Conceptual Variable Design for Scorecards

A Standardized Methodology for the Model-Building Process

Saul Rodrigo Alvarez Zapiain

Apress®

Conceptual Variable Design for Scorecards: A Standardized Methodology for the Model-Building Process

Saul Rodrigo Alvarez Zapiain
Mexico City, Mexico

ISBN-13 (pbk): 979-8-8688-1420-4　　ISBN-13 (electronic): 979-8-8688-1421-1
https://doi.org/10.1007/979-8-8688-1421-1

Copyright © 2025 by Saul Rodrigo Alvarez Zapiain

This work is subject to copyright. All rights are reserved by the Publisher, whether the whole or part of the material is concerned, specifically the rights of translation, reprinting, reuse of illustrations, recitation, broadcasting, reproduction on microfilms or in any other physical way, and transmission or information storage and retrieval, electronic adaptation, computer software, or by similar or dissimilar methodology now known or hereafter developed.

Trademarked names, logos, and images may appear in this book. Rather than use a trademark symbol with every occurrence of a trademarked name, logo, or image we use the names, logos, and images only in an editorial fashion and to the benefit of the trademark owner, with no intention of infringement of the trademark.

The use in this publication of trade names, trademarks, service marks, and similar terms, even if they are not identified as such, is not to be taken as an expression of opinion as to whether or not they are subject to proprietary rights.

While the advice and information in this book are believed to be true and accurate at the date of publication, neither the authors nor the editors nor the publisher can accept any legal responsibility for any errors or omissions that may be made. The publisher makes no warranty, express or implied, with respect to the material contained herein.

　　Managing Director, Apress Media LLC: Welmoed Spahr
　　Acquisitions Editor: Celestin Suresh John
　　Development Editor: James Markham
　　Coordinating Editor: Kripa Joseph

Cover designed by eStudioCalamar

Cover image designed by ©[Just_Super from Getty Images Signature] via Canva.com

Distributed to the book trade worldwide by Springer Science+Business Media New York, 1 New York Plaza, New York, NY 10004. Phone 1-800-SPRINGER, fax (201) 348-4505, e-mail orders-ny@springer-sbm.com, or visit www.springeronline.com. Apress Media, LLC is a Delaware LLC and the sole member (owner) is Springer Science + Business Media Finance Inc (SSBM Finance Inc). SSBM Finance Inc is a **Delaware** corporation.

For information on translations, please e-mail booktranslations@springernature.com; for reprint, paperback, or audio rights, please e-mail bookpermissions@springernature.com.

Apress titles may be purchased in bulk for academic, corporate, or promotional use. eBook versions and licenses are also available for most titles. For more information, reference our Print and eBook Bulk Sales web page at http://www.apress.com/bulk-sales.

Any source code or other supplementary material referenced by the author in this book can be found here: https://github.com/saulzapaiain/ConceptualVariableDesign.

If disposing of this product, please recycle the paper

This book is dedicated to my family: my mother Ernestina, my wife Daniela, our children Rayita and Dexter — and to Dad and Marcus, whose absence I feel every day.

We go through life desperately searching for true love…not realizing that it has always been by our side.

Table of Contents

About the Author xv

About the Technical Reviewers xvii

Preface xix

Introduction xxv

Chapter 1: Conceptual Representations 1

1.1 The Conceptualization Process 2

    1.1.1 Altman Z-Score Model: A Case of Study 4

1.2 Introduction to the Balance Equation 8

1.3 Summary 11

Chapter 2: Conceptual Modeling 13

2.1 Understanding the System at Study 13

    2.1.1 What Is a Loan? A Low-Resolution Analysis 14

    2.1.2 The Magic Ingredient Is Trust 15

    2.1.3 The Contract 16

    2.1.4 The Lender 17

    2.1.5 The Borrower 20

    2.1.6 Integrity 21

    2.1.7 Moral Values 22

    2.1.8 What Does a PD Model Actually Measure? 26

    2.1.9 Consideration of External Factors 28

2.2 Summary 29

TABLE OF CONTENTS

Chapter 3: Balance Equation 31
3.1 Analysis of Contributing and Subtracting Forces 32
3.2 The Married with Children Effect 33
3.2.1 Think of the Children 40
3.2.2 Summary of the Married with Children Effect 41
3.3 Extracting Character from Behavior 42
3.3.1 From Theory to Practice: The Merchant Category Codes 42
3.3.2 Basic Elements of Daily Life 43
3.4 Common Sense and the Inductive, Deductive, and Abductive Reasoning 47
3.5 From Low-Resolution Concepts to Mathematical Expressions 49
3.5.1 Flux Directionality 51
3.6 Summary 58

Chapter 4: Ratios 61
4.1 Dealing with Monetary Variables 61
4.1.1 The Trouble with Normal 62
4.1.2 Wealth Distribution and the Matthew Principle 63
4.1.3 Scale, Dimensionality, and Units 64
4.2 Ratios vs. Magnitudes 70
4.2.1 A Brief History of Ratios 71
4.2.2 A Special Type of Ratio, Dimensionless Numbers 73
4.2.3 Intrinsic Benefits of Using Ratios over Magnitudes 78
4.3 Build Your Own Ratios 79
4.4 Summary 84

Chapter 5: Time and Behavioral Patterns 87
5.1 It Is a Matter of Time 88
5.2 Introducing Time Periodicity into Your Explanatory Variables 90
5.3 Folding Time on Itself 95

5.4 Behavioral Patterns 100

5.5 Summary 101

Chapter 6: Additional Variables 103

6.1 Unconventional Information 104

6.2 Present and Previous Financial Obligations 105

6.3 Customer Personal Information 108

6.4 Customer's Banking Information 114

6.5 Product's Information 114

6.6 Macroeconomic Information 114

6.7 A Word of Caution, Variable Controversy 117

6.8 Summary 118

Chapter 7: Things to Know About ABTs 121

7.1 ABTs Are All About Data Processing 122

7.2 ABTs Require Advanced Programming Skills 123

  7.2.1 SQL Is ABTs' Best Friend 123

7.3 Summary 125

Chapter 8: The Building Plan and Variable Management 127

8.1 Managing Your Variables in Spreadsheets 128

8.2 Code As You Name Your Variables 132

8.3 Nomenclature and Consistency 135

8.4 Summary 138

Chapter 9: Target Population 141

9.1 Defining Your Target Population 141

9.2 Segmentation 142

9.3 Statistical Representativeness and Statistical Power 146

9.4 Business Rules 149

TABLE OF CONTENTS

9.5 Dealing with Keys and Unicity 153
    9.5.1 Synthetic Keys 154
    9.5.2 Correcting Historical Inconsistencies Using Synthetic Keys 155
    9.5.3 Slowly Changing Dimension Type 2 (SCD2) 158
    9.5.4 Pros and Cons of Using Synthetic Keys 159
    9.5.5 The Case of Synthetic Keys for Telcos 160
9.6 Summary 162

Chapter 10: The ABT-Building Process 165

10.1 Overview of the Most Common Types of Input Tables Used
      in the Industry 166
    10.1.1 Snapshot Tables 167
    10.1.2 Transaction Tables 169
    10.1.3 Dimensional Tables and Catalogs 170
10.2 Common Mistakes when Adding New Variables to the ABT 171
    10.2.1 Row Multiplication 172
10.3 Tools for Efficiently and Securely Adding New Variables to the ABT 173
    10.3.1 Oracle Cursor 174
    10.3.2 SAS DATA Step Hash 178
    10.3.3 SELECT Correlated Subqueries 188
10.4 Designing Suitable Inputs for the ABT-Building Process 193
    10.4.1 Input's Building Process and Variable's Addition to the ABT,
          Step by Step 196
10.5 Continuous Variable Integration Cycle (CVIC) 230
    10.5.1 The CVIC for Interval Variables 230
    10.5.2 The CVIC for Nominal Variables 232
    10.5.3 Integrating Information from Different Financial Products 233
    10.5.4 The CVIC via the Account-Customer Bridge 235

TABLE OF CONTENTS

10.6 Event Definition 238
    10.6.1 Other Considerations 243
    10.6.2 Target Population and Event Definition..................................243
10.7 Stacked ABT 245
    10.7.1 The Average Period n 249
    10.7.2 Handling the Stacked ABT Timeline..249
10.8 ABT-Building Recommendations 253
    10.8.1 Build Your Own Data Mart...253
    10.8.2 Represent Your ABT-Building Process As a Diagram 257
10.9 Summary..265

Chapter 11: A Brief Introduction to the Use of SAS® Enterprise Miner™ 269

11.1 Getting Started with SAS® Enterprise Miner™ 270
    11.1.1 The Exploration Panel 271
    11.1.2 The Properties Panel..272
    11.1.3 The Quick Help Panel 272
    11.1.4 The Nodes Panel...272
    11.1.5 The Diagram or Canvas Panel..273
    11.1.6 Startup Code..273
    11.1.7 The Utility Macros...274
    11.1.8 Metadata Properties 284
    11.1.9 EM Diagrams 293
    11.1.10 The Code Node in EM..295
11.2 Data Mining Diagram Design 313
    11.2.1 The Six Basic Steps of the Mining Process 315
    11.2.2 Keep Improvised Tasks Out of Your Mining Diagram 317
11.3 The Importance of Detail 319
11.4 Summary..320

TABLE OF CONTENTS

Chapter 12: Partitioning 323

12.1 Stratified Sampling and Representativeness 324

    12.1.1 Spatial Dimension...324

    12.1.2 Time Dimension 325

    12.1.3 Risk Dimension...325

12.2 Stratified Sampling and Overtraining...326

12.3 Training, Validation, and Test Partitions 328

12.4 Data Proportion Between Training and Validation 328

12.5 Applied Example of a Stratified Sample Analysis......................................329

12.6 Summary...337

Chapter 13: Univariable Analysis 339

13.1 The Univariable Code: A Step-by-Step Explanation.....................................340

    13.1.1 Setting Execution Parameters 341

    13.1.2 Log Rerouting 344

    13.1.3 Parameter Initialization 345

    13.1.4 Obtaining the Reference Category for Nominal Variables 346

    13.1.5 Creating Temporary Storage Tables 349

    13.1.6 Starting the Main Loop and Data Validation 349

    13.1.7 Regression..352

    13.1.8 Estimating Significance and Saving Results 353

    13.1.9 Loop to the Next Variable in the ABT 355

    13.1.10 Save Final Results and Reject Nonsignificant Variables.....................355

13.2 Summary...358

Chapter 14: Collinearity Analysis 361

14.1 Variance Inflation Factor (VIF) 361

14.2 Optimization Issues...363

14.3 Collinearity Diagnostics 369

14.4 Available Methods for Dealing with Collinearity..371
14.5 Iterative Collinearity Reduction Method (ICRM) 372
    14.5.1 Initial Collinearity Diagnostics 373
    14.5.2 Eliminate Perfect Collinearity Issues 375
    14.5.3 Check for Collinearity Issues and Start Main Loop Accordingly 377
    14.5.4 Identifying the Maximum VIF Variable 378
    14.5.5 Identifying the Competing Variable and Selecting the Variable with the Higher Proportion of Variance..380
    14.5.6 Delete the Rejected Variable and Recheck for Collinearity Issues 384
14.6 Applied Example 385
14.7 EM Variable Clustering Node (Varclus) 389
    14.7.1 ICRM or Varclus? 393
14.8 Summary..394

Chapter 15: Weight of Evidence..397

15.1 WoE Definition and Calculation 398
15.2 Advantages of Using the WoE..404
    15.2.1 Controlling for the Presence of Outliers 405
    15.2.2 Standard Units and the Use of Logits 406
    15.2.3 Monotonic Linearization of Interval Variables...................................407
    15.2.4 Dealing with Nominal Variables 415
    15.2.5 Dealing with Missing Values...418
15.3 Using Decision Trees to Calculate the WoE 420
    15.3.1 The Interactive Grouping Node 421
15.4 Setting the Total Number of Groups 433
    15.4.1 The IV and the Gini Index As a Function of the Number of Groups.....436
15.5 WoE Defined As an Optimization Problem..441
    15.5.1 WoE Optimization with the OPTMODEL Procedure 442

15.6 WoEs and Interval Targets 451
  15.6.1 Resolution Issues When Using WoEs for Interval Targets 453
15.7 Summary... 456

Chapter 16: Multivariable Selection Methods 459
16.1 Variable Survival Rate 460
16.2 Multivariable Selection Methods.. 466
  16.2.1 Stepwise Procedures.. 467
  16.2.2 All-Subsets Procedure 478
  16.2.3 Stepwise or All-Subsets Procedure? 490
16.3 Summary... 491

Chapter 17: Experimental Design and Hyperoptimization 495
17.1 Experimental Design 496
  17.1.1 Response (Model's Assessment) 498
  17.1.2 Factors and Factor Levels 507
  17.1.3 Final Specific Factors and Total Number of Variants 515
  17.1.4 Considerations and the Curse of Statistics.................................... 520
  17.1.5 Alternative Factors and Factor Levels 521
  17.1.6 Applied Example 524
  17.1.7 Naming Your Variants 530
17.2 Hyperoptimization 532
  17.2.1 Assessing the Hyperresponse 535
  17.2.2 Assessing the Hyperspace... 541
17.3 Models Are Not Selected, They Are Constructed 562
17.4 Summary... 564

Chapter 18: The Main-Effects Model 567
18.1 The Variable Select Node 569
18.2 Selecting the Best Full Main-Effects Model.. 571

18.3 Results and Discussion of the Best Full Main-Effects Model	577
18.4 The WoE Refinement Process	582
18.4.1 Defining the Expected Trend	584
18.4.2 Interactive Grouping Feature	589
18.4.3 Applied Examples of Visual Assessment, Diagnosis, and Manual Adjustments	592
18.4.4 Conclusion of the WoE Refinement Process	628
18.5 Construction of the Final Main-Effects Model	631
18.5.1 Manual Elimination Based on the χ^2 Statistic	634
18.6 Conclusion	645
18.7 Summary	647

Chapter 19: The Scoring Process 653

19.1 Principal Elements of the Scoring Process	654
19.1.1 Official Data Sources	656
19.1.2 ETLs for the Foundation Mart	656
19.1.3 Foundation Mart	657
19.1.4 CVIC	657
19.1.5 Scoring ABT	657
19.1.6 Scoring Model	659
19.1.7 Scored ABT	666
19.1.8 Scoring Snapshot Table	669
19.1.9 The Observed Default Event and the Validity Window	671
19.1.10 The System Stability Index	674
19.1.11 Follow-Up Statistics	677
19.2 Summary	682

TABLE OF CONTENTS

Chapter 20: Closing Thoughts 685

20.1 The Structure of Magic 686

20.2 The End of a Journey 687

References 689

Online Resources 699

Abbreviations and Acronyms 701

Index 705

About the Author

Saul R. Alvarez Zapiain brings over two decades of diverse professional experience to his role as a principal analytical consultant at SAS. His journey began with a bachelor's degree in industrial biochemical engineering, followed by a master's degree specializing in mathematical modeling of cellular organisms. Having started his career in industries such as pharma, telco, and market research, Saul transitioned to analytics around 20 years ago. Despite the lack of scientific opportunities in his home country, his passion for mathematical models remained unwavering. He sought roles where he could apply his mathematical expertise outside academia.

Around 16 years ago, while working at Santander, Saul discovered data mining, sparking a continuous quest for knowledge in analytical domains. Over the years, he honed his skills through self-education and numerous SAS courses, earning over 10 certifications, including Certified Data Scientist and Certified Professional in AI, Machine Learning, and Data Curation. Delving into specialized statistical techniques, he developed a deep understanding of the modeling process. It wasn't until Saul was invited to present his expertise at a conference organized by the Customer Success division that he fully recognized his niche—building models. This realization, coupled with his wealth of real-life experiences, shaped his approach to analytics and problem-solving.

About the Technical Reviewers

Naeem Siddiqi is the author of *Credit Risk Scorecards: Developing and Implementing Intelligent Credit Scoring* (Wiley and Sons, New York, 2005), *Intelligent Credit Scoring: Building and Implementing Better Credit Risk Scorecards* (Wiley and Sons, 2017), and various papers on credit and climate risk topics.

Naeem is a senior advisor (Risk, Fraud and Compliance Solutions) at SAS Institute Inc. He meets with senior executives and decision-makers worldwide and provides strategic advice to them on areas such as the development and validation of credit scoring models, climate risk, infrastructure planning for analytics, and retail credit risk strategy. He also heads SAS's initiatives on climate risk. He has trained hundreds of bankers in dozens of countries on the art and science of credit scorecard development and helps lenders develop better scorecards. Naeem has an Honours Bachelor of Engineering from Imperial College of Science, Technology and Medicine at the University of London and an MBA from the Schulich School of Business at York University in Toronto. Naeem also serves as volunteer Vice Chair at the Olive Tree Foundation and is a member of the board at Windmill Microlending.

ABOUT THE TECHNICAL REVIEWERS

Raúl Franco is an SAS instructor teaching training courses for a period of approximately 15 years. The main courses he teaches are programming, statistics, and data mining with SAS. He has also been a consultant in the analytics area of SAS.

Edwin Can is a senior statistical programmer with more than 3 years of experience in analyzing and reporting of clinical trial data for various pharmaceutical sponsors for drug study phases I, II, and III in therapeutic areas including cardiovascular, gastrointestinal, and oncology. He also leads programming activities for validation, and delivery of CDISC compliant analysis datasets, tables, listings and figures as well as SDTM domains. Additionally, Edwin has nine years of experience in analytical models, data mining, and data science using SAS, Python, R, and SAP.

Preface

By the end of my studies, I became obsessed with models, with what they could do, what they represented, how they worked, their elegance, in short, with everything about them. I worked on simulations of systems of coupled differential equations integrated by numerical methods. I was fascinated by numerical methods, which led me to one of the passions of my youth—optimization methods.

I became obsessed with solving nonlinear regressions with optimization methods and got the idea to get a graduate degree in optimization algorithms. Life moved on, as it usually does, and I ended up getting a master's degree in biotechnology, but I decided to choose a topic that combined mathematical modeling with optimization methods as my dissertation project.

Ironically, I was more interested in the optimization algorithms for solving mathematical models than in the mathematical models themselves. Fortunately, my dissertation tutor made me realize in time that numerical methods, optimization methods, and mathematical software were all a means to an end and not the end itself.

Sometime after my master's degree in biotechnology, I ended up working for a bank as a low-level database analyst,[1] where my job consisted of performing tedious and monotonous tasks to maintain what to me were trivial and boring corporate reports.

---

[1] It is not uncommon for science graduates to end up working in low-level, unrelated jobs in emerging economies like Mexico.

PREFACE

One day, I heard about this new and fashionable division in the same department I was in, called *data mining*. I was curious about what the data mining division did, so I asked my boss, and he told me that they were in charge of developing models. I was silent for a second, and then I said to him, *"But I know how to develop models too"* to which he replied in a condescending tone, *"Yes, but they build different kinds of models"*.

Hearing about data mining and knowing what they did got me interested in models again. But now I started looking for more information about these mysterious data mining models my boss was talking about.

Many more years passed, and I spent my free time at home and idle time at work learning everything I could about the mysterious world of data mining. At first, I was fascinated after reading about neural networks, decision trees, clustering algorithms, principal components, and the like. I spent most of my time reading about these topics, many afternoons until late at night, and many weekends as well.

But reading data mining books led me to a different place than the one I intended to go, so I suddenly found myself reading more academic books about statistics and became more and more interested in topics like ANOVA, MANOVA, confidence intervals, parametric and nonparametric tests, and so on. But one particular topic caught my eye and changed my course, or rather brought me back to where I was before, to regression analysis, more specifically to *logistic regression analysis*.

During this phase of my life, I read many books, articles, and technical guides, but there were three books that arguably shaped the way I think about statistics today. The first is a book called *Discovering Statistics Using SPSS*, written by Andy Field [20]. There is nothing particularly advanced about this book, but it helped me to complete and strengthen my statistical knowledge, considering that my background was more in the pure sciences than in statistics.

The approach of this book seemed fresh and fun to me, and what stood out to me was the fact that the author is a psychologist and not a statistician as one would expect. So, his book helped me understand the most basic concepts of statistics, and for me, it was like taking a complete undergraduate course in statistics. However, I must say that I found in Field's book understandable and excellent explanations of topics such as what eigenvalues really are that I haven't found anywhere else to this day.

The second is a book called *Intuitive Biostatistics* by Harvey Motulsky [42]. The interesting thing about this book is that it explains complex statistical concepts from a commonsense point of view. The book is intended for students and professionals in the natural sciences who struggle with the proper use of statistical terms, or who want to have a better understanding of statistical concepts that can help them use them properly and fluently. This book may not teach you how to prove complicated mathematical expressions, but it will teach you what the conceptual meaning of statistical significance is, which is something that is not properly explained even at the graduate level.

The third and final book is called *Applied Logistic Regression* by Hosmer and Lemeshow [32]. For me, this book is the cornerstone of logistic regression, and most of what I know about logistic regression comes from this book. But unlike the previous two books, this one has a completely academic approach and presents its content in the most formal way.

Applied Logistic Regression is the first book that introduced me to the topic of the model-building process, and it is also the first book that taught me an ordered, logical, and sequential set of steps for building a model.

I had to give up my passion for models for a while, as all signs pointed to a successful career as a mid-level manager in the telecommunications industry. Fortunately, I was laid off and unemployed for a few months, which led me to my dream job as an analytical consultant at SAS, one of the world's most respected statistical software companies.

PREFACE

My job at SAS was, and still is, pretty straightforward: I build models for a living. I was impressed that a company like SAS could exist, and I must say that at first I had mixed feelings, both excitement and apprehension, because I knew that my modeling skills were a bit rusty and also because I was relatively old compared to the rest of the consulting staff.

I soon realized that I had nothing to worry about, as I excelled in my first project, where I was congratulated and recognized for my good work—building models. Soon after, I was introduced to another type of model that changed my perspective on models forever—I was introduced to *scorecard models*.

I must admit that at first I hated scorecard models because they seemed too simplistic to me, and I was still too academic in my approach to the modeling process. I had the opportunity to work on a credit scoring project where I developed some scorecard models using a data mining node called Interactive Grouping. I did not know it at the time, but I would become one of the biggest proponents of using the Interactive Grouping node, or rather, one of the biggest proponents of using the *weight of evidence* technique, also known as *WoE*.

As I mentioned earlier, I was too academically inclined to understand the scorecard method. At the time, I was too fixated on the use of the fractional polynomials method recommended by Hosmer and Lemeshow as a transformation technique to see the benefit of using the WoE. But once again, something happened that made me reconsider my position on using the WoE—I met Naeem Siddiqi.

Naeem Siddiqi is a senior advisory industry consultant at SAS, an expert in credit risk and credit scoring. He is the author of *Intelligent Credit Scoring* [63], a book that explores the process of building and implementing scorecards for the banking industry from a business perspective.

I had the opportunity to assist in one of Naeem's Business Knowledge courses (known in SAS as the BKS course), where Naeem presented the ins and outs of developing a scorecard model from a business application perspective. During the course, there were many ideas that were at odds with the way I used to think about building models, but instead of rejecting them, I began to get a familiar feeling—I was intrigued again.

I tend to ask a lot of questions (something few people feel fond about) when a topic intrigues me, so I did. I noticed that Naeem was a bit annoyed at being questioned at every point of his presentation, but he kept answering my questions politely, so I kept asking.

Naeem and I are good friends now, but I am sure he has no idea how much he changed the way I think about the modeling process. Perhaps the most striking idea I heard in his class was the idea that you could *arbitrarily* introduce a value into the WoE function to *"correct its trend"*! Being as attached to classical methods as I was, I could not understand how this was possible without supporting data; it was the exact opposite of what I had been learning up to that point.

I immediately raised my hand and asked in shock, *"Can you really do that!?"*, to which Naeem replied in a calm tone, *"Of course you can"*. He then explained to me one of the key ideas that has most influenced my thinking about modeling, that *flawed data cannot outweigh well-known logical trends*. This new radical concept blew my mind and led me to one of the most important conclusions in the modeling process: *"that the absence of evidence does not mean that the evidence does not exist, it just means that the evidence is not available".*

Naeem's course continued, but there was something else that nagged at me, something that kept pounding in my head. It was Naeem's insistence on using *ratios* as independent variables in our scorecard models. He did not go into much detail, but the message was loud and clear—*ratios matter*. Little did he know that I, too, would become a big proponent of using ratios in models.

Among the most enjoyable aspects of Naeem's class were the stories about past experiences he had while developing some of the scorecards. Stories like how buying more expensive food for your pets or buying the right amount of roses for your spouse or girlfriend happened to be

PREFACE

negatively correlated with the likelihood of default. But everything in Naeem's class had a point, and these seemingly anecdotal stories were no exception—because in all of his stories, the underlying variable driving behavior was *character*.

The idea that character is one of the most predictive variables in scorecard models, or in any model that predicts behavior, resonated with me. I had also invested in understanding the psychological aspects that influence our behavior, such as human perception, for a number of years, and I had already applied some of these concepts in my daily work, programming, or in data analysis. However, I had not yet figured out how to incorporate such aspects into my modeling techniques.

A few more years passed, and my thinking about the modeling process continued to evolve as I gained more hands-on experience and took a series of SAS courses on specialized statistical techniques. But it wasn't until I was asked to present a topic of my expertise at a Customer Success conference that it finally dawned on me—my expertise is in building models.

I was overwhelmed with ideas...*you can do this*...I said to myself. I realized that I had been successfully building models for many years without realizing it, trying to teach junior consultants how to do it, but none of them were able to grasp all the little details needed to build a model from scratch, and I could not understand why.

But this revelation also made me realize that not even I could properly articulate what I was doing when I built models. Somehow, I had inadvertently managed to incorporate all the ideas I had learnt during my academic and professional life into a unified methodology for model building, so I thought — maybe I should write a book about it — and so I did.

Introduction

Since the inception of Basel II in 2004 [5], the task of model building has become a routine practice within the banking sector, particularly for developing models such as probability of default (PD), loss given default (LGD), and exposure at default (EAD). Despite the widespread adoption of these models, there is a notable absence of consensus or standardized guidance regarding their methodology, the variables included in them, and the rationale behind their selection. Regulatory bodies provide minimal guidance on these issues, allowing individual banks to devise their own modeling approaches as they see fit. As a result, there is a proliferation of methodologies for model building across various analytics and risk departments, resulting in a vast and diverse body of literature addressing this topic.

The assumption that initial explanatory variables are present is pervasive in the literature on *machine learning* (ML) and *data mining* (DM). In practice, however, the conceptualization of variables and the data manipulation needed for the creation of *analytical base tables* (ABT) demands the majority of the effort required for constructing a predictive model. Despite this fact, current literature focuses its discourse primarily on the statistical methodologies, thereby diverting attention from the complex process of acquiring and structuring the inputs that feed these statistical techniques. Consequently, the variable design process is often neglected, with readers expected to understand the origin, usage, and rationale to include such variables in the model.

The academic literature is notable for its focus on model interpretation and variable design. This underscores the discrepancy between academic and industry literature. The literature concerning industrial applications

INTRODUCTION

of machine learning and data mining places significant emphasis on the delivery of rapid solutions, while simultaneously disregarding the subtleties inherent in individual modeling techniques. Data sources are typically presented for illustrative purposes. Conversely, within academic circles, statistical modeling techniques are routinely subjected to rigorous scrutiny at the graduate level. Academic examples are often grounded in scientific research, where a deep understanding of the phenomenon is needed to understand the technique.

In addition, the industry literature lacks a common methodology for building models in a standardized, repeatable manner. This problem appears to be rooted in the fact that current industry literature covers too many topics and focuses its attention on explaining the intricacies of the accompanying methods, without demonstrating their use in an integrated manner, as part of a unified model-building process.

Nevertheless, there are notable examples of model-building methodologies in the academic literature. An illustrative example is the methodology presented by Hosmer and Lemeshow (H&L) in their book *Applied Logistic Regression* [32]. The book presents an excellent and elegant methodology for the model-building process for logistic regression, where the authors propose a general five-step methodology:

1. Univariable analysis
2. Variable selection for the multivariable model
3. Fit of the first multivariable model
4. Determination of the correct relationship between the logit and the covariates
5. Exploration of possible interactions

The resulting model from these five steps is designated the *preliminary final model* and is subsequently evaluated for its performance and fit.

This methodology is elegant in its simplicity, yet it is not well-suited to the accelerated pace of the contemporary industry, particularly with regard to the development and deployment of increasingly large models. Academic research models are typically smaller and more specific, encompassing no more than one hundred variables during the training process, whereas industry models can include hundreds to thousands of variables.

It is important to acknowledge that academic research models address life-threatening issues like cancer and heart disease. Consequently, academic models require a greater investment of time to develop comprehensive, meticulous models. By contrast, the financial and marketing industries have different needs, driven by short- to medium-term economic goals. Consequently, risk and analytics departments are under pressure to deliver high-quality models in a timely manner. It is therefore unsurprising that the majority of academic researchers have not perceived a necessity to accelerate the modeling process.

Relevance in the Age of Machine Learning and Artificial Intelligence

In the age of machine learning (ML) and artificial intelligence (AI), there is a common misconception that algorithms can discover patterns or solutions on their own without human intervention. This book directly addresses this misconception by emphasizing the importance of *conceptualization* and *abstraction* as fundamental steps in the modeling process. It emphasizes that while AI and ML provide powerful tools for model development, they are no substitute for a deep understanding of the phenomenon being modeled.

It is important to emphasize that while all models operate in a fundamentally similar way, the underlying principle that unites them is the convergence of an error-like function, referred to as *"learning"* in the context of machine learning. However, it is important to note that true learning does not occur in this process and that ML algorithms are not intelligent in themselves.

According to Raymond Catell (1963), there are two types of intelligence, fluid and crystallized intelligence.

Fluid Intelligence: *The ability to solve new problems without referencing prior knowledge*

Crystallized Intelligence: *Refers to the use of previously acquired knowledge, such as specific facts learned in school or specific motor skills or muscle memory*

In this respect, artificial intelligence (AI) is more closely related to crystallized intelligence than to fluid intelligence. This is due to the fact that AI, like all models, is incapable of devising novel solutions to novel problems without prior knowledge or specific training. This is because machines are only good at the specific thing they were designed to do [29].

However, it is abstract thought that truly defines human intelligence. Higher cognitive functions include the capacity for making and employing abstractions, including the formation of judgments, learning from experience, and making inferences.

In contrast, by focusing on *conceptual variable design*, this book demonstrates that the success of predictive models depends not only on the sophistication of algorithms but also on the quality, context, and design of the input variables. This insight is crucial in an era where many practitioners rely excessively on *"black-box"* models without understanding the underlying data.

Standardization in a Fragmented Field

This book also provides a *formal and standardized methodology* for variable design and model building, which is particularly valuable in the AI and ML landscape, where practices often lack consistency. Many industries—especially finance, telecommunications, and retail—face challenges due to the diversity of modeling techniques and the absence of unified approaches. In contrast, the methodology outlined here offers a structured, repeatable process that ensures

- **Transparency** in model development
- **Reproducibility** of results
- **Scalability** across different use cases

In the age of AI, where interpretability and accountability are becoming critical (e.g., due to regulations like GDPR and ethical AI initiatives), this book's emphasis on standardization is a much-needed counterbalance to the often chaotic experimentation seen in ML.

Variable Design As the Foundation of Predictive Models

The central argument of this book is that *variable design is often overlooked* in the rush to implement advanced ML algorithms. In practice, most of the effort in building predictive models lies in acquiring, cleaning, and structuring the inputs that feed these algorithms.

This focus on variable design is particularly relevant today, as ML models are only as good as the data they are trained on. By improving the quality and relevance of input variables, the methodology outlined here may enhance the performance, interpretability, and reliability of AI systems.

Interpretability and Explainability in AI

In the current AI landscape, *explainability* is a growing concern. Black-box models like deep learning often lack transparency, making it difficult to understand how predictions are made. This book addresses this issue by

- Advocating for the use of *interpretable variables*, such as ratios and weights of evidence (WoE), which have clear, logical meanings

- Emphasizing the importance of *conceptual clarity* and *human judgment* in variable selection and model refinement

- Providing tools for *visual assessment* and *diagnostic analysis* of variables, ensuring that models align with both statistical and business logic

Empowering Analysts in the Age of Automation

Finally, this book is a *call to action for analysts* to take control of the modeling process rather than relying blindly on automated tools. It empowers readers by

- Providing a *step-by-step methodology* for building models from scratch

- Encouraging analysts to think critically about the *concepts and assumptions* underlying their models

- Bridging the gap between *academic theory* and *industry practice*

This empowerment is crucial in an era where automation risks reducing analysts to mere operators of software, rather than active participants in the modeling process.

Goals of This Book

This book has two principal goals. The initial goal is to instruct readers in the creation of pertinent variables, specifically tailored to the phenomenon they intend to model, through the utilization of illustrative examples. Secondly, the aim is to present a formal and standardized methodology that can be used as a starting point for the model-building process. This methodology is designed to provide readers with a structured approach to the modeling process, while also allowing for the customization of individual components as deemed necessary. The result of this process is that readers may find both flexibility and commonality among all of their models.

Who This Book Is For

The book is intended for individuals engaged in the practical construction of models in the field of risk and analytics. This target readership includes those employed in the banking, telecommunications, and retail industries, who are tasked with the challenge of meeting the growing demand for more sophisticated and extensive predictive models. The book is also intended for those who have not yet identified a means of standardizing their model-building processes. Additionally, the book is intended for individuals who have multiple and legitimate questions about seemingly trivial issues that bridge the gap between the idealized model-building process and its actual practice.

Benefits of This Book

The initial advantage to be gained from reading this book is the acquisition of transferable skills that will enable the reader to construct any model, regardless of its nature. The second benefit is *the reduction of operational*

INTRODUCTION

risk and *the subsequent increase in productivity*. In order to illustrate the concept of increased productivity, **Table 1** provides statistical data from a recent delivery.

Our team not only met the objective of building nine models, exceeding the original requirement by 29%, but also accomplished this in a significantly reduced timeframe. The total time estimated for this project was 4,584 hours, yet our team completed the task in only 3,307 hours, representing a 28% reduction.

Table 1. Statistical data

Elements developed	Data mart tables	>28
	Data mart ETLs	>28
	ABTs	9
	Explanatory variables	>1,700
	Total number of delivered models	9
	Scoring processes	9
	Monitoring reports	7
Resource management	Available hours	4,584
	Hours actually used	3,307
	Remaining hours	1,277
	Analytical consultants	3
Performance statistics	Time reduction	-28%
	Additional models	+29%
	Average hours per model	367
	Average weeks per model	9

It is noteworthy that each model was constructed in an average of only nine weeks, or less than two and a half months. This is a significant feat, particularly given that the aforementioned timeline does not account for all the elements required during the model-building process.

It is therefore evident that a methodology capable of producing results of the quality demonstrated in this project would be an invaluable tool for any industry that aims to be at the forefront of predictive model development.

Chapters at a Glance

Chapter 1 explains the importance of the conceptualization process for the construction of predictive models, making the case that the first step in the model-building process is the understanding of the phenomenon at a deep level. The Altman Z-score model is employed to illustrate the conceptualization process and the representation of equations at varying resolution levels.

Chapter 2 conducts a conceptual modeling of a loan probability of default (PD) model. It argues that trust is the underlying element that drives the default event and that it is intrinsically linked to concepts like integrity, consistency, moral judgment, and moral values.

Chapter 3's main goal is to explain the mental processes needed to transition from conceptual abstractions to mathematical expressions, and it introduces the concept of "balance equation."

Chapter 4 addresses the drawbacks of dealing with monetary variables and magnitudes in general, making the case for the use of ratios. It argues that ratios convey more information than magnitudes alone.

Chapter 5 underscores the significance of temporal cycles in predicting behavioral patterns and argues that human beings are creatures of habit that structured their activities around yearly, weekly, and daily routines.

INTRODUCTION

Chapter 6 introduces additional data sources pertinent to the construction of explanatory variables, aside from behavioral data. This encompasses the utilization of attributes such as customer demographics, banking information (e.g., maturity, segment), products' attributes (e.g., credit card type, interest rate, credit limit), and macroeconomic indicators (e.g., GDP, unemployment rate, price index).

Chapter 7 highlights the importance of analytical base tables (ABTs) and their intrinsic relationship with the modeling process, arguing that ABTs are the very essence of the model itself, as the consistency of the input data is the foundation upon which the entire analysis rests.

Chapter 8 provides a series of guidelines for the management of an ABT project, emphasizing the critical role of spreadsheets in facilitating variable creation through dynamic formulas and concatenation, which enable the generation of both, variable names and accompanying code, thereby streamlining the model-building process.

Chapter 9 emphasizes the critical importance of meticulously delineating the target population in the development of predictive models. It highlights the segmentation principle, which advocates for the isolation of homogeneous populations based on their characteristics and behaviors, with the aim of enhancing model performance.

Chapter 10 deals with all the topics related with the construction of an ABT, including the various types of input tables commonly used in building ABTs and how to effectively utilize them to create explanatory variables.

Chapter 11 provides an overview and basic guidance on using SAS® Enterprise Miner™ (EM), emphasizing essential elements and functionalities.

Chapter 12 addresses the initial statistical analysis, which precedes the construction of the multivariable model emphasizing the significance of stratified sampling for the generation of representative samples, in both the training and the validation partitions, arguing that the main cause of overtraining in models is due to a lack of representativeness.

Chapter 13 addresses the issue of irrelevancy in the input variables and gives a step-by-step explanation of the process for conducting a univariable analysis, alongside the accompanied code included in this book.

Chapter 14 covers redundancy in the input variables and explains how to conduct a collinearity analysis, through the use of the variance inflation factor (VIF) and the principal component analysis (PCA).

Chapter 15 discusses the application of weight of evidence (WoE) in regression analysis, contrasting its advantages with those of traditional methods.

Chapter 16 explores the different multivariable selection techniques employed in regression analysis, classifying them into two categories: sequential selection based on statistical criteria (such as forward, backward, or stepwise procedures) and fitting all possible models and selecting the optimal one based on information criteria (i.e., all-subsets).

Chapter 17 makes the case that each particular set of options selected during the execution of a regression analysis, including methodologies, parameters, and values, confines a specific *feasible region*, within which an *optimum model* lies in.

Chapter 18 introduces the concept of the *full main-effects model*, which combines variables from different feasible regions into a single model, using various multivariable selection strategies.

Chapter 19 emphasizes the significance of the *scoring process* for the effective deployment of the model. It divides the scoring process into nine principal elements.

Chapter 20 offers a concluding reflection on the meta-structure of the modeling process and provides a concise overview of the key elements covered in the book.

CHAPTER 1

Conceptual Representations

"To abstract is to draw out the essence of a matter."

—Ben Shahn

I started building models in my master's. At that time, my focus was on mathematical models of biological systems, not credit scorecards. I did a dissertation on metabolic engineering with a stoichiometric model of yeast metabolism. The objective was to identify parameters and fluxes to integrate with metabolite measurements. The system of equations was solved using quadratic programming (QP), also called metabolic flux balancing (MFB), which is a form of optimization modeling. Back then, I was more enthusiastic about the optimization software (GAMS) than about modeling yeast reactions. I was insistent with Dr. Kookos that I should work with the optimization model, not the stoichiometric equations; he grew weary and shared a crucial insight that has remained with me:[1]

---

[1] I must explain that Dr. Kookos has a background in chemical engineering, while my background was in biochemistry. So, Dr. Kookos was more of a mathematician than me back then, and I was more of a chemist.

© Saul Rodrigo Alvarez Zapiain 2025
S. R. A. Zapiain, *Conceptual Variable Design for Scorecards*,
https://doi.org/10.1007/979-8-8688-1421-1_1

CHAPTER 1   CONCEPTUAL REPRESENTATIONS

"Defining and executing an optimization routine is the easy part—anyone can do it—but modeling the overall metabolism of a living organism with stoichiometric reactions is the real challenge!"

So the message that I took away from this experience is that the *conceptual representation* of a problem is far more important than the modeling technique or the statistical software you use to solve it.

1.1 The Conceptualization Process

Conceptualizing a problem allows us to understand the essence of the problem, to generalize a solution, and to ignore its peculiarities or specifics. Accordingly, the conceptualization process in modeling means to have *a deep understanding of the system or phenomena under study*. The conceptualization process then takes place before the definition of the mathematical model, and in fact, the mathematical model should be driven by the conceptualization of the problem, rather than the other way around.

Mathematical models, in contrast, are quantitative abstract representations of a phenomenon that takes place in the real world. They are defined by the conceptualization process. Mathematical models serve to simulate the phenomenon at study, thereby facilitating the extraction of knowledge and the making of predictions within a controlled experimentation. In general, mathematical models may help us to

1. Get an insight of which components and interactions are relevant in systems of enormous complexity and size

2. Find patterns and recognize behaviors from intelligible models that will aid us to discover and implement new strategies

3. Give a thorough review and to question the conventional knowledge in order to correct fallacies

4. Understand intangible and complex natural phenomenon qualitatively and quantitatively

Mathematical models are employed in all areas of science, engineering, and business to solve problems, design equipment, interpret data, and communicate information. It is worth noting that reality is of infinite complexity, and therefore any model, including the one in our heads, will be only a modest approximation of the real world. This is a key idea, since even what we perceive as reality is also a simplified mental representation of the world around us (**Figure 1-1**) [67].

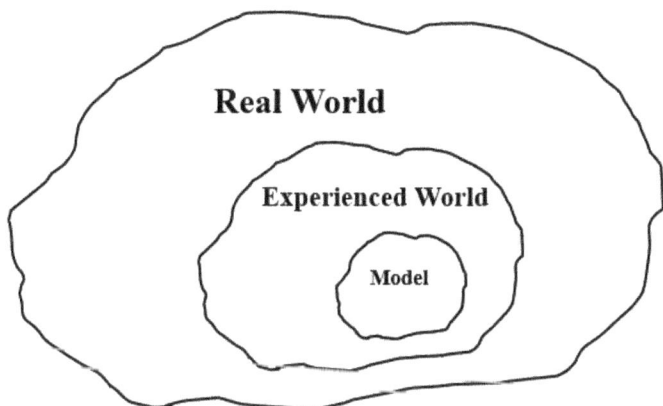

Figure 1-1. *Levels of abstraction in model development.*

A good understanding of the system at study and its proper conceptual representation are then essential for the modeling process, since it not only allows us to incorporate more realistic elements to our model, but it also allows us to represent systems of infinite complexity with simple mathematical expressions.

CHAPTER 1   CONCEPTUAL REPRESENTATIONS

1.1.1 Altman Z-Score Model: A Case of Study

In September 1968, Edward I. Altman published his groundbreaking article, "Financial Ratios, Discriminant Analysis, and the Prediction of Corporate Bankruptcy" [1], introducing his Z-score model. First published over 50 years ago, Altman's Z-score model is still a key tool for academics and professionals worldwide. Furthermore, Altman's model has given rise to literally hundreds of specialized economic papers, articles, master's and doctorate's thesis, books, and publications of all sorts. But what was so special about his original article that made it stand out from the rest? Why is Altman's model still being used? Was it because of the use of the discriminant analysis and the Z-scores? No. The modeling technique was inconsequential. What sets Altman's model apart is the conceptualization process through which he grasped the abstract elements influencing bankruptcy and represented them using straightforward and well-known financial indicators.

Altman's final model was comprised of five different *ratios* (down from 22). Note that Altman did not use hundreds of variables to build his model, nor did he consider variables other than ratios. But Altman's genius was his capacity to express complex abstract ideas in simple terms, e.g., describing an arithmetic operation like the working capital over total assets as the efficiency to convert assets into cash.

$$\frac{\text{Working Capital}}{\text{Total Assets}} =$$

Liquidity: The efficiency or ease with which an asset or security can be converted into ready cash without affecting its market price.

$$\frac{\text{Retained Earnings}}{\text{Total Assets}} =$$

Profitability: Profitability is a measurement of efficiency—and ultimately its success or failure. However, Altman also mentioned that the age of a firm is also implicitly considered in this ratio. Younger firms will tend to show lower RE/TA ratio because they have not had the time to build up their cumulative profits, and therefore, their incidence of failure is much higher in a firm's earlier years.

$$\frac{\text{EBIT}}{\text{Total Assets}} =$$

Productivity: Measures output per unit of input. In Altman's words, it is a measure of the earning power of a firm's assets.

$$\frac{\text{Market Value}}{\text{Total Debt}} =$$

Leverage: Shows the degree to which a company's market value would decline when it declares bankruptcy before the value of liabilities exceeds the value of assets on the balance sheet. A high market value of equity to total liabilities ratio can be interpreted to mean high investor *confidence* in the company's financial strength. Although not mentioned by Altman, you can also interpret this ratio as a measure of how resilient a firm can be under bankruptcy.

$$\frac{\text{Sales}}{\text{Total Assets}} =$$

Activity: It is one measure of management's capability of dealing with competitive conditions.

As an exercise, let us go in the opposite direction of a classical written math problem, so, instead of going from a plain text with simple words to mathematical expressions, we will go from mathematical expressions to a plain text with simple words.

CHAPTER 1 CONCEPTUAL REPRESENTATIONS

Altman's Z-score equation may be described as follows:

$$(x) = \beta_0 + x_1\beta_1 + x_2\beta_2 + x_3\beta_3 + x_4\beta_4 + x_5\beta_5$$

Equation 1-1. Altman's model expressed as a mathematical equation.

Where $\beta_0, \beta_1, ..., \beta_5$ correspond to the model's coefficients.
And

$$x_1 = \frac{\text{Working Capital}}{\text{Total Assets}}$$

$$x_2 = \frac{\text{Retained Earnings}}{\text{Total Assets}}$$

$$x_3 = \frac{\text{EBIT}}{\text{Total Assets}}$$

$$x_4 = \frac{\text{Market Value}}{\text{Total Debt}}$$

$$x_5 = \frac{\text{Sales}}{\text{Total Assets}}$$

We can also define Altman's equation in conceptual terms. Thus, bankruptcy is a function of the firm's liquidity, profitability, productivity, leverage, and activity, which can be defined as follows:

Bankruptcy = Liquidity + Profitability + Productivity + Leverage + Activity

Equation 1-2. Conceptual representation of Altman's model in general financial concepts.

The above equation represents a conceptualization of the bankruptcy phenomenon observed in the real world. However, there are different levels of conceptualization, which are referred to as *resolution levels*.

In this context, "*Resolution*" refers to the level of complexity of a specific element of interest. For instance, **Equation 1-1** is the most detailed representation of the bankruptcy phenomenon. In contrast, the

CHAPTER 1   CONCEPTUAL REPRESENTATIONS

conceptual representation in **Equation 1-2** is a lower-resolution version of **Equation 1-1**. It should be clear then that, in this case, the element of interest is the mathematical representation of the bankruptcy model.

But according to this logic, we can move in both directions, to lower or higher levels of resolution, up to a certain limit, above or below which the resolution cannot be further increased, or decreased, respectively.

In light of these observations, it is possible to take the conceptual representation of the bankruptcy phenomenon one step further to a lower level of resolution. This would allow us to describe financial concepts in simple everyday language.

$$\begin{bmatrix} \text{How Likely is that} \\ \text{your business will} \\ \text{not go bankrupt will} \\ \text{depend upon} \end{bmatrix} = \begin{bmatrix} \text{How much cash you} \\ \text{have available for} \\ \text{contingencies} \end{bmatrix} + \begin{bmatrix} \text{How experienced} \\ \text{you are in your} \\ \text{business} \end{bmatrix} + \begin{bmatrix} \text{How much can you} \\ \text{get with what you} \\ \text{have} \end{bmatrix} + \begin{bmatrix} \text{How resilient you} \\ \text{are in case of} \\ \text{bankruptcy} \end{bmatrix} + \begin{bmatrix} \text{How well can you} \\ \text{deal with} \\ \text{competition} \end{bmatrix}$$

Equation 1-3. Conceptual representation of Altman's model in simple everyday language.

Simple everyday language allows us to understand Altman's model at a commonsense level, but we can reduce the resolution level of **Equation 1-3** even further, by representing Altman's model in one-word concepts.

$$\begin{bmatrix} \text{Bankruptcy} \end{bmatrix} = \begin{bmatrix} \text{Cautious} \end{bmatrix} + \begin{bmatrix} \text{Experienced} \end{bmatrix} + \begin{bmatrix} \text{Resourceful} \end{bmatrix} + \begin{bmatrix} \text{Resilient} \end{bmatrix} + \begin{bmatrix} \text{Persevering} \end{bmatrix}$$

Equation 1-4. Conceptual representation of Altman's model in simple one-word concepts.

Equation 1-4 represents the lowest possible resolution of Altman's model, expressed in simple terms. It can be argued that words such as *"Cautious," "Experienced," "Resourceful," "Resilient,"* and *"Persevering"* are concepts that convey a complex idea that cannot be further simplified. Rather, they can only be explained by simpler words or by other concepts with the same meaning. In conclusion, it can be stated that bankruptcy is associated with a number of characteristics, including *caution, prudence, experience, resourcefulness, resilience,* and *perseverance.*

It is imperative to acknowledge that most of us can understand these concepts, despite their intricate nature. Furthermore, the fact that experience, caution, and resourcefulness are a requirement to achieve a successful business is not surprising, but rather self-evident.

This is the essence of the conceptualization process: the ability to understand a complex phenomenon occurring in the real world by breaking it down into its basic elements, so that it can be described in the simplest way using everyday language with common terms.

1.2 Introduction to the Balance Equation

Thus far, the conceptualization process has been examined in reverse order, with due consideration of the already reported significant variables in Altman's model. However, the primary objective is to ascertain how to approach any given phenomenon with a *cause-and-effect analysis*, as exemplified by the aforementioned approach.

This type of analysis is a standard practice in the sciences and engineering, particularly in chemical engineering, where it is referred to as a *balance equation* (**Figure 1-2**). As its name suggests, a balance equation balances the major *forces that influence a given system*. In this instance, the system under consideration is bankruptcy, and the relevant forces are those pertaining to *financial ratios*. In this specific instance, all the forces have a positive impact on the overall system. However, this is not always the case. Typically, a given system will be influenced by both forces, namely, those that *contribute* to the system and those that *subtract* from it.

CHAPTER 1   CONCEPTUAL REPRESENTATIONS

Figure 1-2. *Balance equation for any given system.*

We can also draw a balance equation diagram for a company considering the most important contributing and subtracting forces, as well as the accumulation factors. To this effect, we will use the conventional components of a balance sheet.

Figure 1-3 illustrates that, through the use of a balance sheet, it is feasible to conceptualize a given company as a system comprising inputs, outputs, and accumulation/consumption factors.

Figure 1-3. *Generic balance equation for any given company.*

Furthermore, an examination of the components illustrated in the diagram reveals that it is possible, at least in theory, to estimate the majority of the components of Altman's ratios.

> **Current Assets** = Sales + Savings + Short-Term Investments + Account Receivables + Inventory

CHAPTER 1　CONCEPTUAL REPRESENTATIONS

Current Liabilities = Short-Term Debt + Account Payables + Taxes

Working Capital = Current Assets − Current Liabilities

Total Assets = Current Assets + No Current Assets

Retained Earnings = Previous Retained Earnings + Net Income − Dividends

Total Liabilities = Total Debt + Other Liabilities

EBIT = Revenue − Cost of Goods Sold − Operating Expenses

Revenue = Sales + Other Income + Interest Income

Net Income = Sales − Operating Expenses − Other Expenses − Taxes − Interest in Debt + Other Income

One component that is absent from the original list of components utilized in Altman's ratios is the market value. *The market value* is distinctive from the other factors in that it is not an intrinsic component of the system; rather, it considers the influence and interaction of all the systems that coexist within the same environment, which is referred to as *the market* (**Figure 1-4**). While market value may appear to be an exception to the balance equation, it is, in fact, a component of the market itself, which can be considered a system, in this case a macro-system, comprising enterprises or subsystems.

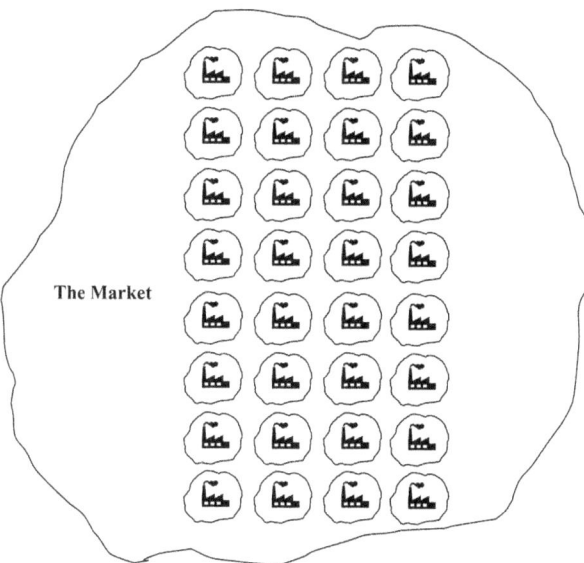

Figure 1-4. *Illustration of the market as a macro-system comprised of subsystems, the companies.*

The market analysis–interesting as it may be–is not our main goal but serves as a good example to show the mental processes of abstraction and conceptualization, intrinsic to the modeling process. The idea here is to show a general conceptualization and abstraction process that might have led to a rationale similar to the one that led Altman to his original model.

1.3 Summary

This chapter serves as an introduction to the conceptualization process, where we begin by abstracting a complex idea in order to extract its essence. The essence of any idea is that it holds true for all possible examples, and it is also what allows us to understand the phenomenon we want to model in simple terms that can be explained in everyday language.

CHAPTER 1   CONCEPTUAL REPRESENTATIONS

Key takeaways:

1. **Essence of Abstraction:** Conceptual representation involves abstracting and distilling the core essence of complex phenomena, as exemplified by the case that quote *"To abstract is to draw out the essence of a matter."*

2. **Conceptualization Precedes Modeling:** The conceptualization process should always drive the mathematical modeling process, and not the other way around.

3. **Levels of Resolution:** Models can be expressed at different resolution levels. For instance, Altman's model can go from detailed mathematical equations to everyday language or even single-word descriptors—each offering a different degree of abstraction.

In the following chapter, we will conduct a conceptual analysis using a probability of default (PD) model as an example. This analysis will help us to answer the question of what does a PD model actually measures. In turn, this answer will allow us to elucidate the potential explanatory variables that might be meaningful to predict the default event, in a cause-and-effect fashion.

CHAPTER 2

Conceptual Modeling

"The purpose of abstraction is not to be vague, but to create a new semantic level in which one can be absolutely precise."

—Edsger Dijkstra

In Section 1.1.1, we presented a potential methodology for the development of Altman's model [1]. We commenced at a high-resolution level, beginning with the final mathematical expression, and subsequently proceeded to a lower-resolution level until we reached an expression that could be elucidated in everyday language. In practice, however, the modeling process progresses from a low-resolution level of analysis, which is conveyed in simple terms, to a high-resolution level of analysis, which is expressed in mathematical expressions. But prior to any analysis, it is important to have a deep understanding of the system or phenomenon at study.

2.1 Understanding the System at Study

From this point forward, our attention will be focused on the default event, specifically the retail financial default. A default is defined as the failure of an obligator (i.e., a customer) to meet its credit obligations to the bank in a timely manner or to meet any material credit obligation to the banking group (Basel, 2004 [5]).

Default predictive models are also referred to as probability of default (PD) models, which represent the primary focus of credit risk modeling. The probability of default (PD) can be defined as the likelihood that a loan will not be repaid and will therefore become defaulted. Credit risk modeling can also be regarded as credit scoring (CS) analysis, which is the most well-known and widely used methodology for measuring default risk in consumer lending.

This brief introduction to credit risk concepts, such as credit scoring and probability of default, is important for the sake of context. However, the reader may find ample literature on the CS methodology elsewhere (Siddiqi [63], Brown [9]), as our main concern is not the technical and specific aspects of the scorecard building process, but the mental processes that lead to the construction of any model of any kind. Therefore, the use of the credit default event as an example is a means to explain the modeling process, not an end itself.

2.1.1 What Is a Loan? A Low-Resolution Analysis

A loan can be defined in different ways depending on the source:

> *"A financial instrument in which one party borrows money from another, such as a mortgage, a credit card debt, or personal line of credit."*
>
> *—Investopedia*

> *"A loan is a sum of money that one or more individuals or companies borrow from banks or other financial institutions so as to financially manage planned or unplanned events. In doing so, the borrower incurs a debt, which he has to pay back with interest and within a given period of time."*
>
> *—CFI*

One may agree or disagree with these definitions. One might also find them too vague or too specific, or one might even possess a personal definition of one's own. It is also possible that one has never considered the matter before. In any case, however, these definitions, or indeed any other formal technical definition, do not define what a loan actually is.

The terminologies utilized to define loans and credit are modern inventions. Nevertheless, the concept of a loan can be traced back to 3,500 BC in the Sumerian civilization [1], where seeds and grain served as collateral. A similar practice was observed in which animals were used as security for loans, with repayment obligations being met through the delivery of a new calf.[1] The question therefore arises as to what the constant factor behind the act of lending and borrowing across all civilizations throughout human history might be—the answer is *trust*.

2.1.2 The Magic Ingredient Is Trust

Trust is a fundamental tenet of our financial system. It is at the heart of modern finance, and without it, the system would not function. As an example, consider the value of money. The value of money is a mental construct. This is because the value of money is not intrinsic to the paper on which it is printed, the purpose of which is merely to transform matter into mind.

Harari [28] suggests that the belief in money as a universal, efficient system of mutual trust is a social construct. The one-dollar bill bears the signature of the US Secretary of the Treasury on one side and "*In God We Trust*" on the other. Harari argues the dollar bill is trusted because of faith in God and the Treasury Secretary. In this respect, it is important to acknowledge that an understanding of trust facilitates an understanding of why financial systems are inextricably linked to politics, social structures,

---

[1] Interesting fact: "mas" is the Sumerian word for interest, and this was the same as the word for "calf." In the Hebrew language today, "mas" is the word used for "tax."

and ideologies. Moreover, financial crises, as Harari [28] observes, are frequently precipitated by political dynamics and can exhibit volatility based on sentiments or capricious inclinations.

Consider, for example, the fact that ancient coins had value not only because of the intrinsic value of gold or silver, but because of the mark imprinted on them. A Roman denarius held value across the known world because people trusted the Roman emperor who was imprinted on it. The continued use of *dinar*, derived from denarii, as the official currency name in numerous countries[2] is an illustration of the influence of trust.

This example sheds light on the role of trust in the contemporary financial system and underscores the limitations of conventional loan definitions. The concept of a loan encompasses more than mere monetary sums[3] or contractual agreements; it is, therefore, essential to delineate its precise definition—so what is a loan then?

> *A loan is a commitment, an obligation that we establish with someone else, that is based on trust, trust that both parties will have the integrity to honor their commitment.*

This definition is what is termed an *archetypical definition of a loan*. This definition is regarded as archetypal in nature, in the sense that it is regarded as holding true across all possible examples of loans; you can also call it *the meta-structure of a loan.*

2.1.3 The Contract

In order to provide a more concrete understanding of the nature of a loan, it is useful to associate each of its components with a tangible element. Consequently, a commitment is formalized in a document that

---

[2] Denarii is still the official name of the currency in Jordan, Iraq, Serbia, Macedonia, and Tunisia, and in Spanish-speaking countries, like Mexico, the word "*dinero*" is also derived from the ancient word denarii.

[3] Or it is, insofar as we consider money as a measure of the amount of trust that we have in our state.

delineates the specifics of the agreed-upon terms, which is then referred to as a contract. To formalize the contract and serve as a testament to our commitment, we put our signature on it. A signature represents a person's word. In essence, one is thereby pledging one's word that the contract will be upheld.

Similarly, in order for the contract to be valid, both parties must demonstrate their trustworthiness by honoring their commitment. In the context of personal loans, the borrower is a person, a flesh-and-blood human being. Conversely, the lender is typically a financial institution. However, financial institutions are not tangible entities; they are merely conceptualized by society. This raises the question of how one can trust such an entity.

2.1.4 The Lender

Financial institutions are *"limited liability companies."* In the United States, a limited liability company is a *"corporation"* (from the Latin *"corpus,"* meaning *"body"* [2]). In the American legal system, these symbolic bodies are regarded as legal persons. Thus, placing trust in financial institutions is analogous to placing trust in the power and integrity of the Roman emperor.

However, in this case, it is not the emperor's face that represents financial institutions; it is their logo [4]. The ten most prominent bank logos, as identified by Arek Dvornechuck,[4] branding expert and graphic designer (Ebaqdesign and Inkbot Design), illustrate the significance of trust in financial institutions. **Table 2-1** and **Figure 2-1** provide the meaning of each logo associated with a bank. After reading these definitions, one will be hard-pressed to deny the fact that trust is still the core element of banks' reputation and credibility and of all corporations for that matter.

---

[4] Arek Dvornechuck
 +1 929 - 245 – 9811
 arek@ebaqdesign.com

CHAPTER 2 CONCEPTUAL MODELING

Table 2-1. *List of the top ten banks' logos, their nationality, and their meaning.*

Bank	Origin	Logo's meaning
Bank of America	United States	Patriotism Farm field Blue = trust
Chase Bank	United States	Durability Simplicity Money flow
Citi Bank	United States	Security Protection Good care
TD Bank, N.A	Canadian-American	Connection Coherence Green = money
Deutsche Bank AG	Dutch	Stable growth Financial gain Sustainability
ING Group	Dutch	Royalty, dignity Courage, strength Orange = energy
Societe General	French	Strength Harmony Balance
UBS Group AG	Swiss	Security Confidence Discretion
Raiffeisen Bank	Romanian	Protection Safety Yellow = optimism
Getin Bank	Polish	Forward moving Call to action Freshness

CHAPTER 2   CONCEPTUAL MODELING

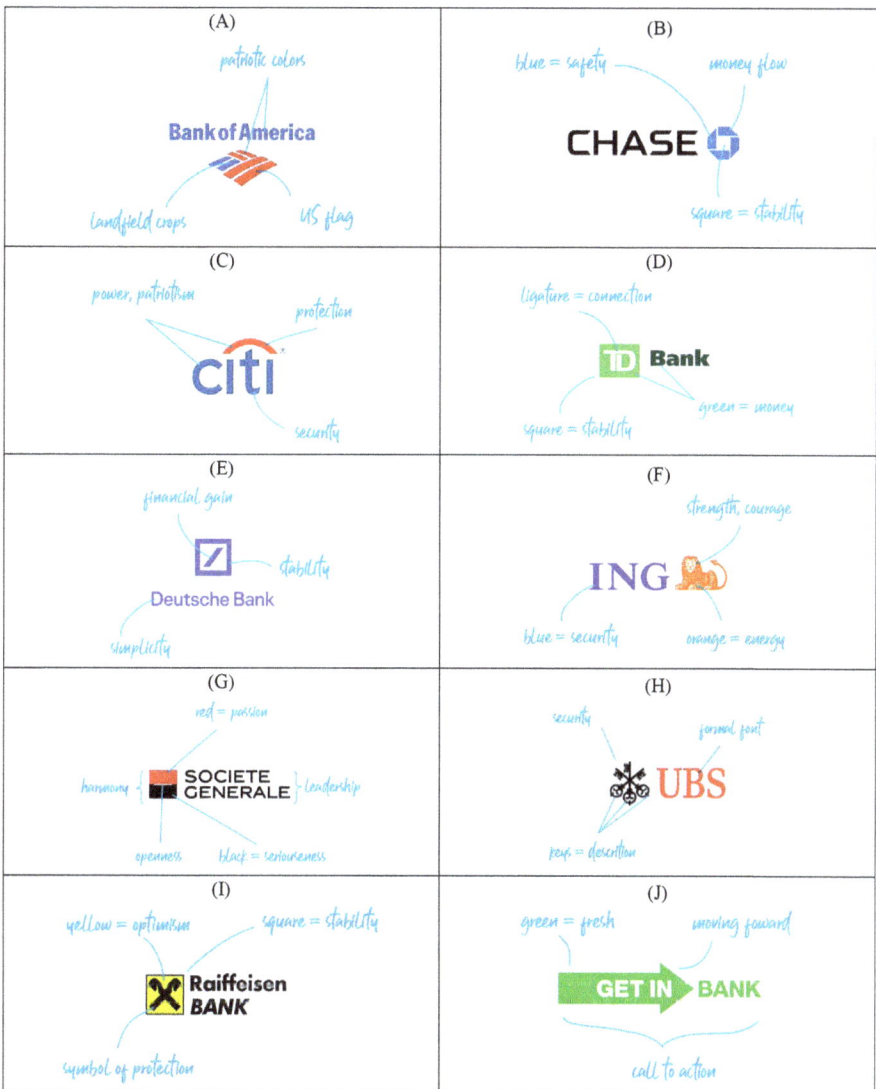

Figure 2-1. *Logos of the ten top banks. (A) Bank of America, (B) Chase Bank, (C) Citi Bank, (D) TD Bank, N.A, (E) Deutsche Bank AG, (F) ING Group, (G) Societe General, (H) UBS Group AG, (I) Raiffeisen Bank, (J) Getin Bank.*

CHAPTER 2   CONCEPTUAL MODELING

It is worth noting that I am not making the case that because banks' logos pretend to convey a message of security, stability, and protection, banks are actually beacons of integrity and trustworthiness, and that you should blindly trust them. The case I am making is that trust is a two-way street, so, for a loan (commitment) to take place, both parties should first trust one another. So, once we have established that there is a reasonable amount of trust on the side of the financial institutions, then we can focus our attention on the other side of the equation–*the borrower*.

2.1.5 The Borrower

So far, we have established that a loan is a commitment and that the lender is a corporation that symbolically represents security and protection, and for the sake of the analysis, we will assume that the lender's *trustworthiness* is somehow granted. But what about the borrower? Can we also assume that their trustworthiness is also granted? After all flesh-and-blood people do not announce themselves with logos on their foreheads or have coins with their faces on them (perhaps not most of them). Nor can we speak of their sense of *integrity* and trustworthiness just as if we had known them for all of their lives, or can we? So, how can we know the trustworthiness of borrowers in advance?–But before we answer this question, we should ask ourselves, what does *integrity* have to do with being trustworthy?

I have been using the word *integrity* deliberately; however, what integrity is, and what do we mean when we use it, is not obvious by any stretch of the imagination. We use the word integrity casually in our everyday conversations and expect everyone to know exactly what we mean by that. It is very likely that not even we really know what we mean when we use it.

2.1.6 Integrity

Becker (2009 [6]) argues that when we refer to someone as a person of integrity, we are really referring to someone's *consistency*. Consistency between one's words and actions, consistency between one's actions and what we claim to honor. But *time* is the key factor here, because consistency unfolds over time as a constant commitment to act in line with one's principles regardless of the circumstances, until such acts become one of the main characteristics that describe us as a person. Petrick and Quinn (2000) [51] describe integrity as follows:

> *"The capability for repeated process alignment of moral awareness, deliberation, character and conduct that demonstrates balanced judgment, enhances sustained moral development and promotes supportive systems for moral decision making."*

From this definition, it can be observed that the term *"moral"* is employed on numerous occasions.

- *Moral* awareness
- *Moral* development
- *Moral* decision-making

In this context, the term *"moral"* or *"morality"* refers to the conduct and/or principles that enable us to distinguish between *good* and *evil*. It can also be defined as *moral judgment*. However, the capacity for moral judgment necessitates the prior establishment of a predefined set of *moral values*. This raises the question of whether moral values are subjective and arbitrary, or where do moral values come from, and whether we all possess such values.

CHAPTER 2 CONCEPTUAL MODELING

2.1.7 Moral Values

In 1958, Lawrence Kohlberg, following Piaget's ideas, proposed that there are six progressive and sequential stages of moral reasoning and understanding [5]:

1. **Obedience and Punishment Orientation:** A simplistic understanding of morality. The individual behaves to avoid punishment and understand that those that have been punished are because they have done something wrong.

2. **Individualism and Exchange:** Realization that morality is relative. The individual understands that there is not only one single point of view when it comes to morality.

3. **Good Interpersonal Relationships:** Realization that the individual is a social being. The individual behaves in order to be seen and perceived as good by others looking for acceptance and a sense of belonging.

4. **Maintaining the Social Order:** Realization that there is a bigger scheme. The individual becomes conscious of the fact that they belong to a higher social order where laws and rules are of outmost importance.

5. **Social Contract and Individual Rights:** Realization that morality is not absolute and can be malleable. The individual understands that rules and laws were made to serve society to achieve a greater good, and not the other way around. When rules and laws do not serve their purpose, they should be modified and not arbitrary imposed.

6. **Universal Principles:** The individual becomes a real individual. People at this stage have developed their own set of complex moral guidelines which may or may not fit the stablished rules and laws. They believe they have reached universal principles, principles that they believe are worth fighting for, perfectly aware of the consequences.

Soon after, Elliot Turiel, a former student of Kohlberg's, continued his legacy. Turiel showed children can recognize universal moral principles. Turiel argues that children construct their moral understanding on the bedrock of the absolute moral truth that *harm is wrong* (Jonathan Haidt, The Righteous Mind [27]). While rules vary across cultures, children can discern between moral and conventional rules. Piaget, Kohlberg, and Turiel are *rationalist/constructivist* thinkers. Rationalists believe moral values are *self-constructed*. They suggest that children learn moral values and differentiate them from *social conventions*, environmental influences, interactions with others, and parental guidance. The central tenet of rationalist theory is that moral values are derived from *reasoning*—unfortunately, recent studies prove these ideas to be incorrect.

In 1975, Edward O. Wilson published a controversial book, *Sociobiology: The New Synthesis* [71]. In his book, Wilson says human behavior is also the result of natural selection and evolution. He asserts the existence of human nature, which may constrain our social interactions. Moreover, Wilson argued that the rationalist argument was a sophisticated rationale for moral intuitions that could be better explained by evolutionary theory. In 1992, a group of economists, philosophers, and neuroscientists developed an alternative approach to morality. It was based on the premise that emotions are the foundation of morality. This peculiar group adopted the term sociobiology and renamed it as *evolutionary psychology*.

2.1.7.1 Evolutionary Psychology

Evolutionary psychology is the application of evolutionary principles to the study of the evolution of mind (Tooby & Cosmids, 1992). It holds that psychological attributes that increased the probability of survival and reproduction are present today in the form of evolved adaptations designed to solve ancestral problems and enhance paternal certainty (Wilson & Daly, 1992), optimizing mate selection (Buss, 1989a), language acquisition (Pinker, 1994), comprehending the mental state of others (Baron-Cohen, 1997), and weighting the cost of risky encounters (Campbell, 1999 [10]).[5]

2.1.7.2 The Origins of Moral Values

Jonathan Haidt, a world-renowned psychologist, presented in his book (*The Righteous Mind* [27]), a comprehensive research, from an evolutionary psychological point of view, about the origin of moral judgment and moral values. In his book, Haidt concluded that when it comes to moral judgment, *intuition come first, strategic reasoning second*. He conducted a series of experiments based on telling volunteers *harmless-taboo stories* and ask them afterwards if they considered the outcome of the story having been morally wrong.

The veracity of these situations is irrelevant to the experiment. The significant finding of Haidt's study was that all subjects exhibited an immediate and emotional moral judgment. Upon prompting, contestants provided more reasons for their responses and demonstrated a reduced tendency to alter their positions.

Haidt found that people make moral judgments quickly and emotionally, using them to justify their opinions. Haidt concluded that moral judgments are cognitive processes that can be divided into two

---

[5] Anne Campbell, A Mind of Her Own: The Evolutionary Psychology of Women.

CHAPTER 2   CONCEPTUAL MODELING

categories: intuition and reasoning. Intuition is the term for the numerous, rapid, and effortless moral judgments and decisions that are made on a daily basis by all individuals.

It is worth mentioning that *moral emotions* are a type of moral intuitions. In *Emotions Revealed*, Paul Ekman [18] says emotions arise when people think something will affect their well-being. Similarly, Ekman agrees with Haidt that emotions are *automatic-appraising mechanisms* and serve a survival function. He believes emotions are survival mechanisms developed over millions of years. Furthermore, humans are born with an evolving sensitivity to events pertinent to our ancestral hunter-gatherer environment. The themes for which the auto-appraisers are constantly scanning our environment, usually without our knowledge, have been selected over the course of our evolution.

The final conclusion of Ekman is that there are six universal emotions across all cultures:

1. Happiness
2. Anger
3. Disgust
4. Sadness
5. Fear
6. Surprise[6]

Similarly, Haidt reached the conclusion that there are six universal moral values:

1. Care/harm
2. Liberty/oppression
3. Fairness/cheating

---

[6] Fear and surprise are sometimes considered as one single emotion.

4. Loyalty/betrayal
5. Authority/subversion
6. Sanctity/degradation

Both, Ekman and Haidt imply that emotions and moral values are innate to human nature, although Haidt makes an important annotation about innateness from the neuroscientist Gary Marcus:

> "Nature bestows upon the newborn a considerably complex brain, but one that is best seen as *prewired*—flexible and subject to change—rather than *hardwired*, fixed, and immutable.... Nature provides a first draft, which experience then revises.... 'Built-in' does not mean unmalleable; it means '*organized in advance of experience.*'"

This annotation is of utmost importance to make the distinction between those who believe that human nature is a blank slate on which any utopian vision can be sketched and those who believe that genetics are destiny. Neither of them is correct; what this annotation entails is that while millions of years of evolution have endowed us with a preset package of basic and vital information that gives us the capabilities that our ancestors needed to strive and survive, evolution has also endowed us with a malleable brain that is capable of evolving for itself over its own lifetime.

2.1.8 What Does a PD Model Actually Measure?

Prior to embarking on any modeling endeavor, it is imperative to ascertain the precise nature of the phenomenon that the model is intended to measure. It was previously suggested that the PD model is measuring trust, or the level of trustworthiness of the account owner being evaluated. Moreover, if the PD model is indeed measuring trustworthiness, it is reasonable to suggest that a PD model can be a function of one's *character*.[7] Character can be understood as the set of qualities and traits that define and distinguish one's behavior from others. It was therefore necessary to provide an explanation of the concept of moral values and their source in order to substantiate the assertion that a borrower's trustworthiness is contingent upon their moral values. Let us now examine this path in a linear sequence of events.

Borrower → Trust → Integrity → Consistency → Moral judgment → Moral values

Postulate 1: The initial step in establishing trust with a borrower is for the borrower to demonstrate integrity.

Postulate 2: In order to demonstrate that one is a person of integrity, it is first necessary to provide evidence of consistency in one's moral judgments.

However, moral judgments are dependent upon the borrower's moral values. This again raises the three questions initially posed at the conclusion of Section 2.1.6, but now, we have the proper answers to each of them:

1. **Are Moral Values Subjective and Arbitrary?:**
   No, they are not. Although different cultures may have variations in the way they worth and express moral values, all of them are based on the same six basic moral values of care, liberty, fairness, loyalty, authority, and sanctity.

---

[7] It is worth noting that the term "character" is employed extensively within the banking sector as a key factor influencing risk assessment. It is, however, important to clarify that the term "character" in and of itself does not encompass any meaning other than the set of characteristics that define one's behavior. Consequently, the term "character" is devoid of any inherent positive or negative connotation. Moral values, on the other hand, have a positive connotation.

2. **Where Do They Come From?:** Moral values are the result of millions of years of evolution that resulted in innate behaviors (knowledge before experience) that increased the probability of survival and reproduction.

3. **Do We All Have Them?:** Yes, all human beings are born with a preset of moral values.

In light of the substantial evidence presented by recent scientific research, it can be reasonably concluded that moral values do indeed exist and are universal. This finding suggests that moral values are not merely illusions, but tangible realities. If these values are to be considered real, it can be posited that they can be quantified and utilized as fundamental building blocks for a comprehensive *scorecard model*.

2.1.9 Consideration of External Factors

It is crucial to underscore that when we make reference to the borrower, we are operating under the assumption that the borrower is a human being. Furthermore, it has been assumed that the borrower is wholly and exclusively responsible for repaying the loan in a timely and appropriate manner. However, this is not entirely accurate.

The objective is to construct a case around the individual and propose that their integrity and consistency will predict the likelihood of their fulfillment of financial obligations. Nevertheless, our assumption is that the borrower has complete control over the situation, which may not be entirely accurate given the potential for unforeseen circumstances beyond anyone's control, such events as wars, pandemics, natural disasters, and economic crises. Such occurrences are not uncommon, and it would be prudent to incorporate them into the model when feasible. Nevertheless, the objective of this book is to present a universal methodology for modeling any phenomenon, regardless of the topic or the specifics.

CHAPTER 2   CONCEPTUAL MODELING

For the sake of simplicity, this study will focus exclusively on the borrower's behavior and motivations to comply with financial obligations, excluding external factors for the time being.

2.2 Summary

Chapter 2 represents the core of conceptual modeling, demonstrating the means of conducting a comprehensive analysis to gain insight into the system under study. In this instance, the system under study is the default event for retail loans. Subsequently, the loan was decomposed into its fundamental constituents, and it was determined that a loan is constituted of three essential elements:

1. A contract
2. A lender
3. A borrower

And a fourth essential element, namely, *trust*. It has been demonstrated that trust is the fundamental basis of our financial system. Without trust, there would be no possibility of trade. In light of this, it is now necessary to define the default event in terms of trust.

> *The likelihood that our borrower will fulfill their financial obligations depends upon how trustworthy they are.*

Given that the probability of default is now a function of trust, it is necessary to develop a means of measuring trust. We then argued that an individual's degree of trustworthiness is contingent upon their level of integrity. We can define integrity as follows:

> *Being a person of integrity means consistently conduct oneself by a good moral compass and having the fortitude to defend one's principles regardless of the circumstances.*

However, in order to possess a moral compass, one must first establish a foundation of moral values. This chapter presents substantial evidence that moral values are not merely subjective and arbitrary concepts. Rather, they are the result of millions of years of evolution, during which innately survival- and reproduction-enhancing behaviors (i.e., knowledge acquired prior to experience) emerged and became fixed. In light of this evidence, we put forth the proposition that integrity—that is, our collective set of morally sound behaviors over time—can be utilized as explanatory variables to construct our scorecard model. In more specific terms, the probability of default can be defined as follows:

> *The likelihood that our borrower will fulfill their financial obligations depends on how consistently they have been conducting themselves in a good moral manner and how many times they have shown the fortitude to defend their principles.*

This definition begins to sound much more like *the balance equation* that we proposed in **Equation 1-3**. And that happens to be the subject of the next chapter.

CHAPTER 3

Balance Equation

> "Lesson not just karate only, lesson for whole life! Whole life have a balance, everything be better."
>
> —Mr. Miyagi, Karate Kid

In this section, the balance equation will be employed as a tool to elucidate the principal elements affecting the system under study. Particular emphasis will be placed on the fact that empirical data demonstrate that the existence of previous commitments may significantly predict the outcome of future commitments, specifically in the context of financial obligations. It is important to note that the use of certain variables presented in this section, such as marital status and economic dependents, may be subject to local regulations and therefore should be used with caution. It is also important to note that the use of the aforementioned variables is optional. They are presented here merely as illustrative examples of previous commitments that demonstrate a strong correlation with the default event. It is important to note, however, that the strength of the association between these variables does not guarantee their inclusion in the final main-effects multivariable model. Similarly, the exclusion of these variables from the final model does not necessarily indicate a lack of association with the target variable. Rather, it suggests that their association was insufficient to justify their inclusion in the model, or that other variables may have emerged to explain the same variation in the data.

CHAPTER 3   BALANCE EQUATION

3.1 Analysis of Contributing and Subtracting Forces

As previously discussed in Section 1.2, a balance equation represents a specific type of analysis, primarily utilized in scientific and engineering contexts. This analytical approach facilitates the identification and decomposition of the primary contributing and subtracting forces influencing the system under investigation. In this case, the system under study is the default event, or, more specifically, *the probability that a borrower will fulfill their financial obligations.*

Figure 3-1 shows a similar diagram as the one presented in **Figure 1-2**, but in this case, trust is at the center of the analysis.

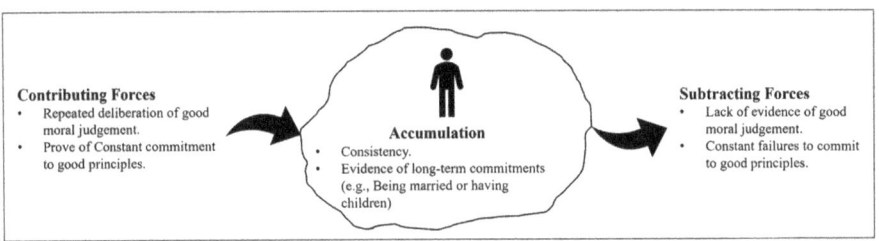

Figure 3-1. *Balance equation for the default event based on a system of trust.*

A close examination of the accumulation factors illustrated in **Figure 3-1** reveals the initial instances of explanatory variables that are directly associated with trust or, more specifically, with *commitment*. It would be challenging to identify a more illustrative representation of commitment than *marriage.*[1] Many young people today hold the view that marriage is not a prerequisite for demonstrating commitment to one's partner; some even claim that marriage is merely a *"piece of paper."* However, is this characterization accurate?

---

[1] It should be noted that the utilization of marriage or marital status as an explanatory variable may be prohibited by local legislations in certain countries.

3.2 The Married with Children Effect

In my experience, marriage or marital status has proven to be an extraordinary predictor of the default rate (DR), and I have seen the same effect on numerous occasions, consistently indicating that married individuals exhibit a reduced risk of default.

As illustrated in **Table 3-1**, the default rate (DR) for single individuals is 3.89%, while the DR for married individuals is 1.97%. This represents a significant disparity, with the DR for married individuals being approximately half that of the DR for single individuals (51%). It is evident that the effect of marital status is significant in reducing the DR. However, it is essential to determine whether this effect is directly related to an individual's level of commitment or if there are other underlying factors at play. In other words, is the DR a function of an individual's capacity to make long-term commitments?

Table 3-1. *Real case of the effect of marital status on default rate*[2].

Marital status	Nondefault	Default	Total	Default rate
Single	28,966	1,171	30,137	**3.89%**
Married	16,860	339	17,199	**1.97%**
	45,826	**1,510**	**47,336**	**3.19%**

One of the advantages of working as a consultant is the opportunity to engage with a diverse range of data sources. One of the most frequently requested pieces of information from prospective customers by banks is their marital status. In the event that customers indicate that they are single, they are occasionally queried as to whether they reside in a union,

---

[2] The information presented belongs to Latin American banks.

are divorced, or are widowed. However, in the case of applicants who declare that they are married, it is uncommon for banks to inquire about the *matrimonial property regime* (MPR) of their prospective customers. This refers to the differentiation between the *separation of goods* and the *community of acquired assets*.[3]

For those unfamiliar with the term matrimonial property regime, such as residents of the UK and select regions of the United States, the matrimonial property regime is a legal framework utilized to ascertain the intended manner of asset ownership in the future. For example, the separation of goods stipulates that the assets held in the names of the married couple remain the property of the respective owners. Conversely, the community of acquired assets regime stipulates that from the date of marriage, all assets purchased from that point forward are deemed to belong to both parties, irrespective of the source of funds used to make the purchase.

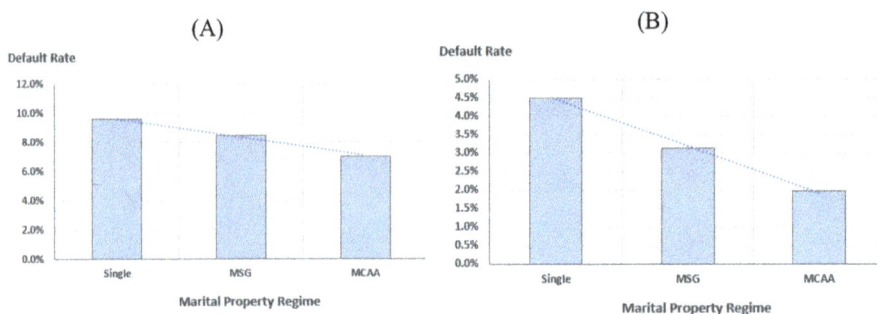

Figure 3-2. *Column bar chart, where MSG = married by separation of goods and MCAA = married by community of acquired assets. (A) corresponds to case 1, and (B) corresponds to case 2.*

---

[3] In countries like France, there is a third matrimonial property regime called universal community property.

CHAPTER 3   BALANCE EQUATION

Tables **3-2** and **3-3** illustrate two instances from Mexican banking institutions where the matrimonial property regime was specified during the application process. It is notable that in both cases, there is a decreasing monotonic linear relationship between the DR and the MPR (**Figure 3-2**). The Mantel-Haenszel χ^2 test for ordinal association (test not shown) yielded extremely significant *p*-values (less than 0.0001) for both cases, thereby strongly suggesting the existence of the association between the DR and the MPR. However, it remains unclear whether this negative correlation is truly attributable to a lack of commitment or trust on the part of the borrower. In other words, does the evidence indicate that *individuals who are reluctant to make long-term commitments may also be more prone to fail in their long-term financial commitments?* The available evidence appears to support this conclusion, but it is crucial to maintain a clear focus on the central issue. The decision to enter into a marital union through the exchange of separate property is not the primary focus of this analysis. Instead, the objective is to demonstrate how the observed data aligns with our hypothesis that previous long-term commitments may predict the behavior of future long-term commitments. The question thus arises as to whether the status of being single is the result of circumstance or of deliberate choice.

Table 3-2. *The effect of marital status on the default rate for different matrimonial property regimes, case 1.*

Marital status	Nondefault	Default	Total	Default rate
Single	22,845	2,426	25,271	**9.60%**
Married by separation of goods	4,764	443	5,207	**8.51%**
Married by community of acquired assets	15,115	1,141	16,256	**7.02%**
	42,724	**4,010**	**46,734**	**8.58%**

CHAPTER 3 BALANCE EQUATION

Table 3-3. *The effect of marital status on the default rate for different matrimonial property regimes, case 2.*

Marital status	Nondefault	Default	Total	Default rate
Single	9,786	462	10,248	**4.51%**
Married by separation of goods	5,467	177	5,644	**3.14%**
Married by community of acquired assets	8,371	168	8,539	**1.97%**
	23,624	**807**	**24,431**	**3.30%**

The available data appear to indicate that the primary reason for being single is that individuals are young. **Figure 3-3** illustrates the percentage of people who are married[4] as a function of age[5] (for the sake of convenience, age ranks were estimated from decile cutoffs). As might be expected, people tend to marry as they grow older. However, is being married the sole characteristic that accompanies the process of aging?

---

[4] Any status other than married, such as divorced, widowed, union, etc., was labeled as single. The reason why all these categories were lumped together into the single category is because there are only two marital statuses recognized by law, which are married or single.

[5] The sample size was of 139,228 applications.

CHAPTER 3   BALANCE EQUATION

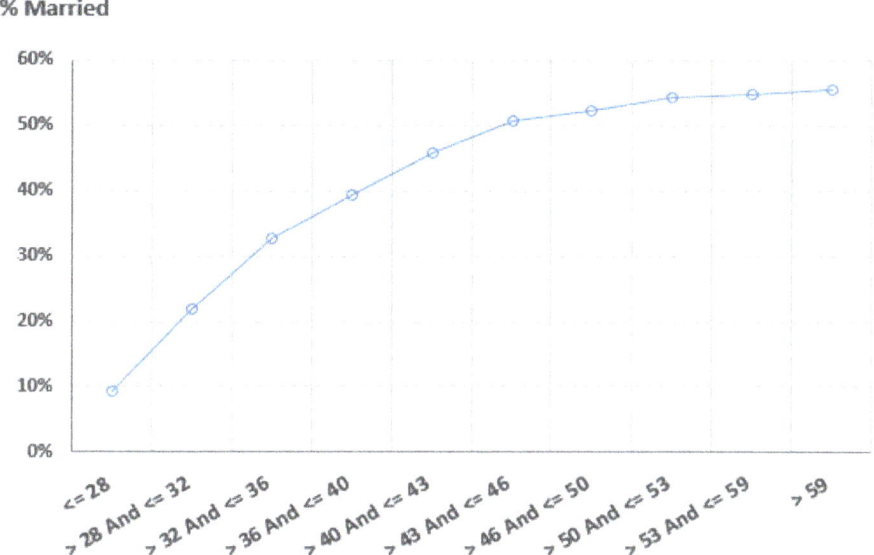

Figure 3-3. *Percentage of married persons by age rank.*

Figure 3-4 illustrates the default rate as a function of age for married and single individuals. As illustrated in the figure, the presence of a marital status appears to exert a mitigating influence on the default rate until the mid-forties. However, beyond this age threshold, the protective effect of marriage appears to dissipate. This is a significant finding because numerous positive characteristics are often linked with advancing age. These include increased maturity, responsibility, stability, and *higher income* (Discrimination and Disparities, Thomas Sowell, page 23 [64]).

37

CHAPTER 3 BALANCE EQUATION

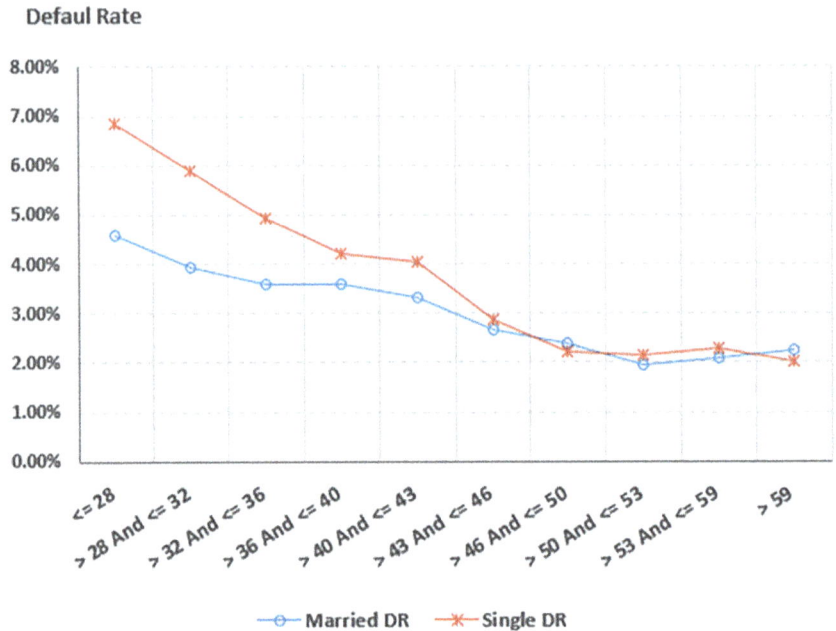

Figure 3-4. *Default rate as a function of age for married and single people.*

The data indicate that age is a significant determinant of marital status. In statistical terms, it indicates that the relationship between the DR, marital status, and age is *confounded*.[6] This finding is expected when one considers that the majority of individuals tend to seek a life partner and establish a long-term relationship as they age. Furthermore, **Figure 3-5** presents an example of the relationship that exists between the average monthly income and age rank.[7] As the figure illustrates, income is also a confounded variable, which is consistent with Sowell's [64] observations, that, on average, advanced age always corresponds to higher incomes. Conversely, higher incomes have been found to be negatively correlated with the DR.

---

[6] The term confounder is used to describe a covariate that is associated with both the outcome variable of interest and a primary independent variable.

[7] The dataset under consideration corresponds to a major banking institution in Mexico. The sample comprises over 100,000 loan applications.

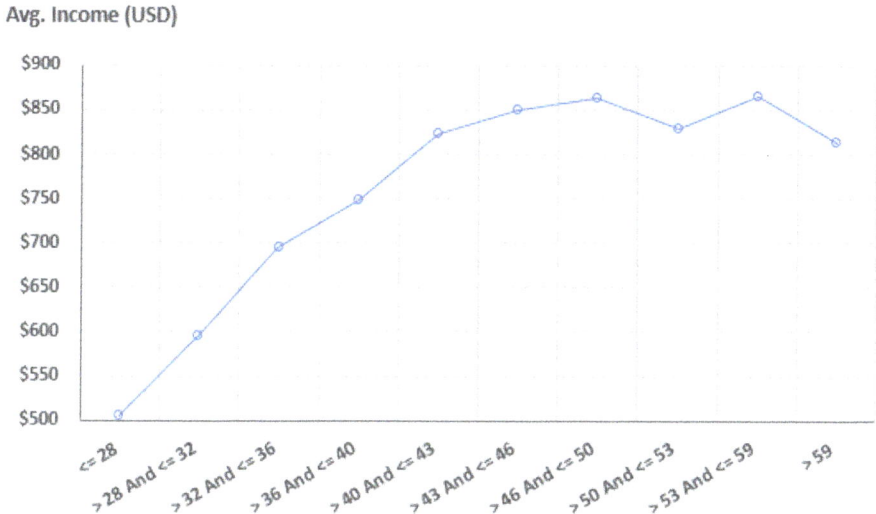

Figure 3-5. *Average monthly income by age rank.*

However, the question remains as to whether singles could also be subdivided into singles by choice and singles by circumstance. The question thus arises as to whether such a phenomenon exists. The present banking institutions do not collect information on whether individuals are single by choice. Nevertheless, the available data suggest that the higher DR reported for singles is mainly due to the fact that singles tend to be younger than married people. However, one should not lose sight of the fact that marital status, income, and number of dependents are confounded with age, which aligns with the idea that integrity plays an important role in predicting the default event. This becomes clear when we consider that integrity is defined as consistency in our moral compass across time. It is clear then that older individuals have a more complete record of their past commitments than young people do. In this respect, it becomes clear why younger individuals tend to have a higher probability of default—*because there is less certainty about what they will do in the future.*

CHAPTER 3   BALANCE EQUATION

3.2.1 Think of the Children

One of the consequences of being married is having children. It is often assumed that the financial obligations associated with raising multiple children are more challenging to fulfill as the number of children increases. However, is this assumption accurate?

Figure 3-6 illustrates the DR as a function of the number of dependent children.[8] As can be observed, the DR does not increase in conjunction with the number of dependent children; rather, it declines as the number of children rises (p-value <0.001 for the Mantel-Haenszel χ^2 test of association). What, then, is the significance of this finding?

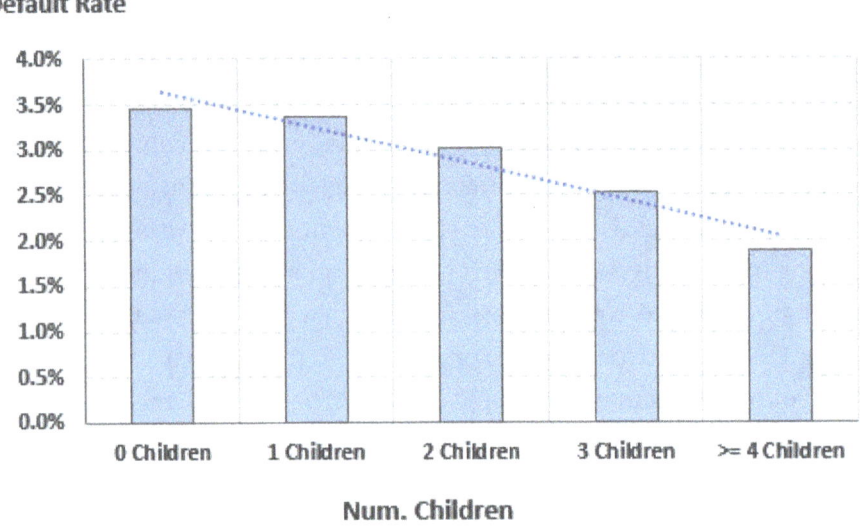

Figure 3-6. Default rate as a function of the number of dependent children.

---

[8] The sample size was of 47,337 applications. The population was comprised of people with access to the financial system, with bureau information available, and all of them were approved during the application process. Therefore, it is very likely that the population, in this particular sample, belonged to middle to upper middle class.

The concept that *"having children is a lifelong responsibility"* is a pervasive one and one that is well-founded. Nevertheless, it is worth questioning whether having children makes individuals more responsible and less likely to default on their financial obligations. An alternative hypothesis is that individuals who are more responsible are more likely to have children. It is possible that two mutually exclusive propositions may be true simultaneously, although the former seems more probable. The crucial point is not whether one becomes a responsible person as a result of having children or whether one was already a responsible person prior to having children. Rather, the point is that *"commitment"* appears to be the common factor in reducing the impact on the DR.

3.2.2 Summary of the Married with Children Effect

The primary objective of demonstrating the inverse relationship between marital status and the DR was to substantiate, with empirical evidence, that the variables that more accurately predict the default event are those that indirectly reflect the borrower's *character*. In this context, character refers to *the borrower's level of commitment, responsibility, moral values, and fortitude in acting in accordance with their principles*. These characteristics are also referred to as integrity. It should be noted, however, that marital status and the presence of dependent children are not the only variables that could reflect an individual's character. The rationale for initially focusing on these variables as examples of explanatory variables is twofold: firstly, they are relatively simple concepts that are easily comprehensible for most people, and secondly, they are concepts that are widely applicable in a variety of contexts.

The following sections will present a methodology for *extracting character from behavior*, regardless of the data source. Furthermore, it will demonstrate other types of variables that are not directly related to character but that also have explanatory power in predicting the default event.

3.3 Extracting Character from Behavior

It has been shown that there is a strong correlation between an individual's marital status, matrimonial property regime, age, and the presence of dependent children and their level of commitment and responsibility. We posit that this relationship is readily discernible due to its reliance on fundamental concepts with which the majority of individuals are intimately familiar. However, variables that reflect character can also be extracted from transactional data. Nevertheless, the process of extracting character from transactional data is not straightforward.

3.3.1 From Theory to Practice: The Merchant Category Codes

Prior to utilizing transactional data, it is imperative to ascertain the meaning of these transactions. The most effective method for interpreting transactions is through the use of the *merchant category code* (MCC). The merchant category code (MCC) is a four-digit numeric identifier that specifies the type of business and the goods and services purchased through a bank transaction, regardless of whether the transaction was conducted using any credit or electronic payment. This catalog is of considerable utility in that it permits the aggregation of disparate categories within a conceptual framework that is grounded in *reality*. The categories available in the MCCs are as follows:

General merchant category codes:
- Agricultural services
- Contracted services
- Transportation services
- Utility services

CHAPTER 3   BALANCE EQUATION

- Retail outlet services
- Clothing stores
- Miscellaneous stores
- Business services
- Professional services and membership organizations
- Government services

Travel and entertainment (T&E) merchant category codes:

- Airlines
- Car rental
- Lodging

It is crucial to acknowledge that the MCCs represent merely one method of categorizing codes. Once the detailed descriptions of each code have been obtained, they can be grouped together in accordance with the requirements of the particular purposes for which they are to be used. It is important to maintain focus on the objective of creating variables that can reflect character traits.

3.3.2 Basic Elements of Daily Life

An alternative method for categorization is to attempt to represent the fundamental elements of the typical person's daily life. The rationale for representing *the basic elements of daily life* is to attempt to encompass the most prevalent aspects that comprise the life of an average person. By doing so, it is possible to ascertain the relative importance of each element for each individual.

CHAPTER 3   BALANCE EQUATION

Figure 3-7 shows an example of the categories that may represent the basic elements of daily life:

1. Education
2. Medical services
3. Communication services
4. Purchase of food and groceries
5. Transportation services
6. Gas, water, and electricity services
7. Cleaning and hair and barber services
8. Home improvements expenses
9. Pets' expenses
10. Clothing expenses
11. Sports-related expenses
12. Entertainment-related expenses
13. The rest, mainly hobbies and luxuries
14. Savings and investments

CHAPTER 3   BALANCE EQUATION

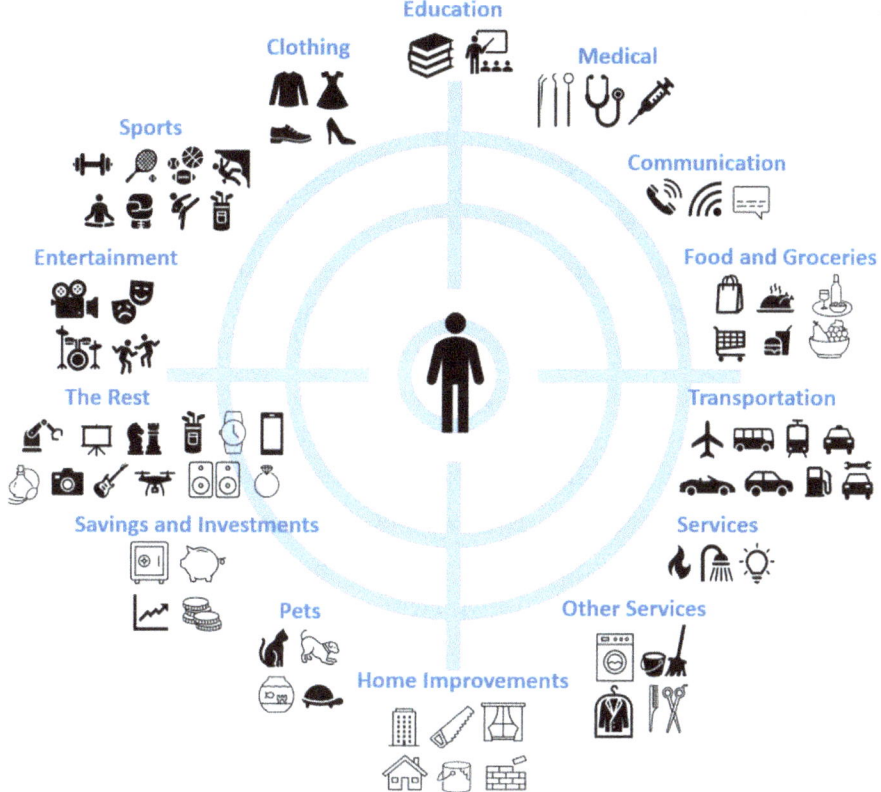

Figure 3-7. Example of basic elements of daily life.

It is notable that with the exception of savings and investments, all credit and debit card transactions can be properly classified within one of the remaining 13 categories. In order to properly classify savings and investments, it is necessary to obtain information regarding savings and current accounts, as well as data on financial instruments, including bonds, funds, and stocks.

It is crucial to highlight that the classification method is fundamental in developing *robust* variables. In this context, *robustness* can be defined as the completeness of information available for each subject of analysis, whether at the application, account, or customer level. To illustrate,

CHAPTER 3 BALANCE EQUATION

consider the creation of a category encompassing airline transactions. If the target population regularly utilizes airline services, for example, due to a proclivity for travel, then the intended variable is likely to have sufficient information and could, in theory, serve as an effective explanatory variable. However, if the target population does not utilize airline services with any regularity, or if the category is further subdivided to encompass every airline available, it is highly probable that the majority of observations within the dataset will lack pertinent information, resulting in the intended variables being discarded due to their lack of predictive power. The question thus arises as to how one might group together the various codes. Furthermore, it is necessary to determine the number of categories that can be combined to form a robust variable. Additionally, there is a need to establish the circumstances under which it is advisable to subdivide a category into more detailed subunits.

Unfortunately, there is no straightforward answer to these questions, as the process of combining categories is inherently a trial-and-error process and requires a degree of experimentation. The initial recommendation is to conduct an exploratory analysis prior to and following the lumping, observing, and comparing the resulting distributions. Nevertheless, conducting an exploratory analysis without a clear objective is not an effective approach. It is important to note that variable design is not an exact science and that it is often not feasible to consult with an expert in this field. Instead, I propose a controversial suggestion—why not rely on your *common sense*?

Recommending the use of common sense to create ones variables is a bit of a dead end, because it assumes that the other person knows exactly what it means to use your common sense. However, common sense is not self-evident. The term encompasses a range of meanings for different individuals, but for the purposes of this discussion, it can be defined as a combination of *intuitive judgment* and *logical reasoning*, including *inductive*, *deductive*, and *abductive* processes.

3.4 Common Sense and the Inductive, Deductive, and Abductive Reasoning

Inductive reasoning [6] is concerned with probability, which is a data-driven concept. One might argue that all scientific research is inductive in nature, as it initially observes a phenomenon of interest and then gathers sufficient information to draw a generalized conclusion. For example, one might wish to ascertain whether the concept of consolidating sports-related expenditures is a viable proposition. Thus, prior to determining the viability of consolidating sports expenses, it is essential to quantify the number of cases with available data and compare it to the total target population to ascertain the potential value of such an endeavor. In the event that only 1% of the observations within the target population have recorded sports expenses, it would be prudent to abandon the idea entirely. Conversely, if a minimum of 20% of observations within the target population have recorded sports expenditures, then the use of sports expenditures as a lumping category may prove to be a viable approach.

In contrast, deductive reasoning [6] is concerned with certainty and is therefore logic-driven. Deductive reasoning begins with a known fact or general rule and proceeds to a specific conclusion. To illustrate, consider the construction of a PD model for a cohort of customers who availed themselves of a promotional offer providing a 30% discount on purchases exceeding $1,000 USD from a curated list of participating retailers over the past year. For the sake of simplicity, we will assume that a flag *"has used the promotion in the last 12 months"* exists and can be used to select the target population. This implies that the sample should only include those customers who purchased at least one item from the participating retailers. An explanatory variable is then constructed as the number of items purchased from the list of selected stores in the past 12 months. It is found that more than 20% of the selected customers have made no purchases over this period. Thus, it can be concluded with absolute certainty that there has been a failure in tracking the use of the promotion, since it is impossible to have used the promotion in the last 12 months without purchasing an item.

Abductive reasoning [6] is akin to the process of detective thinking, whereby a hypothesis is formulated based on a limited amount of information. In other words, abductive reasoning attempts to establish a connection between two seemingly unrelated events based on a synthesis of available evidence—that is, *a sense of causality*.

The design of conceptual variables is an exemplar of this type of reasoning, as evidenced by numerous previously discussed applications. For example, based on historical data and logical reasoning, we have determined that a loan is a personal commitment rather than a simple transaction. If a loan is indeed a personal commitment, then evidence of past commitments fulfilled is evidence of high probability of future commitments fulfilled and also evidence of someone's trustworthiness. In light of these conclusions, it is reasonable to infer that the following elements are all evidence of present commitments, which in turn are evidence of the fulfillment of future commitments, and therefore are evidence of the likelihood of the fulfillment of financial obligations or, in other words, the likelihood that someone will pay their loan.

- How committed is someone with their children's education?
- How committed is someone with their family's health?
- How committed is someone to be in good physical shape and to be clean and presentable to go to work every day to provide for their family?
- How committed is someone to keep their family's home in good conditions?
- How committed is someone with their family's nutrition?

- How committed is someone to going out with their partner and/or family to have a good time in order to strengthen couple/family bonds and improve relationships among them?

- How committed is someone to staying in touch with their loved ones?

- And finally, how committed is someone with their pets?

One can make the case that common sense is a synthesis of inductive, deductive, and abductive reasoning. It draws upon not only one's past experiences but also one's instincts, which, as has been demonstrated in previous chapters, are intrinsic to the human condition. Furthermore, gut instincts are not merely arbitrary whims; rather, they are the product of millions of years of evolution. This includes the ability to distinguish between good and evil, as well as to express the appropriate emotion in a given situation. So, is it really unreasonable to ask someone to rely on their common sense?

3.5 From Low-Resolution Concepts to Mathematical Expressions

As previously discussed in Section 3.1, the initial identification of 14 lumping categories served as the foundation for the subsequent development of our explanatory variables. Additionally, the 14 lumping categories could provide insight into the level of commitment an individual may have toward various aspects of their daily life. **Equation 3-1** presents a streamlined representation of the 14 categories in the form of a balance equation, which may be regarded as a *low-resolution conceptual representation* of the balance equation for a default event. It is important to

CHAPTER 3 BALANCE EQUATION

note that the essential concept underlying each element on the right-hand side of the equation can be distilled as follows:

1. Children
2. Family
3. Job/career
4. Home
5. Loving relationship
6. Friends
7. Pets
8. Yourself

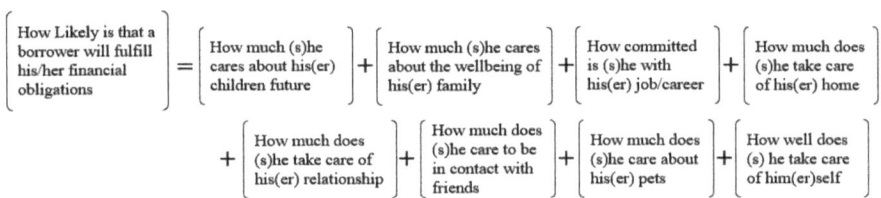

Equation 3-1. Low-resolution conceptual balance equation of a default event.

It should be noted that **Equation 3-1** not only reduces the number of lumping categories into conceptual ideas, but it also provides the first mathematical expression that describes the default event as a function of behaviors that may reflect an individual's character. However, to progress with our balance equation, it is essential to incorporate one additional crucial element—that is, *directionality*.

CHAPTER 3 BALANCE EQUATION

3.5.1 Flux Directionality

One of the fundamental components of a balance equation is the directionality of flux. As has been previously established, a multitude of factors influence the fulfillment of financial obligations. These include the level of commitment to fulfilling other obligations in one's daily life. In practical terms, however, how do these forces interact with our system? If we consider these forces to be monetary fluxes, then it is reasonable to describe our system in terms of input, output, and accumulation/consumption fluxes.

Table 3-4 presents an expanded list of lumping categories, which includes 10 additional categories beyond the previous list of 14 lumping categories.

1. **Cash Withdrawals:** Output flux that includes all types of cash withdrawals, such as ATM, and cashier withdrawals

2. **Account Deposits:** Input flux that considers all types of deposits, such as electronic transfers, cashier, and ATM deposits

3. **Credit Card Payments:** Input flux that considers only credit card payments

4. **Personal Loan Installment:**[9] Sub-input flux that considers only the personal loan installment

5. **Car Loan Installment:** Sub-input flux that considers only the car loan installment

---

[9] The installment we are considering is not the preestablished installment by the bank, but the real payment applied to the outstanding due amount. Most of the time, both quantities will be the same; however, the reason to make the distinction between them is because sometimes people may pay in advance more than one installment at a time.

CHAPTER 3 BALANCE EQUATION

6. **Mortgage Loan Installment:** Sub-input flux that considers only the mortgage loan installment

7. **Credit Card Balance:** Accumulation factor that considers the outstanding credit card balance, reported in the monthly statement

8. **Personal Loan:** Accumulation factor that considers the outstanding balance of all active personal loans, as reported in the monthly statement

9. **Car Loan:** Accumulation factor that considers the outstanding due amount for all active car loans, reported in the monthly statement

10. **Mortgage Loan:** Accumulation factor that considers the outstanding due amount for all active mortgages, reported in the monthly statement

CHAPTER 3   BALANCE EQUATION

Table 3-4. *List of lumping categories influencing the default event with their corresponding flux directionality (input, output, or accumulation/consumption), their conceptual meaning (element of daily life or monetary balance), and their data source (credit, debit transactions, current, savings, and investment accounts, and balance statement).*

Num.	Lumping Category	Code	Data Source	Conceptual Meaning	Direction
1	Education	edu	Credit, Debit card	Element of daily life	Output
2	Medical Services	med	Credit, Debit card	Element of daily life	Output
3	Communication Services	com	Credit, Debit card	Element of daily life	Output
4	Purchase of Food and Groceries	fdg	Credit, Debit card	Element of daily life	Output
5	Transportation Services	Trn	Credit, Debit card	Element of daily life	Output
6	Gas, Water, and Electricity Services	Ser	Credit, Debit card	Element of daily life	Output
7	Cleaning and hair and barber Services	Cln	Credit, Debit card	Element of daily life	Output
8	Home Improvements expenses	Hom	Credit, Debit card	Element of daily life	Output
9	Pets' expenses	Pet	Credit, Debit card	Element of daily life	Output
10	Clothing Expenses	Clt	Credit, Debit card	Element of daily life	Output
11	Sports related expenses	Spr	Credit, Debit card	Element of daily life	Output
12	Entertainment related expenses	Ent	Credit, Debit card	Element of daily life	Output
13	The rest, mainly hobbies and luxuries	Oth	Credit, Debit card	Element of daily life	Output
14	Cash Withdrawals (ATM, and counter)	Csh	Current account, debit card	Monetary Balance	Output
16	Savings and Investments	Svg	Saving, current account, and investments	Element of daily life	Accumulation/Consumption
15	Credit Card Balance	Bal	Statement	Monetary Balance	Accumulation/Consumption
17	Personal Loan (outstanding balance)	Lon	Statement	Monetary Balance	Accumulation/Consumption
18	Car Loan (outstanding balance)	Car	Statement	Monetary Balance	Accumulation/Consumption
19	Mortgage Loan (outstanding balance)	Mor	Statement	Monetary Balance	Accumulation/Consumption
20	Personal Loan instalment	Loi	Statement	Monetary Balance	Input
21	Car Loan instalment	Cai	Statement	Monetary Balance	Input
22	Mortgage Loan instalment	Moi	Statement	Monetary Balance	Input
23	Credit Card Payments	Pay	Statement	Monetary Balance	Input
24	Deposits (All)	Dep	Current account, savings, debit card	Monetary Balance	Input

53

CHAPTER 3 BALANCE EQUATION

It is notable that **Table 3-4** introduces not only additional categories but also a novel classification criterion, encompassing *abbreviated codes, data source, conceptual meaning,* and *directionality*. It is possible to consider the conceptual meaning and directionality of the categories as new dimensions that describe their attributes. Education, for example, is an integral aspect of daily life, yet it also represents a significant monetary output within the system under examination. But more interesting is the fact that the elements presented in **Table 3-4** enable us to present the data pertaining to aggregated categories in the form of mathematical expressions, as it is shown next:

Sets

$J_{Exp} = \{edu, med, com, fdg, trn, ser, cln, hom, pet, clt, spr, ent\}$: Set of all expenses of daily life

$J_{Out} = \{edu, med, com, fdg, trn, ser, cln, hom, pet, clt, spr, ent, csh\}$: Set of all output fluxes

$J_{In} = \{All\ Deposits\}$: Set of all input fluxes

$J_{Liq} = \{Savings, Current\ account, Liquid\ Investments\}$: Set of all assets

$J_{Debt} = \{bal, lon, car, mor,\}$: Set of all outstanding debts (consumption)

$J_{Pay} = \{pay, loi, cai, moi,\}$: Set of all payments

Where **Exp** stands for *daily life expense,* **Out** stands for *output flux,* **In** stands for *input flux,* **Liq** stands for the set of *all liquid assets,* **Debt** stands for *all outstanding debts,* and **Pay** stands for *payment amount*.

Equations

We can now define the total daily expenses as

$$\sum_{i_{exp}=1}^{n_{exp}} C_{i_{exp}}, i_{exp} \in J_{Exp}$$

Equation 3-2. Total daily expenses.

CHAPTER 3   BALANCE EQUATION

Where n_{exp} corresponds to the total number of elements in the set $\mathbf{J}_{Exp}$, and C_{iexp} corresponds to the i_{exp}-th lumping category.

Similarly, we can define the total output flux amount as

$$\sum_{i_{out}=1}^{n_{out}} C_{i_{out}}, i_{out} \in \mathbf{J}_{Out}$$

Equation 3-3. Total output flux amount.

Where n_{out} corresponds to the total number of elements in the set $\mathbf{J}_{Out}$ and C_{iout} corresponds to the i_{out}-th lumping category.

And for the input flux

$$\sum_{i_{In}=1}^{n_{In}} C_{i_{In}}, i_{In} \in \mathbf{J}_{In}$$

Equation 3-4. Total input flux amount.

Where n_{In} corresponds to the total number of elements in the set $\mathbf{J}_{In}$ and C_{iIn} corresponds to the i_{In}-th lumping category.

The total amount of standing debts can be defined as

$$\sum_{i_{debt}=1}^{n_{debt}} C_{i_{debt}}, i_{debt} \in \mathbf{J}_{Debt}$$

Equation 3-5. Total amount of outstanding debts.

Where n_{debt} corresponds to the total number of elements in the set $\mathbf{J}_{Debt}$ and C_{idebt} corresponds to the i_{debt}-th lumping category.

The total amount of payments for the outstanding debts can be defined as

$$\sum_{i_{Pay}=1}^{n_{Pay}} C_{i_{Pay}}, i_{Pay} \in \mathbf{J}_{Pay}$$

Equation 3-6. Total payments for outstanding debts.

CHAPTER 3 BALANCE EQUATION

Where n_{Pay} corresponds to the total number of elements in the set $\mathbf{J}_{Pay}$ and C_{iPay} corresponds to the i_{Pay}-th lumping category.

And finally, the total amount of available liquid assets can be defined as

$$\sum_{i_{Liq}=1}^{n_{Liq}} C_{i_{Liq}}, i_{Liq} \in \mathbf{J}_{Liq}$$

Equation 3-7. Total liquid accounts.

Where n_{Liq} corresponds to the total number of elements in the set $\mathbf{J}_{Liq}$ and C_{iLiq} corresponds to the i_{Liq}-th lumping category.

Notice that all the lumping output categories for the expenses of the daily life can also be described as summations of subsets of MCCs.

$$C_{i_{exp}} = \sum_{j_{i_{exp}}=1}^{m_{i_{exp}}} T_{j_{i_{exp}}}, j_{i_{exp}} \in \mathbf{J}_{MCC_{i_{exp}}}$$

Equation 3-8. All lumping output categories.

Where

$m_{i_{exp}}$ is the total number of MCCs for the i_{exp} th daily life expense,

$T_{j_{i_{exp}}}$ is the $j_{i_{exp}}$ th transaction, and $\mathbf{J}_{MCC_{i_{exp}}}$ is the set of MCCs for the i_{exp}

For explanatory purposes, we may substitute education fees, medical services, and the $m_{n_{exp}}$ expense in **Equation 3-8**.

$\mathbf{J}_{MCC_{edu}} = \{\cdots MCCs\ for\ Education \cdots\}$: Set of MCCs that correspond to transactions related to education fees

$\mathbf{J}_{MCC_{med}} = \{\cdots MCCs\ for\ Medical\ Services \cdots\}$: Set of MCCs that correspond to transactions related to medical services

⋮

CHAPTER 3   BALANCE EQUATION

$$\mathbf{J}_{MCC_{m_{n_{exp}}}} = \{\cdots MCCs \text{ for } m_{n_{exp}} \cdots\}$$: Set of MCCs that correspond to transactions related to the $m_{n_{exp}}$ expense

$$C_{edu} = \sum_{j_{edu}=1}^{m_{edu}} T_{j_{edu}}, j_{edu} \in \mathbf{J}_{MCC_{edu}}$$

Equation 3-9. Total amount of education fees.

$$C_{med} = \sum_{j_{med}=1}^{m_{med}} T_{j_{med}}, j_{med} \in \mathbf{J}_{MCC_{med}}$$

Equation 3-10. Total amount of medical services.

$\vdots$

$$C_{m_{n_{exp}}} = \sum_{j_{m_{n_{exp}}}=1}^{m_{n_{exp}}} T_{j_{m_{n_{exp}}}}, j_{m_{n_{exp}}} \in \mathbf{J}_{MCC_{m_{n_{exp}}}}$$

Equation 3-11. Total amount of the $m_{n_{exp}}$ expense.

We have finally achieved the goal of expressing low-resolution conceptual ideas about the forces influencing the probability of default in proper mathematical expressions. This allows us to quantify the amount of money corresponding to each element of daily life, including the total amount of money corresponding to all daily expenses, all monetary outputs, the total amount of outstanding debt, and the total amount of payments made on a monthly basis. The question that remains is how these monetary values will be used to represent the relative importance of each element of the borrower's daily life.

3.6 Summary

In this chapter, we began our balance equation analysis by identifying factors influencing *the probability of default*, focusing on key data sources such as sociodemographic information and transactional data. We demonstrated that marital status and the presence of children indicate lifetime commitments and are negatively correlated with default rates and that both are confounded with age, which explains why marriage's protective effect diminishes after the mid-forties. Additionally, we highlighted the utility of the MCC catalog in classifying transactions into a proposed scheme of 14 categories representing daily life elements. This grouping, guided by *common sense*—defined as a combination of *inductive*, *deductive*, and *abductive* reasoning—helps assess individual commitments through diverse MCC categories.

In the second section of this chapter, we delved into how to transform broad concepts of daily life expenses and financial commitments into precise mathematical expressions to better understand the factors influencing the probability of default. We began by identifying 14 categories that encompass key aspects of life, such as education, family, and home, and expanded these to include additional financial components like cash withdrawals, loan installments, and outstanding debts. By classifying these categories based on their directionality and linking them to concrete data sources such as credit card transactions or savings accounts, we established a structured approach to quantify financial behaviors. Through the use of equations, we calculated total expenses, monetary flows, outstanding obligations, payments, and liquid assets. By leveraging transaction data grouped through merchant category codes (MCCs), we gained deeper insights into how these financial metrics align with an individual's priorities and commitments. This method enabled us to more effectively analyze the underlying behaviors that contribute to the probability of default.

Key takeaways:

1. **Commitment:** Long-term commitments (e.g., marriage, children) are strong predictors of financial reliability.

2. **Behavioral Insights:** Transactional data can reveal character traits linked to financial behavior.

3. **Mathematical Modeling:** Low-resolution concepts (e.g., daily life elements) are translated into mathematical expressions to quantify financial priorities.

4. **Focus on Character:** Understanding a borrower's character is central to predicting financial outcomes.

In the following chapter, we will provide the missing link needed to represent the relative importance of each element of the borrower's daily life by incorporating the concept of ratios into our variable design, where it will be argued that the magnitudes alone do not convey information because they lack the proper context that gives meaning to our explanatory variables.

CHAPTER 4

Ratios

"The description of this proportion as Golden or Divine is fitting perhaps because it is seen by many to open the door to a deeper understanding of beauty and spirituality in life."

—H.E. Huntley, The Divine Proportion: A Study in Mathematical Beauty

This chapter invites us to reconsider how we approach monetary variables in financial modeling. While these variables are fundamental to understanding financial behavior, they often present challenges that can complicate analysis and fail to convey meaningful information. Here, we explore an alternative perspective—using ratios and frequencies—to better capture the relationships among monetary variables and the dynamics within financial data. This approach ultimately allows us to quantify the relative importance of various aspects of daily life, offering deeper insights into the priorities and commitments that define a customer's financial behavior.

4.1 Dealing with Monetary Variables

The following sections will provide a comprehensive analysis of the most prevalent issues associated with the utilization of monetary variables, with a view to elucidating the rationale for their avoidance.

CHAPTER 4 RATIOS

An alternative approach, namely, ratios and frequencies, will be presented and recommended. It should be noted, however, that the utilization of ratios and frequencies represents merely the initial stages of the modeling process. Chapter 15 represents a continuation of the approach outlined in Chapter 4, whereby the weight of evidence (WoE) for each input will be calculated prior to the construction of the first multivariable scorecard model.

4.1.1 The Trouble with Normal

Two crucial assumptions for linear and logistic regression are linearity and normality (Chaterjee & Hadi [12], Hosmer & Lemeshow [32]). In this context, linearity means that mean values of the outcome variable for each increment of the interval predictor(s) are on a straight line. Normality means the differences between predictions and observed values are randomly distributed and close to 0. The majority of literature cautions against confusing normally distributed residuals with normally distributed predictors. Normal distribution is not a prerequisite for predictors. Nevertheless, it can be argued that this is a statistical euphemism. Interval predictors do not need to be normally distributed, but skewed distributions result in larger residuals and deviate further from normal. It is important to acknowledge that *outliers* are the main cause of skewness in predictors. To further complicate the situation, also consider that some outliers may actually represent *leverage points*.

There are many ways to address outliers. One effective approach is the robust regression (Rousseeuw & Leroy, Robust Regression and Outlier Detection [57]). However, outlier detection and the robust regression are not feasible solutions to deal with outliers, in the context of scorecard models. We will instead focus on the peculiarities of monetary variables and how to address them.

4.1.2 Wealth Distribution and the Matthew Principle

Figure 4-1 illustrates the typical distributions of monetary variables, which are depicted by an L-shaped graph. This type of highly right-skewed distribution represents what is known as Price's law, named after Derek J. de Solla Price, who discovered its applications in 1963. However, the basic principle was discovered much earlier by Vilfredo Pareto (1848-1923), an Italian polymath, who noticed the exact same L-shaped behavior in the distribution of wealth in the early twentieth century (J. Peterson, 12 Rules for Life [50]). It is also known as the Matthew principle, which is derived from the Bible (Matthew 25:29).

"To those who have everything, more will be given; from those who have nothing, everything will be taken."

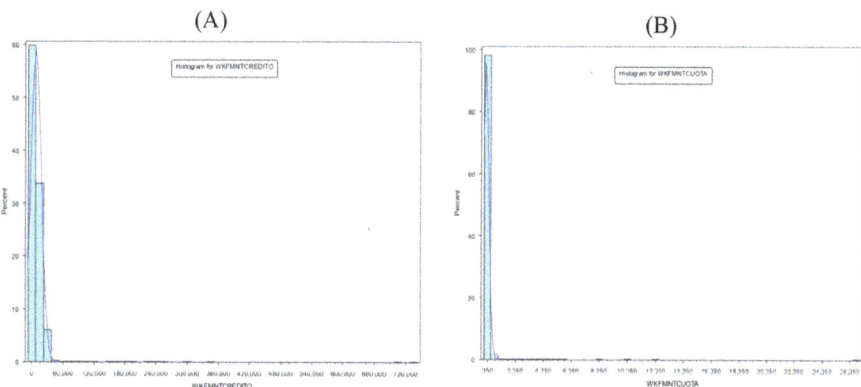

Figure 4-1. *Examples of distribution of monetary variables. (A) Reported requested amount's distribution for retail loans during the application process. (B) Reported estimated installment distribution for retail loans during the application process.*

CHAPTER 4   RATIOS

It would be erroneous, however, to assume that monetary variables are the only examples of L-shaped distributions. In fact, highly right-skewed distributions are prevalent in a multitude of contexts, including across the globe, throughout the universe, and at all points in time.[1]

4.1.3 Scale, Dimensionality, and Units[2]

A further factor to be taken into account in the context of monetary variables is the scale and units employed. This is primarily due to the fact that monetary variables often exhibit a considerable range of values. The issue with *large numerical values* is not one of model correctness or specification; rather, it is the effect that large numerical differences among the explanatory variables have on the computation of parameter estimates and, consequently, on the model's predictions.

To demonstrate the impact of substantial scale disparities on parameter estimates, we present a hypothetical model comprising six explanatory variables:

- Average income in the last 12 months (I_AV_INC_L12M)
- Number of credit card (CC) payments in the last six months (I_FR_CC_PAY_L06M)

---

[1] Just to mention some few examples of phenomena with L-shaped distributions:
All species in all habitats body size (Why are species' body size distributions usually skewed to the right? [34]).
More than 90% of all tornados occur only in the United States (Thomas Sowell [64]).
Lightning occurs far more often in Africa than in Europe and Asia put together not withstanding their superficial size.
Number of species comparing different geographic settings, like number of species found in the America's Amazon vs. the number of species in all Europe.
Number of planets per star in the universe.

[2] The scale and units of the natural monetary variables are not an issue when using the weight of evidence (WoE) transformation since the WoE transformation will precisely deal with this issue and standardized all units to logit units (-ln[Distribution of No Events/Distribution of Events]).

CHAPTER 4 RATIOS

- Number of ATM withdrawals during the last month (I_FR_ATM_WDR_L01M)
- Number of dependent children (I_FR_CHILDREN)
- Max credit card utilization in the last 12 months (I_MX_BAL_CCL_L12M)
- Marital status (N_MARRIED)

Let us now consider the case where the monetary units in question are Costa Rica colones (CRC), with an exchange rate of 590 CRC/USD.

Mathematically speaking, you can describe your model as[3]

$$g(x) = \beta_0$$
$$+ I\_AV\_INC\_L12M \cdot \beta_1$$
$$+ I\_FR\_CC\_PAY\_L06M \cdot \beta_2$$
$$+ I\_FR\_ATM\_WDR\_L01M \cdot \beta_3$$
$$+ I\_FR\_CHILDREN \cdot \beta_4$$
$$+ I\_MX\_BAL\_CCL\_L12M \cdot \beta_5$$
$$+ N\_MARRIED \cdot \beta_6$$

Equation 4-1. Example of a linear model with scale and units issues.

Let us assume the following values for each one of the explanatory variables:

- I_AV_INC_L12M = $10,000 USD.
- I_FR_CC_PAY_L06M = 8.
- I_FR_ATM_WDR_L01M = 2.

---

[3] The nomenclature notation will be explained in Section 8.3.

65

CHAPTER 4   RATIOS

- I_FR_CHILDREN = 3.
- I_MX_BAL_CCL_L12M = 0.82.
- N_MARRIED = 1.

After converting USD into CRC, we have that the average income for the last 12 months is

- I_AV_INC_L12M = $5,900,000 CRC.

As can be observed, the model can be described as a simple linear model by multiplying each explanatory variable by its corresponding regression coefficient. Nevertheless, this approach may also result in *numerical stability issues* [11].

Variables such as the number of payments, the number of withdrawals, the number of children, and so forth are classified as *frequencies*. Frequencies are integer variables that typically range from 0 to tens or occasionally hundreds of units, with rare instances of values falling into the thousands. This characteristic renders frequencies a *controllable* and *robust* variable from a numerical standpoint.

The credit card utilization variable utilized in this example can be classified as a *percentage, proportion,* or *ratio.*[4] It can be defined as a ratio, as it represents the quotient of the outstanding balance divided by the credit limit. In this case, the credit card utilization is represented as a fraction, with values typically ranging from 0 to 1, though exceptional values below 0 or above 1 are possible.

---

[4] See Section 4.2.

Marital status is considered a *nominal* variable due to its qualitative nature, with binary values of 0 or 1. It should be noted that in some cases, marital status may encompass more than two categories, which makes it evident that it will not always be of a binary nature.

The final variable is income, which can be classified as an *interval* variable due to its potential to assume any value between 0 and $+\infty$. It is evident that this income will never reach an infinite value, nor will it have fractional values smaller than a cent unit (in terms of physical money). However, what are the implications of including such a diverse range of units in a single equation?

The inclusion of variables with significant discrepancies in their scale units in the same equation will directly impact the numerical differentiation required to estimate the regression coefficients. It should be noted that all modeling techniques rely on optimization methods in order to estimate the coefficients, or weights, of the model.

It should also be noted that optimization methods are only capable of optimizing a single objective function at a time. This may be either a *minimization* of the sum of squared errors or a *maximization* of the -2 log-likelihood function. Consequently, the significance of identifying a suitable coefficient for a specific explanatory variable is directly correlated with the extent to which that variable influences the reduction or increase of the objective function, in numerical terms.

In other words, variables with *larger magnitudes* will have greater influence over variables with *smaller magnitudes* during the optimization process, not because of any inherent conceptual significance, but simply because of their numerical magnitude.

Furthermore, the numerical value of parameter estimates of large magnitude variables will be negligible, notwithstanding whether the variable is statistically significant or not, not to mention the loss of *significant digits* [11] in your parameter estimates. The loss of significant digits is also known as a *round-off error* [11]. This type of error is due to the fact that computers can only represent quantities with a finite number of digits.

CHAPTER 4　RATIOS

Let us illustrate the round-off error with the following example. Suppose that you ran a logistic regression on the model described in **Equation 4-1**. Now consider that β_1, which corresponds to the average income over the last 12 months, takes the value of -2.23×10^{-6}, which means that $\beta_1 = -0.00000223$. Notice that in this case, the zeros are not a significant figure because they are only needed to locate the decimal point. Therefore, the number -0.00000223 has only three significant figures. The resulting odds ratio (OR) for β_1 will be

$$OR = e^{\beta_1} = 0.99999777$$

This is a meaningless numerical result, caused by a significant discrepancy in the scale units during the optimization process. In contrast, consider β_6, which corresponds to the marital status, and assume that it has a value of -0.7344232. Unlike β_1, β_6 has seven significant figures, which gives us more certainty about the parameter estimate of the marital status. The OR now yields

$$OR = e^{\beta_6} = 0.479782118$$

The OR for marital status, β_6, now has an interpretable result of 0.48[5] in contrast with β_1. However, there is a more significant drawback to utilizing unbalanced units than merely the loss of significant digits and the interpretability of the parameter estimate. This is the *instability*, or *volatility*, of the model's predictions.

---

[5] Which means that the occurrence of the default event is one half as likely to occur among those who are married than among those who are not.

It is important to recall that one of the primary objectives when developing a scorecard is to achieve the most accurate prediction of the default event. This is because erroneous predictions will result in financial losses, either due to the misallocation of funds for loan loss provision or the rejection or acceptance of customers during the application process based on inaccurate assessments.

The issue with unstable or volatile models is that their predictions are highly influenced by parameter estimates such as β_1. It is important to note that the numerical value of β_1 was only due to the large difference in scale between the average annual income, which is expressed in CRC, and the rest of the explanatory variables, which are expressed as frequencies and fractions.

This large difference forces the optimization algorithm to minimize the value of β_1 in order to balance all the elements present within **Equation 4-1**. In conclusion, β_1 is not a meaningful coefficient with a conceptual interpretation; it is merely a *numerical artifice*.

There are multiple approaches to addressing scaling issues. One potential solution is to modify the monetary units of CRCs, either by converting them to thousands, or millions of CRCs, or even US dollars, with the aim of reducing the average income scale and enhancing the number of significant figures. Additionally, the two most commonly utilized scaling techniques, *unit-length scaling* and *standardization*, could have been employed.[6] However, the most significant challenge when working with monetary variables is not the scale or units used to express them, but rather the inherent nature of these variables as *magnitudes*.

---

[6] For a comprehensive overview of these two methods, one may refer to the book *Regression Analysis by Example*, 5th Edition, Chapter 3, by Samprit Chatterjee and Ali S. Hadi [12].

4.2 Ratios vs. Magnitudes

The examination of the prevalent issues pertaining to monetary variables, as outlined in Section 4.1, was a crucial undertaking, not only because the majority of our data sources are of a monetary nature but also because these issues are common to *magnitudes* in general. The definition that we will use when referring to magnitudes is as follows:

> *"A continuous or interval variable in nature used to quantify the size of things (whether physical or virtual) in their given units, and that it is capable of taking values between $-\infty$ and $+\infty$, that may take fractional values but that in most cases will have absolute values greater than one unit."*

It is important to distinguish between frequencies and magnitudes. In a sense, all numerical variables can be considered magnitudes. However, frequencies are a particularly special kind of magnitude, as they can only take integer values. Furthermore, from a statistical and conceptual standpoint, frequencies are often used to quantify *events*. Events represent a distinct category of measurement, as they imply *intention* or *fate*. Therefore, the quantification of events, i.e., frequencies, has an intrinsic meaning by itself, without the need for *context*. To illustrate this idea, consider the following examples:

1. John has withdrawn cash from an ATM 20 times in the past month.
2. John withdrew 200 USD from an ATM in the last month.

From the first example, it can be inferred that John's visits to the ATM are becoming too frequent. Of course, that additional information will be required, such as John's average monthly ATM withdrawals, or the specific number of ATM withdrawals for each month of the past year, to rule out a seasonal or cyclical effect. But just as a matter of *common sense*, do you think that 20 ATM visits in a month is a normal number of withdrawals for an average person? You would be hard-pressed to say "yes" to this question, and that is because you made a *snap judgment* just as soon as you finished reading the first example.

Furthermore, you may be inclined to consider several explanations for the situation, even before you have any additional information at hand; could it be that John made a payroll payment last month? But why in cash? Could it be that John, or his family, had some kind of accident? Or could it be that John is in serious debt and has lost his regular income?

Notice how the first sentence, simple as it is, can trigger multiple responses without the need of any additional context.[7] Now let us turn our attention to the second example and repeat the same exercise by trying to come up with sensible reasons to why John withdrew $200 USD in the last month.

If one is like most people, one will be unable to infer or provide a specific answer to the second example. This is because magnitudes, in and of themselves, do not *convey* any information. Therefore, we can refer to magnitudes as *inert* variables.

Unlike magnitudes, *ratios* convey a substantial amount of information about the underlying phenomenon, but before delving into the benefits of using ratios, let us first define what a ratio is and where it comes from.

4.2.1 A Brief History of Ratios

The Latin word *ratio* was often used as a translation of the Greek word *logos*, which meant *"word," "reason,"* or *"plan"* {2}. In ancient Greek philosophy and early Christian theology, logos was used to describe the divine reason implicit in the cosmos, ordering it and giving it form and meaning {3}. Moreover, ratios are the cornerstone of calculus since the

---

[7] This kind of response is normal for any human beings, and this is due to the fact that our brains evolved to find patterns everywhere and all the time.

CHAPTER 4 RATIOS

very definition of a derivative, which serves as the fundamental vehicle for differentiation, coincides with the definition of a ratio, which is

"The rate of change of a dependent variable with respect to an independent variable."

$$\frac{\Delta y}{\Delta x} = \frac{f(x_i + \Delta x) - f(x_i)}{\Delta x}$$

Equation 4-2. Derivative definition as a difference approximation.

Equation 4-2 illustrates the mathematical definition of the derivative as a difference approximation. If we allow the value of Δx to approach zero, we arrive at the formal mathematical definition of a derivative:

$$\frac{dy}{dx} = \lim_{\Delta x \to 0} \frac{f(x_i + \Delta x) - f(x_i)}{\Delta x}$$

Equation 4-3. Derivative definition.

Subsequently, if we differentiate a simple linear equation, as a function of x, such as

$$(x) = \beta_0 + \beta_1 x$$

Equation 4-4. Simple linear equation.

we obtain

$$\frac{d(\beta_0 + \beta_1 x)}{dx} = \beta_1$$

Equation 4-5. Linear regression coefficient as a derivative.

This implies that all parameter estimates are, in fact, ratios. Consequently, in linear models, all parameter estimates possess meaningful units of measurement that describe the rate of change of the target (dependent variable) with respect to the independent variables in the model. To be more precise, parameter estimates provide the rate of change in the dependent variable in response to an increase of one unit in the independent variable.

4.2.2 A Special Type of Ratio, Dimensionless Numbers

Dimensionless numbers are another type of ratio, and their importance lies in the fact that dimensionless numbers reduce the number of variables needed to describe the system at study. They are most commonly used in engineering as a collection of variables that provide order-of-magnitude estimates of the behavior of a system. They are often derived by combining coefficients from differential equations and are often a ratio between two physical quantities (Membrane Science and Technology Chapter 1 – Correlations [46]).

$$\text{Reynolds number} = \frac{Lv\rho}{\mu}$$

Equation 4-6. The Reynolds number.

Equation 4-6 shows the most common dimensionless number employed in *fluid mechanics*, the *Reynolds number* (Re),[8] named after Osborne Reynolds who published a series of papers describing flow in pipes (Reynolds, 1883). It represents the ratio of inertial forces to viscous forces, which can be expressed conceptually as

---

[8] Where L is the cross-sectional length, v is the fluid velocity, ρ is the fluid density, and μ is the fluid viscosity.

CHAPTER 4 RATIOS

$$\frac{\text{Inertial Force}}{\text{Viscous Force}}$$

Equation 4-7. Reynolds number definition.

The key idea to understand is that dimensionless numbers can express the relationship between the contributing and subtracting forces in any given system. In this case, it is expressed in **Equation 4-7** as the struggle between the force required to keep a fluid in motion (*inertia*) vs. the resistance of that fluid to flow (*viscosity*). Moreover, dimensionless numbers can also help us to identify *critical points* at which the system changes from one state to another.

The idea here is not to become an expert in fluid mechanics, but to show the pervasiveness and importance of ratios in all fields. For instance, all the financial ratios that we saw in Section 1.1.1 are in fact dimensionless numbers, and they also explain the relationship between contributing and subtracting forces, but instead of doing it in a mixing reactor system, they do it in a monetary system, namely, the business enterprise.

It is interesting to mention that the very reason that motivated Altman to write his groundbreaking paper[9] was to defend the use of ratio analysis. In his own words,

> *"Academics seem to be moving toward the elimination of the ratio analysis as an analytical technique in assessing the performance of the business enterprise. Theorist downgrade arbitrary rules of thumb, such as company ratio comparisons, widely used by practitioners...Can we bridge the gap, rather than sever the link, between traditional ratio 'analysis' and more rigorous statistical techniques which have become popular among academicians in recent years?"*

---

[9] Financial Ratios, Discriminant Analysis and The Prediction of Corporate Bank, 1968.

CHAPTER 4   RATIOS

Table 4-1. *Selected list of financial ratios described as the relationship of contributing/subtracting forces.*

Financial ratios	Ratio analysis	Accounting ratio category
$\dfrac{\text{Equity (Book Value)}}{\text{Total Liabilities}}$	$\dfrac{\textit{Worth}}{\textit{Obligations}}$	**Leverage** Your present worth vs. all you owe
$\dfrac{\text{Liabilities}}{\text{Total Assets}}$	$\dfrac{\textit{Immediate Obligations}}{\textit{Resources}}$	
$\dfrac{\text{Working Capital}}{\text{Total Assets}}$	$\dfrac{\textit{Cash}}{\textit{Resources}}$	**Liquidity** Your capacity to produce cash with what you have
$\dfrac{\text{Cash}}{\text{Net Sales}}$	$\dfrac{\textit{Cash}}{\textit{Earnings}}$	
$\dfrac{\text{Net Income}}{\text{Total Assets}}$	$\dfrac{\textit{Real Earnings}}{\textit{Resources}}$	**Profitability** How much wealth have you accumulated with the resources that you have
$\dfrac{\text{Retained Earnings}}{\text{Total Assets}}$	$\dfrac{\textit{Accumulated wealth and growth}}{\textit{Resources}}$	
$\dfrac{\text{Sales}}{\text{Total Assets}}$	$\dfrac{\textit{Profitability in the present economic climate}}{\textit{Resources}}$	**Activity** How profitable you presently are with the resources that you have

CHAPTER 4 RATIOS

It is therefore reasonable to consider whether there are any similarities between Altman's financial ratios and the dimensionless numbers used in *chemical engineering*. To prove this similarity, let us turn our attention to **Table 4-1** that presents us a selected list of financial ratios used by Altman and Sabato (Credit Risk model for SMEs from the US market, 2005 [2]), with the addition of a conceptual interpretation for each ratio.

From this example, it can be observed that the Reynolds number, which represents the ratio of internal forces to viscous forces, is not significantly different from the liquidity indicator, which represents the ability of a company's infrastructure to convert assets into cash, a measure of readiness, if you will.

As an example, consider another dimensionless number, the single most important predictor for the residential mortgage LGD (loss given default), *the current loan-to-value ratio*, better known as the *CLTV ratio*.[10]

$$CLTV = \frac{House\ Market\ Price - Down\ Payment}{House\ Market\ Price} = \frac{Borrowed\ Money\ (Exposure)}{House\ Market\ Price}$$

Equation 4-8. CLTV.

The CLTV represents the extent to which a borrower is willing to lend, or has already lent, against the market value of a property. Nevertheless, despite its apparent simplicity, the CLTV has been demonstrated to be one of the most predictive variables for LGD models, as evidenced by the findings of Min Qi and Xiaolong Yang (2007) [52] and Yanan Zhang, Lu Ji, and Fei Liu (2010) [72]. Furthermore, the CLTV ratio exhibits analogous

---

[10] Working papers: Loss Given Default of High Loan-to-Value Residential Mortgages by Min Qi and Xiaolong Yang, 2007. Local Housing Market Cycle and Loss Given Default: Evidence from Sub-Prime Residential Mortgages, by Yanan Zhang, Lu Ji, and Fei Liu, 2010.

characteristics to the Reynolds number in terms of identifying pivotal points at which the system transitions from one state to another. For example, as documented by Qi and Yang (2007) [52], for loans originated in New England between 1990 and 1994 with a CLTV of $80 \leq CLTV \leq 90$, the mean LGD was 5.3% higher (19.3%) than the mean LGD of the region (14%).

Figure 4-2 presents the findings of Qi and Yang (2007 [52]), which illustrate the sudden rise in the mean LGD values for MICA[11] mortgages at CLTV levels exceeding 80%. It is noteworthy that the system under examination undergoes a notable transition as a function of the CLTV ratio, shifting from a low-loss regime to a high-loss regime at an identifiable cutoff point.

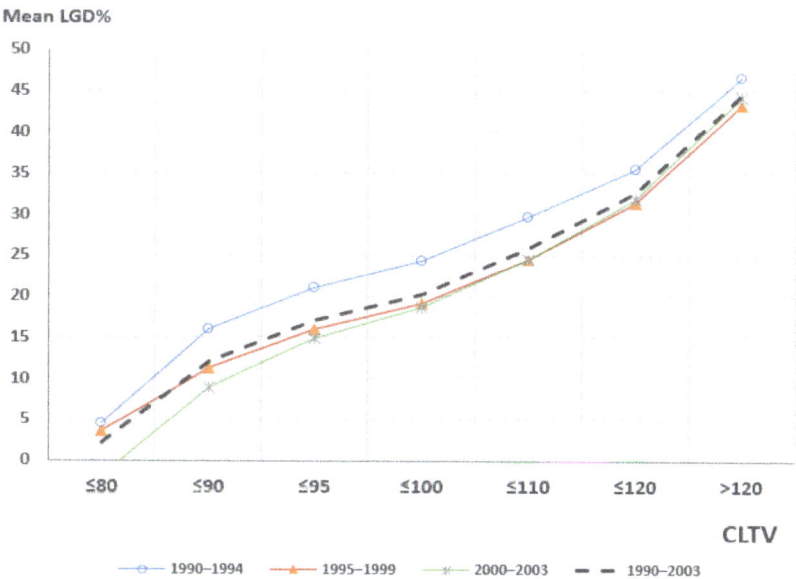

Figure 4-2. National mean loss given default (%) at different time periods and different CLTV cut points.

---

[11] Mortgage Insurance Companies of America.

As can be observed, the utilization of dimensionless numbers is not confined to the domain of chemical engineering. In fact, it is very likely that you have been using dimensionless numbers all along without realizing it.

4.2.3 Intrinsic Benefits of Using Ratios over Magnitudes

As previously demonstrated, ratios offer a more comprehensive representation of data than mere magnitudes. However, the primary reason that ratios convey more information is that ratios confer context to magnitudes.

For instance, knowing that someone spent $500 dollars on a dinner last month tells us nothing about how to interpret that event. In contrast, knowing that someone spent 25% of their monthly income on a single dinner last month tells us a story that can not only be interpreted but also evaluated, even *judged* as good or bad behavior from a credit risk point of view—that is, ratios help us to find critical points from which our system at study changes from one state to another (just as the Reynolds number does), in this case from high-risk behavior to low-risk behavior.

In general, the advantages of employing ratios over magnitudes can be summarized as follows:

- Ratios give context to magnitudes, giving meaning to seemingly unrelated magnitudes.

- Ratios are inherently capable of identifying critical points at which a system undergoes a transition from one state to another. This intrinsic capacity enables ratios to discriminate between defaults and nondefaults.

- Ratios are capable of meaningful interpretation, which lends them an inherent suitability for incorporation into scorecard models.

- Ratios facilitate the standardization of units and scaling of variables, thereby eliminating numerical inconsistencies and enabling the construction of more stable models.

- The use of ratios enables the reduction of the impact of outliers, thereby facilitating the modeling process and eliminating the necessity for the utilization of sophisticated outlier detection techniques.

4.3 Build Your Own Ratios

Section 4.2 was of significant importance, not only to illustrate the intrinsic advantages of employing ratios but also to provide a clear and comprehensive explanation of the nature, origin, and implications of ratios. This enabled a deeper comprehension of the rationale behind the pivotal role of ratios in the modeling process.

Table 4-2 presents a list of selected ratios that can be derived from the elements presented in **Table 3-4**[12] (*elements of daily life* and *monetary balance*). It is evident that the elements previously presented in **Table 3-4** now have the missing context needed to assess the relative importance that a particular customer assigns to each element of daily life. The greater the amount a person allocates to a particular expenditure, in comparison to the total amount of expenditure, the more important it must be for their daily life.

---

[12] This is just an example and by no means represents the whole range of possible combinations to create ratios.

CHAPTER 4 RATIOS

*Table 4-2. Selected list of ratios created from the elements presented in **Table 3-4**.*

Notation	Descriptive ratio	Full Description
$R_{i_{exp}}$	$\dfrac{ith \text{ Expense}}{\text{Total Expenses}}$	The amount of the *ith* expense over the sum of all expenses
$R_{Output/Input}$	$\dfrac{\text{Total Outputs}}{\text{Total Inputs}}$	The sum of all outputs over the sum of all inputs
$R_{i_{debt}}$	$\dfrac{ith \text{ Debt}}{\text{Total Debts}}$	The amount of the *ith* debt over the sum of all debts
$R_{Pay/Debt}$	$\dfrac{\text{Total Payments}}{\text{Total Debts}}$	The sum of all payments (obligations) over the sum of all debts
$R_{Exp/Input}$	$\dfrac{\text{Total Expenses}}{\text{Total Inputs}}$	The sum of all expenses over the sum of all inputs
$R_{i_{exp/input}}$	$\dfrac{ith \text{ Expense}}{\text{Total Inputs}}$	The amount of the *ith* expense over the sum of all inputs
$R_{Liabilities/Assets}$	$\dfrac{\text{Total Expenses + Total Debts}}{\text{Total Inputs + Total Liquid Assets}}$	The sum of all expenses plus the sum of all debts over the sum of all inputs plus the sum of all liquid assets, such as savings, current account, and liquid investments

The equations presented in Section 3.5.1 can now be employed to describe the ratios that have been conceptually delineated in mathematical notation. The resulting equations, from **Equation 4-9** through **Equation 4-15**, are the result of this process.

CHAPTER 4 RATIOS

$$R_{i_{exp}} = \frac{C_{i_{exp}}}{\sum_{i_{exp}=1}^{n_{exp}} C_{i_{exp}}}, i_{exp} \in J_{exp}$$

Equation 4-9. Expense over total expenses ratio.

$$R_{i_{exp/input}} = \frac{C_{i_{exp}}}{\sum_{i_{in}=1}^{n_{in}} C_{i_{in}}}, i_{exp} \in J_{exp}, i_{In} \in J_{In}$$

Equation 4-10. Expense over total inputs ratio.

$$R_{i_{debt}} = \frac{C_{i_{debt}}}{\sum_{i_{debt}=1}^{n_{debt}} C_{i_{debt}}}, i_{debt} \in J_{Debt}$$

Equation 4-11. Debt over total debt ratio.

$$R_{Output/Input} = \frac{\sum_{i_{out}=1}^{n_{out}} C_{i_{out}}}{\sum_{i_{in}=1}^{n_{in}} C_{i_{in}}}, i_{out} \in J_{Out}, i_{In} \in J_{In}$$

Equation 4-12. Output over input ratio.

$$R_{Pay/Debt} = \frac{\sum_{i_{Pay}=1}^{n_{Pay}} C_{i_{Pay}}}{\sum_{i_{debt}=1}^{n_{debt}} C_{i_{debt}}}, i_{debt} \in J_{Debt}, i_{Pay} \in J_{Pay}$$

Equation 4-13. Payment over debt ratio.

81

CHAPTER 4 RATIOS

$$R_{exp/Input} = \frac{\sum_{i_{exp}=1}^{n_{exp}} C_{i_{exp}}}{\sum_{i_{in}=1}^{n_{in}} C_{i_{in}}}, i_{exp} \in J_{exp}, i_{In} \in J_{In}$$

Equation 4-14. Total expenses over total inputs ratio.

$$R_{Curr\ Liabilities/Curr\ Assets} = \frac{\sum_{i_{exp}=1}^{n_{exp}} C_{i_{exp}} + \sum_{i_{debt}=1}^{n_{debt}} C_{i_{debt}}}{\sum_{i_{in}=1}^{n_{in}} C_{i_{in}} + \sum_{i_{liq}=1}^{n_{liq}} C_{i_{liq}}},$$

$$i_{exp} \in J_{exp}, i_{debt} \in J_{debt}, i_{In} \in J_{In}, i_{liq} \in J_{Liq}$$

Equation 4-15. Liabilities over assets.

Table 4-3 shows the expanded list of 36 ratios derived from the seven ratios proposed in Table 4-2.

CHAPTER 4   RATIOS

Table 4-3. Expanded list of ratios derived from the seven ratios proposed in Table 4-2.

Num.	Category	Code	Ratio	Description
1	Education	edu	Education / Expenses	Expense over Total Expenses
2	Medical	med	Medical / Expenses	Expense over Total Expenses
3	Communication	com	Communication / Expenses	Expense over Total Expenses
4	Food and Groceries	fdg	Food and Groceries / Expenses	Expense over Total Expenses
5	Transportation	trn	Transportation / Expenses	Expense over Total Expenses
6	Governmental	ser	Governmental / Expenses	Expense over Total Expenses
7	Personal	cln	Personal / Expenses	Expense over Total Expenses
8	Home Improvements	hom	Home Improvements / Expenses	Expense over Total Expenses
9	Pets	pet	Pets / Expenses	Expense over Total Expenses
10	Clothing	ct	Clothing / Expenses	Expense over Total Expenses
11	Sports	spr	Sports / Expenses	Expense over Total Expenses
12	Entertainment	ent	Entertainment / Expenses	Expense over Total Expenses
13	Hobbies and luxuries	oth	Hobbies and luxuries / Expenses	Expense over Total Expenses
14	Cash Withdrawals	csh	Cash Withdrawals / Expenses	Expense over Total Expenses
15	Education	edu	Education / (Payments+Deposits)	Expense over Total Inputs
16	Medical	med	Medical / (Payments+Deposits)	Expense over Total Inputs
17	Communication	com	Communication / (Payments+Deposits)	Expense over Total Inputs
18	Food and Groceries	fdg	Food and Groceries / (Payments+Deposits)	Expense over Total Inputs
19	Transportation	trn	Transportation / (Payments+Deposits)	Expense over Total Inputs
20	Governmental	ser	Governmental / (Payments+Deposits)	Expense over Total Inputs
21	Personal	cln	Personal / (Payments+Deposits)	Expense over Total Inputs
22	Home Improvements	hom	Home Improvements / (Payments+Deposits)	Expense over Total Inputs
23	Pets	pet	Pets / (Payments+Deposits)	Expense over Total Inputs
24	Clothing	ct	Clothing / (Payments+Deposits)	Expense over Total Inputs
25	Sports	spr	Sports / (Payments+Deposits)	Expense over Total Inputs
26	Entertainment	ent	Entertainment / (Payments+Deposits)	Expense over Total Inputs
27	Hobbies and luxuries	oth	Hobbies and luxuries / (Payments+Deposits)	Expense over Total Inputs
28	Cash Withdrawals	csh	Cash Withdrawals / (Payments+Deposits)	Expense over Total Inputs
29	Credit Card (outstanding balance)	bal	Credit Card (outstanding balance) / Debt	Debt over Total Debt
30	Personal (outstanding balance)	lon	Personal (outstanding balance) / Debt	Debt over Total Debt
31	Car (outstanding balance)	car	Car (outstanding balance) / Debt	Debt over Total Debt
32	Mortgage (outstanding balance)	mor	Mortgage (outstanding balance) / Debt	Debt over Total Debt
33	Aggregated Categories	loi	Payments / Debt	Payments over Debt
34	Aggregated Categories		(Expenses+Instalments+Others) / (Payments+Deposits)	Total Outputs over Total Inputs
35	Aggregated Categories		Expenses/(Payments+Deposits)	Total Expenses over Total Inputs
36	Aggregated Categories		(Expenses+Debt) / (Payments+Deposits+Savings+Others)	Liabilities over Assets

83

CHAPTER 4   RATIOS

Thirty-six variables may not sound like a lot for a scorecard model, but our modeling work is far from over. We will leave our ratios aside for now, and we will move on to the next stage of our modeling process—and that will only be a matter of *time*.

4.4 Summary

Chapter 4 invited us to reflect on the limitations of relying solely on raw monetary variables in financial modeling. These variables, while essential, often came with challenges such as skewed distributions, scaling issues, and numerical instability, all of which disrupted the accuracy of predictions. To address these issues, the chapter introduced ratios and frequencies as powerful tools that provided the context and interpretability raw magnitudes lacked.

We also explored the rich mathematical and historical foundations of ratios, uncovering their ability to reveal critical system transitions and standardize variables for more stable and reliable models. Finally, the chapter guided us through the process of constructing custom ratios tailored to specific financial modeling needs, showing deeper insights into the financial priorities and behaviors of individuals.

Key takeaways:

1. **Limitations of Monetary Variables:** Raw monetary variables, despite their importance, often fell short due to skewed distributions, scaling issues, and a lack of meaningful context.

2. **Power of Ratios and Frequencies:** Ratios uncovered the relationships between variables, while frequencies revealed patterns and behaviors over time.

3. **Mathematical Foundations:** Exploring the historical and mathematical roots of ratios revealed how they exposed critical system transitions and brought stability to models.

4. **Custom Ratios:** The chapter demonstrated how tailored ratios enhanced accuracy and provided deeper insights into financial behavior, helping to better understand priorities and commitments.

We now have most of the principal elements needed to create powerful explanatory variables; however, understanding behavior requires more than just static data points. To truly enhance our variables, we must introduce the time dimension, uncovering how patterns emerge not only from what is done but also from when it is done. In the next chapter, we will explore in detail how to unfold our variables over time and capture the *behavioral patterns* essential for predicting the default event.

CHAPTER 5

Time and Behavioral Patterns

"On a long enough timeline, the survival rate for everyone drops to zero."

—Tyler Durden, *Fight Club (1999)*

"All those moments will be lost in time, like tears in rain."

—Roy Batty, *Blade Runner (1982)*

Section 3.5 introduced the process of replacing commonsense concepts with mathematical expressions, particularly through ratios. By the end of Section 4.3, we had defined 36 ratios, forming 36 independent variables for our scorecard model. Chapter 5 shifts focus from constructing ratios to adding another dimension to our variables—that is, *time*. Time plays a crucial role in shaping behavior and outcomes, and this chapter explores how temporal patterns enhance predictive modeling by emphasizing the importance of when actions occur.

CHAPTER 5 TIME AND BEHAVIORAL PATTERNS

5.1 It Is a Matter of Time

Behavior is not solely determined on the actions undertaken, but also on the temporal context in which they are performed. For example, one of our studies[1] revealed that watching television between 24:00 and 5:59 increased the likelihood of defaulting on a telecommunications or cable television subscription. Conversely, having a higher ratio of out-calls compared to in-calls between 18:00 and 23:59 had the opposite effect, reducing the risk of default.

The increase in risk associated with TV from 24:00 to 5:59 was attributed to unemployment or a disorganized lifestyle, while the decrease in risk associated with a high ratio of out-calls/in-calls from 18:00 to 23:59 was attributed to high-performing individuals who continued to work after hours. It is noteworthy that a 12.5% increase in hours worked has been reported to result in a 42.3% increase in earnings [7].

The aforementioned examples were based on an hourly basis; however, scorecards tend to focus their attention on monthly events, primarily due to the fact that the banking industry has a financial month-end close, which necessitates the closure of all payments and balances on a monthly basis. It is therefore unsurprising that the majority of scorecard variables also refer to a monthly periodicity.[2]

The most common periodicities used are *months back* and *last months*. Months back indicates a point in time, e.g., one, two, or more months prior to the reference date. Then, *the balance amount from one month back* means the balance from the bank statement from one month prior to a given point. On the other hand, *the average balance amount*

---

[1] Telecommunication industry, sample of 3 million subscribers, Mexico City.
[2] It is important to acknowledge that in an increasingly digitalized and real-time data-driven world, some lenders are utilizing real-time data, including daily or hourly information, for decision-making processes related to fraud and transaction authorizations.

over the last three months means the average of the last three months' statements, up to the most recent. It is worth noting that scorecards can also include information on a nonmonthly basis, since most transactions are recorded with a timestamp showing the exact day and time.

As an illustration of the usage of timestamps, in one of our studies, we had the opportunity to develop a scorecard model for a telecommunications and cable provider. The findings of this study suggested that the consumption of television programs, the number of telephone calls made, and the amount of internet usage can be used to predict the risk of default when the time of occurrence is taken into account. For instance, in this same study, the day of the week was also identified as a significant factor. A higher ratio of uploaded to downloaded megabytes[3] on Mondays was associated with a reduced probability of default. It was postulated that this phenomenon may be prevalent among remote workers and those in higher-level white-collar professions. However, our assumptions could not be confirmed, which represents a secondary issue. Our primary objective is to demonstrate that humans are creatures of habit and that incorporating this into our explanatory variables enhances their predictive power.

In my personal experience, I have not had the opportunity to analyze banking transactions by day of the week or time of day. Nevertheless, I have the inkling that purchasing items on Amazon after hours, grocery shopping on Sundays, and the purchase of expensive gifts for loved ones at Christmas may also be correlated with the probability of default.

---

[3] In telecommunications, data-transfer rate is the average number of bits per unit time passing through a communication link in a data-transmission system.

5.2 Introducing Time Periodicity into Your Explanatory Variables

Scorecard models and time series are separate topics. Time series forecast future values of the same series, while scorecards predict event likelihood as a function of time-varying variables. Time series identify patterns related to specific time periods whereas scorecards identify patterns related to customer behavior regardless of the reference date.

When introducing time periods into a scorecard, it is essential to bear in mind that they must be *relative* and not *specific*. It is important to note that a predictive model should be capable of making predictions at any given point in time. To illustrate, consider a scenario where one aims to build a PD model, with a 12-month window for the occurrence of the event, using the number of credit card balance increments in the previous six months, and the credit card utilization ratio from the previous month as explanatory variables.

It should be noted that no specific date was provided in the preceding example, as the objective is to enable the model to identify relative trends and cycles from any reference date. Let us now consider the scenario in which the model is trained using a timeframe from April 2021 to April 2023, with the reference date set to April 2022.

Figure 5-1 illustrates an example of a 25-month window, wherein the reference date is April 2022, the event period[4] is May 2022 through April 2023 (12 months), and the performance period[5] is April 2021 through March 2022 (12 months). In the case of this specific dataset, the last six months align with the period between October 2021 and March 2022, while one month back corresponds to March 2022.

---

[4] The event period is the timeframe during which the payment performance of the accounts is monitored and evaluated to ascertain whether a default event or target has occurred.

[5] The performance period is the timeframe during which the accounts' transactional activity (payments, purchases, past due days, etc.) is monitored and the explanatory variables are introduced.

CHAPTER 5   TIME AND BEHAVIORAL PATTERNS

Figure 5-1. Example of a timeframe window with specific dates.

Figure 5-2. Conceptual representation of a stacked ABT, with each reference date aligned on top of each other.

CHAPTER 5 TIME AND BEHAVIORAL PATTERNS

Let us now consider a scenario in which the present date is December 2023. Would it be appropriate for the model to make predictions regarding the probability of default for 2024, given that the original training reference date was April 2022?

The utilization of particular calendar dates for the training of the model has constrained the model's applicability to solely those specified dates. The issue arises from the fact that the aforementioned six-month period does not refer to a generic set of the last six months, but rather to specific dates that correspond to October 2021 through March 2022.

It is evident that the economic conditions, employment, inflation, and rates of change in March of 2022 will not be comparable to those of any other March in previous years. The issue with using specific dates is that the model is anchored to a specific point in time, rendering it unusable for any period other than the one used for training. This raises the question of how to refer to relative periods.

Figure 5-2 illustrates a stacked ABT with aligned reference dates, with one date on top of the other. Twelve periods are used as reference dates, so now the one month back (n-1) is a broader span of 12 months. Consequently, when referencing the last six months, the specific dates of October 2021 to March 2022 are no longer the focus of analysis.

Calendar seasons, trends, and cycles cannot be explicitly used. They can, however, be used relative to a reference date. The goal then is to find a periodic pattern capable of correlating with the default event without calendar date influence. For example, a number close to six increments on your credit card balance over the last six months means that you are spending more than you are bringing in.

It should be noted that the number of increments should be used in conjunction with the credit card utilization rate, as a six-month increment streak does not necessarily indicate an inability to repay the credit card. An individual may exhibit a six-month increment streak in their outstanding balance and a 26% utilization level in their most recent statement. This does not indicate an imminent default, as it is highly probable that any individual could recover from this level of debt.

Nevertheless, monitoring the utilization level at various points in time would allow one to determine not only the level of debt beyond which recovery is no longer possible, but it would also allow one to determine the point in time beyond which recovery is no longer feasible.

To illustrate, assume that it was determined that a utilization level of 65% three months back to the reference date was a significant predictor of a default event. One might be inclined to interpret this result as an indication that a 65% utilization level will result in default. However, the correct interpretation is that a 65% utilization level with only three months to recover has a high probability of default. Therefore, by examining various points in time, it is possible to identify the critical utilization level and determine the point in time at which it becomes critical.

Table 5-1 presents a number of potential monthly periodicities that could be incorporated into a scorecard model. Monthly periodicities can be employed either as point-in-time or as aggregated values. The remaining periodicities may only be utilized as aggregated values, provided that they are derived from either monthly bank statements or transactional data. Aggregated values must be the result of an aggregation function, such as averages, sums, maximums, minimums, and/or frequencies.

CHAPTER 5   TIME AND BEHAVIORAL PATTERNS

Table 5-1. *Selected list of recommended periodicities for use in scorecard models.*

Periodicity	Row num.	Period num.	Description	Usage
Monthly	1	1	1 month(s) back	Point in time/aggregation
	2	2	2 month(s) back	Point in time/aggregation
	3	3	3 month(s) back	Point in time/aggregation
	4	4	4 month(s) back	Point in time/aggregation
	5	5	5 month(s) back	Point in time/aggregation
	6	6	6 month(s) back	Point in time/aggregation
	7	7	7 month(s) back	Point in time/aggregation
	8	8	8 month(s) back	Point in time/aggregation
	9	9	9 month(s) back	Point in time/aggregation
	10	10	10 month(s) back	Point in time/aggregation
	11	11	11 month(s) back	Point in time/aggregation
	12	12	12 month(s) back	Point in time/aggregation
Quarterly	13	1	Last 3 months	Aggregation
	14	2	Last 4 to 6 months	Aggregation
	15	3	Last 7 to 9 months	Aggregation
	16	4	Last 10 to 12 months	Aggregation
Cumulative 3-, 6-, 9-, and 12-month intervals	17	1	Last 6 months	Aggregation
	18	2	Last 9 months	Aggregation
	19	3	Last 12 months	Aggregation
	20	4	Last 13 to 18 months	Aggregation
	21	5	Last 18 to 24 months	Aggregation
	22	6	Last 13 to 24 months	Aggregation

CHAPTER 5   TIME AND BEHAVIORAL PATTERNS

We will now extend the list of 36 ratios presented in **Table 4-3** using all periodicities in **Table 5-1**. For simplicity, we will assume that only the average aggregation function is used. **Table 5-1** shows 22 different periodicities, meaning the scorecard model will generate 792 explanatory variables (22×36). Adding different periodicities creates a *time series* for each observation. This allows us to measure the level of the 36 ratios presented in **Table 4-3** at various points in time. We must next identify the time series' components, which include *seasons, cycles, trends,* and *streaks*.

5.3 Folding Time on Itself

Time series demonstrate the manner in which ratios evolve over time. Each point in time represents a distinct measurement. However, a mere aggregation of individual measurements does not provide a comprehensive account of the evolution of ratios over time. In order to achieve this, it is necessary to introduce an additional layer of complexity to the individual measurements by means of a comparison of different points in time. The magnitude and direction of the evolution of ratios over time can be measured by means of deviations. The calculation of streaks is achieved by the addition of the number of positive or negative deviations that occur within a specified timeframe. Such streaks are referred to as the sum of increments or decrements.

A deviation is defined as the rate of change of a measurement from one point in time to another (or percent change over time). For example, the monthly and yearly deviation of our total payment over total debt, $R_{Pay/Debt}$, is defined as follows:

$$D^{month}_{R_{Pay/Debt}} = \frac{R_{\frac{Pay}{Debt},t} - R_{\frac{Pay}{Debt},t-1}}{R_{\frac{Pay}{Debt},t-1}}$$

Equation 5-1. Monthly deviation of payment over debt.

CHAPTER 5 TIME AND BEHAVIORAL PATTERNS

$$D^{Annual}_{R_{Pay/Debt}} = \frac{\frac{R_{Pay}}{Debt}_{,t} - \frac{R_{Pay}}{Debt}_{,t-12}}{\frac{R_{Pay}}{Debt}_{,t-12}}$$

Equation 5-2. Annual deviation of payment over debt.

The general definition will yield

$$D^{month}_R = \frac{R_{,t} - R_{,t-1}}{R_{t-1}}$$

Equation 5-3. Generalized monthly deviation.

$$D^{Annual}_R = \frac{R_t - R_{t-12}}{R_{t-12}}$$

Equation 5-4. Generalized annual deviation.

The sum of increments and decrements between a lower time interval, t_l, and an upper time interval, t_u, can be defined as follows:

$$S_{I,t_1-t_n} = \sum_{t_l}^{t_u} \left[\text{if } D^{month}_{R,t_i} > 0 \text{ then } 1 \text{ else } 0 \right]$$

Equation 5-5. Sum of increments.

And respectively, the sum of decrements can be defined as follows:

$$S_{D,t_1-t_n} = \sum_{t_l}^{t_u} \left[\text{if } D^{month}_{R,t_i} < 0 \text{ then } 1 \text{ else } 0 \right]$$

Equation 5-6. Sum of decrements.

CHAPTER 5   TIME AND BEHAVIORAL PATTERNS

It is important to note that the definition of time intervals is not limited to a specific point in time. However, it is this author's recommendation that the reader utilize the quarter (**Table 5-1**, rows from 13 to 16) and cumulative 3-, 6-, 9-, and 12-month intervals (**Table 5-1**, rows from 17 to 22).

It should be noted that the changes over time can also be measured using different grouping functions, such as MAX and MIN, or by measuring the deviation from the mean over the last 12 months, for example, by identifying the biggest percentage difference in deposits over the mean deposit for the last 12 months, which is a measure of income stability.

Tables 5-2 and **5-3** provide a list of periods and times that can be employed in the calculation of ratios and deviations across a range of periodicities and time intervals. **Table 5-2** presents 45 distinct point-in-time variables, comprising 11 periodicities representing 11 variations, 11 increments, and 11 decrements (yielding 33 variants), along with 12 additional annual periodicities (yielding 45 variants, 33 + 12 = 45). **Table 5-3**, in turn, presents 20 different aggregation time intervals. The two tables collectively present an additional 65 metrics, which, when multiplied by the 36 ratios presented in **Table 4-3**, result in a total of 2,340 variables. The incorporation of the 2,340 variables into the existing set of 792 variables results in a total of 3,132 variables.

Table 5-2. *Selected list of recommended time intervals for variations and derived monthly increment/decrement flags.*

Periodicity	Periodicity num.	Interval	Usage
Monthly	1	From 2 to 1 months	Variation/increment/decrement
	2	From 3 to 2 months	Variation/increment/decrement
	3	From 4 to 3 months	Variation/increment/decrement
	4	From 5 to 4 months	Variation/increment/decrement
	5	From 6 to 5 months	Variation/increment/decrement
	6	From 7 to 6 months	Variation/increment/decrement
	7	From 8 to 7 months	Variation/increment/decrement
	8	From 9 to 8 months	Variation/increment/decrement
	9	From 10 to 9 months	Variation/increment/decrement
	10	From 11 to 10 months	Variation/increment/decrement
	11	From 12 to 11 months	Variation/increment/decrement
Annual	1	From 13 to 1 months	Variation
	2	From 14 to 2 months	Variation
	3	From 15 to 3 months	Variation
	4	From 16 to 4 months	Variation
	5	From 17 to 5 months	Variation
	6	From 18 to 6 months	Variation
	7	From 19 to 7 months	Variation
	8	From 20 to 8 months	Variation
	9	From 21 to 9 months	Variation
	10	From 22 to 10 months	Variation
	11	From 23 to 11 months	Variation
	12	From 24 to 12 months	Variation

Table 5-3. *Selected list of increments/decrements sums at different cumulative time intervals.*

Periodicity	Row num.	Periodicity num.	Description	Usage
Quarterly	1	1	Last 3 months	Sum increments
	2	2	Last 4 to 6 months	Sum increments
	3	3	Last 7 to 9 months	Sum increments
	4	4	Last 10 to 12 months	Sum increments
Cumulative 3, 6, 9, and 12 months	5	1	Last 6 months	Sum increments
	6	2	Last 9 months	Sum increments
	7	3	Last 12 months	Sum increments
	8	4	Last 13 to 18 months	Sum increments
	9	5	Last 18 to 24 months	Sum increments
	10	6	Last 13 to 24 months	Sum increments
Quarterly	11	1	Last 3 months	Sum decrements
	12	2	Last 4 to 6 months	Sum decrements
	13	3	Last 7 to 9 months	Sum decrements
	14	4	Last 10 to 12 months	Sum decrements
Cumulative 3, 6, 9, and 12 months	15	1	Last 6 months	Sum decrements
	16	2	Last 9 months	Sum decrements
	17	3	Last 12 months	Sum decrements
	18	4	Last 13 to 18 months	Sum decrements
	19	5	Last 18 to 24 months	Sum decrements
	20	6	Last 13 to 24 months	Sum decrements

CHAPTER 5　TIME AND BEHAVIORAL PATTERNS

Remarkably, the inquiry that began with the question *"What is a loan?"* has evolved into a comprehensive set of 3,132 explanatory variables for our scorecard model. This expansion requires reflection on the process that led to it. The expansion of variables is mainly due to the introduction of temporal elements into the model, but introducing time itself was not the final objective. We added time to our model to help us identify *behavioral patterns* in the 36 variables presented in **Table 4-3**. The term *"behavioral patterns"* is pervasive in the fields of artificial intelligence (AI) and machine learning (ML). Nevertheless, the term is seldom defined. One of the key promises of machine learning algorithms is the ability to identify and extract behavioral patterns from data, thereby enhancing predictive capabilities. Although this may facilitate modeling, it is unlikely that ML algorithms will discover these patterns on their own, which begs the question: *What are behavioral patterns?*

5.4 Behavioral Patterns

Behavior is set of conducts and actions that we perform every day. In fact, we have already grouped and categorized different behaviors embedded in our transactional data sources. We grouped these behaviors into 14 categories in Section 3.3.2, and we called them *basic elements of daily life*. We further explained that these 14 groups are in fact a manifestation of our core values, such as our children's future, our family's well-being, our career commitment, how much we care about our home, how much we care about our romantic relationship, how much do we value friendship, how much do we care about our pets, and, finally, how much do we care about ourselves.[6] Therefore, behavior does not exist as an isolated phenomenon.

---

[6] Although core values do not change across countries and cultures (see Section 2.2.7.2), it is important to acknowledge that some variables will differ from country to country, cultures, legal frameworks, etc.

Patterns can be defined as repeated, systematic occurrences of behaviors. Patterns may also be defined as consistent behaviors occurring over time, which is also the definition of *integrity*. This is the primary rationale behind the unfolding of our 14 explanatory variables over time and the subsequent folding of our time series into different timeframes and periodicities. The objective is to identify and differentiate between distinct risk populations based on their patterns and trends. It is noteworthy that the periodicities were not randomly assigned; rather, they were selected to align with human behavior, specifically to correspond with yearly cycles.

5.5 Summary

Chapter 5 highlighted the critical role time plays in predictive modeling, demonstrating how temporal patterns can provide deeper insights into behavior and outcomes. By incorporating the dimension of time into our independent variables, we moved beyond static measurements to a more dynamic understanding of actions and their contexts. This chapter emphasized the importance of timing in shaping behavior and improving the accuracy of our models.

Key takeaways:

1. **Time as a Behavioral Dimension**: Adding a temporal perspective allows us to uncover not just what actions occur, but when they happen, providing richer insights into behavior.

2. **Significant Variable Expansion**: By applying time-based transformations to our original 36 ratios, we dramatically increased the number of independent variables, enabling the model to capture behavioral changes and trends over time.

3. **Relative Periods over Fixed Dates**: Using relative measures (e.g., *"days since last transaction"*) ensures the variables remain adaptable and relevant across different time horizons.

4. **Temporal Patterns and Trends**: Patterns such as periodicities, seasonality, or irregularities over time reveal critical behavioral dynamics that static variables cannot capture.

5. **Enhanced Predictive Power**: The inclusion of time-driven variables significantly strengthens the model's ability to predict outcomes by accounting for the evolving nature of behavior.

So far, we have seen that incorporating the time dimension into our 36 ratios not only drastically expands the number of independent variables but also adds a higher level of sophistication and complexity to our model. However, it is important to recognize that many other variables can further enrich our model's predictability by capturing diverse aspects of the same account or customer that are not necessarily tied to the dimensions explored so far. Chapter 6 will delve into these additional variables, presenting an extensive range of options to consider during the construction of our scorecard model.

CHAPTER 6

Additional Variables

"Better to have, and not need, than to need, and not have."

—Franz Kafka

A total of 3,132 explanatory variables have been theoretically designed to encompass consumption, payment, and accumulation behavior over time. Nevertheless, there are numerous additional variables that have yet to be taken into account.

Table 6-1 presents a list of eight distinct categories of variables, classified according to their conceptual meaning. Thus far, the initial two categories have been presented: elements of daily life and monetary balance. But we will now introduce additional variables that capture new dimensions of information. These may include demographic attributes, historical financial information, account-level characteristics, or external data sources. By incorporating such variables, we can explore aspects like customer segmentation, debt trends, or geographic influences, all of which complement the existing variables and add depth to the model.

CHAPTER 6 ADDITIONAL VARIABLES

Table 6-1. *Categorization of variables based on their conceptual meaning.*

Num.	Conceptual meaning	Data source
1	Elements of daily life	Statement and transactional information
2	Monetary balance	Statement and transactional information
3	Unconventional information	Available online data
4	Current and past financial obligations	Historical financial information
5	Customer's personal information	Application information
6	Customer's banking information	Application information
7	Product's information	Application information
8	Macroeconomic information	Public information

6.1 Unconventional Information

The term *"unconventional information"* refers to online data. It is important to acknowledge that the digitalization era has opened the door to exciting new sources brimming with behavioral data. To illustrate, one might consider the plethora of streaming companies currently in existence, including prominent names such as Netflix, Max, Amazon Prime, and Crunchyroll. It is irrefutable that the consumption of streaming content will have explanatory power. However, this may also represent a challenge in technical terms, as well as in ethical and legal issues.

Nevertheless, the advent of new online sources, including eBay, Amazon, Etsy, Uber, YouTube, X (formerly Twitter), and others, has led to the emergence of a vast trove of information that was previously inaccessible. It is probable that analysts will encounter greater challenges in exploiting this data than in addressing its scarcity.

6.2 Present and Previous Financial Obligations

Table 6-2 presents a list of the most commonly utilized characteristics and features as explanatory variables in scorecard models. It should be noted that a considerable number of these characteristics are expressed in magnitude. Consequently, in accordance with the methodology delineated in the preceding sections, it is necessary to first convert these magnitudes to ratios before incorporating them into the model.

Table 6-3 presents a list of recommended variables derived from the information presented in the current and previous credits.[1] It is notable that **Table 6-3** suggests the potential for the creation of hundreds of additional variables through the unfolding of group variables over time. Additionally, it is notable that some of these variables pertain to characteristics that may not be expected to undergo temporal change. These include the total number of installments, the original approved amount, the interest rates associated with the loans, and the loan-to-value ratio of the applications. Conversely, variables such as the maximum number of days past due may only be meaningful when considered in an aggregate form. Ultimately, some of the variables can be unfolded as a time series, with sequential backward points in time. However, they can also be aggregated at different timeframes and periodicities, such as the percentage of recurrent payments and the percentage of outstanding debt.

---

[1] For the sake of simplicity, we will not describe in detail the mathematical expressions of these variables.

CHAPTER 6   ADDITIONAL VARIABLES

Table 6-2. List of possible attributes per product.

Product	Type of Information
Auto	Approved amount
	Collateral
	Recurrent payment
	Interest rate
	down payment
	Loan to Value (LTV)
	Total installments
	Due amount
	Paid amount
	Installments left
	Payment Frequency
	Past due days
Mortgage	Approved amount
	Collateral
	Recurrent payment
	Interest rate
	down payment
	Loan to Value (LTV)
	Total installments
	Due amount
	Paid amount
	Installments left
	Payment Frequency
	Past due days
Loan	Approved amount
	Recurrent payment
	Interest rate
	Total installments
	Due amount
	Paid amount
	Installments left
	Payment Frequency
	Past due days
Credit Card	Credit Limit
	Interest rate
	Type (Gold, Platinum, etc.)
	Past due days

CHAPTER 6   ADDITIONAL VARIABLES

Table 6-3. List of recommended variables based on past and present credits.

Variable group	Product	Description	Periodicity	Periodicity description
% Recurrent Payment	Auto	Payment/Total Debt, n months back, average in the last n months	Aggregation/Point-in-Time	Recurrent Point-in-Time
	Mortgage	Payment/Total Debt, n months back, average in the last n months	Aggregation/Point-in-Time	Recurrent Point-in-Time
	Loan	Payment/Total Debt, n months back, average in the last n months	Aggregation/Point-in-Time	Recurrent Point-in-Time
	Credit Card	Payment/Total Debt, n months back, average, maximum, increments, decrements, deviations in the last n months	Aggregation/Point-in-Time	Recurrent Point-in-Time
% Outstanding Debt	Auto	Due amount/Approved amount, n months back, average in the last n months	Aggregation/Point-in-Time	Recurrent Point-in-Time
	Mortgage	Due amount/Approved amount, n months back, average in the last n months	Aggregation/Point-in-Time	Recurrent Point-in-Time
	Loan	Due amount/Approved amount, n months back, average in the last n months	Aggregation/Point-in-Time	Recurrent Point-in-Time
interest rate	Auto	i_car	Point-in-Time	Constant across time
	Mortgage	i_Mtg	Point-in-Time	Constant across time
	Loan	i_Lon	Point-in-Time	Constant across time
	Credit Card	i_CC	Point-in-Time	Recurrent Point-in-Time
Total Installments	Auto	Installments	Point-in-Time	Constant across time
	Mortgage	Installments	Point-in-Time	Constant across time
	Loan	Installments	Point-in-Time	Constant across time
Installments left	Auto	Installments left	Point-in-Time	Recurrent Point-in-Time
	Mortgage	Installments left	Point-in-Time	Recurrent Point-in-Time
	Loan	Installments left	Point-in-Time	Recurrent Point-in-Time
% Instalments left	Auto	Installments left/Total Installments	Point-in-Time	Recurrent Point-in-Time
	Mortgage	Installments left/Total Installments	Point-in-Time	Recurrent Point-in-Time
	Loan	Installments left/Total Installments	Point-in-Time	Recurrent Point-in-Time
LTV	Auto	(Market Value-down payment)/Approved Amount	Point-in-Time	Constant across time
	Mortgage	(Market Value-down payment)/Approved Amount	Point-in-Time	Constant across time
Ever flag	Auto	Has ever had a car loan	Aggregation	Constant across time
	Mortgage	has ever had a mortgage loan	Aggregation	Constant across time
	Loan	has ever had a loan	Aggregation	Constant across time
Active product	Auto	Active loan in the last 12, 24, 36 months	Aggregation	Variable across time
	Mortgage	Active loan in the last 12, 24, 36 months	Aggregation	Variable across time
	Loan	Active loan in the last 12, 24, 36 months	Aggregation	Variable across time
Past due days	Auto	Max past due days in the last 6, 12, and 13 to 24 months	Aggregation	Variable across time
	Mortgage	Max past due days in the last 6, 12, and 13 to 24 months	Aggregation	Variable across time
	Loan	Max past due days in the last 6, 12, and 13 to 24 months	Aggregation	Variable across time
	Credit Card	Max past due days in the last 6, 12, and 13 to 24 months	Aggregation	Variable across time

107

CHAPTER 6   ADDITIONAL VARIABLES

6.3 Customer Personal Information

The customer's personal information is collected during the application process, which has led some analysts to conclude that it is not representative of the customer's current situation. In contrast with the aforementioned viewpoint, I am inclined to disagree regardless of the potential obsolescence of the customer's information. My disagreement with this viewpoint is based on a fundamental set of concepts. It is essential to first establish a hypothesis and then to discard information that does not align with it. Furthermore, the only evidence that can invalidate a data source is the rejection of the alternative hypothesis. This hypothesis is tested at different statistical levels to determine whether the observed effect is due to chance. This is the essence of statistical rigor.

Table 6-4 illustrates the various characteristics that could be employed as explanatory variables in a scorecard model. It should be noted that the customer's personal information has been divided into the following categories:

1. Immutable characteristics
2. Education
3. Home
4. Work
5. Family
6. Geolocation

Table 6-4. Example of customer's personal information that can be used as an explanatory variable.

Trait category	Specific trait
Immutable characteristics	Sex
	Birthdate (age)
Education	Level: elementary school, high school, undergraduate, or graduate education
	If graduate degree, main field, or area
Home	Type: house or department
	Rental or owner
	Time living at current address
Work	If employed, then: industry, position
	Time working at present job
	Salary If self-employed, then Field or industry Main position Time as independent Income
Family	Single or married
	Children
	Number of children
	Number of economical dependents
Geolocation	Home ZIP code and/or district
	Work ZIP code and/or district

CHAPTER 6   ADDITIONAL VARIABLES

An immutable characteristic is defined as a personal attribute that is unchanging, such as one's sex[2] or date of birth. It is important to note that while an individual's age may fluctuate over time, their date of birth remains constant. It is important to note that a negative correlation between age and the default event is to be expected; that is to say, the older an individual is, the lower their risk should be.

Education is defined as the academic degree or qualification attained by an individual following the completion of a specified number of years of study. Two distinct, additive effects are anticipated. Firstly, a higher level of education may be negatively correlated with the default event; however, this may not always be the case, particularly for successful entrepreneurs. Secondly, the effect is linked to difficulty, which is related to an individual's character. Bachelor's degrees in challenging fields like *science, technology, engineering, and mathematics* (STEM) should show a negative correlation with the default event.

The Pew Research Center's findings that 50% of Americans believe that a greater number of individuals are deterred from pursuing STEM degrees due to their perceived difficulty substantiate the preceding assertion. Furthermore, the National Center for Education Statistics[3] reports that of the 1.8 million bachelor's degrees awarded between 2015 and 2016, only 331,000 (18%) were in STEM fields,[4] and the enrollment rate was only 15%[5] in 2003-2004.

---

[2] Be aware that use of sex (which is not the same as gender) as an explanatory variable might be prohibited by local regulations in some countries and is presented here for the sake of completeness.

[3] The National Center for Education Statistics (NCES) is the primary federal entity for collecting and analyzing data related to education in the United States and other nations. NCES is located within the US Department of Education and the Institute of Education Sciences. NCES fulfills a Congressional mandate to collect, collate, analyze, and report complete statistics on the condition of American education; conduct and publish reports; and review and report on education activities internationally.

[4] https://nces.ed.gov/programs/raceindicators/indicator_reg.asp
[5] https://nces.ed.gov/pubs2009/2009161.pdf

CHAPTER 6 ADDITIONAL VARIABLES

Moreover, the employment situation remains largely unchanged. The US Bureau of Labor Statistics reports that out of a total of 158,134,000 occupations in 2021, it is estimated that only 80,000 individuals (representing 5% of the total) will be employed in STEM occupations. This is particularly noteworthy given that the median annual wage for those in STEM occupations was $95,420, compared to $40,120 for all individuals employed in non-STEM occupations. In light of the aforementioned evidence, it seems reasonable to posit that individuals with a background in STEM fields should, in theory, present a lower risk than those with a background in non-STEM fields.

The term *"home"* is used to describe the characteristics of the place where an individual resides. It would be ideal to have detailed knowledge of the square meters, number of bedrooms, and number of floors in a person's home. However, such detailed data are seldom available. The information that is more commonly collected is whether an individual resides in an apartment or a house, and whether they rent or own their residence. The extant evidence is inconclusive as to whether one group is at a greater risk than the other. For instance, it has been traditionally assumed that homeowners present a lower risk in North America. However, this may not be the case in locations where housing is unaffordable, as renting is the norm and ownership is uncommon. But as a general rule, it seems reasonable to suggest that those who rent are at a higher risk than those who own.

In this context, *"work"* refers to one's occupation or profession. The initial differentiation will be between those who are employed by another individual or entity and those who are self-employed. In my experience, individuals who are self-employed tend to exhibit a higher risk profile than those who are employed by another party. This seems to be related to the debt to service ratio, or DSR, since employed DSRs tend to be more stable. To illustrate, consider an entrepreneur who owns and operates their own business. Regardless of the size of the enterprise, the owner bears complete responsibility for their business. In contrast, an employee's

CHAPTER 6 ADDITIONAL VARIABLES

risk is limited to maintaining their position and performing their duties satisfactorily. Therefore, the former example entails a significantly greater degree of exposure than the latter, which in turn gives rise to a greater risk of default.

The length of time spent in one's current position should have a similar effect to age, indicating stability and experience. Therefore, it would be reasonable to suggest a negative correlation between time in current job and risk.

The correlation between industry and risk may be influenced by two factors: employee turnover and macroeconomic conditions. For employees, salary and position will have the same effect; the higher the position, the higher the income, and higher incomes tend to correlate with lower default risk. However, salary alone is not a reliable predictor, and thus, the ratio of salary to expenses would be a preferable measure to salary alone.

The term *family* is used to denote the state of being married and having children. As previously discussed in Section 3.2, both marital status and having children have been correlated with the risk of default. It is noteworthy that the inverse relationship between the number of children and risk is particularly counterintuitive. In other words, the higher the number of children, the lower the risk.

The term *geolocation* is used to describe not only the place where an individual resides but also the place where they are employed. Based on my experience, geographic information is highly predictive of the default event. However, there are several important considerations to keep in mind before using it. It should first be noted that no other geographic level has been mentioned, aside from ZIP code and district.[6]

---

[6] It should be noted that the utilization of ZIP code and/or district as explanatory variables may be prohibited in certain countries by virtue of local regulations. In such instances, it would be advisable to employ an alternative variable that offers a reasonable degree of detail, such as that pertaining to a banking branch. In the event that this is not feasible, it may be advisable to refrain from utilizing any geographical variable.

This is primarily due to the fact that the predictive capacity of geographic variables increases in direct proportion to their level of detail. It is not uncommon for analysts to include state as a geographic variable, only to remove it from the final analysis due to its lack of significance. It is not my intention to suggest that the state in which one resides is never a significant variable. Rather, I am proposing that even if it is, there is a question as to how it can be interpreted. One might be bold enough to assert that the collective default risk of the Arizona population is greater than that of the California population. It is possible that this is the case, but this generalization is of no use since it lacks both *interpretability* and *resolution*. The term *"resolution,"* in this context, can be understood to imply a degree of specificity. To illustrate this point, consider the following example: To what extent do you believe the population of an entire state is homogeneous? Do you believe that the majority of individuals within the population will exhibit similar values and behaviors? In addition, it is reasonable to inquire whether the majority of individuals within a given population possess similar levels of education and income. The answer to all these questions is in the negative. There are more differences within the population of the same state than between the populations of different states.

To illustrate, consider the neighborhood in which you were raised. Was there a predominant establishment frequented by the majority of residents? To what extent did the students at your educational institution reside in the same neighborhood as you, or in a nearby area? It is reasonable to inquire whether the majority of individuals frequented the same shopping mall, movie theater, and even church on Sundays. Do you believe that the majority of individuals residing in your neighborhood adhere to a similar set of values? It can be reasonably deduced that the *proximity* of individuals to one another is a significant factor influencing their habits and values. This does not imply that all neighborhoods are inherently distinct; rather, it suggests that two disparate neighborhoods are more likely to exhibit similarities in terms of values and socioeconomic status with other neighborhoods in different states than with those in the same state. This is why the ZIP code is a superior variable to state, as it has a higher resolution. The principal conclusion to be drawn is not that the ZIP code can be employed as the sole explanatory variable, but rather, that in the case of geographic variables, the higher the resolution, the more beneficial the result.

6.4 Customer's Banking Information

Customer's banking information refers to specific characteristics as a customer, such as maturity (time as a customer of the bank since the first product application), segment, application branch, etc.

6.5 Product's Information

The product information pertains to the specific characteristics of the product that is being modeled. To illustrate, the credit card type may be either Visa or MasterCard, while the credit card category may be Gold, Platinum, Black, and so forth. Details such as the interest rate, credit limit, points awarded and redeemed, and benefits used may also be considered part of the product information.

6.6 Macroeconomic Information

Macroeconomic information is primarily comprised of public data, including but not limited to price levels, rates of economic growth, national income, gross domestic product (GDP), and changes in unemployment.

Table 6-5 illustrates a mere sampling of the economic indicators that could be employed in a scorecard model. However, the primary challenge associated with economic indicators is their level of granularity. The majority of economic indicators are population-level metrics, which precludes the possibility of analyzing distributions of varying values per account or customer within a given sample.

Table 6-5. Example of economic indicators.

Indicator	Description
GDP	National and by economic sector
Employment/unemployment	National and by economic sector
Exchange rates	USD/GBP, USD/EUR
Price index	National and by consumer good
Inflation	National and/or specific
IDEA (Index of Economic Activity)	Index of Economic Activity
Imports/exports	National, by product, by state, etc.
Debt	National, state, and/or local
Population growth	National, state, and/or local
Housing	Index, price, units, etc.
Home ownership	National, state, and/or local
Interest rates	Regular, prime, etc.

The issue is that a change in the gross domestic product, even at the industry level, has a ripple effect across the entire national economy. One potential solution to this issue is the use of a *link function* [7]. A link function is an additional model that, in lieu of predicting the probability of default (PD) at the *account level*, predicts the *default rate* (DR) at the *subportfolio level*.

The fundamental concept is to categorize the modeling population based on criteria that extend beyond the boundaries of the natural portfolio segmentation. As an illustration, one may choose to further subdivide the credit card population into distinct geographical regions, such as the Pacific Region (PR), the Rocky Mountain Region (RMR), the Mid-West Region (MWR), the Southwest Region (SWR), the Southeast Region (SER), and the Northeast Region (NR). The DR for each region can now be estimated. This is defined as the ratio of the number of defaults that have occurred over a given time horizon (as defined by the default definition) to the number of live accounts at the beginning of the reference period. The DR for each region can be calculated on a monthly basis over an extended period of time, such as ten years. This will result in a DR time series for each region, which can then be used to fit a regression model for each region. In this model, the DR will serve as the dependent variable, while macroeconomic indicators will act as the independent variables. The resulting regression models will serve as the *link functions*. Once the link functions have been established, it is then possible to estimate a *PD shift*. The PD shift represents the application of a variation based on macroeconomic information to the one-year PD scorecard. The complete process of correctly implementing the link function concept is considerably more intricate than what has been outlined here. However, for those interested in applying link functions to their scorecard models, it is recommended to consult Tiziano Bellini's book, *IFRS 9 and CECL Credit Risk Modelling and Validation* [7].

Two additional methods for incorporating macroeconomic data into scorecard models are available. One such method necessitates the utilization of a stacked ABT, thereby facilitating a vertical configuration of disparate reference periods. The next step is to align the macroeconomic data with the corresponding reference period. In the case of a reference period of January 2018, the GDP from three months prior would be that of November 2017. Conversely, if the reference period is March 2018, the three months prior GDP would be the January 2018 GDP. The alternative approach necessitates the availability of region-specific economic indicators, including, but not limited to, house price indices and crime indices. In such a case, the region-specific indices may be employed by matching the region indicated in the index with the region associated with the account or customer in question.

6.7 A Word of Caution, Variable Controversy

In one of our projects, we employed *"bank branch"* as one of the explanatory variables of our model. However, the company's management expressed concerns that incorporating this variable into the model could lead to significant controversy among its managerial personnel and might even be perceived as discriminatory.[7] Nevertheless, it was evident that the opposition to utilizing the banking branch application as an explanatory variable was not rooted in statistical or personal sensitivity concerns, but rather in political considerations. In this particular instance, our customer was not a single financial institution, but rather a consortium of disparate financial entities with competing interests. As a result, we discovered a significant issue within our customer's organization. Consequently, the problematic variable was removed without further consideration.

In contrast to the aforementioned example, other customers have expressed reservations about the use of certain explanatory variables due to concerns about subjective considerations. In this case, the issue of geography was raised, and our customer argued that the use of their customers' district as explanatory variables would create the impression that some marginalized areas were being discriminated against as opposed to more privileged areas.

In order to address these concerns, it is advisable to first familiarize oneself with the local regulations governing the utilization of specific

---

[7] This line of reasoning is perplexing to those trained in statistics, as variables are designed to be discriminatory. Otherwise, they would be of no consequence. Indeed, from a statistical standpoint, the greater the discriminatory power of a variable, the more valuable it is.

variables. Furthermore, it is imperative to take into account prevailing ethical and accepted moral norms and their implications for the interpretation of statistical data. If the audience is deemed to be receptive, it may be beneficial to engage in an open discourse on the nuances of statistical modeling. For instance, it could be explained to the audience that the final outcome of the scorecard model is not determined by the result of a single variable, but rather by the additive effect of multiple variables, hence the use of multivariable models.

6.8 Summary

Chapter 6 emphasized the importance of broadening our model by incorporating additional variables that go beyond the original ratios and time-based transformations. By introducing new dimensions of information, such as demographic attributes, historical financial obligations, and macroeconomic factors, we expanded the scope and depth of our analysis. These variables not only complement the existing ones but also provide unique perspectives that enhance the predictive power of the model.

Key takeaways:

1. **Diverse Sources of Information:** The additional variables were drawn from various sources, including transactional data, application details, public information, and online data, offering a wide range of insights.

2. **Conceptual Categorization:** Variables were grouped into seven categories based on their conceptual meaning: elements of daily life, monetary balance, unconventional information, financial obligations, personal information, banking details, and macroeconomic data.

3. **Enriching the Model:** These variables introduced new dimensions, such as customer segmentation, spending behavior, geographic influences, and financial trends, which complement the previously explored ratios and time-based variables.

4. **Improved Predictive Power:** By diversifying the inputs, the model becomes more robust, capable of capturing a broader range of behaviors and patterns.

5. **Strategic Selection:** While the number of potential variables is vast, careful selection ensures that only those with meaningful contributions are incorporated into the final model.

In the preceding chapters, we established a strong theoretical foundation for variable design, covering daily life elements, ratios, time-based transformations, and the predictive power of past commitments. Chapter 6 expanded this framework by incorporating additional variables, enriching our perspective with diverse data sources. Now, we shift from theory to practice: Chapter 7 focuses on constructing analytical base tables (ABTs), where these designed variables are organized, transformed, and integrated into a structured format, bridging the gap between conceptual design and practical implementation.

CHAPTER 7

Things to Know About ABTs

> "In God, we trust. All others must bring data."
>
> —W. Edwards Deming

ABT, or "analytical base table," is a specialized dataset designed for purposes far beyond mere data storage. Unlike transactional tables, which are optimized for relational databases, ABTs are fully *denormalized*; that is, each record in an ABT is organized into a single horizontal vector per observation (or subject), making it unsuitable for traditional relational database structures but ideal for analytical modeling.

In an ABT, each variable—including the dependent variable—is represented as a column. While all variables are columns, not all columns are variables. What sets an ABT apart is its purpose: variables in an ABT are not simply "data"; they embody the hypotheses we aim to test for statistical significance (see Chapter 13). This distinction elevates the ABT from being a simple table to a critical component of the modeling process.

Moreover, it can be argued that an ABT is not merely a "table" or a "dataset." Rather, it represents the essence of the model itself. The success of any model depends heavily on the design of the ABT, often more so than on the mathematical models, algorithms, or techniques employed.

CHAPTER 7   THINGS TO KNOW ABOUT ABTS

The validity of a model hinges on the robustness of its underlying hypotheses, as exemplified by Altman's model. This underscores the pivotal role of the ABT in determining the efficacy of a model.

At this juncture, it is imperative to highlight the fact that while machine learning (ML) algorithms may facilitate the construction of models, it is ultimately the responsibility of the modeler to ensure the soundness of the hypotheses upon which the model is based.

7.1 ABTs Are All About Data Processing

The terms *"data mining,"* *"machine learning,"* and *"data science,"* among others, are currently in vogue. However, they tend to ignore a crucial aspect of the modeling process, i.e., *"data processing."* Consequently, when we encounter references to *state-of-the-art algorithms* and related ML terminology, such as deep learning, neural networks, gradient boosting, and so forth, the topic of data processing is seldom discussed. It is noteworthy that data processing is not a topic of discussion in academic circles either. It is as if data processing never occurs during the model-building process; however, it does, and in fact, it is the most demanding and arduous part of the entire modeling process. This is because it requires a hands-on approach, which many people find challenging.

Notwithstanding the growing complexity of modeling techniques, a considerable number of individuals continue to regard *data processing* and *modeling* as two distinct activities. However, this perception is erroneous, as the two aspects in question are, in fact, two facets of the same methodology. This is because the accurate materialization of explanatory variables is contingent upon the correct processing of data, which in turn is dependent upon the proficiency of the programmer. It is therefore essential that the appropriate extraction rules are applied and that the spatial configuration of the data is accurately visualized in order to facilitate the transformation and manipulation of the data in a manner that accurately reflects the reality being modeled. It is crucial to recognize that no individual is more familiar with the data in question than oneself. However, in order to effectively manipulate the data, it is first necessary to possess *advanced programming skills*.

7.2 ABTs Require Advanced Programming Skills

It is imperative that programming skills be a prerequisite for those seeking to become modelers; they should be viewed not as a desirable attribute but as an essential qualification. It is, however, not essential for a modeler to have knowledge of a specific programming language, provided that they have previously exhibited a high level of proficiency in programming and, more importantly, have demonstrated an aptitude and willingness to learn. There are numerous programming languages in existence; however, the most important for data manipulation purposes is the *Structured Query Language* (SQL).

7.2.1 SQL Is ABTs' Best Friend

In 1680, the Levant, a region on the eastern shores of the Mediterranean Sea, was the common ground for merchants from all over the world to trade and do business. But the biggest challenge they faced had nothing to do with the deals they wanted to make with other merchants, nor with the inherent risks of sailing the seas—no, the biggest challenge was *communication*. But to overcome this problem, half a century ago, a common language emerged among the merchants trading in the Levant; it was a combination of Italian, Spanish, French, Greek, Arabic, and Turkish. This language was called *lingua franca*, and it became the standard in many ports of the time (Oracle SQL by Example [56]).

CHAPTER 7   THINGS TO KNOW ABOUT ABTS

The Structured Query Language (SQL) is today's *lingua franca* for the information technology (IT) industry, and it is currently used to effectively communicate with the database in a completely *standard* way. A standard language means that the code written is easy to understand and can be fully supported and quickly extended. The SQL was originally developed by the IBM Research Laboratory (based on a paper by Dr. E. F. Codd at IBM in 1970) as a standard language for use with relational databases.

There are several reasons to prefer SQL over other *data manipulation languages*, such as follows:

- SQL is used by most (if not all) database applications.

- Although the language has evolved over the past 50 years, most of the basic functionality remains the same.

- Even after half a century, there is no indication that the SQL will be replaced by another language any time soon.

It should be noted that the term *"programming languages"* was deliberately omitted in reference to SQL. Instead, the term *"data manipulation languages"* was employed, as the suitability of different programming languages depends on the task at hand. One might argue that Python could be used to construct the ABT; however, this is not a widely adopted approach within the IT industry. In fact, it is not a commonly used language in this field. It is not this author's intention to suggest that Python should not be employed for data processing. Rather, the argument is that a language like Python would be better suited to modeling tasks than to data processing. The recommendation, therefore, is that the most appropriate course of action would be to utilize the strengths of each tool in the context of its respective forte.

7.3 Summary

In this chapter, we examined the critical role of analytical base tables (ABTs) in the modeling process and their unique importance in bridging raw data and statistical analysis. ABTs are not just data storage tools; they are the foundation upon which successful models are built. By embodying hypotheses and organizing data in a fully denormalized structure, ABTs enable the testing of statistical significance and directly influence the validity and efficacy of models.

Additionally, we highlighted that while machine learning algorithms facilitate model construction, the ultimate success of a model depends on the expertise of the modeler. Advanced programming skills are indispensable in this context, as they empower modelers to design robust ABTs, manipulate data effectively, and implement sophisticated techniques to align with the hypotheses being tested.

Key takeaways:

1. **Definition and Structure:** ABTs are specialized, fully denormalized datasets organized into one horizontal vector per observation, making them unsuitable for relational databases but ideal for analytical purposes.

2. **Role of Variables:** Each variable, including the dependent variable, occupies a column in the ABT, with variables representing hypotheses rather than mere data.

3. **Significance of Design:** The success of a model depends more on the design of the ABT than on the mathematical models or algorithms used.

CHAPTER 7   THINGS TO KNOW ABOUT ABTS

4. **Modeler's Responsibility:** While machine learning algorithms facilitate model construction, it is ultimately the modeler's responsibility to ensure the soundness of the hypotheses upon which the ABTs—and the model—are based.

5. **Programming Skills as a Key Asset:** Advanced programming skills are crucial for modelers to design ABTs effectively, handle complex data transformations, and implement techniques that align with the hypotheses and modeling goals.

With the foundational importance of analytical base tables (ABTs) established in Chapter 7, we now turn to the practical aspects of managing variables in the modeling process. Chapter 8 emphasizes the importance of organization, consistency, and attention to detail when working with large numbers of variables. It introduces the use of spreadsheets as a powerful tool to standardize and streamline workflows, helping modelers avoid common mistakes and maintain precision.

CHAPTER 8

The Building Plan and Variable Management

"You know what I've noticed? Nobody panics when things go 'according to plan.'"

—The Joker, *The Dark Knight (2008)*

In any modeling process, managing a large number of variables can quickly become overwhelming without a clear and organized approach. This chapter introduces practical strategies to ensure consistency and precision when working with variables. Just as the construction of any building relies on a well-designed plan to guide its progress, constructing an effective ABT requires its own carefully crafted building plan. By implementing structured workflows and leveraging simple yet powerful tools, modelers can streamline their processes, maintain organization, standardize the assembly of SQL code, and avoid common pitfalls.

CHAPTER 8 THE BUILDING PLAN AND VARIABLE MANAGEMENT

8.1 Managing Your Variables in Spreadsheets

As illustrated in **Figure 8-1**, a spreadsheet can be utilized for the creation and administration of explanatory variables. It is imperative to acknowledge that the final variable name, upon completion, does not constitute mere text; rather, it is a dynamic formula. This underscores the efficacy of spreadsheets as a tool for the extension, creation, and management of explanatory variables, particularly in the context of developing a comprehensive *building plan*.

A building plan should include a comprehensive and explicit account of the characteristics of each variable, as well as facilitate a *meta-analysis* of the variables along multiple dimensions, including data source, type, periodicity, operation, and other relevant elements. For example, the number of variables created by data source can be quantified in order to determine the primary data source for the scorecard model. Additionally, the number of variables created by each block can be quantified in order to identify inconsistencies and to balance *experiments*.[1] As illustrated in **Figure 8-1**, there are 29 variants (column M) of the base ratio, balance/credit limit. Consequently, by organizing the variables according to their base ratio, the same number of variants should be obtained for each base ratio.

---

[1] In this context, experiment balancing refers to the process of designing and adjusting an experiment to ensure that the treatment groups are equivalent with respect to key variables or covariates. In essence, experiment balancing is a critical aspect aimed at reducing bias and enhancing the validity of causal inferences drawn from the experiment. More about experimental design in Section 17.1.

CHAPTER 8   THE BUILDING PLAN AND VARIABLE MANAGEMENT

Variable Name	Length	Ratio	Var. Type	Operation	Code	Period Label	Start Period	End Period	Num
I_R_BAL_LM_01MB	15	BALANCE / CREDIT LIMIT	I	R	R_BAL_LM	01MB	1	1	1
I_R_BAL_LM_02MB	15	BALANCE / CREDIT LIMIT	I	R	R_BAL_LM	02MB	2	2	2
I_R_BAL_LM_03MB	15	BALANCE / CREDIT LIMIT	I	R	R_BAL_LM	03MB	3	3	3
I_R_BAL_LM_06MB	15	BALANCE / CREDIT LIMIT	I	R	R_BAL_LM	06MB	6	6	4
I_R_BAL_LM_08MB	15	BALANCE / CREDIT LIMIT	I	R	R_BAL_LM	08MB	8	8	5
I_R_BAL_LM_10MB	15	BALANCE / CREDIT LIMIT	I	R	R_BAL_LM	10MB	10	10	6
I_R_BAL_LM_12MB	15	BALANCE / CREDIT LIMIT	I	R	R_BAL_LM	12MB	12	12	7
I_AV_R_BAL_LM_L03M	18	BALANCE / CREDIT LIMIT	I	AV	R_BAL_LM	L03M	1	3	8
I_AV_R_BAL_LM_L06M	18	BALANCE / CREDIT LIMIT	I	AV	R_BAL_LM	L06M	1	6	9
I_AV_R_BAL_LM_L12M	18	BALANCE / CREDIT LIMIT	I	AV	R_BAL_LM	L12M	1	12	10
I_AV_R_BAL_LM_L13_24M	21	BALANCE / CREDIT LIMIT	I	AV	R_BAL_LM	L13_24M	13	24	11
I_IN_R_BAL_LM_01MB	18	BALANCE / CREDIT LIMIT	I	IN	R_BAL_LM	01MB	1	1	12
I_IN_R_BAL_LM_02MB	18	BALANCE / CREDIT LIMIT	I	IN	R_BAL_LM	02MB	2	2	13
I_IN_R_BAL_LM_03MB	18	BALANCE / CREDIT LIMIT	I	IN	R_BAL_LM	03MB	3	3	14
I_IN_R_BAL_LM_06MB	18	BALANCE / CREDIT LIMIT	I	IN	R_BAL_LM	06MB	6	6	15
I_IN_R_BAL_LM_12MB	18	BALANCE / CREDIT LIMIT	I	IN	R_BAL_LM	12MB	12	12	16
I_SIN_R_BAL_LM_L03M	19	BALANCE / CREDIT LIMIT	I	SIN	R_BAL_LM	L03M	1	3	17
I_SIN_R_BAL_LM_L06M	19	BALANCE / CREDIT LIMIT	I	SIN	R_BAL_LM	L06M	1	6	18
I_SIN_R_BAL_LM_L12M	19	BALANCE / CREDIT LIMIT	I	SIN	R_BAL_LM	L12M	1	12	19
I_SIN_R_BAL_LM_L13_24M	22	BALANCE / CREDIT LIMIT	I	SIN	R_BAL_LM	L13_24M	13	24	20
I_DC_R_BAL_LM_01MB	18	BALANCE / CREDIT LIMIT	I	DC	R_BAL_LM	01MB	1	1	21
I_DC_R_BAL_LM_02MB	18	BALANCE / CREDIT LIMIT	I	DC	R_BAL_LM	02MB	2	2	22
I_DC_R_BAL_LM_03MB	18	BALANCE / CREDIT LIMIT	I	DC	R_BAL_LM	03MB	3	3	23
I_DC_R_BAL_LM_06MB	18	BALANCE / CREDIT LIMIT	I	DC	R_BAL_LM	06MB	6	6	24
I_DC_R_BAL_LM_12MB	18	BALANCE / CREDIT LIMIT	I	DC	R_BAL_LM	12MB	12	12	25
I_SDC_R_BAL_LM_L03M	19	BALANCE / CREDIT LIMIT	I	SDC	R_BAL_LM	L03M	1	3	26
I_SDC_R_BAL_LM_L06M	19	BALANCE / CREDIT LIMIT	I	SDC	R_BAL_LM	L06M	1	6	27
I_SDC_R_BAL_LM_L12M	19	BALANCE / CREDIT LIMIT	I	SDC	R_BAL_LM	L12M	1	12	28
I_SDC_R_BAL_LM_L13_24M	22	BALANCE / CREDIT LIMIT	I	SDC	R_BAL_LM	L13_24M	13	24	29

Figure 8-1. Example of how to use a spreadsheet to create and manage your explanatory variables.

CHAPTER 8   THE BUILDING PLAN AND VARIABLE MANAGEMENT

It is notable that no code has been written thus far. This is because the building plan should serve as the primary driver for the coding process, rather than the other way around. Accordingly, the building plan should explicitly indicate all of the variables that are to be included in the scorecard model. This will enable the precise determination of the number of variables to be constructed, the anticipated number of variants for each base variable, and the periodicities to be employed, which will subsequently inform the modeling timeframe. Furthermore, the building plan enables the visual *assessment* of the *symmetry* of the *experiments*. It should be noted that each variable represents an independent hypothesis test, which is equivalent to an independent experiment. This process can also be referred to as *experiment design* or *experiment balancing*.

Furthermore, the building plan should facilitate the identification of any potential errors or inconsistencies in the final ABT. **Table 8-1** illustrates the output of a report generated from a building plan. In this instance, a cross-tabulation was constructed by crossing base variables with periodicity.

The resulting report demonstrates the efficacy of detecting errors and inconsistencies through a simple observation: the total number of variables to be created for the periodicity of five months back (5MB) is 18, while the total number of variables to be created for the remaining periodicities between 01MB and 11MB should be 19. Similarly, the inconsistency can be identified by row. It can be observed that the root variable, ratio food expenses/total expenses, produces 11 variables, whereas the remaining ratios yield 12. This leads to the conclusion that the periodicity 05MB is absent for the ratio of food expenses to total expenses. However, the report contains a wealth of additional information. As an illustration, it can be observed that the sum of all increment/decrement flags is limited to a maximum of 11 variables within a 12-month interval. Conversely, the point-in-time ratios should yield a total of 12 within the same interval. Additionally, it is notable that this concise report encompasses the information of over 200 variables.

CHAPTER 8   THE BUILDING PLAN AND VARIABLE MANAGEMENT

Table 8-1. Illustration of the use of your building plan to detect errors and inconsistencies.

Root Variable	01MB	02MB	03MB	04MB	05MB	06MB	07MB	08MB	09MB	10MB	11MB	12MB	Grand Total
Decrement Flag Purchases/Balance	1	1	1	1	1	1	1	1	1	1	1		11
Decrement Flag Payments/Purchases	1	1	1	1	1	1	1	1	1	1	1		11
Decrement Flag Payment/Balance	1	1	1	1	1	1	1	1	1	1	1		11
Decrement Flag Balance/Credit Limit	1	1	1	1	1	1	1	1	1	1	1		11
Increment Flag Purchases/Balance	1	1	1	1	1	1	1	1	1	1	1		11
Increment Flag Payments/Purchases	1	1	1	1	1	1	1	1	1	1	1		11
Increment Flag Payment/Balance	1	1	1	1	1	1	1	1	1	1	1		11
Increment Flag Balance/Credit Limit	1	1	1	1	1	1	1	1	1	1	1		11
Ratio Purchase/Balance	1	1	1	1	1	1	1	1	1	1	1	1	12
Ratio Balance/Credit Limit	1	1	1	1	1	1	1	1	1	1	1	1	12
Ratio Payments/Balance	1	1	1	1	1	1	1	1	1	1	1	1	12
Ratio Payments/Purchases	1	1	1	1	1	1	1	1	1	1	1	1	12
Ratio Food Expenses/Total Expenses	1	1	1	1		1	1	1	1	1	1	1	11
Ratio Retail expenses/Total Expenses	1	1	1	1	1	1	1	1	1	1	1	1	12
Ratio Education Expenses/Total Expenses	1	1	1	1	1	1	1	1	1	1	1	1	12
Ratio Luxuries Expenses/Total Expenses	1	1	1	1	1	1	1	1	1	1	1	1	12
Ratio Health Expenses/Total Expenses	1	1	1	1	1	1	1	1	1	1	1	1	12
Ratio Services Expenses/Total Expenses	1	1	1	1	1	1	1	1	1	1	1	1	12
Ratio Other Services Expenses/Total Expenses	1	1	1	1	1	1	1	1	1	1	1	1	12
Total Number of Variables Created	19	19	19	19	18	19	19	19	19	19	19	11	219

131

8.2 Code As You Name Your Variables

Another benefit of utilizing spreadsheet concatenation formulas is that they can be employed not only for the generation of names but also for the generation of code. **Figure 8-2** provides a visual representation of the process of generating Oracle SQL code while simultaneously naming variables.

In this example, the conditions for extraction, as defined by columns G (Measure) and I (Industry), are used to construct the WHERE clause. Conversely, column L, designated "End Period," defines the extraction period by subtracting the number of periods from the reference period.

It is evident from this example that the ADD_MONTHS function offers a convenient means of moving in any direction from the reference date. However, a minor issue arises due to the fact that the PERIOD column in both tables, A and B, is represented by a numerical value in the year-month format, expressed as 'YYYYMM'. This can be readily addressed by employing the TO_DATE and TO_CHAR functions.

It is important to note that in this example, each variable is created using a *correlation subquery* (see Section 10.3.3 [56]), which is only possible if certain conditions are met. For example, the input data source, designated as INPUT_TABLE, must exhibit a vertical conformation, thereby indicating that INPUT_TABLE is a transactional table. Secondly, it is imperative that the result of the subquery is a single record; otherwise, the SQL code will not execute. Consequently, the operation performed in the SELECT clause is an AVG aggregation. It is important to exercise caution when utilizing aggregation operations, as the selection of one over another is contingent upon the interpretation of the variable in question and the structural configuration of the data source.

In this particular case, it can be stated with certainty that the set of conditions placed in the WHERE clause will return a single record. The rationale behind utilizing the AVG aggregation operation is to mitigate the potential for duplicate records.

CHAPTER 8 THE BUILDING PLAN AND VARIABLE MANAGEMENT

	C	D	E	F	G	H	I	J	K	L	M
	Var Name	Length	Measure	Var Type	Operation	Code	Industry	Periodicity	Start Period	End Period	Variable SQL
1											
2	I_R_GDPs_GDPt_PRO_01MB	22	GDPi / GDPt	I	R	R_GDPi_GDPt	GOV	01MB	0	1	SELECT AVG(VALUE) FROM INPUT_TABLE B WHERE B.PERIOD=TO_CHAR(ADD_MONTHS(TO_DATE(A.PERIOD,'YYYYMM'),-1),'YYYYMM') AND MEASURE='GDPi' / GDPt' AND INDUSTRY='PRO') AS I_R_GDPs_GDPt_PRO_01MB
3	I_R_GDPs_GDPt_PRO_02MB	22	GDPi / GDPt	I	R	R_GDPi_GDPt	GOV	02MB	0	2	SELECT AVG(VALUE) FROM INPUT_TABLE B WHERE B.PERIOD=TO_CHAR(ADD_MONTHS(TO_DATE(A.PERIOD,'YYYYMM'),-2),'YYYYMM') AND MEASURE='GDPi' / GDPt' AND INDUSTRY='GOV') AS I_R_GDPs_GDPt_PRO_02MB
4	I_R_GDPs_GDPt_PRO_03MB	22	GDPi / GDPt	I	R	R_GDPi_GDPt	GOV	03MB	0	3	SELECT AVG(VALUE) FROM INPUT_TABLE B WHERE B.PERIOD=TO_CHAR(ADD_MONTHS(TO_DATE(A.PERIOD,'YYYYMM'),-3),'YYYYMM') AND MEASURE='GDPi' / GDPt' AND INDUSTRY='GOV') AS I_R_GDPs_GDPt_PRO_03MB

Figure 8-2. Example of how to create your SQL code as you create and name your variables.

CHAPTER 8   THE BUILDING PLAN AND VARIABLE MANAGEMENT

To further illustrate this point, we can consider a scenario in which the AVG operation is replaced with the SUM operation. Additionally, let us assume that the second condition, INDUSTRY, is overlooked for some reason. It is important to note that the variable being extracted is a ratio, specifically the GDP of the "GOV" industry (which stands for *government*) divided by the total GDP. Thus, by neglecting the INDUSTRY condition, the ratios of all industries are aggregated, resulting in a value of 1 for all columns, which is not the desired outcome. In certain instances, it may be desirable to aggregate a variable, such as when working with transactional data, such as the total amount of purchases or the total number of withdrawals from an ATM.

A third factor to consider when employing subqueries is the time required for execution and the resulting performance. To illustrate this point, consider the specific instance where it is known in advance that the INPUT_TABLE within the FROM clause comprises fewer than 100,000 records. By contrast, the transactional data tables, prevalent in the modern era, generally contain billions of records and are therefore significantly more substantial. Consequently, it is recommended that correlation subqueries be avoided in such cases unless the information has been preaggregated.

While correlation subqueries are undoubtedly a convenient tool, they may not be the optimal choice for building explanatory variables, as the suitability of this approach hinges on the specifics of each case. It is possible that a more complex and sophisticated process may be a more suitable fit for your particular circumstances (see Chapter 10). In such cases, it may be preferable to utilize PL/SQL programming instead. In any case, it is recommended to utilize spreadsheets to their fullest potential when designing and building variables, as they are an invaluable tool that will greatly streamline the model-building process.

8.3 Nomenclature and Consistency

It is important to acknowledge that prior to initiating the construction plan, it is essential to address two fundamental aspects: nomenclature and consistency. The term *"nomenclature"* is used to describe the names and codes that are employed to refer to variables and the various elements that constitute them. The concept of *"consistency"* pertains to the adherence to these names and codes. To illustrate, consider the following scenario: You and your team (assuming you have one) have agreed upon the abbreviation of the word *"average,"* and you have chosen to use the three-letter word "AVG." Subsequently, as the creation of additional variables is undertaken, it becomes apparent that certain names exceed the permitted length for the system in use (e.g., >32 characters). One might then decide to use the two-letter word "AV" for *"average"* and the three-letter word "AVG" on occasion, concluding that this is inconsequential. However, is this truly the case?

Table 8-2 provides a list of recommended codes for the various elements that constitute the variable names. It is noteworthy that in certain instances, multiple codes are recommended for a given element. For example, the grouping function *"count"* can be coded as either CN or CNT, as it refers to an SQL grouping function. However, since counting always returns a frequency, it is also possible to code the count grouping function as FR or FRQ or even FREQ. The key consideration is not whether one code is preferable to another, but rather to maintain consistency in the coding chosen.

CHAPTER 8   THE BUILDING PLAN AND VARIABLE MANAGEMENT

Table 8-2. *List or recommended coding for the different elements of your explanatory variables.*

Category	Object name	Recommended code
Grouping Functions	Average	Av or Avg
Grouping Functions	Maximum	Mx or Max
Grouping Functions	Minimum	Mn or Min
Grouping Functions	Count	Cn or Cnt or Frq or Fr
Grouping Functions	Standard Deviation	St or Std
Variable Type	Interval	I or Int
Variable Type	Nominal	N or Nom
Variable Type	Binary	B or Bin
Variable Type	Ordinal	O or Ord
Operation	Ratio	R
Operation	Decrement	DC or Dec
Operation	Increment	IN or Inc
Operation	Variation/Deviation	V or Vr or Var or D or Dv or Dev
Operation	Sum of Increments	Sin
Operation	Sum of Decrements	Sdc
Banking Portfolios	Credit Card	CC or CrCr
Banking Portfolios	Retail Loan	Ln or Lon or Loan
Banking Portfolios	Mortgage	Mg or Mor or Mrtg
Banking Portfolios	Auto	Au or Aut or Auto or Car
Banking Portfolios	Small and Medium Enterprises	Sm or SME
Banking Portfolios	Corporations	Crp or Corp
Periodicity	Months Back	MB
Periodicity	Last Months	L##M
Periodicity	Next Months	N##M
Event	Known Good and Bad	KGB
Event	Target	Tgt
Event	Default	Df or Def or Dflt
Banking Information	Balance	Bl or Bal
Banking Information	Credit Limit	CL or Lim
Banking Information	Payments	Py or Pay
Banking Information	Purchases/Expenses	Pr or Pur or Ex or Exp
Banking Information	Deposits	Dp or Dep
Banking Information	Withdrawals	Wdr or Wdrw
Banking Information	Savings	Sv or Svg
Banking Information	Installment	Ins or Instl
Miscellaneous	Credit	Crd or Cred
Miscellaneous	Debit	Db or Deb
Miscellaneous	Transaction	Tr or Trn
Miscellaneous	Snapshot	Sn or Snp

CHAPTER 8   THE BUILDING PLAN AND VARIABLE MANAGEMENT

Figure 8-3 provides a visual representation of the various elements contributing to the variable's nomenclature, where

> I = Interval
>
> AV = Average[2]
>
> R = Ratio operation
>
> ATM_EXP = ATM cash withdrawal/total expenses
>
> CSN = Credit card snapshot table
>
> L03M = Last three months

Accordingly, I_AV_R_ATM_EXP_CSN_L03M is read as the *interval variable of the average ratio of the ATM withdrawals to total expenses over the last three months.*

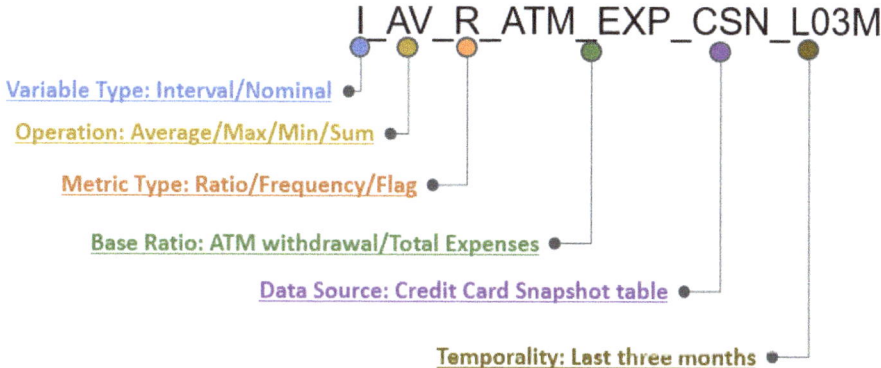

Figure 8-3. *Diagram exemplifying the different elements conforming the variable's nomenclature.*

---

[2] It is imperative to acknowledge the complexities inherent to all calculations. For instance, in the case of the average calculation, how should the presence of null values be addressed? Should the average be deflated by considering null values in the calculation, or should it be inflated by disregarding null values? Thus, when feasible, it is recommended to incorporate coding to identify the type of calculation being performed.

8.4 Summary

In this chapter, we explored the importance of having a structured approach to managing variables when constructing an analytical base table (ABT). By adopting a "building plan" for variable management, we highlighted how organization, consistency, and precision are essential to avoiding errors and ensuring a smooth modeling process. Beyond standardizing the nomenclature of variables, we also emphasized the importance of standardizing the code itself. This approach not only enhances clarity and reduces errors but also streamlines the transition from the planning phase to the production phase—facilitating the seamless construction of the ABT.

Key takeaways:

1. **The Importance of a Building Plan:** A clear and organized "building plan" is essential for managing large numbers of variables. It provides structure and guidance, ensuring that complex processes are approached systematically.

2. **The Value of Standardization:** Standardizing both the variable nomenclature and the code ensures consistency, reduces the risk of errors, and simplifies collaboration across teams.

3. **The Role of Tools in Streamlining Workflows:** Leveraging spreadsheets as tools for standardization helps bridge the gap between planning and production, making workflows more efficient and manageable.

4. **The Power of Structured Workflows:** Structured workflows streamline the process of constructing the ABT, ensuring that every step is deliberate, organized, and aligned with the overall objectives.

CHAPTER 8   THE BUILDING PLAN AND VARIABLE MANAGEMENT

5. **The Foundation of Consistency:** Consistency in planning and execution lays a strong foundation for effective data modeling, ensuring that each phase of the process supports the next.

With a solid plan for managing and standardizing variables and code in place, we are now ready to focus on the foundation of any analytical base table: the **target population**. Chapter 9 will explore how to define the specific entities—whether individuals, transactions, or other units of analysis—that will serve as the focus of your ABT. By clearly identifying the target population, we set the stage for aligning data with business objectives and ensuring that all subsequent steps in the ABT construction process are purposeful and precise.

CHAPTER 9

Target Population

"If you aim at everything, you hit nothing."

—Strategic planning and marketing literature

This chapter focuses on identifying *who* or *what* that will serve as the cornerstone of your analytical base table (ABT). The target population forms the backbone of your analysis, encompassing the specific entities—be it individuals, customers, accounts, or other units of analysis—that your model seeks to understand or make predictions about. A clear and precise definition of this population is not just a technical step; it is a strategic one, ensuring the scope of your data aligns seamlessly with the business objective. But the influence of the target population extends far beyond the number of records in your ABT—it shapes the selection of variables, their importance, and, ultimately, the accuracy and relevance of your final model.

9.1 Defining Your Target Population

It is of the utmost importance that the target population is correctly defined in order to guarantee the accuracy and efficacy of the model. This is due to a segmentation principle, which aims to isolate homogeneous populations. One may construct a single PD model for the portfolio using product classification as an explanatory variable (e.g., credit card, mortgage, loan, etc.). Nevertheless, it would be more prudent to construct a different model for each product.

CHAPTER 9 TARGET POPULATION

The use of independent models for each product is often necessary due to the distinct characteristics and behaviors exhibited by these products. To illustrate, the CLTV variable is applicable solely to mortgages and auto loans, and not to personal loans or credit cards. Similarly, the utilization variable, which gauges credit card usage, is only applicable to credit cards and credit lines. The construction of a single ABT encompassing all products will result in the dissemination of information across a multitude of variables in a disorganized manner.

It is not uncommon for customers to have multiple products active; however, it is also the case that many do not. It would be preferable to have distinct models for mortgages and credit cards, with a flag for other products. For example, the addition of flags could facilitate the identification of active credit cards and mortgages. This approach will facilitate the isolation of product characteristics and behaviors, thereby enhancing the discriminatory power of the explanatory variables.

9.2 Segmentation

In general, it is always preferable to segment the population in question according to the default rate. This is because populations with homogeneous DRs will tend to exhibit similar characteristics among themselves, as opposed to populations with significantly disparate DRs. For example, it is well documented that mortgage default rates are significantly lower than credit card default rates[1] [9]. **Table 9-1** provides a summary of the average DR in the United States reported for the first quarter of 2020 and 2021. It is notable that although the values reported in 2020 are considerably larger, there is nevertheless a marked difference between products. However, regardless of the year, mortgage consistently represents the lower risk (1.17% and 0.42%, for 2020 and 2021,

---

[1] https://www.creditcards.com/statistics/credit-card-delinquency-statistics-1276/

respectively), in comparison to auto loan (2.37% and 1.72%, for 2020 and 2021, respectively) and credit card (5.31% and 3.78%, for 2020 and 2021, respectively). Similarly, credit cards consistently represent a higher risk, irrespective of the year.

Table 9-1. Average risk reported for the first quarter of 2020 and 2021 for mortgages, auto loans, and credit cards.

Product	Q1 2020	2021
Mortgage	1.17%	0.42%
Auto loan	2.37%	1.72%
Credit card	5.31%	3.78%

The issue of population segmentation raises the question of whether further segmentation of the population in question will necessarily lead to improved results. The response to this question is contingent upon a number of variables. For instance, it may be advantageous to subdivide the mortgage population into even more granular categories, such as Pacific, Rocky Mountain, Midwest, Southwest, Southeast, and Northeast regions. This approach could prove beneficial. Nevertheless, prior to undertaking further segmentation, it would be prudent to consider the following:

1. Does the level of risk by region justify a further segmentation?

2. Is there enough resources to build six more models instead of just one?

3. Would it be better to just add a nominal variable with the regional classification?

4. Are there enough events (defaults) for each region to ensure statistical representativeness?

CHAPTER 9 TARGET POPULATION

As you can see, there is no simple answer to the question of whether or not to continue segmenting your population.

For the sake of argument, let us assume that the resources are available to construct six distinct mortgage models, one for each region in the United States. In this case, it would be advisable to initially undertake a univariable analysis of the region variable to ascertain whether region, as a risk factor, is a significant predictor of different levels of risk across different regions. In such a case, it would be advisable to conduct a *post hoc* analysis, with the objective of comparing all the different combinations of the treatment groups (in this instance, the regions) in order to ascertain which regions exhibit significantly disparate risk levels. This method is known as *category lumping*.

Category lumping is also referred to as clustering. As its name implies, category lumping specifically reduces the number of categories present in a nominal variable by lumping together categories that have no significant difference in their target response.

This technique permits the population to be properly segmented based on a mere two variables: the nominal variable and the response variable. The following three methods may be employed as category lumping techniques:

1. Greenacre's method [26,47]
2. Decision trees
3. Weight of evidence (WoE)

Let us assume that the region variable can significantly predict different levels of risk. Furthermore, let us assume that four significant risk groups have been identified among the six different regions in the United States, with corresponding DRs of

1. Pacific region, 1.5%
2. Rocky-Mountain-Midwest region, 0.5%

3. Southwest-Southeast region, 2%
4. And the Northeast region, 1%

The question that now arises is whether the number of subjects in each of the aforementioned segments is sufficient to construct the model. **Table 9-2** presents a hypothetical illustration, assuming a total mortgage population of 900,000 accounts, with a fictitious population distribution by lumped region.

Table 9-2. *Example of expected independent variables for mortgage models segmented by region.*

Region	DR	Accounts	Defaults	Expected variables
Pacific	1.50%	212,636	3,189	80
Midwest	0.50%	185,752	929	23
South	2%	347,581	6,952	174
Northeast	1%	154,030	1,540	39
	1.4%	900,000	12,611	315

It should be noted that the final column is labelled *"Expected variables."* The term "Expected variables" refers to the total number of independent variables that can be used during the modeling process, given the total number of events (defaults) that are available. The total number of independent variables is defined by the following expression [42]:

$$\frac{n}{40} > m$$

Equation 9-1. Relationship between the total number of independent variables and the total number of events in a logistic regression.

Where *n* refers to the total number of events, and not to the total number of observations in your sample, and *m* refers to the total number of independent variables you start with, regardless of the number of significant variables in your final model.

It is notable that with a total number of 900,000 observations and a default rate of 1.4%, only 315 explanatory variables can be utilized during the modeling process, as illustrated in **Equation 9-1**. It is therefore important to consider that in subsequent stages of the process, the information will be partitioned into two distinct groups: a training partition and a validation partition. This will inevitably result in a further reduction in the number of defaults and, consequently, the number of input variables, in accordance with the principles of *statistical representativeness* and *statistical power*.

9.3 Statistical Representativeness and Statistical Power

The concept of statistical representativeness pertains to the sufficiency of observations to accurately depict the default risk profile of a given population group. To illustrate, the Midwest segment comprises a total of 185,752 subjects and a total of 929 defaults. Let us now consider a scenario in which there are six explanatory variables: sex, economic class, education, marital status, age, and profession. For each of these variables, we can define the following categories:

1. Sex:
    a. Male
    b. Female

2. Economic class:

    a. A/B

    b. C+

    c. C-

    d. D

    e. E,F

3. Education:

    a. Elementary school

    b. High school

    c. Bachelor's degree

    d. Graduate degree

4. Marital status:

    a. Single

    b. Married by separation of goods

    c. Married by community of acquired assets

5. Profession:

    a. Farming

    b. Blue collar

    c. White collar

    d. Businessman

    e. Independent

    f. Government

CHAPTER 9   TARGET POPULATION

6. Age by bin:

    a. 18 to 25

    b. 25 to 35

    c. 35 to 45

    d. 45 to 55

    e. 55 to 65

    f. >65

Assuming a multiplicative combination, the result is as follows:

$2 \times 5 \times 4 \times 3 \times 6 \times 6 = 4{,}320$ different groups

If subjects are distributed uniformly across groups, we can expect 43 subjects per group. But the 929 defaults must also be distributed across the 4,320 groups (assuming all combinations exist). Consequently, even if one group had one default case, 3,391 would have 0% risk and 929 would have 2.33% risk. It is therefore pertinent to inquire as to whether these groups are representative of the population as a whole. Consider a hypothetical scenario in which single women in economic class A/B with a university education, who are independent and between 35 and 45 years of age, have a 0% risk of defaulting. How reliable is this result? This leads to the question of whether a larger sample might have yielded a different result.

It should be noted that not all combinations of the six explanatory variables are real-world possibilities. It is highly unlikely that default risk will be homogeneously distributed among the 4,320 groups. However, this is not the focus of this discussion. The objective is to ascertain whether the number of cases is sufficient to achieve statistical representativeness and to illustrate the relationship between the total number of default cases and the total number of input variables in a given model.

In contrast, *statistical power*, *statistical representativeness*, and *sample size* are interrelated, and over-segmentation will impact all three in varying ways (as illustrated in the previous example). But before proceeding, it is essential to define statistical power.

> *The probability of detecting a statistically significant difference in a given sample, assuming that a difference of a given magnitude exists in the population.*

If we consider the cohort of college-educated, single women in economic class A/B between the ages of 35 and 45, what is the likelihood of detecting a significant effect on the default rate if we focus solely on the Midwest population? What are the implications for the probability of a given outcome if the entire population is considered instead?

It is evident that the practice of over-segmentation not only leads to an inaccurate measurement of default risk for specific groups but also reduces the likelihood of identifying an existing effect. This is also referred to as a *type I statistical error*.

In general, it is preferable to segment populations by default rate, as populations with homogeneous DRs tend to exhibit similar characteristics. Nevertheless, it is imperative that the *equilibrium* between the total number of independent variables that can be employed at the start of the model-building process and the total number of events required to achieve statistical representativeness and statistical power is not compromised by over-segmentation.

9.4 Business Rules

It has been previously assumed that the definition of a target population can be achieved through the selection of a specific category from a nominal variable, such as a financial product. However, this is rarely, if ever, the case. In practice, defining the target population is often a more complex undertaking than merely applying a straightforward condition

CHAPTER 9   TARGET POPULATION

such as PRODUCT = 'MORTGAGE'. Indeed, it is not uncommon for such a classification to be absent, necessitating the implementation of intricate *business rules* to accurately identify and categorize each product within the portfolio.

For example, it may be the case that the financial industry in question lacks a formal definition of a *mortgage*. Instead, the financial product may be identified by the marketing campaign coding that created it, or even by the marketing campaign description, such as follows:

1. H123-The home of your dreams!
2. H124-Mortgage loans 2023
3. H125-Housing 2023
4. H126-Housing+30
5. H127-Mor+30
6. H128-Home+30
7. H129-House 2023
8. H111-Mortgage loan

The following list provides an overview of the potential codes and/or descriptions that may be employed to identify the mortgage product in question. It should be noted that the term *"mortgage"* alone is insufficient for identifying mortgage loans. Instead, a combination of words is typically used, including *"home," "mor," "housing,"* and *"house,"* among others. It bears noting that this is not a hypothetical scenario; rather, it is a genuine account of a real-world experience, and not even the most extreme case I have encountered during my professional career. Let us now consider a further complication, as follows:

- There is no formal financial product definition in your DB, so there is no classification column available.

CHAPTER 9 TARGET POPULATION

- There is, however, the commercial campaign code and description of the original loan application.

- Unfortunately, codes are recycled for future campaigns, so you cannot rely on the campaign codes.

- You can rely on the campaign descriptions; however, the text is messy, and they cannot always be used to properly classify the financial products.

- However, you can use multiple columns and nested rules and regular expressions to determine the financial product with an acceptable level of error; let us say with a misclassification rate of less than 2%.

Listing 9-1 illustrates a brief excerpt of 150 lines of business rules code utilized for the classification of financial products. It is evident that four distinct variables (PORTFOLIOTYPE, USAGE, ACTIVITY, and COMMERCIAL_CAMPAIGN) are necessary to differentiate between mortgage (housing) and personal retail personal loans. It is also important to note that in order to properly identify mortgage loans, multiple keywords and regular expressions are required, such as follows:

- HOUSING
- HOUSE
- HOUSE OUTLET
- MORTGAG
- MORTG
- MORTGAGE
- HOME IMPROVEMENT
- REAL ESTATE
- REAL EST

CHAPTER 9 ■ TARGET POPULATION

Listing 9-1. Extract from a subset of nested rules used to classify financial products

```
CASE
    WHEN PORTFOLIOTYPE=0 AND USAGE=9 AND UPPER(REGEXP_REPLACE(ACTIVITY, '\s{2,}', ' ')) NOT LIKE '%HOME IMPROVEMNTS%'   THEN 'Housing'
    WHEN PORTFOLIOTYPE IN (3,8) AND UPPER(REGEXP_REPLACE(ACTIVITY, '\s{2,}', ' ')) NOT LIKE '%HOME IMPROVEMNTS%'        THEN 'Housing'
    WHEN UPPER(REGEXP_REPLACE(COMMERCIAL_CAMPAIGN, '\s{2,}', ' ')) LIKE '%REAL ESTATE%'                                 THEN 'Housing'
    WHEN UPPER(REGEXP_REPLACE(COMMERCIAL_CAMPAIGN, '\s{2,}', ' ')) LIKE '%REAL EST%'                                    THEN 'Housing'
    WHEN UPPER(COMMERCIAL_CAMPAIGN) LIKE '%HOUSING%'                                                                    THEN 'Housing'
    WHEN UPPER(COMMERCIAL_CAMPAIGN) LIKE '% HOUSE %'                                                                    THEN 'Housing'
    WHEN UPPER(COMMERCIAL_CAMPAIGN) LIKE '% HOUSE OUTLET %'                                                             THEN 'Housing'
    WHEN UPPER(COMMERCIAL_CAMPAIGN) LIKE '% MORTGAG. %'                                                                 THEN 'Housing'
    WHEN UPPER(COMMERCIAL_CAMPAIGN) LIKE '% MORTG. %'                                                                   THEN 'Housing'
    WHEN UPPER(COMMERCIAL_CAMPAIGN) LIKE '% MORTGAGE%'                                                                  THEN 'Housing'
    WHEN UPPER(REGEXP_REPLACE(ACTIVITY, '\s{2,}', ' ')) LIKE '%HOME IMPROVEMNTS%'                                       THEN 'Housing'
    WHEN UPPER(REGEXP_REPLACE(ACTIVITY, '\s{2,}', ' ')) NOT LIKE '%HOME IMPROVEMNTS%' AND UPPER(ACTIVITY) LIKE '% HOUSING %' THEN 'Housing'
    ELSE 'Loan'
END
```

CHAPTER 9   TARGET POPULATION

Additionally, it is worth noting the utilization of sophisticated text processing functions, such as Oracle's REGEXP (regular expressions), for the purpose of eliminating multiple consecutive spaces between characters or words. It is also noteworthy that this is a common occurrence when attempting to identify the target population that will be utilized during the construction of your ABT.

Furthermore, a multitude of sophisticated business rules are employed to identify financial products. Nevertheless, it is crucial to acknowledge that the process of identifying and extracting the target population is inherently complex.

9.5 Dealing with Keys and Unicity

The target population represents the initial column of the ABT and is typically comprised of numeric codes that delineate the *subject level of analysis*. The subject level of analysis pertains to the degree of detail or granularity of the data. As previously stated, the most prevalent subjects of analysis within the financial industry are the account, customer, and application levels. It should be noted, however, that there are numerous additional subjects of analysis that may be employed, including, but not limited to, phone number, member number/ID, subscriber number/ID, social security number/ID, and so forth. It is anticipated that the subject level of analysis will be represented by an ID of a numeric nature. However, this will not always be the case.

The subject level of analysis, therefore, encompasses not only the target population but also the primary key that will serve as a driver for the sequential addition of information to the ABT. Consequently, the primary key, as any other primary key, should have the characteristics that allow for the proper cross-referencing of the target population with the input data sources utilized during the construction of the ABT. However, in most cases, the native account number, for instance, does not provide a robust and secure means for the cross-referencing of information. This may be due to various reasons, including historical inconsistencies resulting from debt restructurings, deficiencies in database operations, and the risk of record multiplication. These issues are often encountered during the development of any ABT, underscoring the necessity for a *synthetic key* replacement to enhance data security and integrity.

9.5.1 Synthetic Keys

The use of a numeric ID will facilitate the cross-referencing of the target population with the input data sources utilized during the ABT-building process. From a database perspective, numbers require less storage space than characters, which reduces the computational burden during join operations.

In some industries, the *native keys* (the original IDs) are not utilized, and instead, *system-generated keys* or *synthetic keys* are employed. System-generated keys, also referred to as synthetic keys, are consistently numeric and are therefore particularly useful when cross-referencing information from one source to another. This is due to the fact that synthetic keys restart their sequence at a certain point within the database, which significantly reduces the size of the key used for cross-referencing information.

For example, consider account numbers, which often have 10 to 16 digits, which are too large for efficient join operation performance. An alternative is to replace the original account numbers with a consecutive number, commencing at one.

Table 9-3 illustrates the manner in which synthetic keys are allocated to a distinctive account number. It is notable that account numbers are typically lengthy, encompassing distinctive information about the account in question. For example, it is not uncommon for banks to concatenate the code of the branch where the account was originally opened, together with the numeric code that identifies the specific product for which the account is intended, and a consecutive number to uniquely identify the current account.

Table 9-3. Example of a system-generated key.

Account number	Synthetic key
11223335548489	1
11223335548490	2
11223335548491	3
11223335548492	4
11223335548493	5
11223335548494	6
11223335548495	7
11223335548496	8
11223335548497	9

9.5.2 Correcting Historical Inconsistencies Using Synthetic Keys

Synthetic keys offer a convenient solution that is not only fast but also enables the correction of inconsistencies in information. One of the most prevalent challenges encountered when consolidating data from disparate sources is the presence of *historical inconsistencies*. A historical inconsistency can be defined as the issue of having *truncated time series* in the available information. The phenomenon of time series truncation occurs when data abruptly ceases to be available at a specific point in time, whether in the forward or backward direction.

CHAPTER 9   TARGET POPULATION

Table 9-4 illustrates a case of historical inconsistency. From this example, it is evident that Account-1 and Account-2 are, in fact, the same account. This is due to the fact that they both belong to the same customer, John Adams, and also because the balance amount of Account-1 appears to follow the same decreasing trend that is evident in the balance amount of Account-2. It is therefore reasonable to assume that some event must have occurred between June and July of 2019 that updated the original account number from 11223335548489 to 11223335548490.

There are numerous reasons for updating an account number, with *debt restructuring* being the most prevalent. Debt restructuring refers to the process by which a customer renegotiates the terms of their contract with the aim of reducing the interest rate, increasing the number of payments, etc., in order to avoid default. Debt restructuring is a relatively common occurrence that frequently results in the formation of a new contract, which also culminates in the generation of a new account number.

Banks are well aware of this issue, and most, if not all, of them already have corrective measures in place to mitigate this effect. The most common corrective process used is *account fusion or account linkage*. Account fusion, or account linkage, is the process of creating a third ID in order to link together multiple accounts that share a common predecessor.

Table 9-4. *Example of historical inconsistency.*

Period	Customer ID	Customer name	Account-1	Balance 1	Account-2	Balance 2
201901	55446622	Jhon Adams	11223335548489	$2,321.00		
201902	55446622	Jhon Adams	11223335548489	$2,100.00		
201903	55446622	Jhon Adams	11223335548489	$2,150.00		
201904	55446622	Jhon Adams	11223335548489	$1,500.00		
201905	55446622	Jhon Adams	11223335548489	$1,700.00		
201906	55446622	Jhon Adams	11223335548489	$1,200.00		
201907	55446622	Jhon Adams			11223335548490	$1,000.00
201908	55446622	Jhon Adams			11223335548490	$800.00
201909	55446622	Jhon Adams			11223335548490	$1,500.00
201910	55446622	Jhon Adams			11223335548490	$1,200.00
201911	55446622	Jhon Adams			11223335548490	$1,000.00
201912	55446622	Jhon Adams			11223335548490	$800.00

9.5.3 Slowly Changing Dimension Type 2 (SCD2)

Table 9-5 shows a new version of **Table 9-3**, but unlike **Table 9-3**, **Table 9-5** shows a version where the system-generated key tracks all different account numbers that belong to a common predecessor account, by linking them to the same ID. For instance, using our previous example, we can see that the account numbers 11223335548489, 11223335548490, and 11223335548491 have been given the same synthetic key, 1, which means that the original account 11223335548489 has gone through two different restructures from its origination to the present day.

Table 9-5. Example of a system key implementing the slowly changing dimension 2.

Account number	System key	Start period	End period
11223335548489	1	1/1/2019	6/30/2019
11223335548490	1	7/1/2019	12/31/2019
11223335548491	1	1/1/2020	12/31/4747
11223335548492	2	1/1/2018	12/31/2019
11223335548493	2	1/1/2020	12/31/4747
11223335548494	3	3/1/2018	12/31/4747
11223335548495	4	9/1/2019	12/31/4747
11223335548496	5	7/1/2017	12/31/4747
11223335548497	6	1/1/2020	12/31/4747

Notice also that **Table 9-5** shows you the expiration date of each account, so you can see that account 11223335548489 had a start date of January 1, 2019, and an expiration date of June 30, 2019. Also note that the next account number, 11223335548490, starts exactly one day after the previous account ends, on July 1, 2019, and ends on December 31.

Likewise, the last occurrence of the system key, 1, corresponding to the account number 11223335548491, begins on January 1, 2020, but unlike the previous two occurrences, this one projects a date into the distant future, with an end date of December 31, 4747, meaning that the last account number is still active today.

This way of storing information is also known as *slowly changing dimension type 2*, or just *SCD2* for short. The SCD2 is a convenient way to store and track historical data by creating multiple records for a given natural key, but keeping the same value for the synthetic key, also called *retained key*.[2] As the name implies, a retained key *retains* the same key value across all versions of the natural key, or in this case, across all versions of the account number.

9.5.4 Pros and Cons of Using Synthetic Keys

Switching keys from native to synthetic represents an additional effort to your data processing activities, because it requires you to

1. Create and maintain a new Account × Account_rk (account retained key) catalog.
2. Propagate the account_rk through all of your input tables.
3. From this point onward, the only key allowed during join operations is the account_rk, not the account number.
4. The only column that can be used for grouping operations should be the account_rk, not the account number.

---

[2] Some literature refers to the synthetic key as surrogate key, so you may find both terms interchangeably.

However, this approach guarantees the uniqueness of records in the ABT and ensures *historical consistency*.

Historical consistency represents a fundamental requirement for a scorecard model, as it assesses consistent moral judgments over time. Furthermore, the introduction of inconsistencies into the data introduces errors into the model, resulting in discrepancies between predicted and actual default events. While it is not feasible to eliminate all errors, it is imperative to minimize those that impact historical consistency, given that *the quality of a model is dependent upon the quality of its explanatory variables, which in turn are dependent upon the consistency of the data sources from which they originate*. Therefore, the historical consistency of our ABT model justifies the additional data processing effort required to implement synthetic keys in our data sources.

9.5.5 The Case of Synthetic Keys for Telcos

Telecommunications companies (telcos) face a unique challenge when it comes to managing keys. This is because the inherent key for all telcos is, in theory, the phone number. However, telephone numbers are an inherently unreliable key because they are recycled over time, particularly mobile telephone numbers. This implies that a single phone number may have been associated with different users at different points in time within the same temporal frame.

Table 9-6 illustrates the recycling of a mobile number, 212-639-7575, which was in use by John Adams from January to June of 2019. Thereafter, the number changed ownership to Kevin Smith, who retained it from July to December of that year. It is notable that this is the exact opposite of the issue reported by banks when a loan changes its account number due to a debt restructure. For banks, the issue is that a loan belonging to the same person is reported with two different account numbers. In contrast, for telecommunications companies, the same mobile number is associated with two different individuals.

Table 9-6. *Example of a mobile number belonging to two different subscribers within the same time window.*

Period	Mobile number	Customer name	Subscriber number
201901	212-639-7575	John Adams	1122367899
201902	212-639-7575	John Adams	1122367899
201903	212-639-7575	John Adams	1122367899
201904	212-639-7575	John Adams	1122367899
201905	212-639-7575	John Adams	1122367899
201906	212-639-7575	John Adams	1122367899
201907	212-639-7575	Kevin Smith	11223679888
201908	212-639-7575	Kevin Smith	11223679888
201909	212-639-7575	Kevin Smith	11223679888
201910	212-639-7575	Kevin Smith	11223679888
201911	212-639-7575	Kevin Smith	11223679888
201912	212-639-7575	Kevin Smith	11223679888

To address this challenge, telecommunications companies employ the subscriber number in preference to the mobile number, thereby ensuring historical consistency in their analytical processes. It is worthwhile to consider the consequences of using the mobile number in place of the subscriber number in the ABT. This would result in the erroneous combination of the behaviors of different subscribers into a single record, which would in turn alter the variable selection process and lead to misleading interpretations of the variables present in the final model.

CHAPTER 9   TARGET POPULATION

The fundamental conclusion of this section is the realization that inconsistencies in historical data are not unique to the financial sector; they are pervasive in various industries to varying extents. This underscores the critical importance of leveraging synthetic keys in order to ensure the integrity and validity of data and to maintain the reliability of subsequent inferences derived therefrom.

9.6 Summary

In this chapter, we explored the critical role of defining the target population as the foundation of your analytical base table (ABT). The target population is the pillar from which all other components of the ABT—such as variables and the target variable—are built, making its definition essential for aligning your data with business objectives and ensuring meaningful insights.

A key challenge in this process is addressing historical inconsistencies, which can undermine the detection of behavioral patterns over time. Synthetic keys play a vital role here, helping to resolve such inconsistencies. Thus, it is important to keep in mind that the entity level of analysis in your ABT might be replaced by synthetic keys to ensure consistency and accuracy. Ignoring these complexities can lead to flawed models and unreliable outcomes, especially when adding new information to your ABT.

Additionally, the statistical representativeness of your target population drives the level of segmentation you can achieve. Striking the right balance between representativeness and granularity ensures that your ABT is both precise and capable of supporting robust analysis.

CHAPTER 9 TARGET POPULATION

Key takeaways:

1. **The Target Population as the Foundation:** The target population serves as the pillar of the ABT, influencing the addition of variables, the target variable itself, and the overall structure of the table.

2. **Synthetic Keys and Historical Inconsistencies:** Addressing historical inconsistencies is essential for detecting behavioral patterns over time, with synthetic keys often replacing the entity level of analysis.

3. **Statistical Representativeness:** A well-defined target population must balance representativeness and segmentation to support the desired level of analytical precision.

4. **Alignment with Business Objectives:** The target population must align with the business goals to ensure the ABT is relevant and actionable.

5. **Impact on Variables and Model Accuracy:** The definition of the target population directly affects the selection of variables, their weights, and the overall reliability of the model.

With the target population defined and its complexities addressed, the next step is to construct your analytical base table (ABT). Chapter 10 will guide you through the ABT-building process, where raw data is transformed into a structured table ready for analysis. From selecting and engineering variables to ensuring data integrity, this chapter will provide the practical steps needed to build an ABT that captures behavioral patterns, respects historical consistency, and aligns with your business objectives. Let us move forward and begin the construction of your ABT.

CHAPTER 10

The ABT-Building Process

"Hard work beats talent when talent doesn't work hard."

—Tim Notke

"Great things are not done by impulse, but by a series of small things brought together."

—Vincent Van Gogh

With the *target population* defined, we now turn to one of the most challenging and time-demanding steps in the modeling process: building the *analytical base table* (ABT). This phase involves transforming raw, often messy, data into a structured and cohesive table that serves as the foundation for analysis. While demanding, this meticulous process ensures the ABT can support meaningful insights and reliable predictions.

In this chapter, we move away from *conceptual and analytical thinking* to a more *practical and technical hands-on approach*. We will explore the most common types of inputs, the tools and techniques for safely integrating information, and the use of *intermediate structures*, such as data marts, to streamline variable creation and integration. Recognizing the *live nature of ABT construction*, we also introduce the *continuous variable integration cycle (CVIC)*, a framework for seamlessly adding new data sources and variables without significant reprogramming.

Finally, we address two critical components: the *definition of the target variable* and the *stacking of periodic snapshots*, both essential for capturing patterns and enabling robust predictive modeling. Building an ABT requires precision, consistency, and creativity, but it is this effort that lays the groundwork for analytical success.

10.1 Overview of the Most Common Types of Input Tables Used in the Industry

Prior to commencing the ABT-building process, it is essential to undertake a review of the most prevalent types of input tables, accompanied by a formal definition of each and guidance on how to optimize their utility for the generation of explanatory variables, and an examination of the factors that should be taken into account before they are employed as data sources for the ABT.

In general, all input tables fall into four categories:

1. Snapshot tables
2. Transactional tables
3. Dimensional tables
4. Catalog tables

It can be reasonably assumed that any given industry will include one or more of the aforementioned tables within its official database, data mart, or data warehouse.

10.1.1 Snapshot Tables

A snapshot table stores a point-in-time view of data at periodic intervals. Common intervals are monthly and daily. Snapshot tables are often redundant, storing complete chunks of information without regard to alterations over time. In banking, the *credit card statement table* is a common snapshot table. A bank statement shows financial transactions and the credit card balance at the end of a billing cycle. It should have one record per account, with a summary of transactions and the final balance from the first statement ever issued to the last billing cycle.

Statement tables cover all accounts for a set period, regardless of status. Some regulations require banking institutions to retain closed or canceled accounts for a minimum period (e.g., five or ten years) before permanently deleting them. Therefore, it is important to consider the status of the accounts before using a snapshot table as a data source for our ABT. The following list outlines the most significant attributes of a snapshot table:

1. Snapshot tables should have one unique record per period.

2. Snapshot tables should have a status column that indicates whether a record is active, inactive, canceled, blocked, etc.

3. Snapshot tables can have both, a summarized information per period, consisting of sums and/or averages, and point-in-time information, usually comprised of the last billing cycle information, such as the credit card balance.

4. Snapshot tables have consecutive periods; therefore, all consecutive periods of a single record should be present for the entire validity period of the record.

CHAPTER 10   THE ABT-BUILDING PROCESS

Snapshot tables are ideal for creating explanatory variables for our ABT. A single record per period ensures uniqueness, facilitates joining datasets, and reduces grouping operations. Snapshot tables can serve as an excellent source of data for selecting the target population, as they often include multiple filter columns, such as status, number of days past due, product descriptions, and other characteristics. It is worth noting that snapshot tables can contain dozens or hundreds of columns, including summaries of transactional data, such as purchases (global and/or specific), payments, or promotions/points.

Snapshot tables are also used to aggregate data from multiple sources. This enables the computation of crucial business metrics and facilitates the aggregation of data across time, facilitating the observations of behavioral patterns such as trends, seasonality, and cycles. In general, it can be stated that snapshot tables are an excellent data source because of the following:

1. They regularly summarize the most critical information about the company.

2. They serve as an official and reliable source of information within any institution or industry.

3. Maintain consistency over time due to their official status.

4. They can be used to derive the target population because they typically encompass all records, allowing the selection of different subpopulations of interest through multiple built-in filters.

5. Because of the sequential periodicity of snapshot tables, time-based operations can be easily implemented using lag or lead functions.

6. The wealth of information contained in snapshot tables often eliminates the need for transactional tables.

In summary, snapshot tables are pervasive in databases, data marts, and data warehouses globally, rendering them an exemplary source of information. It is advisable to seek out these tables regardless of the industry in question. It is also noteworthy that snapshot tables are commonly referred to by other designations, including periodic tables and historical tables. In the event that the database administrator is unaware of the existence of snapshot tables, it may be beneficial to provide an explanation of their purpose. Such tables are likely to exist, but may be referred to by another name.

10.1.2 Transaction Tables

Unlike snapshot tables, transaction tables store events, usually in real time, rather than subjects, such as accounts, or customers. Therefore, we do not expect unique records at any particular periodicity, but rather a series of events that may or may not occur sequentially over time.

In contrast to snapshot tables, transaction tables record events in real time; as a consequence, they do not have specific periodicities. Transactional and snapshot tables also differ greatly in row-to-column ratio; since snapshot tables tend to be wider in terms of columns, transactional tables tend to be narrower and longer, in terms of columns and rows, respectively.

One of the most significant advantages of utilizing transactional tables is the capability to generate customized groupings based on the merchant category code or MCC (see Section 3.3.1), which is not feasible with snapshot tables due to the fact that their data has already been aggregated and structured.

It is important to note that the subjects (accounts) of the snapshot table may not be present in the transaction table during the same period. The aggregation of transactional data by period and subject (account) is an inadequate method for achieving historical consistency, as accounts without transactions in a given period will be absent from the aggregated form. It is not possible to transform a transactional table into a snapshot table through the process of grouping alone, as the two types of table are not interchangeable; rather, they are complementary.

The following are the main characteristics of transactional tables:

1. They store information at the lowest atomic level, capturing real-time events as they occur.

2. Transaction tables typically exhibit a significant row-to-column ratio, containing a large number of rows (often numbering in the millions, billions, or even trillions) relative to their relatively modest number of columns, which is usually less than 20.

3. These tables establish a one-to-many relationship between the subject of analysis and the events recorded, resulting in multiple events associated with each subject.

4. They facilitate the creation of custom grouping categories using transaction identifiers such as the merchant category code (MCC).

Consequently, transaction tables are capable of accommodating the creation of a multitude of variables, in addition to facilitating the generation of meaningful variables that can be linked to authentic human behavior.

10.1.3 Dimensional Tables and Catalogs

Dimensional tables and catalogs are often used interchangeably. They store one record per unique identifier (ID), also called a *foreign key*. Foreign keys normalize stored info by replacing descriptions with IDs in catalogs to avoid redundancy and save space. An example is transaction type (MCC). As previously stated, transactional tables comprise billions or even trillions of rows, capturing events in real time. However, each transaction must be typified to be useful. To circumvent replication of transaction labels, it is possible to substitute the description with a specific ID in a catalog table. Foreign keys are accompanied by implicit constraints that ensure integrity. They prohibit deletion or updating of records in both tables, including the transactional table.

Dimension tables may have multiple records per ID, which can be used to identify unique records. Dimension tables can be classified as SCD2-type tables (Section 9.5.3). An illustrative example is a customer information table. Customer tables store characteristics like date of birth, profession, and address. These can change over time, so dimensional tables may be structured in an SCD2 fashion with a new record inserted each time an update occurs.

Bridge tables are dimensional or catalog tables with entity relationships. Examples include account-customer number and account-application number relationships. Bridge tables are important because they store the relationship between natural and synthetic keys, maintaining referential integrity and historical consistency. They are also structured like SCD2, tracking the effective date of each ID change.

10.2 Common Mistakes when Adding New Variables to the ABT

Once historical inconsistencies have been addressed (either through the use of synthetic keys or other means) and the target population has been properly identified, explanatory variables can be added to the ABT in a sequential manner. However, prior to this, it is essential to ensure that the process does not fall prey to the most common mistake made when cross-referencing data sources, namely, *row multiplication*.

10.2.1 Row Multiplication

One of the most prevalent errors when transferring data from one source to another is the replication of rows. Row multiplication is typically the result of erroneous join operations, whereby data is added by sequentially creating tables through repetitive join operations. The fundamental concept of this technique is to incrementally incorporate data into the ABT by constructing multiple intermediate tables, each with a greater number of columns than the preceding table, until the ABT is fully populated. Regrettably, this methodology is fundamentally flawed. This is due to its failure to grasp the essential concept of an ABT. ABTs should maintain the exact number of rows from start to finish; furthermore, the initial table should never be replaced, dropped, or deleted. One secure method for incorporating new variables into the original ABT is to utilize the ALTER TABLE statement.

```
ALTER TABLE table_name ADD column_name data_type constraint;
```

This approach offers two possible methods for populating a new ABT: the first is to create an initial ABT with all the necessary columns from the outset, and the second is to add new columns as the ABT is populated with further information.

It is imperative that the initial column populated at the beginning of the building process be designated as the primary key,[1] otherwise known as the target population (Chapter 9). The level of granularity in the target population is inconsequential, provided that each observation is unique within the ABT, which is also referred to as *unicity*.

---

[1] The primary key can also be comprised of two or more columns. Such is the case of stack ABTs, where the primary key is the concatenation of the reference period and the subject ID.

10.3 Tools for Efficiently and Securely Adding New Variables to the ABT

The ALTER TABLE statement is widely regarded as the most secure method for adding new columns to the ABT. However, the challenge lies in determining how to cross-reference the ABT with input tables to populate these new columns without compromising the ABT's structure.

The optimal approach involves temporarily cross-referencing the ABT with the input data source. This process fetches the value or values of interest from the input data source, along with the matching *primary key* obtained from the join operation. A second process then captures this data vector and updates the ABT, ensuring that the update operation is constrained to match the captured primary key. By separating the fetching and updating steps, this method eliminates any risk of record multiplication and maintains the integrity of the ABT.

One example of a tool capable of performing this operation[2] is Oracle's *PL/SQL cursor*, which sets a benchmark for securely and efficiently adding new variables to the ABT. The cursor enables precise control over the cross-referencing process, ensuring that only the intended values are captured and updated. An alternative to the PL/SQL cursor is the SAS *hash DATA step*; however, it is important to acknowledge that although the hash DATA step is a plausible alternative, it works by overwriting the driver table (ABT) just as any other DATA step would do, which makes the PL/SQL cursor a preferable option.

---

[2] The SAS DS2 procedure also allows for a dynamic update of records via the SQLSTMT package. However, the DS2 procedure is beyond the scope of this book.

CHAPTER 10   THE ABT-BUILDING PROCESS

10.3.1 Oracle Cursor[3]

Once the primary key has been correctly established, the next step is to sequentially populate the explanatory variables on the right-hand side of the primary key. The optimal methodology for accomplishing this task is to utilize the *PL/SQL cursor*. PL/SQL cursors permit the addition of data to a base table by dynamically calling the update statement, thereby eliminating the potential for row multiplication.

```
UPDATE
    table_name
SET
    column1 = value1,
    column2 = value2,
    column3 = value3,
    ...
WHERE
    condition;
```

Listing 10-1 shows you a standard Oracle cursor structure, which consist of three basic sections:

1. Declare section
2. Cursor
3. Begin

---

[3] All codes presented in this section can be downloaded from https://github.com/saulzapiain/ConceptualVariableDesign. SQL examples are in the file: **Listing 9 and 10 SQL Examples.sql**.

Listing 10-1. Base Oracle SQL cursor structure.

```
                  1   DECLARE
  Declare Section 2      V_REGS NUMBER:=0;
                  3      CURSOR C1 IS
                  4          SELECT
                  5              A.ROWID
                  6           , B.VARIABLE1
  Cursor Section  7          FROM
                  8              ABT_TABLE          A
                  9             ,DATA_SOURCE_TABL   B
                 10          WHERE
                 11              A.ACCOUNT_KEY=B.ACCOUNT_KEY
                 12          ;
                 13   BEGIN
                 14      UPDATE ROW_COUNTER SET COUNTER = 0;
                 15      COMMIT;
                 16      FOR B IN C1 LOOP
                 17          update ABT_TABLE A
                 18             set
                 19                 VARIABLE1=B.VARIABLE1
                 20             WHERE ROWID = B.ROWID;
  Begin Section   21          IF MOD(C1%ROWCOUNT,10000) = 0 THEN
  Loop Section    22              UPDATE ROW_COUNTER SET COUNTER = COUNTER + 10000;
                 23              COMMIT;
                 24          END IF;
                 25          V_REGS := C1%ROWCOUNT;
                 26      END LOOP;
                 27      COMMIT;
                 28   END;
                 29   /
```

10.3.1.1 Declare Statement

As its name suggests, the DECLARE statement enables the user to specify the variables that will be employed throughout the process.

10.3.1.2 Cursor Statement

A cursor is a pointer that points to a result of a query. There are two types of cursors, implicit and explicit, but in this book, we will only refer to explicit cursors. Explicit cursors are a SELECT statement that is explicitly declared in the declaration section of the current block. For our purposes, it is in this particular section where we will define the join operation between our ABT and our input data source.

Listing 10-1, spanning lines 4 to 11, illustrates a hypothetical scenario in which our ABT is linked to a DATA_SOURCE_TABLE through an equijoin on the ACCOUNT_KEY. It is notable that the SELECT statement references only two columns: the ROWID from the ABT and VARIABLE1 from the

DATA_SOURCE_TABLE. From this query, we can infer that both tables are at the same granularity level, which means that the query will return either one row indicating that the account exists or no rows indicating that the account does not exist in the DATA_SOURCE_TABLE. Nevertheless, as will be demonstrated, the question of whether the DATA_SOURCE_TABLE possesses a unique ACCOUNT_KEY is inconsequential with regard to row multiplication.

10.3.1.3 Begin Statement

A begin statement is employed to construct compound statements that are delimited by the BEGIN and END keywords. In this particular case, the BEGIN statement encloses a cursor FOR LOOP statement that fetches a record for each resulting row from the CURSOR statement. It is important to note that the UPDATE statement is enclosed by the cursor FOR LOOP. This implies that the UPDATE statement is executed as many times as rows are fetched from the parent cursor. Additionally, it should be noted that the UPDATE statement modifies the VARIABLE1 column, which is present in both the ABT and the DATA_SOURCE_TABLE tables. It is also noteworthy that the UPDATE statement is constrained by ROWID rather than ACCOUNT_KEY, which is integral to the PL/SQL cursor. This is because the ROWID referenced in both the cursor and the update statement refers to the same table, namely, ABT. In Oracle, a ROWID is a unique identifier for a row that is not limited to a specific table. Instead, it is a global identifier that can be used to reference the same row across different tables within the same database. This is the most expedient method for joining and retrieving data from the database, thereby ensuring optimal performance during execution.

It should be noted that the PL/SQL cursor illustrated in **Listing 10-1** comprises a number of additional elements, the inclusion of which is optional. To illustrate, the table designated as ROW_COUNTER serves as a monitoring mechanism, providing the total number of rows processed in a given process. It is noteworthy that the ROW_COUNTER variable is referenced

CHAPTER 10   THE ABT-BUILDING PROCESS

twice within the PL/SQL code. The initial instance of the counter is observed on line 13, where it is reset to zero. The second occurrence of the ROW_COUNTER is on line 21, where it is updated every 1,000 records. It should be noted that in the second occurrence, the ROW_COUNTER is enclosed by an IF THEN ELSE statement. The purpose of the IF THEN ELSE statement in this context is to regulate the number of instances in which the ROW_COUNTER table is updated via the following function:

MOD(C1%ROWCOUNT,1000) = 0

The Oracle MOD function returns the remainder of a quotient. In this case, C1%ROWCOUNT refers to the total number of rows fetched so far from the cursor, divided by 1,000. Consequently, each time the total number of rows retrieved by the cursor is a multiple of 1,000, the condition is satisfied, and the ROW_COUNTER table is updated. It is essential to condition the update of the counter to prevent the cursor from performing suboptimally by continuously updating the counter with each record fetched by the cursor statement.

10.3.1.4 Considerations

Oracle cursors are a powerful tool for the secure addition of new data to an ABT. However, it is important to note that the *performance of the underlying query* is a critical factor in the efficacy of the cursor. The term *"query performance"* is used to describe the time taken for a database to respond to a request for data retrieval and processing. There are multiple approaches that can be employed to enhance the performance of a query. These include the use of indexes, hints, data restructuring, and query optimization. However, a relatively straightforward strategy would be to initially prioritize the use of indexes.

The creation of indexes can markedly reduce query time; thus, it is advisable to consider indexing all relevant tables on the appropriate columns. For example, it is strongly advised that not only the primary key of the ABT (whether natural or synthetic) be indexed, but also that a `UNIQUE INDEX` be created. The use of unique indexes is recommended whenever feasible, as they are significantly faster than simple indexes.

Conversely, it is highly probable that the majority of the inputs will be of a historic nature. Therefore, indexing solely by subject ID will prove inadequate. The creation of a composite index may prove a superior solution, for instance, using the information date and subject ID columns. A superior alternative would be to create a partitioned index, although this would require the use of partitioned tables. The subject of partitioned indexes and tables is beyond the scope of this book. However, further information can be found in other sources.

10.3.2 SAS DATA Step Hash[4]

SAS offers its own SQL language, designated as PROC SQL. However, SAS SQL is not sufficiently robust when confronted with substantial amounts of data. The second option when using SAS is the DATA step syntax. However, cross-referencing information using the traditional DATA step is often a convoluted and cumbersome process. This is due to the fact that SAS is not a database engine; it does not have the capacity to manage actual tables. Instead, it manages datasets. Datasets are files containing simple data that are stored in an operating system directory. Prior to cross-referencing two data sources, it is necessary to sort each dataset separately in order to facilitate the joining of the data.

---

[4] All Hash examples can be downloaded from https://github.com/saulzapiain/ConceptualVariableDesign. Files: Listing <InternalRef RefID="PC5">10-2 Hash Outer join DATA step.sas, Listing <InternalRef RefID="PC6">10-3 Improved hash outer join DATA step.sas, Listing <InternalRef RefID="PC9">10-4 The hash inner join DATA step.sas, and Listing <InternalRef RefID="PC9">10-4 Improved hash inner join DATA step.sas.

Fortunately, SAS offers a far superior alternative in the form of the *hash DATA step* [14, 15, 16, 17, 38, 55, 69]. In contrast to the conventional DATA step merge and SQL join, which are disk-based lookups, the hash DATA step is memory-based. This implies that the table is read into memory on a single occasion and then accessed repeatedly in memory. Given that memory-based operations are faster than disk-based operations, it is reasonable to conclude that lookups based on a hash DATA step will be faster than the disk-based alternatives. Indeed, they may even be comparable to Oracle cursors in terms of speed (The SAS® Hash Object: It's Time To .find() Your Way Around – Tutorial [17]).

The following list enumerates the advantages of the hash DATA step (Another Helping of HASH [55]):

- Key lookups are done in memory, not on disk, which is usually faster.

- Key lookups only search a small subset of records.

- The key can contain multiple unconcatenated variables, and they do not need to have the same variable names.

- The hash table (object) allocates memory only as records are added and the number of records that can be stored is limited only by the amount of memory available to SAS.

- The dataset used to load the hash table does not need to be sorted (or indexed).

The hash DATA step statement has a variety of applications; however, this book will focus on two key areas: the addition of new data to an ABT via an outer join hash DATA step and the extraction of data using an inner join hash DATA step.

10.3.2.1 The Hash Outer Join Statement[5]

Listing 10-2 illustrates the four fundamental components of the hash outer join DATA step statement:

1. Hash objects
2. Input table loop
3. ABT loop
4. Iterator loop

It should be noted that the most significant components of the hash DATA step statement are indicated by a blue rectangle and are numbered from 0 to 6.

> **0:** Helps to stablish the properties of the variables in the input dataset without actually reading any record.
>
> **1:** Declares the hash table object. Notice that the HASHEXP and the ORDERED arguments are also used. The HASHEXP argument serves to specify the internal hash table sized (the maximum value is 16), and the ORDERED argument specifies that the hash table is ordered in ascending order—hence the 'A'.
>
> **2:** The second declare statement declares the hash iterator object, which is used to move the pointer through the hash table and bring the values from the hash table into the SAS program data vector.

---

[5] The hash DATA step statement will not be reviewed in depth in this book; for more information about the hash DATA step, you can read the article "How Do I Love Hash Tables? Let Me Count The Ways!" by Judy Loren and the article "I cut my processing time by 90% using hash tables - You can do it too!" By Jennifer K. Warner-Freeman.

CHAPTER 10 THE ABT-BUILDING PROCESS

3: Defines the name of the key column, which must be the same in both tables, the input table and the base table. Notice that the key column should always be enclosed in single quotation marks ''. To define a composite index, simply add as many columns as needed, separated by a comma "," in the following fashion: HH.DEFINEKEY ('KEY1','KEY2', ...,'KEY(n)').

4: Defines the data to be added from the input table to the base table. Notice that the definedata statement has the same logic as the definekey statement, so you can also define as many columns as you need from your hash table, here is an example, HH.DEFINEDATA ('COLUMN1','COLUMN2'..,'COLUMN (n)','_F');. Finally, notice that the '_F' argument is not a physical column in the hash table, and it is actually a variable calculated on the fly that is used as a flag to identify whether or not a record was found during the iteration process or not.

5: See the SET statement for the input data source table, which is also called the lookup table, but the important thing to notice here is the KEEP statement which is used to control the key and the input columns that will be present in the hash DEFINEDATA statement.

6: The last element worth noting is the CALL MISSING statement, which is key to making the outer join process work properly. The reason why the CALL MISSING statement is indispensable for a hash outer join is because the hash table works in conjunction with the SAS program data vector, in the same way

CHAPTER 10   THE ABT-BUILDING PROCESS

as a retain statement does when there is no match between the base and the hash table, causing the last data vector to repeat itself until new information matches a new record.

Listing 10-2. Hash outer join DATA step.

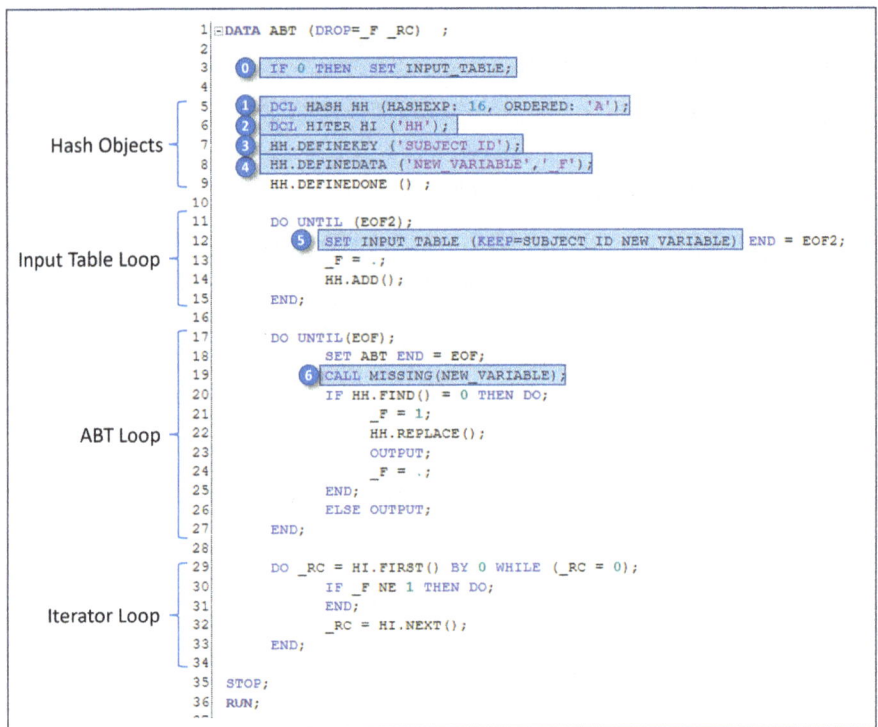

Figure 10-1 illustrates the impact of the CALL MISSING statement. Figure (A) illustrates the outcome of executing a hash DATA step in the absence of the CALL MISSING statement. It is evident that records with the identifiers 2 and 4 are present in the base table, yet they lack a matching entry in the input data source table. In the absence of the CALL MISSING statement, record ID = 1 is retained until a subsequent match is identified, and similarly, record ID = 3 is retained until a new match is established.

182

(A)

Base Table			Input Data Source		
Record Id	Data		Record Id	Data	
1	$ 12,305		1	$	12,305
2	$ 12,305		3	$	11,115
3	$ 11,115		5	$	5,300
4	$ 11,115				
5	$ 5,300				

(B)

Base Table			Input Data Source		
Record Id	Data		Record Id	Data	
1	$ 12,305		1	$	12,305
2	-		3	$	11,115
3	$ 11,115		5	$	5,300
4	-				
5	$ 5,300				

Figure 10-1. *Illustrative example of the effect that the* `CALL MISSING` *statement has over a hash operation. (A) When the* `CALL MISSING` *statement is not present. (B) When the* `CALL MISSING` *statement is present.*

In contrast, example (B) illustrates the anticipated outcome of an outer join operation. It can be observed that records with the identifiers 2 and 4 have been assigned the null value, while only records 1, 3, and 5 have been matched with the lookup table, as anticipated. This is due to the fact that the `CALL MISSING` statement initiates the reinitialization of the new variables at each instance of data being read from the base table.

CHAPTER 10   THE ABT-BUILDING PROCESS

10.3.2.2 Improve Your Hash Outer Join Using the DOSUBL Function

The hash outer join statement illustrated in **Listing 10-2** can be enhanced by incorporating the DOSUBLE *function*. The DOSUBLE function enables the immediate execution of SAS code following the passing of a text string, thereby facilitating the real-time updating and reporting of the number of records undergoing processing.

Listing 10-3 illustrates the enhanced version of the hash outer join DATA step. It should be noted that three additional segments have been incorporated into the original code presented in **Listing 10-2**: the initial execution time, the successful match counter, and the DOSUBLE segment.

Listing 10-3. Improved hash outer join DATA step.

```
 1  DATA ABT (DROP=_F _RC) ;
 2
 3      RETAIN TIMESTART;                                          ─┐ Initial time
 4      TIMESTART=DATETIME();                                      ─┘
 5
 6      IF 0 THEN  SET INPUT_TABLE;
 7
 8      DCL HASH HH (HASHEXP: 16, ORDERED: 'A');
 9      DCL HITER HI ('HH');
10      HH.DEFINEKEY ('SUBJECT_ID');
11      HH.DEFINEDATA ('NEW_VARIABLE','_F');
12      HH.DEFINEDONE () ;
13
14      DO UNTIL (EOF2);
15          SET INPUT_TABLE (KEEP=SUBJECT_ID NEW_VARIABLE) END = EOF2;
16          _F = .;
17          HH.ADD();
18      END;
19
20      DO UNTIL (EOF);
21          SET ABT END = EOF;
22          CALL MISSING(NEW_VARIABLE);
23          IF HH.FIND() = 0 THEN DO;
24              _F = 1;
25              HITS+1;                                            ─┐ Successful match counter
26              HH.REPLACE();
27              OUTPUT;
28              _F = .;
29          END;
30          ELSE OUTPUT;
31  /*_____START_COUNTER_____*/
32          K+1;
33          IF MOD(K,200000)=0 THEN
34              DO;
35                  TIMEEND = DATETIME();
36                  TIMEDIFF = SUM(TIMEEND, -TIMESTART);
37                  ELAPSED = PUT(TIMEDIFF, TIMES.);               ─┐ DOSUBLE segment
38                  START = PUT(TIMESTART,DATETIME14.);
39                  END = PUT(TIMEEND,DATETIME14.);
40                  COUNTER=PUT(K,COMMA24.);
41                  HITSF=PUT(HITS,COMMA24.);
42                  RC03 = DOSUBL(CATX(' ','SYSECHO "R:',COUNTER,' H:',HITSF,' E:',ELAPSED,'"'));
43              END;
44          DROP TIMESTART TIMEEND TIMEDIFF ELAPSED START END K COUNTER HITS RC03;
45  /*_____END_COUNTER_____*/
46      END;
47
48      DO _RC = HI.FIRST() BY 0 WHILE (_RC = 0);
49          IF _F NE 1 THEN DO;
50          END;
51          _RC = HI.NEXT();
52      END;
53
54      STOP;
55  RUN;
```

CHAPTER 10 THE ABT-BUILDING PROCESS

The enhanced DOSUBLE hash DATA step operates as follows: initially, the initial execution time is captured and retained by the RETAIN TIMESTAR and TIMESTART=DATETIME(). The TIMESTART variable enables the calculation of the elapsed time in subsequent steps. Secondly, it can be observed that two distinct counters have been incorporated into the code in order to monitor the number of join matches and the total number of records processed. These counters are designated as HITS+1 and K+1, respectively. It was observed that in line 33, the statement

`IF MOD(K,200000)=0 THEN`

is analogous to the Oracle MOD function and serves to regulate the number of instances in which the DOSUBLE information is to be updated. Let us now turn our attention to the DOSUBLE statement.

`RCO2 = DOSUBL(CATX(' ','SYSECHO "R:',COUNTER,' H:',HITSF,' E:',ELAPSED,'";'));`

It should be noted that three distinct variables are updated at each 200,000 records, in accordance with the MOD condition:

1. COUNTER: Returns the total number of records processed from the base table by the hash DATA step at multiples of 200,000 records
2. HISTF: Returns the total number of successful matches, or hits from the outer join
3. ELAPSED: Returns the elapsed time in the format HH:MM:SS

It should be noted that the objective of lines 35 through 41[6] is to format the variables into a more readily comprehensible format.

---

[6] There are two additional variables, Start and End, but I decided to not include them in the final output.

Figure 10-2 shows the final output generated by the DOSUBLE function, with the legend "R: 600,000 H: 25,000 E: 0:00:01", where R is the total number of records processed so far, H is the total number of hits so far, and E is the elapsed time, so far.

Task Status	
Task	Status
Program (5)	"R: 600,000 H: 25,000 E: 0:00:01 "

Figure 10-2. Example of a real-time status bar showing the total number of records processed so far, R:, the total number of hits from the outer join so far, H:, and the elapsed time so far, E:.

It is noteworthy that the DOSUBLE function is exclusively applicable within the SAS® Enterprise Guide (EG) environment. This is due to the fact that the DOSUBLE function is only capable of returning updated processing information on the Task Status panel of EG.

10.3.2.3 Hash Table Consideration

The hash outer join is the best way to add new data to an ABT, but there is a potential limitation: it cannot effectively address duplicate entries. The DATA step allows duplicates in the base table but not in the lookup table since duplicates in the input data source will prevent successful execution. Therefore, the input table's granularity must match that of the ABT before adding any information. All ABT records must be uniquely referenced, even when not present in both tables.

10.3.2.4 The Hash Inner Join Statement

There is yet another type of hash DATA step that you can use to add information to your ABT, and that is the hash inner join DATA step. The hash inner join works in a similar way to the hash outer join, and it is actually faster than the hash outer join. However, you would rarely have the opportunity to use a hash inner join to add new variables to your ABT, because hash inner joins do reduce the number of records during the join operation when the number of records in the lookup table is less than the number of records in the base table, which defeats the purpose of not altering the initial structure of the ABT, which begs the question—why would anyone use a hash inner join then?

While hash inner joins are not suitable for adding information to an ABT, they are well-suited for extracting information from the original data sources. This is due to the intrinsic property of hash inner joins to reduce the number of records in a base table when the number of records in the lookup table is smaller, which makes them an excellent data extraction tool.

Listing 10-4 illustrates the utilization of a hash inner join DATA step for the purpose of data extraction. It is important to note the following differences between the hash outer join and other forms of joins:

- The code is apparently simpler, and it does not have a hash iterator object.

- It has two different declare statements, one for each one of the tables involved in the inner join.

- No actual data is added to the base table.

- A new column, called FLG, is created on the fly on the target population side that will be used as a selection flag that may or may not be dropped later.

- The base table is not the ABT, but the data source from which the input table will be derived.

- The data source is constrained to a specific time window, or extraction window.

Notice that the resulting table from executing this hash inner join will be a table with the exact same structure (columns) as the original data source, but restricted to the IDs of the subjects present in the target population and restricted to the extraction window. This table will be the precursor to the input table containing the new variable(s) to be added to the ABT.

Listing 10-4. The hash inner join DATA step.

```
1  DATA TARGET_DATA_SOURCE (COMPRESS=NO DROP=FLG);
2
3      DCL HASH HH (HASHEXP: 16,ORDERED: 'A');         ← Hash Object for
4          HH.DEFINEKEY ('SUBJECT_ID', '_N_');            the lookup table
5          HH.DEFINEDATA ('FLG') ;
6          HH.DEFINEDONE () ;
7      DCL HASH HS (HASHEXP: 16,ORDERED: 'A');         ← Hash Object for the
8          HS.DEFINEKEY ('SUBJECT_ID');                   data source table
9          HS.DEFINEDATA ('_N_');
10         HS.DEFINEDONE ();
11
12     DO UNTIL ( END_LOOKUP ) ;
13         SET TARGET_POPULATION (KEEP=SUBJECT_ID) END = END_LOOKUP;
14         FLG=1;
15         IF HS.FIND() NE 0 THEN _N_ = 0;              ← Target population loop
16         _N_ ++ 1;
17         HS.REPLACE();
18         HH.ADD();
19     END ;
20
21     DO UNTIL ( END_DRIVER );
22         SET DATA_SOURCE ( WHERE=(START_PERIOD<=PERIOD<=END_PERIOD)) END = END_DRIVER;
23         IF HS.FIND() = 0 THEN
24             DO _N_ = 1 TO _N_;                        ← Data source loop
25                 HH.FIND();
26                 OUTPUT;
27             END;
28     END;
29
30     STOP;
31 RUN;
```

10.3.3 SELECT Correlated Subqueries[7]

The SELECT correlated subqueries method represents a secure approach to incorporating new data into your ABT, as the subquery is situated within the SELECT clause, thereby ensuring that

---

[7] All SQL examples are in Listing 9 and 10 SQL Examples.sql, https://github.com/saulzapiain/ConceptualVariableDesign.

CHAPTER 10 THE ABT-BUILDING PROCESS

1. The original number of records in the base table is not altered

2. Only one record can be retrieved from the subquery, eliminating the possibility of duplicating the base table

3. If there is no match within the subquery, a null value is returned, just like an outer join

Listing 10-5 illustrates the use of a SELECT correlated subquery. It should be noted that the table specified in the FROM clause can be either the target population or the ABT. However, in both cases, there must be a unique SUBJECT_ID per record. This is what is referred to as a *backbone table*.[8] It is important to note the following elements of the subquery:

1. A grouping operation takes place in the inner query.

2. Both, the inner and the outer tables, are joined by the SUBJECT_ID, hence the name, correlated subquery.

3. The inner query also restricts the time window of the extraction.

---

[8] A backbone table may serve as a data shuttle that can help us to extract and carry data from the data source to the base table.

Listing 10-5. SELECT correlated subquery.

```
    SELECT
      SUBJECT_ID
    , (
        SELECT
        SUM/MIN/MAX/AVG(INTERVAL_VARIABLE)
        FROM DATA_SOURCE B
        WHERE
            A.SUBJECT_ID=B.SUBJECT_ID
        AND PERIOD BETWEEN START_PERIOD AND END_PERIOD
      ) AS INPUT_VARIABLE
    FROM TARGET_POPULATION/ABT A
```

It should be noted that in this specific instance, we are dealing with an interval variable. Therefore, a grouping operation may be necessary, given that subqueries within the SELECT clause are only capable of retrieving a single record per match. The decision of whether or not to utilize a grouping function is dependent upon the specific characteristics of the data source in question and the desired outcome.

The following examples are worthy of consideration. In the first scenario, the data source is a catalog of sociodemographic information, and the objective is to extract a single characteristic from it. In the second case, the inner table references transaction data from the current account, and the objective is to ascertain the total amount withdrawn from ATMs over the last three-month period prior to the reference date.

Listing 10-6 illustrates a realistic query for each of the aforementioned cases. The initial case is illustrated in figure (A), which depicts a query wherein the marital status is extracted from the sociodemographic catalog. It should be noted that this example assumes that the SOCIODEMOGRAPHIC_CATALOG contains only one record per SUBJECT_ID.

CHAPTER 10 THE ABT-BUILDING PROCESS

The second case is illustrated in figure (B), which depicts a query that extracts the total amount withdrawn from ATMs over the past three months from a transactional table. The dataset comprises five columns: subject ID, transaction ID, transaction timestamp, transaction code, and transaction amount. In order to obtain the desired information, it is first necessary to define a set of conditions in the WHERE clause:

- A.SUBJECT_ID=B.SUBJECT_ID: Specifies that only the target population should be extracted

- B.TRN_PERIOD BETWEEN ADD_MONTHS(B.TRN_PERIOD,-3) AND A.REF_PERIOD: Defines that only the last three months from the reference date (present in the target population table) should be extracted

- B.TRN_CODE='ATM': Specifies that only the transactions coded as "ATM" should be extracted

Given that we are dealing with an interval variable and that we are asking for the total amount withdrawn from ATMs in the last three months, we use the SUM grouping function in the inner SELECT clause. Upon executing the code in figure (B), a new variable, TOTAL_ATM_WDRW_AMOUNT_L03M, will be generated.

Listing 10-6. Applied examples of a SELECT correlated subqueries. (A) An example where only the marital status is being extracted from a catalog with sociodemographic information. (B) An example where the total amount withdrawn from an ATM in the last three months is being extracted from the transactional information of the current account.

(A)

```
SELECT
    SUBJECT_ID
    , (
        SELECT
            MARITAL_STATUS
            FROM SOCIODEMOGRAPHIC_CATALOG B
            WHERE
                A.SUBJECT_ID=B.SUBJECT_ID
        ) AS N_MARITAL_STATUS
FROM TARGET_POPULATION/ABT A
```

(B)

```
SELECT
    SUBJECT_ID
    , (
        SELECT
            SUM(B.CASH_AMOUNT)
            FROM TRN_CURRENT_ACCOUNT B
            WHERE
                A.SUBJECT_ID=B.SUBJECT_ID
                AND B.TRN_PERIOD BETWEEN ADD_MONTHS(B.TRN_PERIOD,-3) AND A.REF_PERIOD
                AND B.TRN_CODE='ATM'
        ) AS TOTAL_ATM_WDRW_AMOUNT_L03M
FROM TARGET_POPULATION/ABT A
```

10.4 Designing Suitable Inputs for the ABT-Building Process[9]

One aspect that is rarely addressed during the process of ABT construction is the suitability of the initial data sources for the construction of explanatory variables. To illustrate this issue, we will present an applied example. **Table 10-1** illustrates 26 distinct variants derived from the base ratio ATM/EXP, where ATM represents total monthly credit card cash withdrawals from ATMs and EXP represents total monthly credit card spending. The coding can be summarized as follows:

- Data type
  - I = Interval variable
- Operation
  - R = Ratio
  - MV = Monthly variation
  - QV = Quarterly variation
  - SV = Semestral variation
  - AV = Average
  - SIN = Sum of increments
  - SDC = Sum of decrements

---

[9] All SQL codes presented in this section can be downloaded from https://github.com/saulzapiain/ConceptualVariableDesign. File: Listing 9 and 10 SQL Examples.sql. In addition, two complete examples of the CVIC process, at account and customer level, are also provided. Files: Bonus TBT and CVIC example at account level.sql and Bonus TBT and CVIC example at customer level.sql.

CHAPTER 10 THE ABT-BUILDING PROCESS

- Time period type
  - P = Point-in-time variable
  - A = Aggregate variable

In essence, the objective is to compute seven distinct operations at varying temporalities for the ATM/EXP ratio, which serves as the base metric. Furthermore, the same 26 variants must be computed for nine additional metrics, as outlined in **Table 10-2**, which provides a summary of the ten credit card spending categories, their corresponding codes, and their historical distribution.

CHAPTER 10 THE ABT-BUILDING PROCESS

Table 10-1. *Example of 26 different variants derived from the ratio ATM/EXP, where ATM refers to the total monthly cash withdrawals of the credit card from ATMs and EXP refers to the total monthly spending of the credit card.*

Num.	Data Source Cd.	Var Name	Length	Base Metric	Data Type	Operation	Base Code	Subcategory	Period Label	Time Period Type	Start Period	End Period
1	TRX	I_R_ATM_EXP_TRX_00MB	20	ATM / EXP	I	R	R_ATM_EXP	ATM	00MB	P	0	0
2	TRX	I_R_ATM_EXP_TRX_01M3	20	ATM / EXP	I	R	R_ATM_EXP	ATM	01MB	P	1	1
3	TRX	I_R_ATM_EXP_TRX_02M3	20	ATM / EXP	I	R	R_ATM_EXP	ATM	02MB	P	2	2
4	TRX	I_R_ATM_EXP_TRX_03MB	20	ATM / EXP	I	R	R_ATM_EXP	ATM	03MB	P	3	3
5	TRX	I_R_ATM_EXP_TRX_06VlB	20	ATM / EXP	I	R	R_ATM_EXP	ATM	06MB	P	6	6
6	TRX	I_R_ATM_EXP_TRX_09MB	20	ATM / EXP	I	R	R_ATM_EXP	ATM	09MB	P	9	9
7	TRX	I_R_ATM_EXP_TRX_12MB	20	ATM / EXP	I	R	R_ATM_EXP	ATM	12MB	P	12	12
8	TRX	I_MV_R_ATM_EXP_TRX_L01_00	25	ATM / EXP	I	MV	R_ATM_EXP	ATM	L01_00	P	1	0
9	TRX	I_MV_R_ATM_EXP_TRX_L02_01	25	ATM / EXP	I	MV	R_ATM_EXP	ATM	L02_01	P	2	1
10	TRX	I_MV_R_ATM_EXP_TRX_L03_02	25	ATM / EXP	I	MV	R_ATM_EXP	ATM	L03_02	P	3	2
11	TRX	I_QV_R_ATM_EXP_TRX_L06_03	25	ATM / EXP	I	QV	R_ATM_EXP	ATM	L06_03	P	6	3
12	TRX	I_QV_R_ATM_EXP_TRX_L09_06	25	ATM / EXP	I	QV	R_ATM_EXP	ATM	L09_06	P	9	6
13	TRX	I_QV_R_ATM_EXP_TRX_L12_09	25	ATM / EXP	I	QV	R_ATM_EXP	ATM	L12_09	P	12	9
14	TRX	I_SV_R_ATM_EXP_TRX_L12_06	25	ATM / EXP	I	SV	R_ATM_EXP	ATM	L12_06	P	12	6
15	TRX	I_AV_R_ATM_EXP_TRX_L03M	23	ATM / EXP	I	AV	R_ATM_EXP	ATM	L03M	A	1	3
16	TRX	I_AV_R_ATM_EXP_TRX_LC6M	23	ATM / EXP	I	AV	R_ATM_EXP	ATM	L06M	A	1	6
17	TRX	I_AV_R_ATM_EXP_TRX_L12M	23	ATM / EXP	I	AV	R_ATM_EXP	ATM	L12M	A	1	12
18	TRX	I_AV_R_ATM_EXP_TRX_L13_24	25	ATM / EXP	I	AV	R_ATM_EXP	ATM	L13_24	A	13	24
19	TRX	I_SIN_R_ATM_EXP_TRX_L03M	24	ATM / EXP	I	SIN	R_ATM_EXP	ATM	L03M	A	1	3
20	TRX	I_SIN_R_ATM_EXP_TRX_L06M	24	ATM / EXP	I	SIN	R_ATM_EXP	ATM	L06M	A	1	6
21	TRX	I_SIN_R_ATM_EXP_TRX_L12M	24	ATM / EXP	I	SIN	R_ATM_EXP	ATM	L12M	A	1	12
22	TRX	I_SIN_R_ATM_EXP_TRX_L13_24	26	ATM / EXP	I	SIN	R_ATM_EXP	ATM	L13_24	A	13	24
23	TRX	I_SDC_R_ATM_EXP_TRX_L03M	24	ATM / EXP	I	SDC	R_ATM_EXP	ATM	L03M	A	1	3
24	TRX	I_SDC_R_ATM_EXP_TRX_L06M	24	ATM / EXP	I	SDC	R_ATM_EXP	ATM	L06M	A	1	6
25	TRX	I_SDC_R_ATM_EXP_TRX_L12M	24	ATM / EXP	I	SDC	R_ATM_EXP	ATM	L12M	A	1	12
26	TRX	I_SDC_R_ATM_EXP_TRX_L13_24	26	ATM / EXP	I	SDC	R_ATM_EXP	ATM	L13_24	A	13	24

Table 10-2. Observed distribution of the credit card spending.

Expense category	Code	Distribution
Retail stores, supermarket, and home	RSH	21.2%
Hobbies, luxuries, and entertainment	HLE	20.9%
Cash withdrawals	ATM	13.8%
Transportation	TRN	13.0%
Equipment and specialized supplies and services	ESS	6.8%
Others	OTH	6.0%
Vacations and trips	VTR	5.4%
Personal care, health, and education	PHE	5.3%
Governmental expenses and services	GES	4.2%
Restaurants and fast food	RFF	3.3%
		100%

10.4.1 Input's Building Process and Variable's Addition to the ABT, Step by Step

In this example, three assumptions are made regarding the initial input. The initial data source will be the credit card transactional information. Additionally, the merchant category code (MCC) and transaction type code will be included in the transaction information. Finally, a catalog that classifies each MCC into one of the ten desired expense categories presented in **Figure 10-3** will already be available. From this point onward, the process can be completed in approximately eight steps, from point A (the input) to point B (the output).

CHAPTER 10   THE ABT-BUILDING PROCESS

1. Add grouping columns to the original transactional data source.
2. Add customer ID column for additional grouping operations (optional).
3. Group transactions to the desired level of analysis.
4. If there are no consecutive dates, then:
   a. Create a transactional base table (TBT; see Section 10.4.1.4).
   b. Add grouped transactional information to a TBT via cursor.
5. Compute base metrics using Oracle's analytical functions:
   a. Base ratio.
   b. Ratio's lags.
   c. Monthly, quarterly, and semestral variations.
   d. Monthly increment and decrement flag.
6. Add new variables to the ABT.
7. Create ABT's input variables from TBT using *conditional grouping functions* [56].
8. Add input variables to ABT via cursor.

CHAPTER 10 THE ABT-BUILDING PROCESS

MCC	MCC_NAME	MCC_GENERAL	Expense Category	Expense Code
4815	Specialty Cleaning, Polishing and Sanitation Preparations	Utility Services	Governmental expenses and services	GES
2842	Accounting, Auditing, and Bookkeeping Services	Contracted Services	Retail Stores, Supermarket, and Home	RSH
8931	Advertising Services	Professional Services and Membership Organizations	Equipment, and specialized Supplies and services	ESS
7311	Agricultural Co-operatives	Business Services	Equipment, and specialized Supplies and services	ESS
0763	Air Conditioning and Refrigeration Repair Shops	Agricultural Services	Others	OTH
7623	Ambulance Services	Business Services	Retail Stores, Supermarket, and Home	RSH
4119	Amusement Parks, Circuses, Carnivals, and Fortune Tellers	Transportation Services	Personal care, health and education	PHE
7996	Antique Reproductions	Business Services	Hobbies, Luxuries, and Entertainment	HLE
5937	Antique Shops – Sales, Repairs, and Restoration Services	Miscellaneous Stores	Hobbies, Luxuries, and Entertainment	HLE
5932	Aquariums, Seaquariums, Dolphinariums, and Zoos	Miscellaneous Stores	Hobbies, Luxuries, and Entertainment	HLE
7998	Architectural, Engineering, and Surveying Services	Business Services	Hobbies, Luxuries, and Entertainment	HLE
8911	Art Dealers and Galleries	Professional Services and Membership Organizations	Equipment, and specialized Supplies and services	ESS
5971	Artist's Supply and Craft Shops	Miscellaneous Stores	Hobbies, Luxuries, and Entertainment	HLE
5970	Automated Fuel Dispensers	Miscellaneous Stores	Hobbies, Luxuries, and Entertainment	HLE
5542	Automated Referral Service	Retail Outlet Services	Transportation	TRN
9700	Automobile Associations	Government Services	Governmental expenses and services	GES
8675	Automobile Rental Agency	Professional Services and Membership Organizations	Others	OTH
7512	Automotive Body Repair Shops	Business Services	Transportation	TRN
7531	Automotive Paint Shops	Business Services	Transportation	TRN
7535	Automotive Parts and Accessories Stores	Business Services	Transportation	TRN
5533	Automotive Service Shops (Non-Dealer)	Retail Outlet Services	Transportation	TRN
7538	Automotive Tire Stores	Business Services	Transportation	TRN
5532	Bail and Bond Payments	Retail Outlet Services	Transportation	TRN
9223	Bakeries	Government Services	Governmental expenses and services	GES
5462	Bands, Orchestras, and Miscellaneous Entertainers (Not Elsewhere Classified)	Retail Outlet Services	Retail Stores, Supermarket, and Home	RSH
7929	Beauty and Barber Shops	Business Services	Hobbies, Luxuries, and Entertainment	HLE
7230	Betting, including Lottery Tickets, Casino Gaming Chips, Off-Track Betting, Wagers	Miscellaneous Stores	Personal care, health and education	PHE
7995	Bicycle Shops – Sales and Service	Business Services	Hobbies, Luxuries, and Entertainment	HLE
5940	Billiard and Pool Establishments	Miscellaneous Stores	Hobbies, Luxuries, and Entertainment	HLE
7932	Blueprinting and Photocopying Services	Business Services	Hobbies, Luxuries, and Entertainment	HLE
7332	Boat Dealers	Business Services	Equipment, and specialized Supplies and services	ESS
5551	Boat Rentals and Leasing	Retail Outlet Services	Transportation	TRN
4457	Book Stores	Transportation Services	Hobbies, Luxuries, and Entertainment	HLE
5942	Books, Periodicals and Newspapers	Miscellaneous Stores	Hobbies, Luxuries, and Entertainment	HLE
5192	Bowling Alleys	Retail Outlet Services	Hobbies, Luxuries, and Entertainment	HLE
7933	Bus Lines	Business Services	Hobbies, Luxuries, and Entertainment	HLE
4131	Business and Secretarial Schools	Transportation Services	Transportation	TRN
8244		Professional Services and Membership Organizations	Personal care, health and education	PHE

Figure 10-3. Example of a MCC catalog and its corresponding customized expense classification.

10.4.1.1 Adding Grouping Columns to the Transactional Data Source

Listing 10-7 presents a CREATE table syntax where the transactional data source, CC_TRANSACTIONS, is cross-referenced with the MCC_CATALOG and the TRX_CATALOG. The MCC_CATALOG contains the new expense type and its corresponding MCC code, while the TRX_CATALOG contains the transaction type and its classification according to its directionality:

- Current interest (accumulation)
- Extra charges (output)
- Cash withdrawal (output)
- Purchases (output)
- Payments (input)
- Reverse charges (input)

CHAPTER 10　THE ABT-BUILDING PROCESS

Listing 10-7. Syntax to create the credit card transactional input at the account level.

```
CREATE TABLE TRX_INPUT_CC_ACC NOLOGGING STORAGE (INITIAL 2M NEXT 1M)
AS
SELECT /*+FIRST_ROWS(10000) INDEX(C, INX_TRX_CAT) INDEX(B, INX_MCC) INDEX(A, INX_TRX_MCC)*/
       TO_NUMBER(TO_CHAR(A.TRX_DATE,'YYYYMM'),'999999') AS PERIOD
     , A.ACCOUNT_ID
     , A.TRX_DATE
     , A.TRX_AMOUNT
     , A.MCC
     , A.TRX_CODE
     , B.GENERAL_MCC
     , B.MCC_NAME
     , B.MCC_DESCRIPTION
     , B.EXPENSE_DESC
     , B.EXPENSE_CODE
     , C.TRX_TYPE
     , C.TRX_DIRECTION
     , CASE
           WHEN C.TRX_DESCRIPTION LIKE 'WITHDRAW%' THEN 'ATM'
           ELSE B.EXPENSE_CODE
       END AS EXPENSE_TYPE_CODE
     , CASE
           WHEN C.TRX_DESCRIPTION LIKE 'WITHDRAW%' THEN 'CASH WITHDRAW'
           ELSE B.EXPENSE_DESC
       END AS EXPENSE_TYPE_DESC
  FROM
       CC_TRANSACTIONS A
     , MCC_CATALOG B
     , TRX_CATALOG C
  WHERE
       A.MCC = B.MCC (+)
   AND A.TRX_CODE = C.TRX_CODE
;
```

The resulting table after executing **Listing 10-7** will be the initial credit card's transactional input at account level, named TRX_INPUT_CC_ACC.

10.4.1.2 Add Customer ID for Additional Grouping Operations (Optional)

As an alternative, it may be advisable to incorporate the CUSTOMER_ID into the transactional input to facilitate additional grouping operations, particularly if the intention is to utilize the transactional data for models other than the credit card model, such as loans or mortgages.

Listing 10-8 presents the ALTER TABLE, and the CURSOR syntax needed to add the CUSTOMER_ID into the transactional input, TRX_INPUT_CC_ACC. It should be noted that this code is predicated on the assumption that a bridge table, designated as BRD_CUSTOMER_X_ACCOUNT_CC, exists which establishes the relationship between the credit card account number and the customer number.

Listing 10-8. *CURSOR syntax to add the CUSTOMER_ID to the TRX_INPUT_CC_ACC table.*

```sql
ALTER TABLE TRX_INPUT_CC_ACC ADD CUSTOMER_ID VARCHAR2(50);

DECLARE
T1 INTEGER:=DBMS_UTILITY.GET_TIME;T2 INTEGER:=0;V_REGS NUMBER:=0;
    CURSOR C1 IS
        SELECT    /*+ FIRST_ROWS(10000)*/
            A.ROWID
          , B.CUSTOMER_ID
        FROM
         TRX_INPUT_CC_ACC A
        ,BRD_CUSTOMER_X_ACCOUNT_CC B
        WHERE
            A.ACCOUNT_ID=B.ACCOUNT_ID
        ;
BEGIN
UPDATE COUNTER SET COUNTER = 0;
    COMMIT;
    FOR B IN C1 LOOP

    UPDATE TRX_INPUT_CC_ACC A
        SET
          CUSTOMER_ID= B.CUSTOMER_ID
        WHERE ROWID = B.ROWID;

        IF MOD(C1%ROWCOUNT,10000) = 0   THEN
                UPDATE COUNTER SET COUNTER = COUNTER + 10000;
                COMMIT;
        END IF;
        V_REGS := C1%ROWCOUNT;
    END LOOP;
    COMMIT;
    T2 :=DBMS_UTILITY.GET_TIME;
    DBMS_OUTPUT.PUT_LINE('-- Records Processed: ' ||TO_CHAR(V_REGS,'999,999,999.99'));
    DBMS_OUTPUT.PUT_LINE('-- Elapsed: ' ||TO_CHAR((T2-T1)/100/60,'999,999,999.99'));
END;
/
```

10.4.1.3 Grouping Transactions at the Desired Level of Analysis

The classification of expenses and transactions allows for the grouping of transactions at the desired level of analysis. In this case, the level of analysis is as follows:

- Period, defined as year-month in the format of YYYYMM
- ACCOUNT_ID
- EXPENSE_TYPE_CODE

Listing 10-9 presents the CREATE syntax for creating the grouped table TRX_INPUT_CC_ACC_GRP. Notice that the column EXPENSE_TYPE_DESC is only added to the GROUP clause for clarity purposes; also notice that a WHERE clause is being added to select only purchases and cash withdrawals outputs.

Listing 10-9. CREATE syntax for creating the grouped table TRX_INPUT_CC_ACC_GRP.

```
41  CREATE TABLE TRX_INPUT_CC_ACC_GRP
42  AS
43  SELECT
44    PERIOD
45    ,ACCOUNT_ID
46    ,EXPENSE_TYPE_CODE
47    ,EXPENSE_TYPE_DESC
48    ,SUM(TRX_AMOUNT) AS  TOT_TRX_AMOUNT
49    ,COUNT(1)        AS  TRANSACTIONS
50  FROM TRX_INPUT_CC_ACC
51  WHERE
52  TRX_TYPE IN (
53    'PURCHASE'
54    ,'CASH WITHDRAW'
55  )
56  AND TRX_DIRECTION='OUTPUT'
57  GROUP BY
58    PERIOD
59    ,ACCOUNT_ID
60    ,EXPENSE_TYPE_CODE
61    ,EXPENSE_TYPE_DESC
62  ;
```

10.4.1.4 Creating a Transactional Base Table When No Consecutive Dates Exist

One challenge in the analysis of transactional data is the lack of uniformity in the occurrence of expenses. Expenses may be classified as either regular or irregular. For instance, the typical individual will likely incur supermarket expenses on a monthly basis, whereas expenses related to hobbies, luxuries, and entertainment are not expected to occur with the same regularity. The issue arises when attempting to make comparisons between expenses within the same category at different points in time, as there is no guarantee that the same expense will be present regularly in the transactional data.

To illustrate this issue, consider the following example. The reference date is January 2024. The objective is to estimate the average transaction amount of the last three months (Avg. L03M), the monthly variation (MV), the quarterly variation (QV), the semestral variation (SV), and the number of increments and decrements of the last three months (SIN L03M and SDC L03M). It is further assumed that only expenses related to hobbies, luxuries, and entertainment are to be considered.

Table 10-3 presents a fictitious case comprising the preceding 12 months' transaction information classified by category. Only three months reported HLE expenses: $100 in January 2023, $300 in August 2023, and $500 in December 2023. The average and variations have been estimated on a moving basis, and increments and decrements have been flagged from January 2023. Using January 2024 as the reference date, we obtain the following results:

- The average transaction amount for HLE expenses for the last three months is $166.67.

- The monthly, quarterly, and semestral variations cannot be computed due to the lack of information from previous months.

- The sum of the increments of the last three months equals one increment, and the sum of the decrements of the last three months equals zero.

CHAPTER 10 THE ABT-BUILDING PROCESS

Table 10-3. *Fictitious transactional information for the last 12 months of hobbies, luxuries, and entertainment expenses.*

Period	Expense type	Trx. amount	Avg. L03M	MV	QV	SV	IN_ FLG	DC_ FLG
202312	HLE	$500.00	$166.67				1	0
202311	HLE	0	$-		−100.0%		0	0
202310	HLE	0	$100.00				0	0
202309	HLE	0	$100.00	−100.0%			0	1
202308	HLE	$300.00	$100.00				1	0
202307	HLE	0	$-			0.0%	0	0
202306	HLE	0	$-				0	0
202305	HLE	0	$-				0	0
202304	HLE	0	$-		−100.0%		0	0
202303	HLE	0	$33.33				0	0
202302	HLE	0		−100.0%			0	1
202301	HLE	$100.00						

Unfortunately, in reality, transactional data does not account for the absence of transactions in the intermediate months. Instead, these months are not represented in the transactional table. **Table 10-4** illustrates a more realistic layout for a transactional table, wherein only existing transactions are represented. It should be noted that this layout precludes the possibility of *explicit periodic calculations*, resulting in the generation of erroneous data.

CHAPTER 10 THE ABT-BUILDING PROCESS

Table 10-4. Realistic transactional layout, where only the existing transactions are recorded in the transactional table.

Period	Expense type	Trx. amount	Avg. LO3M	MV	QV	SV	IN_FLG	DC_FLG
202301	HLE	$500.00	$300.00	66.7%			1	0
202305	HLE	$300.00	$200.00	200.0%			1	0
202312	HLE	$100.00	$100.00				1	0

It should be noted that the term *"explicit periodic calculations"* has been used in preference to *"periodic calculations."* This is because periodic calculations are still possible without intermediate periods; however, unnecessarily complex queries would be required to do so. A preferable alternative would be the establishment of a *transactional base table* (TBT), which would ensure the accurate and chronological ordering of time periods for each of the ten expense types.

Listing 10-10 illustrates the four steps necessary for the creation of the TBT. The initial three steps of the process entail the creation of three distinct catalogs: the TIME_PERIOD_CATALOG, which encompasses all potential time periods; the ACCOUNT_TIME_PERIOD_CATALOG, which includes all possible ACCOUNT_IDs and their corresponding time periods; and the EXPENSE_TYPE_CATALOG, which contains all identified expense types. Step 4 creates the final TBT with a semi-cartesian join. It should be noted that the condition in the WHERE clause A.PERIOD>=B.PERIOD avoids including periods prior to the account's inception.

CHAPTER 10 THE ABT-BUILDING PROCESS

Listing 10-10. Steps required to create the transactional base table, or TBT.

```
65      --Step 1
66      CREATE TABLE TIME_PERIOD_CATALOG
67      AS
68      SELECT
69      DISTINCT
70      PERIOD
71      FROM
72      TRX_INPUT_CC_ACC_GRP  A
73      ORDER BY
74      PERIOD
75      ;
76      --Step 2
77      CREATE TABLE ACCOUNT_TIME_PERIOD_CATALOG
78      AS
79      SELECT
80      DISTINCT
81       ACCOUNT_ID
82      ,PERIOD
83      FROM
84      TRX_INPUT_CC_ACC_GRP  A
85      ORDER BY
86       ACCOUNT_ID
87      ,PERIOD
88      ;
89      --Step 3
90      CREATE TABLE EXPENSE_TYPE_CATALOG
91      AS
92      SELECT
93      DISTINCT
94      EXPENSE_TYPE_CODE
95      FROM
96      TRX_INPUT_CC_ACC_GRP  A
97      ORDER BY
98      EXPENSE_TYPE_CODE
99      ;
100     --Step 4
101     CREATE TABLE TBT_INPUT_CC_ACC_GRP NOLOGGING STORAGE(INITIAL 2M NEXT 1M)
102     AS
103     SELECT
104     DISTINCT
105      A.PERIOD
106     ,B.ACCOUNT_ID
107     ,C.EXPENSE_TYPE_CODE
108     ,0 TOT_TRX_AMOUNT
109     ,0 TRANSACTIONS
110     FROM
111      TIME_PERIOD_CATALOG A
112     ,ACCOUNT_TIME_PERIOD_CATALOG B
113     ,EXPENSE_TYPE_CATALOG C
114     WHERE
115         A.PERIOD>=B.PERIOD
116     ORDER BY
117      A.PERIOD
118     ,B.ACCOUNT_ID
119     ,C.EXPENSE_TYPE_CODE
120     ;
```

CHAPTER 10 THE ABT-BUILDING PROCESS

The subsequent step is illustrated in **Listing 10-11**, wherein the documented transactional data is incorporated into the TBT via a CURSOR. It should be noted that **Listing 10-11** also includes the syntax for the creation of a UNIQUE INDEX. It is strongly recommended that unique indexes be created prior to the execution of a CURSOR to enhance its performance.

Listing 10-11. Adding reported transactional information to the TBT via CURSOR.

```
126    CREATE UNIQUE INDEX INX_TBT_INPUT_CC_ACC_GRP ON
127    TBT_INPUT_CC_ACC_GRP(PERIOD,ACCOUNT_ID,EXPENSE_TYPE_CODE) TABLESPACE SC_SASIDX;
128
129    CREATE UNIQUE INDEX INX_TRX_INPUT_CC_ACC_GRP    ON
130    TRX_INPUT_CC_ACC_GRP(PERIOD,ACCOUNT_ID,EXPENSE_TYPE_CODE) TABLESPACE SC_SASIDX;
131
132    DECLARE
133    t1 integer:=dbms_utility.get_time;t2 integer:=0;v_regs number:=0;
134    CURSOR C1 IS
135      SELECT    /*+ INDEX(A,INX_TBT_TRX_INPUT_CC_ACC_GRP) INDEX(B, INX_TRX_INPUT_CC_ACC_GRP) FIRST_ROWS(10000) */
136                A.ROWID
137              , B.TOT_TRX_AMOUNT
138              , B.TRANSACTIONS
139        FROM
140          TBT_TRX_INPUT_CC_ACC_GRP  A
141         ,TRX_INPUT_CC_ACC_GRP      B
142        WHERE
143            A.ACCOUNT_ID=B.ACCOUNT_ID
144        AND A.PERIOD=B.PERIOD
145        AND A.EXPENSE_TYPE_CODE=B.EXPENSE_TYPE_CODE
146        ;
147    BEGIN
148     UPDATE COUNTER SET COUNTER = 0;
149     COMMIT;
150     FOR B IN C1 LOOP
151      update TBT_TRX_INPUT_CC_ACC_GRP A
152         set
153            TOT_TRX_AMOUNT=     B.TOT_TRX_AMOUNT
154           ,TRANSACTIONS= B.TRANSACTIONS
155        where rowid = B.rowid;
156        IF MOD(C1%ROWCOUNT,50000) = 0  THEN
157              UPDATE COUNTER SET COUNTER = COUNTER + 50000;
158              COMMIT;
159        END IF;
160        v_regs := c1%rowcount;
161     END LOOP;
162     commit;
163     t2 :=dbms_utility.get_time;
164     dbms_output.put_line('-- Registros Procesados: '||to_char(v_regs,'999,999,999.99'));
165     dbms_output.put_line('-- Tiempo: '||to_char((t2-t1)/100/60,'999,999,999.99'));
166    END;
167    /
```

10.4.1.5 Computing Base Metrics Using Oracle's Analytical Functions

One of the most effective tools for calculating base metrics is Oracle's analytical functions (OAF). It is beyond the scope of this book to provide an exhaustive account of the characteristics of each of the OAFs. Consequently, for the time being, our attention will be focused on just three of them:

1. RATIO_TO_REPORT
2. LAG
3. ROW_NUMBER

Listing 10-12 illustrates the SQL syntax that generates all the requisite metrics in a single step through the use of nested subqueries. It should be noted that nested subqueries are executed in an inside-out fashion; therefore, attention should be focused on the innermost query, as shown in **Listing 10-13**. The initial query reads the entirety of the TBT_INPUT_CC_ACC_GRP table resulting from the previous step, but adds an analytical function called RATIO_TO_REPORT. For the time being, ignore the NVL function that surrounds the RATIO_TO_REPORT and let us focus on the RATIO_TO_REPORT function first.

CHAPTER 10 ■ THE ABT-BUILDING PROCESS

Listing 10-12. *CREATE syntax in a nested series of queries for the creation of the base metrics from the TBT, using Oracle's analytical functions.*

```
CREATE TABLE OAF_TBT_INPUT_CC_ACC_GFP NOLOGGING STORAGE(INITIAL 2M NEXT 1M)
AS
SELECT
  A.*
, CASE WHEN N > 1 THEN DECODE(NVL(R_EXPI_EXPT_LAG01,0),0,0,NVL(R_CNSM_CNSMI,0)/NVL(R_EXPI_EXPT_LAG01,0)-1) ELSE 0 END AS MV_R_EXPI_EXPT
, CASE WHEN N > 2 THEN DECODE(NVL(R_EXPI_EXPT_LAG03,0),0,0,NVL(R_CNSM_CNSMI,0)/NVL(R_EXPI_EXPT_LAG03,0)-1) ELSE 0 END AS QV_R_EXPI_EXPT
, CASE WHEN N > 5 THEN DECODE(NVL(R_EXPI_EXPT_LAG06,0),0,0,NVL(R_CNSM_CNSMI,0)/NVL(R_EXPI_EXPT_LAG06,0)-1) ELSE 0 END AS SV_R_EXPI_EXPT
, CASE WHEN NVL(R_EXPI_EXPT,0) > NVL(R_EXPI_EXPT_LAG01,0)            THEN 1 ELSE 0 END            AS IN_R_EXPI_EXPT
, CASE WHEN NVL(R_EXPI_EXPT,0) < NVL(R_EXPI_EXPT_LAG01,0)            THEN 1 ELSE 0 END            AS DC_R_EXPI_EXPT
FROM
(
  SELECT
    A.*
  , ROW_NUMBER() OVER (PARTITION BY ACCOUNT_ID,EXPENSE_TYPE_CODE ORDER BY ACCOUNT_ID,EXPENSE_TYPE_CODE,PERIOD) AS N
  , LAG(R_EXPI_EXPT,1,0) OVER (PARTITION BY ACCOUNT_ID,EXPENSE_TYPE_CODE ORDER BY ACCOUNT_ID,EXPENSE_TYPE_CODE,PERIOD) AS R_EXPI_EXPT_LAG01
  , LAG(R_EXPI_EXPT,3,0) OVER (PARTITION BY ACCOUNT_ID,EXPENSE_TYPE_CODE ORDER BY ACCOUNT_ID,EXPENSE_TYPE_CODE,PERIOD) AS R_EXPI_EXPT_LAG03
  , LAG(R_EXPI_EXPT,6,0) OVER (PARTITION BY ACCOUNT_ID,EXPENSE_TYPE_CODE ORDER BY ACCOUNT_ID,EXPENSE_TYPE_CODE,PERIOD) AS R_EXPI_EXPT_LAG06
  FROM
  (
    SELECT
      A.*
    , NVL(RATIO_TO_REPORT(TOT_TRX_AMOUNT) OVER (PARTITION BY PERIOD,ACCOUNT_ID),0) AS R_EXPI_EXPT
    FROM
      TBT_INPUT_CC_ACC_GRP  A
  )
) A
;
```

Listing 10-13. *Innermost query of* ***Listing 10-12.***

```
SELECT
  A.*
, NVL(RATIO_TO_REPORT(TOT_TRX_AMOUNT) OVER (PARTITION BY PERIOD,ACCOUNT_ID),0) AS R_EXPI_EXPT
FROM
  TBT_INPUT_CC_ACC_GRP  A
```

CHAPTER 10 THE ABT-BUILDING PROCESS

The RATIO_TO_REPORT function is a reporting function that calculates the ratio of a value to the sum of a set of values. Its syntax is as follows:

RATIO_TO_REPORT(magnitude) OVER ([query partition clause])

In this case, the objective is to calculate the ratio of each expense to the total sum of expenses (**Equation 4-9**) for each month period, for each account. An illustrative example of the anticipated output from this query is presented in **Figure 10-4**.

Row#	PERIOD	ACCOUNT_ID	EXPENSE_TYPE_CODE	TOT_TRX_AMOUNT	TRANSACTITIONS	R_EXPI_EXPT
1	201811	03-103362	ATM	0	0	0
2	201811	03-103362	CPS	0	0	0
3	201811	03-103362	ESE	0	0	0
4	201811	03-103362	GSC	11.99	1	0.107121
5	201811	03-103362	HLE	14.95	2	0.133566
6	201811	03-103362	OTR	0	0	0
7	201811	03-103362	RFF	0	0	0
8	201811	03-103362	RSH	84.99	1	0.759314
9	201811	03-103362	TRP	0	0	0
10	201811	03-103362	VCN	0	0	0
11	201812	03-103362	ATM	0	0	0
12	201812	03-103362	CPS	0	0	0
13	201812	03-103362	ESE	0	0	0
14	201812	03-103362	GSC	11.99	1	0.051685
15	201812	03-103362	HLE	10	2	0.043107
16	201812	03-103362	OTR	0	0	0
17	201812	03-103362	RFF	0	0	0
18	201812	03-103362	RSH	209.99	1	0.905207
19	201812	03-103362	TRP	0	0	0
20	201812	03-103362	VCN	0	0	0

Figure 10-4. *Expected output from executing the innermost query.*

Notice in this output that 20 records are returned corresponding to the ten possible expense types for November 2018 and December 2018. Also notice that the new variable R_EXPI_EXPT[10] adds up to 1 for each corresponding month. From this example, we can see that 10.7% of the expenses in November 2018 corresponded to government expenses and

---

[10] EXPI corresponds to the *ith* expense, and EXPT corresponds to the sum of total expenses.

CHAPTER 10 THE ABT-BUILDING PROCESS

services; 13.3% corresponded to hobbies, luxuries, and entertainment; and 75.9% corresponded to retail stores, supermarkets, and household-related expenses. Similarly, for December 2018, we can see that the monthly expenses were again distributed among governmental expenses and services; hobbies, luxuries, and entertainment; and retail stores, supermarkets, and home-related expenses, with 5.1%, 4.3%, and 90.5%, respectively.

Listing 10-14 illustrates the second innermost query, presented in a sequential manner, employing two additional analytical functions: LAG and ROW_NUMBER. The LAG function enables the retrieval of values from other rows in relation to the position of the current row. Its syntax is as follows:

```
LAG(variable, offset, default) OVER ([query partition clause]
order by clause)
```

In this case, the variable is R_EXPI_EXPT, which is calculated from the previous innermost query. The partition is by ACCOUNT_ID and EXPENSE_TYPE_CODE. Unlike the RATIO_TO_REPORT function, the LAG function must specify the ordering column, which in this case is the month period.

The ROW_NUMBER function is the simplest analytical function that does not take an argument. The function assigns a unique number to each row in a sequential manner, commencing with the value of 1 and continuing based on the ORDER BY clause. At each block defined by the PARTITION BY clause, the counter is reset. The syntax is as follows:

```
ROW_NUMBER() OVER ([query partition clause] order by clause)
```

The ROW_NUMBER function is employed in a similar fashion to the LAG function, with the same partition and order parameters. However, its purpose is merely to serve as a reference number for the LAG function.

CHAPTER 10 THE ABT-BUILDING PROCESS

Listing 10-14. Second innermost query from the inside out.

```
SELECT
A.*
, ROW_NUMBER() OVER (PARTITION BY ACCOUNT_ID,EXPENSE_TYPE_CODE ORDER BY ACCOUNT_ID,EXPENSE_TYPE_CODE,PERIOD)       AS N
, LAG(R_EXPI_EXPT,1,0) OVER (PARTITION BY ACCOUNT_ID,EXPENSE_TYPE_CODE ORDER BY ACCOUNT_ID,EXPENSE_TYPE_CODE,PERIOD) AS R_EXPI_EXPT_LAG01
, LAG(R_EXPI_EXPT,3,0) OVER (PARTITION BY ACCOUNT_ID,EXPENSE_TYPE_CODE ORDER BY ACCOUNT_ID,EXPENSE_TYPE_CODE,PERIOD) AS R_EXPI_EXPT_LAG03
, LAG(R_EXPI_EXPT,6,0) OVER (PARTITION BY ACCOUNT_ID,EXPENSE_TYPE_CODE ORDER BY ACCOUNT_ID,EXPENSE_TYPE_CODE,PERIOD) AS R_EXPI_EXPT_LAG06
FROM
```

CHAPTER 10 THE ABT-BUILDING PROCESS

Figure 10-5 illustrates the anticipated outcome of executing **Listing 10-14**. It is notable that the three lags, at 1, 3, and 6, return the requisite monthly, quarterly, and semester lags for the computation of the corresponding variations. Additionally, it should be noted that the N column provides the relative position of each lag in relation to the initial row within the partitioned block.

Row#	PERIOD	ACCOUNT_ID	EXPENSE_TYPE_CODE	R_EXPI_EXPT	N	R_EXPI_EXPT_LAG01	R_EXPI_EXPT_LAG03	R_EXPI_EXPT_LAG06
187	201811	03-103362	GSC	0.107121	1	0	0	0
188	201812	03-103362	GSC	0.051685	2	0.107121	0	0
189	201901	03-103362	GSC	0.099892	3	0.051685	0	0
190	201902	03-103362	GSC	0.125563	4	0.099892	0.107121	0
191	201903	03-103362	GSC	0.09718	5	0.125563	0.051685	0
192	201904	03-103362	GSC	0.25271	6	0.09718	0.099892	0
193	201905	03-103362	GSC	0.067282	7	0.25271	0.125563	0.107121
194	201906	03-103362	GSC	0.666508	8	0.067282	0.09718	0.051685
195	201907	03-103362	GSC	0.053283	9	0.666508	0.25271	0.099892
196	201908	03-103362	GSC	0.030116	10	0.053283	0.067282	0.125563
197	201909	03-103362	GSC	1	11	0.030116	0.666508	0.09718
198	201910	03-103362	GSC	1	12	1	0.053283	0.25271
199	201911	03-103362	GSC	0.123271	13	1	0.030116	0.067282
200	201912	03-103362	GSC	0.064069	14	0.123271	1	0.666508
201	202001	03-103362	GSC	1	15	0.064069	1	0.053283
202	202002	03-103362	GSC	0.053275	16	1	0.123271	0.030116

Figure 10-5. *Expected output from executing the second innermost query from the inside out.*

The base metrics are calculated in the outermost query, which is illustrated in **Listing 10-15**. At this stage, no analytical functions are required; instead, the CASE clause and DECODE function will be employed to prevent errors throughout the overall execution. The final stage of computation yields the base metrics, which are as follows:

1. MV_R_EXPI_EXPT: Monthly variation of the Expi/ExpT ratio

2. QV_R_EXPI_EXPT: Quarterly variation of the Expi/ExpT ratio

3. SV_R_EXPI_EXPT: Semestral variation of the Expi/ExpT ratio
4. IN_R_EXPI_EXPT: Monthly increment flag of the Expi/ExpT ratio
5. DC_R_EXPI_EXPT: Monthly decrement flag of the Expi/ExpT ratio

Listing 10-12 illustrates the complete SQL code needed to compute the base metrics, and **Figure 10-6** shows the expected output from the created table OAF_TBT_INPUT_CC_ACC_GRP for ACCOUNT_ID 03-103362 and the EXPENSE_TYPE_CODE HLE.

The NVL is Oracle's null value function. The NVL function allows you to replace a null with an alternate value. Notice throughout this process we have used the NVL function to replace possible nulls values with a zero value to avoid possible errors during the process execution.

CHAPTER 10 THE ABT-BUILDING PROCESS

Listing 10-15. *Outermost query.*

```
SELECT
A.*
, CASE WHEN N > 1 THEN DECODE(NVL(R_EXPI_EXPT_LAG01,0),0,0,NVL(R_EXPI_EXPT,0)/NVL(R_EXPI_EXPT_LAG01,0)-1) ELSE 0 END AS MV_R_EXPI_EXPT
, CASE WHEN N > 2 THEN DECODE(NVL(R_EXPI_EXPT_LAG03,0),0,0,NVL(R_EXPI_EXPT,0)/NVL(R_EXPI_EXPT_LAG03,0)-1) ELSE 0 END AS QV_R_EXPI_EXPT
, CASE WHEN N > 5 THEN DECODE(NVL(R_EXPI_EXPT_LAG06,0),0,0,NVL(R_EXPI_EXPT,0)/NVL(R_EXPI_EXPT_LAG06,0)-1) ELSE 0 END AS SV_R_EXPI_EXPT
, CASE WHEN NVL(R_EXPI_EXPT,0) > NVL(R_EXPI_EXPT_LAG01,0) THEN 1 ELSE 0 END AS IN_R_EXPI_EXPT
, CASE WHEN NVL(R_EXPI_EXPT,0) < NVL(R_EXPI_EXPT_LAG01,0) THEN 1 ELSE 0 END AS DC_R_EXPI_EXPT
FROM
```

PERIOD	ACCOUNT_ID	EXPENSE_TYPE_CODE	TOT_TRX_AMOUNT	TRANSACTIONS	R_EXPI_EXPT	N	R_EXPI_EXPT_LAG01	R_EXPI_EXPT_LAG03	R_EXPI_EXPT_LAG06	MV_R_EXPI_EXPT	QV_R_EXPI_EXPT	SV_R_EXPI_EXPT	IN_R_EXPI_EXPT	DC_R_EXPI_EXPT
201811	03-103362	HLE	14.95	2	0.133566	1	0	0	0	0	0	0	0	0
201812	03-103362	HLE	10	2	0.043107	2	0.133566	0	0	-0.677259	0	0	0	1
201901	03-103362	HLE	8.05	2	0.067067	3	0.043107	0	0	0.55581	0	0	1	0
201902	03-103362	HLE	0	0	0	4	0.067067	0.133566	0	-1	-1	0	0	1
201903	03-103362	HLE	129.97	3	0.90282	5	0	0.043107	0	19.943624	0	0	1	0
201904	03-103362	HLE	1.96	2	0.035405	6	0.90282	0.067067	0	-0.960784	-0.472097	0	0	1
201905	03-103362	HLE	3.95	1	0.018997	7	0.035405	0	0.133566	-0.463438	0	-0.857772	0	1
201906	03-103362	HLE	7	2	0.333492	8	0.018997	0.90282	0.043107	16.555195	-0.630611	6.736351	1	0
201907	03-103362	HLE	4.99	1	0.019005	9	0.333492	0.035405	0.067067	-0.943012	-0.463201	-0.716622	0	1
201908	03-103362	HLE	353.81	2	0.761635	10	0.019005	0.018997	0	39.075136	39.092861	0	1	0
201909	03-103362	HLE	0	0	0	11	0.761635	0.333492	0.90282	-1	-1	-1	0	1
201910	03-103362	HLE	0	0	0	12	0	0.019005	0.035405	0	-1	-1	0	0
201911	03-103362	HLE	199	1	0.876729	13	0	0.761635	0.018997	0	0.151114	45.15147	1	0
201912	03-103362	HLE	0	0	0	14	0.876729	0	0.333492	-1	0	-1	0	1
202001	03-103362	HLE	0	0	0	15	0	0	0.019005	0	0	-1	0	0
202002	03-103362	HLE	0	0	0	16	0	0.876729	0.761635	0	-1	-1	0	0
202003	03-103362	HLE	0	0	0	17	0	0	0	0	0	0	0	0
202004	03-103362	HLE	0	0	0	18	0	0	0.876729	0	0	-1	0	0
202005	03-103362	HLE	0	0	0	19	0	0	0	0	0	0	0	0

Figure 10-6. *The expected output for OAF_TBT_INPUT_CC_ACC_GRP with real data.*

215

10.4.1.6 Delete Empty Rows from the TBT (Optional Step)

The final TBT, OAF_TBT_INPUT_CC_ACC_GRP, is an augmented version of the original transactional data, expanded to accommodate missing intermediate periods and expense types for infrequent transaction. However, now that the periodic computations have been performed correctly, we can delete empty observations with no impact on subsequent steps.

10.4.1.7 Adding New Variables to the ABT

As previously stated in Section 10.2.1, new columns should be added using the ALTER TABLE syntax. In accordance with the recommendations outlined in Section 8.2, the addition of numerous variables, up to the thousands, should be straightforward if the ALTER TABLE code is generated directly within the building plan spreadsheet.

Cnt	Data Source Cd.	Var Name	Length	Base Metric	Data Type	Operation	Base Code	Sub Category	Period Label	Time Period Type	Start Period	End Period	ALTER TABLE ADD
													="ALTER TABLE ABT ADD "&G6&" NUMBER;"
1	TRN	I_R_ATM_EXP_TRN_00MB	20	ATM / EXP	I	R	R_ATM_EXP	ATM	00MB	P	0	0	="ALTER TABLE ABT ADD "&G6&" NUMBER;"
2	TRN	I_R_ATM_EXP_TRN_01MB	20	ATM / EXP	I	R	R_ATM_EXP	ATM	01MB	P	1	1	ALTER TABLE ABT ADD I_R_ATM_EXP_TRN_01MB NUMBER;
3	TRN	I_R_ATM_EXP_TRN_02MB	20	ATM / EXP	I	R	R_ATM_EXP	ATM	02MB	P	2	2	ALTER TABLE ABT ADD I_R_ATM_EXP_TRN_02MB NUMBER;
4	TRN	I_R_ATM_EXP_TRN_03MB	20	ATM / EXP	I	R	R_ATM_EXP	ATM	03MB	P	3	3	ALTER TABLE ABT ADD I_R_ATM_EXP_TRN_03MB NUMBER;
5	TRN	I_R_ATM_EXP_TRN_06MB	20	ATM / EXP	I	R	R_ATM_EXP	ATM	06MB	P	6	6	ALTER TABLE ABT ADD I_R_ATM_EXP_TRN_06MB NUMBER;
6	TRN	I_R_ATM_EXP_TRN_09MB	20	ATM / EXP	I	R	R_ATM_EXP	ATM	09MB	P	9	9	ALTER TABLE ABT ADD I_R_ATM_EXP_TRN_09MB NUMBER;
7	TRN	I_R_ATM_EXP_TRN_12MB	20	ATM / EXP	I	R	R_ATM_EXP	ATM	12MB	P	12	12	ALTER TABLE ABT ADD I_R_ATM_EXP_TRN_12MB NUMBER;
8	TRN	I_VM_R_ATM_EXP_TRN_L01_00	25	ATM / EXP	I	MV	R_ATM_EXP	ATM	L01_00	P	1	0	ALTER TABLE ABT ADD I_VM_R_ATM_EXP_TRN_L01_00 NUMBER;
9	TRN	I_VM_R_ATM_EXP_TRN_L02_01	25	ATM / EXP	I	MV	R_ATM_EXP	ATM	L02_01	P	2	1	ALTER TABLE ABT ADD I_VM_R_ATM_EXP_TRN_L02_01 NUMBER;
10	TRN	I_VM_R_ATM_EXP_TRN_L03_02	25	ATM / EXP	I	MV	R_ATM_EXP	ATM	L03_02	P	3	2	ALTER TABLE ABT ADD I_VM_R_ATM_EXP_TRN_L03_02 NUMBER;
11	TRN	I_VT_R_ATM_EXP_TRN_L06_03	25	ATM / EXP	I	QV	R_ATM_EXP	ATM	L06_03	P	6	3	ALTER TABLE ABT ADD I_VT_R_ATM_EXP_TRN_L06_03 NUMBER;
12	TRN	I_VT_R_ATM_EXP_TRN_L09_06	25	ATM / EXP	I	QV	R_ATM_EXP	ATM	L09_06	P	9	6	ALTER TABLE ABT ADD I_VT_R_ATM_EXP_TRN_L09_06 NUMBER;
13	TRN	I_VT_R_ATM_EXP_TRN_L12_09	25	ATM / EXP	I	QV	R_ATM_EXP	ATM	L12_09	P	12	9	ALTER TABLE ABT ADD I_VT_R_ATM_EXP_TRN_L12_09 NUMBER;
14	TRN	I_VS_R_ATM_EXP_TRN_L12_06	25	ATM / EXP	I	SV	R_ATM_EXP	ATM	L12_06	P	12	6	ALTER TABLE ABT ADD I_VS_R_ATM_EXP_TRN_L12_06 NUMBER;
15	TRN	I_AV_R_ATM_EXP_TRN_L03M	23	ATM / EXP	I	AV	R_ATM_EXP	ATM	L03M	A	1	3	ALTER TABLE ABT ADD I_AV_R_ATM_EXP_TRN_L03M NUMBER;
16	TRN	I_AV_R_ATM_EXP_TRN_L06M	23	ATM / EXP	I	AV	R_ATM_EXP	ATM	L06M	A	1	6	ALTER TABLE ABT ADD I_AV_R_ATM_EXP_TRN_L06M NUMBER;
17	TRN	I_AV_R_ATM_EXP_TRN_L12M	23	ATM / EXP	I	AV	R_ATM_EXP	ATM	L12M	A	1	12	ALTER TABLE ABT ADD I_AV_R_ATM_EXP_TRN_L12M NUMBER;
18	TRN	I_AV_R_ATM_EXP_TRN_L13_24	25	ATM / EXP	I	AV	R_ATM_EXP	ATM	L13_24	A	13	24	ALTER TABLE ABT ADD I_AV_R_ATM_EXP_TRN_L13_24 NUMBER;
19	TRN	I_SIN_R_ATM_EXP_TRN_L03M	24	ATM / EXP	I	SIN	R_ATM_EXP	ATM	L03M	A	1	3	ALTER TABLE ABT ADD I_SIN_R_ATM_EXP_TRN_L03M NUMBER;
20	TRN	I_SIN_R_ATM_EXP_TRN_L06M	24	ATM / EXP	I	SIN	R_ATM_EXP	ATM	L06M	A	1	6	ALTER TABLE ABT ADD I_SIN_R_ATM_EXP_TRN_L06M NUMBER;
21	TRN	I_SIN_R_ATM_EXP_TRN_L12M	24	ATM / EXP	I	SIN	R_ATM_EXP	ATM	L12M	A	1	12	ALTER TABLE ABT ADD I_SIN_R_ATM_EXP_TRN_L12M NUMBER;
22	TRN	I_SIN_R_ATM_EXP_TRN_L13_24	26	ATM / EXP	I	SIN	R_ATM_EXP	ATM	L13_24	A	13	24	ALTER TABLE ABT ADD I_SIN_R_ATM_EXP_TRN_L13_24 NUMBER;
23	TRN	I_SDC_R_ATM_EXP_TRN_L03M	24	ATM / EXP	I	SDC	R_ATM_EXP	ATM	L03M	A	1	3	ALTER TABLE ABT ADD I_SDC_R_ATM_EXP_TRN_L03M NUMBER;
24	TRN	I_SDC_R_ATM_EXP_TRN_L06M	24	ATM / EXP	I	SDC	R_ATM_EXP	ATM	L06M	A	1	6	ALTER TABLE ABT ADD I_SDC_R_ATM_EXP_TRN_L06M NUMBER;
25	TRN	I_SDC_R_ATM_EXP_TRN_L12M	24	ATM / EXP	I	SDC	R_ATM_EXP	ATM	L12M	A	1	12	ALTER TABLE ABT ADD I_SDC_R_ATM_EXP_TRN_L12M NUMBER;
26	TRN	I_SDC_R_ATM_EXP_TRN_L13_24	26	ATM / EXP	I	SDC	R_ATM_EXP	ATM	L13_24	A	13	24	ALTER TABLE ABT ADD I_SDC_R_ATM_EXP_TRN_L13_24 NUMBER;

Figure 10-7. Illustrative example of how to manufacture your ALTER TABLE syntax directly in your building plan spreadsheet.

10.4.1.8 Creating ABT Input Variables from TBT via Conditional Grouping Functions

The TBT named OAF_TBT_INPUT_CC_ACC_GRP can be employed in conjunction with the ABT to create the final input variables, which can then be added directly to the ABT via Oracle's CURSOR. **Listing 10-16** illustrates the generation of 10 of the 26 variants displayed in **Table 10-1** through the application of conditional grouping functions.

Listing 10-16 shows that the ABT and the OAF_TBT_INPUT_CC_ACC_GRP table are joined solely by ACCOUNT_ID, with the additional condition B.PERIOD<=A.PERIOD. This results in a semi-cartesian join, where the transaction info is constrained to be less than or equal to the reference period in the ABT. All grouping variables are associated with the ABT, so the number of records generated by the query should be less than or equal to the number of rows in the ABT.

Listing 10-16. Example of a SELECT transposition via conditional grouping functions.

```sql
CREATE TABLE ABT_INPUT_VARIABLES_FROM_IBI NOLOGGING STORAGE(INITIAL 2M NEXT 1M)
AS
SELECT
  A.PERIOD
, A.REF_DATE
, A.ACCOUNT_ID
, AVG(CASE WHEN EXPENSE_TYPE_CODE='ATM' AND B.PERIOD=TO_CHAR(ADD_MONTHS(A.REF_DATE,-0),'YYYYMM') THEN R_EXPI_EXPT END)  AS I_R_ATM_EXP_TRX_00MB
, AVG(CASE WHEN EXPENSE_TYPE_CODE='ATM' AND B.PERIOD=TO_CHAR(ADD_MONTHS(A.REF_DATE,-12),'YYYYMM') THEN R_EXPI_EXPT END) AS I_R_ATM_EXP_TRX_12MB
, AVG(CASE WHEN EXPENSE_TYPE_CODE='ATM' AND B.PERIOD=TO_CHAR(ADD_MONTHS(A.REF_DATE,-0),'YYYYMM') THEN VM_R_EXPI_EXPT END) AS I_MV_R_ATM_EXP_TRX_L01_00
, AVG(CASE WHEN EXPENSE_TYPE_CODE='ATM' AND B.PERIOD=TO_CHAR(ADD_MONTHS(A.REF_DATE,-1),'YYYYMM') THEN VM_R_EXPI_EXPT END) AS I_MV_R_ATM_EXP_TRX_L02_01
, AVG(CASE WHEN EXPENSE_TYPE_CODE='ATM' AND B.PERIOD=TO_CHAR(ADD_MONTHS(A.REF_DATE,-6),'YYYYMM') THEN VT_R_EXPI_EXPT END) AS I_QV_R_ATM_EXP_TRX_L09_06
, AVG(CASE WHEN EXPENSE_TYPE_CODE='ATM' AND B.PERIOD=TO_CHAR(ADD_MONTHS(A.REF_DATE,-9),'YYYYMM') THEN VT_R_EXPI_EXPT END) AS I_QV_R_ATM_EXP_TRX_L12_09
, SUM(CASE WHEN EXPENSE_TYPE_CODE='ATM' AND B.PERIOD=TO_CHAR(ADD_MONTHS(A.REF_DATE,-6),'YYYYMM') THEN VS_R_EXPI_EXPT END) AS I_SV_R_ATM_EXP_TRX_L12_06
, AVG(CASE WHEN EXPENSE_TYPE_CODE='ATM' AND B.PERIOD BETWEEN TO_CHAR(ADD_MONTHS(A.REF_DATE,-3),'YYYYMM')
                                            AND TO_CHAR(ADD_MONTHS(A.REF_DATE,-1),'YYYYMM') THEN R_EXPI_EXPT END) AS I_AV_R_ATM_EXP_TRX_L03M
, SUM(CASE WHEN EXPENSE_TYPE_CODE='ATM' AND B.PERIOD BETWEEN TO_CHAR(ADD_MONTHS(A.REF_DATE,-3),'YYYYMM')
                                            AND TO_CHAR(ADD_MONTHS(A.REF_DATE,-1),'YYYYMM') THEN IN_R_EXPI_EXPT END) AS I_SIN_R_ATM_EXP_TRX_L03M
, SUM(CASE WHEN EXPENSE_TYPE_CODE='ATM' AND B.PERIOD BETWEEN TO_CHAR(ADD_MONTHS(A.REF_DATE,-3),'YYYYMM')
                                            AND TO_CHAR(ADD_MONTHS(A.REF_DATE,-1),'YYYYMM') THEN DC_R_EXPI_EXPT END) AS I_SDC_R_ATM_EXP_TRX_L03M
FROM
  ABT A
, OAF_IBI_INPUT_CC_ACC_GRP B
WHERE
  A.ACCOUNT_ID=B.ACCOUNT_ID
AND B.PERIOD<=A.PERIOD
GROUP BY
  A.PERIOD
, A.REF_DATE
, A.ACCOUNT_ID
;
```

CHAPTER 10 THE ABT-BUILDING PROCESS

The key element of this particular query is the set of conditions in each grouping function, which serve not only to transpose transactional information but also to compute the specific timeframes of each variant presented in **Table 10-1**. To illustrate, we may consider the conditional grouping function for the I_R_ATM_EXP_TRX_12MB variable (**Listing 10-17**).

Listing 10-17. Conditional grouping function for the I_R_ATM_EXP_TRX_12MB variable.

```
AVG (CASE
        WHEN EXPENSE_TYPE_CODE='ATM'
        AND B.PERIOD=TO_CHAR(ADD_MONTHS(A.REF_DATE,-12),'YYYYMM') THEN R_EXPI_EXPT
    END)     AS I_R_ATM_EXP_TRX_12MB
```

It is important to note that conditional grouping functions read all records during the grouping operation. However, they then select a specific value or range of values that meet a specific set of conditions. Consequently, the CASE WHEN statement lacks an alternative ELSE clause.

Table 10-5 provides an illustrative example of the conditional grouping process for the I_R_ATM_EXP_TRX_12MB point-in-time variable as presented in the aforementioned code. Notice that due to the absence of the ELSE clause, all values returned by the CASE WHEN statement are NULL, with the exception of the specific value requested by the EXPENSE_TYPE_CODE='ATM' and the B.PERIOD=TO_CHAR(ADD_MONTHS(A.REF_DATE,-12),'YYYYMM') conditions. These conditions specify the selection of the ratio of ATM cash withdrawals to total expenditures, reported 12 months back from the December 2023 reference date.

CHAPTER 10 THE ABT-BUILDING PROCESS

Table 10-5. *Illustrative example of the conditional grouping process for the point-in-time* $I\_R\_ATM\_EXP\_TRX\_12MB$ *variable.*

A.ACCOUNT_ID	B.PERIOD	A.REF_DATE	B.EXPENSE_TYPE_CODE	B.R_EXPI_EXPT	CASE WHEN
111223344	202312	Dec-23	ATM	0.12	NULL
111223344	202311	Dec-23	ATM	0.05	NULL
111223344	202310	Dec-23	ATM	0.08	NULL
111223344	202309	Dec-23	ATM	0.10	NULL
111223344	202308	Dec-23	ATM	0.09	NULL
111223344	202307	Dec-23	ATM	0.12	NULL
111223344	202306	Dec-23	ATM	0.15	NULL
111223344	202305	Dec-23	ATM	0.00	NULL
111223344	202304	Dec-23	ATM	0.03	NULL
111223344	202303	Dec-23	ATM	0.12	NULL
111223344	202302	Dec-23	ATM	0.05	NULL
111223344	202301	Dec-23	ATM	0.08	NULL
111223344	**202212**	**Dec-23**	**ATM**	**0.10**	**0.10**
111223344	202312	Dec-23	TRN	0.20	NULL
111223344	202311	Dec-23	TRN	0.16	NULL
111223344	202310	Dec-23	TRN	0.18	NULL
111223344	202309	Dec-23	TRN	0.15	NULL
111223344	202308	Dec-23	TRN	0.17	NULL
111223344	202307	Dec-23	TRN	0.21	NULL
111223344	202306	Dec-23	TRN	0.20	NULL
111223344	202305	Dec-23	TRN	0.16	NULL
111223344	202304	Dec-23	TRN	0.18	NULL
111223345	202303	Dec-23	TRN	0.15	NULL
111223346	202302	Dec-23	TRN	0.17	NULL
111223347	202301	Dec-23	TRN	0.21	NULL
111223348	202301	Dec-23	TRN	0.21	NULL
111223349	202212	Dec-23	TRN	0.20	NULL

CHAPTER 10   THE ABT-BUILDING PROCESS

It is noteworthy that the AVG grouping function in this particular example does not, in fact, perform any average calculations. This is due to the fact that the specific set of conditions for this variable will return a unique match from the TBT.

Let us now examine the conditional grouping function for the I_AV_R_ATM_EXP_TRX_L03M variable.

Listing 10-18. Conditional grouping function for the I_AV_R_ATM_EXP_TRX_L03M variable.

```
AVG (CASE
        WHEN EXPENSE_TYPE_CODE='ATM'
            AND B.PERIOD BETWEEN TO_CHAR(ADD_MONTHS(A.REF_DATE,-3),'YYYYMM')
            AND TO_CHAR(ADD_MONTHS(A.REF_DATE,-1),'YYYYMM') THEN R_EXPI_EXPT
        END)    AS I_AV_R_ATM_EXP_TRX_L03M
```

It should be noted that in this case, the distinguishing factor between a point-in-time and an aggregate variant is the use of the BETWEEN clause. The BETWEEN clause permits the inclusion of a range rather than a specific date, which will not yield a unique match but rather a series of records within the range covered by the BETWEEN clause.

Table 10-6 illustrates the aforementioned example with respect to the aggregation variable I_AV_R_ATM_EXP_TRX_L03M. It should be noted that, in contrast to the preceding example, the conditional grouping function returns multiple records, thereby enabling the calculation of the average of ATM cash withdrawals relative to total expenditures. This encompasses the three-month period preceding the December 2023 reference date.

221

CHAPTER 10 THE ABT-BUILDING PROCESS

Table 10-6. *Illustrative example of the conditional grouping process for the aggregation I_AV_R_ATM_EXP_TRX_LO3M variable.*

A.ACCOUNT_ID	B.PERIOD	A.REF_DATE	B.EXPENSE_TYPE_CODE	B.R_EXPI_EXPT	CASE WHEN
111223344	202312	Dec-23	ATM	0.12	NULL
111223344	202311	Dec-23	ATM	0.05	0.05
111223344	202310	Dec-23	ATM	0.08	0.08
111223344	202309	Dec-23	ATM	0.10	0.10
111223344	202308	Dec-23	ATM	0.09	NULL
111223344	202307	Dec-23	ATM	0.12	NULL
111223344	202306	Dec-23	ATM	0.15	NULL
111223344	202305	Dec-23	ATM	0.00	NULL
111223344	202304	Dec-23	ATM	0.03	NULL
111223344	202303	Dec-23	ATM	0.12	NULL
111223344	202302	Dec-23	ATM	0.05	NULL
111223344	202301	Dec-23	ATM	0.08	NULL
111223344	202212	Dec-23	ATM	0.10	0.10
111223344	202312	Dec-23	TRN	0.20	NULL
111223344	202311	Dec-23	TRN	0.16	NULL
111223344	202310	Dec-23	TRN	0.18	NULL
111223344	202309	Dec-23	TRN	0.15	NULL
111223344	202308	Dec-23	TRN	0.17	NULL
111223344	202307	Dec-23	TRN	0.21	NULL
111223344	202306	Dec-23	TRN	0.20	NULL
111223344	202305	Dec-23	TRN	0.16	NULL
111223345	202304	Dec-23	TRN	0.18	NULL
111223346	202303	Dec-23	TRN	0.15	NULL
111223347	202302	Dec-23	TRN	0.17	NULL
111223348	202301	Dec-23	TRN	0.21	NULL
111223349	202212	Dec-23	TRN	0.20	NULL

CHAPTER 10   THE ABT-BUILDING PROCESS

The next step is to create 260 input variables for the 26 variants and ten expense categories via concatenation formulas. **Figure 10-8** shows the complete spreadsheet formula that assembles the SQL code. Notice that when L2="P", the time period condition takes the form of the generic point-in-time condition of B.PERIOD=TO_CHAR(ADD_MONTHS(A.REF_DATE,-#),'YYYYMM'). Otherwise, it takes the form of the generic range time period condition of

B.PERIOD BETWEEN TO_CHAR(ADD_MONTHS(A.REF_DATE,-#),'YYYYMM')
AND TO_CHAR(ADD_MONTHS(A.REF_DATE,-#),'YYYYMM')

K	L	M	N	P
Period Label	Time Period Type	Start Period	End Period	SQL Time Period Condition
00MB	P	0	0	=IF(L2="P","B.PERIOD=TO_CHAR(ADD_MONTHS(A.REF_DATE,-"&N2&"),'YYYYMM')","B.PERIOD BETWEEN TO_CHAR(ADD_MONTHS(A.REF_DATE,-"&N2&"),'YYYYMM') AND TO_CHAR(ADD_MONTHS(A.REF_DATE,-"&M2&"),'YYYYMM')")
12MB	P	12	12	

Figure 10-8. *Spreadsheet formula that assembles the code for the time period condition.*

After assembling the time period condition code, we proceed to assemble the conditional grouping function. **Figure 10-9** shows the spreadsheet formula that concatenates A2, D2, J2, and P2 into the SQL code. Notice that H2 is not a concatenation variable, and it is used as a condition to choose between an AVG, or a SUM aggregation function. The condition is only true when MID(H2,1,1)="S", and this is the only case when the SUM aggregation function is selected.

The final result is shown in **Figure 10-10**, which shows a simple way to create 260 input variables in a spreadsheet, along with the minimum number of columns needed to manufacture the SQL time period condition and the SQL conditional grouping function.

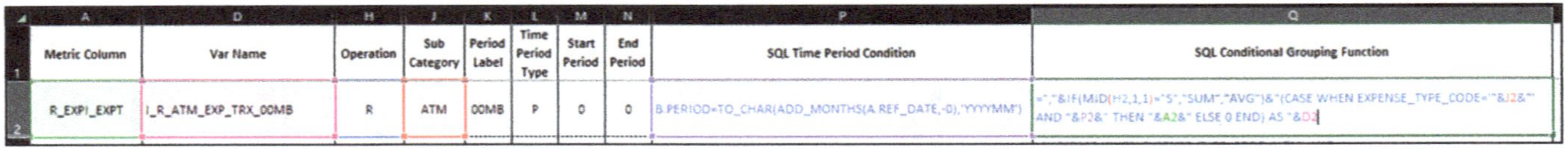

CHAPTER 10   THE ABT-BUILDING PROCESS

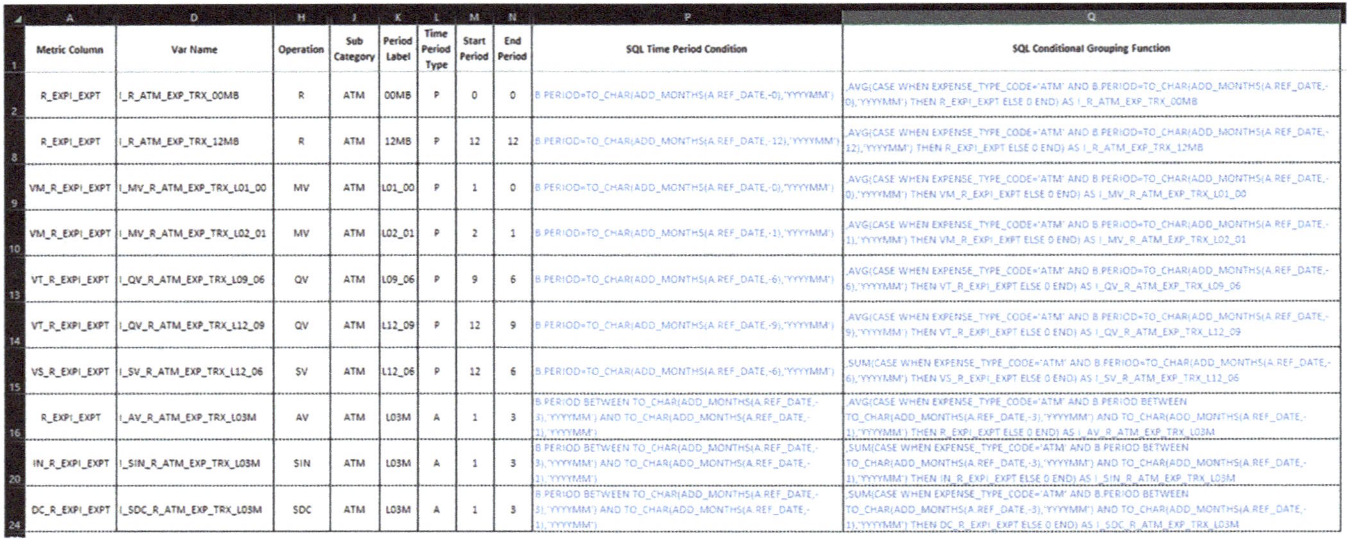

Figure 10-9. Spreadsheet formula that assembles the code for the conditional grouping function.

Figure 10-10. Illustrative example of how to manufacture conditional grouping functions in your building plan spreadsheet.

CHAPTER 10   THE ABT-BUILDING PROCESS

The final table is created with a CREATE syntax identical to that shown in **Listing 10-16**, with the only difference being that the final table will have 260 variables corresponding to the 26 variants for the ten expense types derived from the MCC.

10.4.1.9 Adding Input Variables to the ABT via Cursor

Listing 10-19 shows an example of a CURSOR that adds ten variables previously shown in **Listing 10-16** and **Figure 10-10**. Such a CURSOR could be used to add the 260 variables to the ABT. Notice that adding the input variables from the TBT, ABT_INPUT_VARIABLES_FROM_TBT, is fairly straightforward because both the ABT and the TBT are at the same granularity level, PERIOD and ACCOUNT_ID, which allows the joining operation to be performed by an inner equijoin. Also notice that the SELECT clause consists only of the ROWID from the ABT and the input variables from the TBT; it is worth noting that the A.PERIOD and the A.ACCOUNT_ID are not really necessary in the SELECT clause, and they are included only for clarity.

225

Listing 10-19. *Example of input variables being added to the ABT via CURSOR.*

```
DECLARE
T1 INTEGER:=DBMS_UTILITY.GET_TIME;T2 INTEGER:=0;V_REGS NUMBER:=0;
    CURSOR C1 IS
    SELECT
     A.ROWID
    ,A.PERIOD
    ,A.ACCOUNT_ID
    ,B.I_R_ATM_EXP_TRX_00MB
    ,B.I_R_ATM_EXP_TRX_12MB
    ,B.I_MV_R_ATM_EXP_TRX_L01_00
    ,B.I_MV_R_ATM_EXP_TRX_L02_01
    ,B.I_QV_R_ATM_EXP_TRX_L09_06
    ,B.I_QV_R_ATM_EXP_TRX_L12_09
    ,B.I_SV_R_ATM_EXP_TRX_L12_06
    ,B.I_AV_R_ATM_EXP_TRX_L03M
    ,B.I_SIN_R_ATM_EXP_TRX_L03M
    ,B.I_SDC_R_ATM_EXP_TRX_L03M
    FROM
     ABT A
    ,ABT_INPUT_VARIABLES_FROM_TBT B
    WHERE
        A.ACCOUNT_ID=B.ACCOUNT_ID
    AND B.PERIOD=A.PERIOD
    ;
BEGIN
 UPDATE COUNTER SET COUNTER = 0;
 COMMIT;
     FOR B IN C1 LOOP
     UPDATE ABT A
        SET
         I_R_ATM_EXP_TRX_00MB=B.I_R_ATM_EXP_TRX_00MB
        ,I_R_ATM_EXP_TRX_12MB=B.I_R_ATM_EXP_TRX_12MB
        ,I_MV_R_ATM_EXP_TRX_L01_00=B.I_MV_R_ATM_EXP_TRX_L01_00
        ,I_MV_R_ATM_EXP_TRX_L02_01=B.I_MV_R_ATM_EXP_TRX_L02_01
        ,I_QV_R_ATM_EXP_TRX_L09_06=B.I_QV_R_ATM_EXP_TRX_L09_06
        ,I_QV_R_ATM_EXP_TRX_L12_09=B.I_QV_R_ATM_EXP_TRX_L12_09
        ,I_SV_R_ATM_EXP_TRX_L12_06=B.I_SV_R_ATM_EXP_TRX_L12_06
        ,I_AV_R_ATM_EXP_TRX_L03M=B.I_AV_R_ATM_EXP_TRX_L03M
        ,I_SIN_R_ATM_EXP_TRX_L03M=B.I_SIN_R_ATM_EXP_TRX_L03M
        ,I_SDC_R_ATM_EXP_TRX_L03M=B.I_SDC_R_ATM_EXP_TRX_L03M
        WHERE ROWID = B.ROWID;

        IF MOD(C1%ROWCOUNT,10000) = 0 THEN
                UPDATE COUNTER SET COUNTER = COUNTER + 10000;
                COMMIT;
        END IF;
        V_REGS := C1%ROWCOUNT;
     END LOOP;
 COMMIT;
 T2 :=DBMS_UTILITY.GET_TIME;
 DBMS_OUTPUT.PUT_LINE('-- Records Processed: ' ||TO_CHAR(V_REGS,'999,999,999.99'));
 DBMS_OUTPUT.PUT_LINE('-- Elapsed: ' ||TO_CHAR((T2-T1)/100/60,'999,999,999.99'));
END;
/
```

Finally, it is important to highlight the fact that both the SELECT clause and the SET clause of the UPDATE syntax can be easily manufactured by assembling the SQL code using spreadsheet formulas.

10.4.1.10 Adding Input Variables via PL/SQL (Alternative Method)

There is yet another way to add your input variables to the ABT, namely, via PL/SQL. PL/SQL is Oracle's programming language, and it works in a similar way to SAS macros do. The main idea of using a PL/SQL is to automate a piece of code by adding a variable clause, argument, or syntax. This enables the code to be executed repeatedly, with a distinct value in each iteration, until the desired task or routine is completed.

Listing 10-20 provides an illustrative example of a PL/SQL code that can also be utilized to incorporate the input variables into the ABT. It is notable that this specific CURSOR employs the use of a control table to facilitate the automated execution of another CURSOR. The control table is designated as CTRL_CURSOR, and **Figure 10-11** illustrates its contents. As its name suggests, the control table is responsible for regulating the number of times that the inner cursor is invoked and executed. It can be observed that the inner cursor is executed 26 times, corresponding to the number of variants described in **Table 10-1**. Additionally, the control table functions in a manner analogous to the spreadsheet building plan, but in this case, the SQL code to be concatenated into the inner CURSOR is specified by the CTRL_CURSOR table.

It is notable that, in contrast to other CURSORs, which add information record by record from top to bottom, the inner CURSOR adds information from top to bottom and from left to right. This is due to the fact that the inner query returns a transactional output that is transposed during the PL/SQL execution. In this sense, one step is saved in comparison to the previous methodology, which involved first creating the transpose input variable table and then adding it to the ABT. However, there is a caveat.

Listing 10-20. Example of a PL/SQL to add the input variables to the ABT.

```plsql
DECLARE
    V_CURSOR VARCHAR2(30000);
    V_PROCESS_NUMBER NUMBER;
    V_VAR_PREFIX VARCHAR2(50);
    V_SQL_FUNCTION VARCHAR2(100);
    V_PERIOD_LABEL VARCHAR2(50);
    V_WHERE VARCHAR2(500);
    CURSOR C1 IS
    SELECT
        PROCESS_NUM
        , VAR_PREFIX||'_'   as VAR_PREFIX
        , GRP_FUNCTION
        , PERIOD_CD
        , SQL_WHERE
    FROM  CTRL_CURSOR
    ORDER BY  PROCESS_NUM
    ;
BEGIN
    FOR B IN C1 LOOP
        V_PROCESS_NUMBER:=B.PROCESS_NUM;
        V_VAR_PREFIX:=  B.VAR_PREFIX;
        V_SQL_FUNCTION:=B.GRP_FUNCTION;
        V_PERIOD_LABEL:=B.PERIOD_CD;
        V_SQL_WHERE:=   B.SQL_WHERE;
        V_CURSOR:='
                    DECLARE
                    VSQL VARCHAR2(1000);
                    VVARIABLE_NAME VARCHAR2(50);
                    VPERIOD_EXE NUMBER;
                    VDATE_EXE DATE:=SYSDATE;
                    VPROCESS INTEGER;
                    VELAPSED NUMBER;
                    V_REGS NUMBER:=0;
                    T1 DATE:=SYSDATE;
                    T2 DATE;
                    CURSOR C2 IS
                    SELECT /*+FIRST_ROWS(5000)*/
                        A.ROWID
                        ,A.PERIOD
                        ,A.ACCOUNT_ID
                        ,B.EXPENSE_TYPE_CODE
                        ,'''''''||V_VAR_PREFIX||''''''||'||TRIM(B.EXPENSE_TYPE_CODE)||''''''||'_EXP_TRX_'||V_PERIOD_LABEL||''''''|| AS VAR_NM
                        ,'||V_SQL_FUNCTION||' AS METRIC
                    FROM
                        ABT A
                        ,ABT_INPUT_VARIABLES_FROM_TBT B
                    WHERE
                        A.ACCOUNT_ID=B.ACCOUNT_ID
                    AND '||V_SQL_WHERE||'
                    GROUP BY
                        A.ROWID
                        ,A.PERIOD
                        ,A.ACCOUNT_ID
                        ,B.EXPENSE_TYPE_CODE
                        ,'''''''||V_VAR_PREFIX||''''''||'||TRIM(B.EXPENSE_TYPE_CODE)||''''''||'_EXP_TRX_'||V_PERIOD_LABEL||''''''||'
                    ;
                    BEGIN
                    UPDATE COUNTER COUNTER = 0;
                    COMMIT;
                    FOR B IN C2 LOOP
                        VVARIABLE_NAME := B.VAR_NM;
                        VSQL := '||''''''||'UPDATE ABT  A
                        SET '||''''''||'||VVARIABLE_NAME||''''''|| = :VMETRIC WHERE  ROWID = :VROWID'||''''''||'
                        ;
                        EXECUTE IMMEDIATE VSQL USING B.METRIC, B.ROWID;
                        IF MOD(C2%ROWCOUNT,5000) = 0  THEN
                            UPDATE COUNTER SET COUNTER = COUNTER + 5000;
                            VPERIODO_EXE:=TO_CHAR(VDATE_EXE,''YYYYMMDD'');
                            VPROCESS:='||V_PROCESS_NUMBER||';
                            T2:=SYSDATE;
                            VELAPSED:=TO_NUMBER(T2-T1)*1440*60;
                            V_REGS := C2%ROWCOUNT;
                            UPDATE CTRL_CURSOR
                                SET
                                    EXE_PERIOD=VPERIOD_EXE
                                    ,EXE_DATE=VDATE_EXE
                                    ,ELAPSED=VELAPSED
                                    ,ROW_COUNT=V_REGS
                                WHERE PROCESS_NUM=VPROCESS
                            ;
                            COMMIT;
                        END IF;
                    END LOOP;
                    COMMIT;
                    END;';
        EXECUTE IMMEDIATE V_CURSOR;
        COMMIT;
    END LOOP;
END;
/
```

CHAPTER 10 THE ABT-BUILDING PROCESS

EXE_PERIOD	EXE_DATE	PROCESS_NUM	ELAPSED	ROW_COUNT	VAR_PREFIX	GRP_FUNCTION	PERIOD_CD	SQL_WHERE
20231214	14/12/2023 18:20:49	1	1135.99...	6880000	I_R	AVG(R_CNSM_CNSMT)	00MB	B.PERIODOMES=TO_CHAR(ADD_MONTHS(A.REF_DATE,-0),'YYYYMM')
20231214	14/12/2023 18:39:45	2	1292	6740000	I_R	AVG(R_CNSM_CNSMT)	01MB	B.PERIODOMES=TO_CHAR(ADD_MONTHS(A.REF_DATE,-1),'YYYYMM')
20231214	14/12/2023 19:01:17	3	1140.99...	6645000	I_R	AVG(R_CNSM_CNSMT)	02MB	B.PERIODOMES=TO_CHAR(ADD_MONTHS(A.REF_DATE,-2),'YYYYMM')
20231214	14/12/2023 19:20:19	4	1344.00...	6590000	I_R	AVG(R_CNSM_CNSMT)	03MB	B.PERIODOMES=TO_CHAR(ADD_MONTHS(A.REF_DATE,-3),'YYYYMM')
20231214	14/12/2023 19:42:44	5	1181.00...	6410000	I_R	AVG(R_CNSM_CNSMT)	06MB	B.PERIODOMES=TO_CHAR(ADD_MONTHS(A.REF_DATE,-6),'YYYYMM')
20231214	14/12/2023 20:02:25	6	1454	6190000	I_R	AVG(R_CNSM_CNSMT)	09MB	B.PERIODOMES=TO_CHAR(ADD_MONTHS(A.REF_DATE,-9),'YYYYMM')
20231215	15/12/2023 13:20:57	7	1358	5975000	I_R	AVG(R_CNSM_CNSMT)	12MB	B.PERIODOMES=TO_CHAR(ADD_MONTHS(A.REF_DATE,-12),'YYYYMM')
20231215	15/12/2023 13:43:35	8	3190	6880000	I_vM_R	AVG(vM_R_CNSM_CNSMT)	L01_00	B.PERIODOMES=TO_CHAR(ADD_MONTHS(A.REF_DATE,-0),'YYYYMM')
20231215	15/12/2023 14:36:46	9	1490.99...	6740000	I_vM_R	AVG(vM_R_CNSM_CNSMT)	L02_01	B.PERIODOMES=TO_CHAR(ADD_MONTHS(A.REF_DATE,-1),'YYYYMM')
20231215	15/12/2023 15:01:38	10	981.999...	6645000	I_vM_R	AVG(vM_R_CNSM_CNSMT)	L03_02	B.PERIODOMES=TO_CHAR(ADD_MONTHS(A.REF_DATE,-2),'YYYYMM')
20231216	16/12/2023 12:06:49	11	460	6410000	I_v5_R	AVG(v5_R_CNSM_CNSMT)	L12_06	B.PERIODOMES=TO_CHAR(ADD_MONTHS(A.REF_DATE,-6),'YYYYMM')
20231216	16/12/2023 12:14:30	12	1103	6590000	I_VT_R	AVG(VT_R_CNSM_CNSMT)	L06_03	B.PERIODOMES=TO_CHAR(ADD_MONTHS(A.REF_DATE,-3),'YYYYMM')
20231216	16/12/2023 12:32:53	13	1093.00...	6410000	I_VT_R	AVG(VT_R_CNSM_CNSMT)	L09_06	B.PERIODOMES=TO_CHAR(ADD_MONTHS(A.REF_DATE,-6),'YYYYMM')
20231216	16/12/2023 12:51:06	14	1066.99...	6190000	I_VT_R	AVG(VT_R_CNSM_CNSMT)	L12_09	B.PERIODOMES=TO_CHAR(ADD_MONTHS(A.REF_DATE,-9),'YYYYMM')
20231216	16/12/2023 13:08:53	15	1263.00...	7905000	I_aV_R	AVG(R_CNSM_CNSMT)	L03M	B.PERIODOMES BETWEEN TO_CHAR(ADD_MONTHS(A.REF_DATE,-3),'YYYYMM') AND TO_CHAR(ADD_MONTHS(A.REF_DATE,-1),'YYYYMM')
20231216	16/12/2023 13:29:56	16	1444.99...	8730000	I_aV_R	AVG(R_CNSM_CNSMT)	L06M	B.PERIODOMES BETWEEN TO_CHAR(ADD_MONTHS(A.REF_DATE,-6),'YYYYMM') AND TO_CHAR(ADD_MONTHS(A.REF_DATE,-1),'YYYYMM')
20231219	19/12/2023 11:21:44	17	3099.99...	9545000	I_aV_R	AVG(R_CNSM_CNSMT)	L12M	B.PERIODOMES BETWEEN TO_CHAR(ADD_MONTHS(A.REF_DATE,-12),'YYYYMM') AND TO_CHAR(ADD_MONTHS(A.REF_DATE,-1),'YYYYMM')
20231219	19/12/2023 12:13:24	18	2740.00...	7905000	I_aV_R	AVG(R_CNSM_CNSMT)	L13_24	B.PERIODOMES BETWEEN TO_CHAR(ADD_MONTHS(A.REF_DATE,-24),'YYYYMM') AND TO_CHAR(ADD_MONTHS(A.REF_DATE,-13),'YYYYMM')
20231219	19/12/2023 12:59:04	19	1324.99...	7905000	I_SDC_R	SUM(NVL(DC_R_CNSM_CNSMT,0))	L03M	B.PERIODOMES BETWEEN TO_CHAR(ADD_MONTHS(A.REF_DATE,-3),'YYYYMM') AND TO_CHAR(ADD_MONTHS(A.REF_DATE,-1),'YYYYMM')
20231220	20/12/2023 11:05:11	20	997.999...	8730000	I_SDC_R	SUM(NVL(DC_R_CNSM_CNSMT,0))	L06M	B.PERIODOMES BETWEEN TO_CHAR(ADD_MONTHS(A.REF_DATE,-6),'YYYYMM') AND TO_CHAR(ADD_MONTHS(A.REF_DATE,-1),'YYYYMM')
20231220	20/12/2023 11:21:49	21	1329.99...	9545000	I_SDC_R	SUM(NVL(DC_R_CNSM_CNSMT,0))	L12M	B.PERIODOMES BETWEEN TO_CHAR(ADD_MONTHS(A.REF_DATE,-12),'YYYYMM') AND TO_CHAR(ADD_MONTHS(A.REF_DATE,-1),'YYYYMM')
20231220	20/12/2023 11:43:59	22	1065.00...	8080000	I_SDC_R	SUM(NVL(DC_R_CNSM_CNSMT,0))	L13_24	B.PERIODOMES BETWEEN TO_CHAR(ADD_MONTHS(A.REF_DATE,-24),'YYYYMM') AND TO_CHAR(ADD_MONTHS(A.REF_DATE,-13),'YYYYMM')
20231220	20/12/2023 12:01:44	23	1034	8060000	I_SIN_R	SUM(NVL(IN_R_CNSM_CNSMT,0))	L13_24	B.PERIODOMES BETWEEN TO_CHAR(ADD_MONTHS(A.REF_DATE,-24),'YYYYMM') AND TO_CHAR(ADD_MONTHS(A.REF_DATE,-13),'YYYYMM')
20231220	20/12/2023 12:18:38	24	962.000...	7905000	I_SIN_R	SUM(NVL(IN_R_CNSM_CNSMT,0))	L03M	B.PERIODOMES BETWEEN TO_CHAR(ADD_MONTHS(A.REF_DATE,-3),'YYYYMM') AND TO_CHAR(ADD_MONTHS(A.REF_DATE,-1),'YYYYMM')
20231220	20/12/2023 12:35:00	25	1073.00...	8730000	I_SIN_R	SUM(NVL(IN_R_CNSM_CNSMT,0))	L06M	B.PERIODOMES BETWEEN TO_CHAR(ADD_MONTHS(A.REF_DATE,-6),'YYYYMM') AND TO_CHAR(ADD_MONTHS(A.REF_DATE,-1),'YYYYMM')
20231220	20/12/2023 12:52:53	26	1146.00...	9545000	I_SIN_R	SUM(NVL(IN_R_CNSM_CNSMT,0))	L12M	B.PERIODOMES BETWEEN TO_CHAR(ADD_MONTHS(A.REF_DATE,-12),'YYYYMM') AND TO_CHAR(ADD_MONTHS(A.REF_DATE,-1),'YYYYMM')

Figure 10-11. Contents of the CTRL_CURSOR control table, which controls the execution of the PL/SQL code in Listing 10-20.

While the PL/SQL option streamlines the process by eliminating one step, it also significantly increases the execution time. This is because each inner CURSOR execution iterates over the ABT not once, but ten times, which is the desired outcome. It is therefore probable that the two-step option will prove to be more time-efficient than the PL/SQL option. This is due to the fact that a CREATE syntax, when combined with a regular CURSOR, is less complex in terms of database operations than a PL/SQL program.

10.5 Continuous Variable Integration Cycle (CVIC)[11]

Now that the target population has been defined and the tools for safely adding information to the ABT have been mastered, a process for continuously integrating new variables into the ABT is possible. I have named this the continuous variable integration cycle or CVIC (reads as *"civic"*), for short.

10.5.1 The CVIC for Interval Variables

Figure 10-12 illustrates the conceptual diagram of the CVIC for interval variables at the account level. The process can be described by three steps. The initial step, designated as steps 1 and 1', pertains to the extraction process. The extraction process entails the retrieval of the target population from a historical data source, which may be either a transactional or snapshot-type table. It should be noted that during the extraction process, not only is the target population filtered from the data source table, but the analysis timeframe is also filtered.

---

[11] Two complete examples of the CVIC process, at account, and customer level, are provided at https://github.com/saulzapiain/ConceptualVariableDesign. Files: Bonus TBT and CVIC example at account level.sql and Bonus TBT and CVIC example at customer level.sql.

CHAPTER 10 THE ABT-BUILDING PROCESS

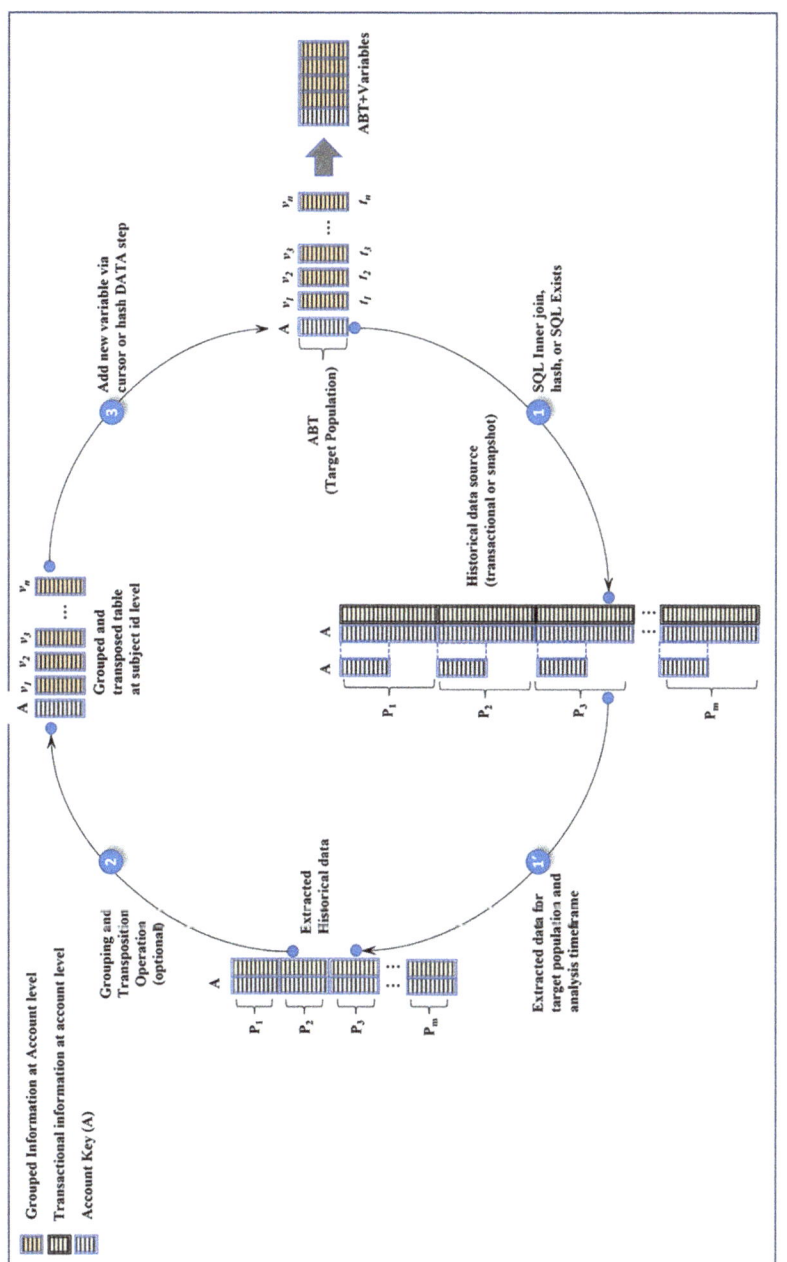

Figure 10-12. *Conceptual diagram of the continuous variable integration cycle for interval variables, where v refers to variables, t refers to time, and P refers to period.*

231

At this step, a single point in time for the extraction can be defined, in the case of variables that require backward extraction (e.g., two months back, three months back, etc.), or a time range for aggregation variables (e.g., the last n periods, etc.). It is important to note that, in any case, the final table must be at the account level (in this example) before it can be added to the ABT.

Once the requisite information has been extracted, the second step, the grouping and transposition operation, may be initiated. In this phase, the data is grouped according to the subject ID, and the variables are calculated using the conditional grouping functions described in Section 10.4.1.8. The resulting table should be at the account level and may contain one or more variables calculated at different periodicities, dependent on the characteristics of the initial input. It should be noted that the second step is optional. This is because, in the case of point-in-time variables, there is no need to summarize the information, given that it is already at the subject ID level.

The final step is the continuous integration of variables into the ABT. It should be noted that this process is repeated n times, with the option of adding multiple variables or one variable at a time, depending on the characteristics of the input data and the design of the variables. The final ABT should be comprised of n interval explanatory variables and should have a unique record for each subject ID in the target population.

10.5.2 The CVIC for Nominal Variables

The CVIC for nominal variables functions identically for interval variables, with the exception that the grouping operation step is excluded from the process for nominal variables.

10.5.3 Integrating Information from Different Financial Products

The aforementioned CVIC is predicated on the assumption that all information is available at the same level of analysis and that the subject ID is consistent across the target population and all input data sources. However, what are the implications when this is not the case?

To solve this problem, we need to find a common identifier that will allow us to cross-reference all the account numbers—that is, that we need the customer's number. The customer's number is what ties all the financial information together, and it can be used as a bridge that allows us to combine information with different keys in the same ABT.

10.5.3.1 Account-Customer Relationship

It is imperative to exercise caution when cross-referencing information utilizing the customer number. This is due to the fact that the relationship between accounts and customers is often complex and characterized by inconsistencies, both in the account and in the customer number. The use of synthetic keys can assist in the elimination of some inconsistencies; however, it is inevitable that some inconsistencies will persist in the data.

Table 10-7 illustrates a problematic case that would be exceedingly challenging to resolve. In this case, three distinct accounts, 1122334455, 1122334488, and 1122334477, are observed. In all three instances, the customer number has undergone a change at some point in the future. Upon closer examination, it becomes evident that the three accounts in question actually belong to the same customer. The customer's account history reveals that they initially opened an account in January 2016 with the number 556677 and subsequently changed it to 556678 in July of the same year. In 2018, the customer in question decided to open a new account in January 2019, subsequently modifying the customer number

CHAPTER 10 THE ABT-BUILDING PROCESS

in August 2020 to 556679. On top of that, the customer initiated another account in December 2020, modifying the customer number once more in September 2021 to 556680.

Table 10-7. *Illustrative example of a problematic case where a single customer changed their number within different accounts at different points in time.*

Account num.	Customer num.	Start_Period	End_Period
1122334455	556677	1/1/2016	6/30/2018
1122334455	556678	7/1/2018	12/31/4747
1122334488	556678	1/1/2019	9/30/2020
1122334488	556679	10/1/2020	12/31/4747
1122334477	556679	12/1/2020	8/31/2021
1122334477	556680	9/1/2021	12/31/4747

The issue at hand is that the customer number has been altered on multiple occasions across various accounts, rendering it impossible to create a single key based solely on the consistency of the account number. One might consider the following scenario: if a single synthetic key were to be created for each account containing more than one customer number, the result would be as illustrated in **Table 10-8**.

Table 10-8. *Example of the relationship between the Account_Rk and the Customer_Rk and its effective period.*

Account num.	Customer_Rk	Start_Period	End_Period
1122334455	1	1/1/2016	12/31/4747
1122334488	2	1/1/2019	12/31/4747
1122334477	3	12/1/2020	12/31/4747

It is evident that not only has a new key been generated, but an SCD2 has also been applied to the table, thereby extending the effective period for each account's account-customer relationship. Nevertheless, it is regrettable that the three accounts cannot be associated with a single customer number. Furthermore, **Table 10-7** still indicates that the three accounts belong to a single customer. However, the replacement of the customer number with Customer_Rk prevents the identification of the three accounts as belonging to the same customer.

The resolution of this issue will necessitate the establishment of a sequential relationship between the ostensibly adjacent customer numbers, continuing until no further alterations are identified. This is by no means a trivial problem. However, it would be best left unresolved and the focus shifted to historical inconsistencies from debt restructuring. This recommendation is based on the fact that the issue presented in **Table 10-7** is relatively rare and may represent only a small fraction of all cases.

10.5.4 The CVIC via the Account-Customer Bridge

It has been demonstrated that interval and nominal variables can be added to an ABT via the CVIC in an iterative manner. Furthermore, it has been shown that variables from disparate products can be incorporated by cross-referencing through the customer number. This naturally gives rise to the question of how variables from different products can be added in an iterative manner.

Figure 10-13 illustrates an alternative CVIC diagram that considers the use of products other than the one currently under consideration. It should be noted that the sequence of steps remains identical, with the exception of the fact that in this diagram, all variables are expressed in terms of the customer number, rather than the account number. The objective is to initially generate a new input derived from the original data source, which will be grouped at the customer level. This constitutes steps 0 and 0'.

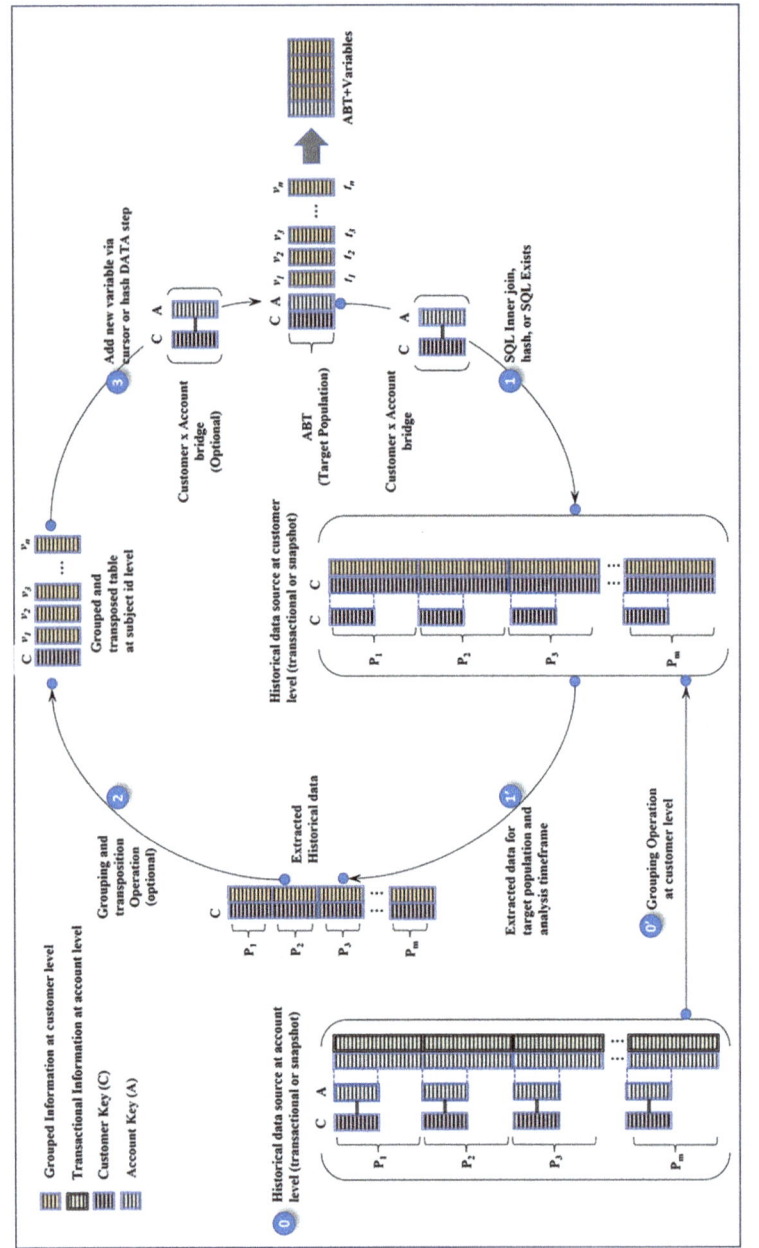

Figure 10-13. Conceptual diagram of the continuous variable integration cycle of different products over a customer-account bridge.

CHAPTER 10 THE ABT-BUILDING PROCESS

To illustrate, consider the case of a personal loan model. In this scenario, the target population and subject ID will be defined at the loan account level. Now, consider the scenario in which the objective is to incorporate the customer's credit card data. However, given that a single customer may possess multiple credit card accounts, it is necessary to initially aggregate the information at the customer level. The underlying assumption is that all of the elements that comprise the credit card statement can be aggregated to create a single credit line. Consequently, the credit line's utilization can be defined as follows:

$$\frac{\sum_{i=1}^{n} Balance_i}{\sum_{i=1}^{n} Credit\ Limit_i}$$

Equation 10-1. Credit line utilization.

Likewise, we can also define the following ratios in terms of a single credit line:

$$\frac{\sum_{i=1}^{n} Payments_i}{\sum_{i=1}^{n} Balance_i} \qquad \frac{\sum_{i=1}^{n} Payments_i}{\sum_{i=1}^{n} Purchases_i} \qquad \frac{\sum_{i=1}^{n} Purchases_i}{\sum_{i=1}^{n} Balance_i}$$

As a result, the previous historical data source at the account level is replaced by the historical data source at the customer level (middle and lower part of the diagram). From here, the process continues in the same way, except that now the extraction process takes place at the customer level instead of the account level.

Now that we have extracted the historical data at the customer level (center-left side of the diagram), we can proceed with a second grouping operation to transpose our information using conditional grouping functions (Section 10.4.1.8). The resulting table should be at the customer level and may contain one or more variables calculated at different periodicities, depending on the characteristics of your initial input.

In step 3, we will continuously add information to the ABT via Oracle's cursor or hash DATA step. Notice that although all steps within the cycle are at the customer level, the ABT remains at the loan account level (in this example). It is important to mention that using the customer-account bridge is optional, depending on whether your ABT already has the customer key or not—but what happens if a single account belongs to more than one customer number?

As previously observed, a customer's number may undergo alteration for a number of reasons. In theory, this should not present an issue if the customer's number has previously been homologated via synthetic keys. Consequently, there is no risk of duplicates occurring when the accounts are cross-referenced with their corresponding customer numbers.

However, there are cases, such as the one illustrated in **Table 10-7**, in which this does not apply. In certain instances, a single account may be associated with two distinct customer IDs. In such instances, the account-customer bridge may also be used to specify the effective date of each account-customer relationship. It must be acknowledged that this will not address the issue of historical consistency. Nevertheless, this will permit the joining operation to be performed for the specified reference date.

10.6 Event Definition

The target population is representative of the sample and was extracted using complex rules. Synthetic keys have been introduced to overcome historical inconsistencies caused by debt restructurings. Furthermore, interval and nominal variables have been added from all available data sources, and the account number of the target population has been cross-referenced with the corresponding customer number. However, we still need to define the event that we want to predict.

The definition of the event to be predicted is a challenging undertaking, potentially requiring as much or more effort than the work already completed. It is imperative, however, to differentiate between the conceptual interpretation of the event and the measurement of the event itself.

Conceptually, we saw in Section 2.1 that the event that we actually want to predict is

> *The likelihood that a borrower will fulfill their financial obligations based on the fortitude of their character or the lack thereof.*

But from a practical point of view, we need to be able to define what it means not *to fulfill one's financial obligations*—also called *default*.

One widely accepted method for defining the default event is by the number of days past due. The days past due is the number of days that have elapsed since the last payment date on a loan. But the important thing to note in this definition is not the total number of days past due on an account, but the number of days after which an account becomes *uncollectible*.

In order to comprehend the concept of uncollectible, it is first necessary to recognize that the mere fact of an account being a certain number of days past due does not inherently signify that the account is uncollectible or that it cannot be brought back into compliance with its payment schedule. Conversely, an account is deemed uncollectible when it reaches a *point of no return*. The point of no return is defined as the elapsed time in days past due at which the probability of repayment is significantly reduced.

The point of no return is determined through the *roll rate analysis* (for further details, please refer to *Intelligent Credit Scoring* by Naeem Siddiqi [63]). The roll rate analysis is based on a two-year observation period, during which the accounts in the portfolio are ranked according to the number of days past due reported during the initial 12-month analysis

CHAPTER 10   THE ABT-BUILDING PROCESS

window.[12] The analysis entails a comparison of the change in ranking with regard to the number of days past due from the initial 12-month period to the subsequent 12-month interval of the analysis window. To illustrate, if the maximum number of days past due from month 1 to month 12 is employed, the following ranking can be utilized:

- Current
- 1 to 30 days
- 31 to 60 days
- 61 to 90 days
- 91 to 120 days
- 121 to 150 days
- 150 to 180 days

Table 10-9 shows a customer's genuine roll rate analysis. In the initial case, 70% of accounts remained current in the subsequent year. The remaining 30% were divided between two delinquency categories: 30 to 59 days and 60 days or greater. The recovery rate declines as the delinquency rate rises in the second year. As the maximum number of days past due shifts from 30-59 to 60-89 and then 90-119 days, the recovery rate declines from 28%, to 18%, to 13%.

---

[12] It is important to highlight the fact that all accounts included in the analysis most be up-to-date in their payments, with zero days past due, and they should also be present along the observation timeframe of 24 months.

CHAPTER 10 THE ABT-BUILDING PROCESS

Table 10-9. *Example of the dynamics of the roll rate analysis. (A) Example of the expected distribution of the change in status if the account was reported current in the first 12 months of observation. (B) Example of the expected distribution of the change in status if the account was reported between 31 and 60 days past due in the first 12 months of observation. (C) Example of the expected distribution of the change in status if the account was reported between 90 and 120 days past due in the first 12 months of observation.*

First 12 months	Next 12 Months	Recovery Rate
01 Current	**01 Current**	**68.79%**
01 Current	02 From 30 to 59 Days	15.06%
01 Current	03 From 60 to 89 Days	7.66%
01 Current	04 From 90 to 119 Days	3.49%
01 Current	05 From 120 to 149 Days	1.94%
01 Current	06 From 150 to 179 Days	1.38%
01 Current	180 or More Days	1.67%
02 From 30 to 59 Days	**01 Current**	**28.36%**
02 From 30 to 59 Days	02 From 30 to 59 Days	25.35%
02 From 30 to 59 Days	03 From 60 to 89 Days	23.44%
02 From 30 to 59 Days	04 From 90 to 119 Days	9.72%
02 From 30 to 59 Days	05 From 120 to 149 Days	4.54%
02 From 30 to 59 Days	06 From 150 to 179 Days	2.84%
02 From 30 to 59 Days	180 or More Days	5.75%
03 From 60 to 89 Days	**01 Current**	**17.85%**
03 From 60 to 89 Days	02 From 30 to 59 Days	17.94%
03 From 60 to 89 Days	03 From 60 to 89 Days	21.55%
03 From 60 to 89 Days	04 From 90 to 119 Days	17.57%
03 From 60 to 89 Days	05 From 120 to 149 Days	8.26%
03 From 60 to 89 Days	06 From 150 to 179 Days	5.10%
03 From 60 to 89 Days	180 or More Days	11.74%
04 From 90 to 119 Days	**01 Current**	**13.22%**
04 From 90 to 119 Days	02 From 30 to 59 Days	10.82%
04 From 90 to 119 Days	03 From 60 to 89 Days	14.88%
04 From 90 to 119 Days	04 From 90 to 119 Days	15.87%
04 From 90 to 119 Days	05 From 120 to 149 Days	13.82%
04 From 90 to 119 Days	06 From 150 to 179 Days	9.11%
04 From 90 to 119 Days	180 or More Days	22.28%
05 From 120 to 149 Days	**01 Current**	**12.17%**
05 From 120 to 149 Days	02 From 30 to 59 Days	8.67%
05 From 120 to 149 Days	03 From 60 to 89 Days	9.25%
05 From 120 to 149 Days	04 From 90 to 119 Days	6.52%
05 From 120 to 149 Days	05 From 120 to 149 Days	7.79%
05 From 120 to 149 Days	06 From 150 to 179 Days	13.53%
05 From 120 to 149 Days	180 or More Days	42.06%

CHAPTER 10   THE ABT-BUILDING PROCESS

It is important to understand the difference between *"recovery"* and *"improvement."* With accounts 60-89 days past due, four potential outcomes are possible in the first year: recovery, improvement, no change, or deterioration.

Having distinguished between recovery and improvement, we may now examine the change in the recovery rate from bucket 4 (90 to 119 days past due) to bucket 5 (120 to 149 days past due). This change is approximately 1% (13.2%-12.2%), which differs from the change in the recovery rate from the previous bucket, 3 (60 to 89 days past due), which is 4.6% (17.8%-13.2%).

Figure 10-14 presents a bar chart illustrating the change in recovery rates between delinquency buckets in ascending order. It is notable that the recovery rate declines exponentially and attains its lowest point between bucket 4 (90 to 119 days past due) and bucket 5 (120 to 149 days past due). In light of these findings, it can be concluded that a delinquency of 90 days or more represents a suitable default definition, given that no notable decline in the recovery rate is likely to be observed upon transitioning to subsequent delinquency buckets.

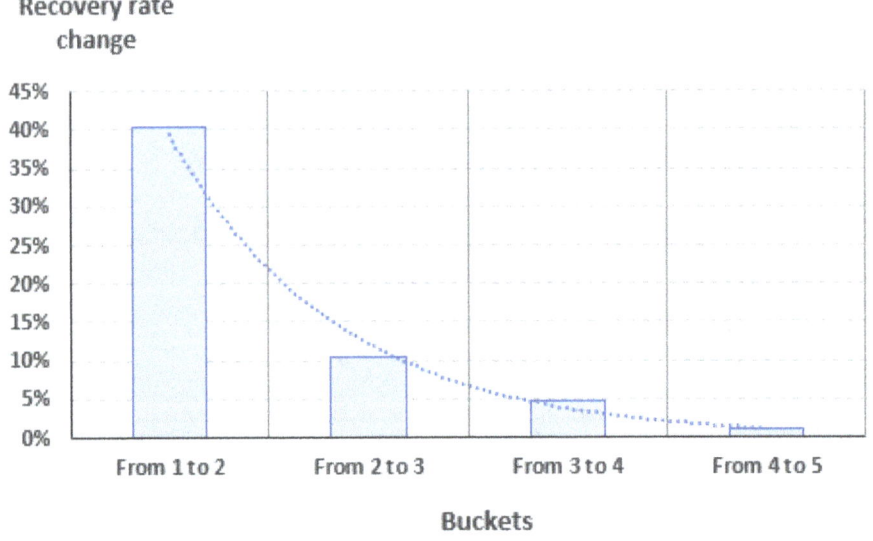

Figure 10-14. *Bar chart showing the exponential decay of the change in the recovery rate across incremental delinquency buckets.*

CHAPTER 10 THE ABT-BUILDING PROCESS

10.6.1 Other Considerations

While the default event is typically defined by the maximum number of days past due within a 12-month period, exceptions to this rule do exist. This is due to the fact that on occasion the default event is a combination of multiple and complex business rules that may necessitate cross-referencing of multiple data sources. It is therefore important to be aware that additional data processing may be necessary in order to properly define the default event.

10.6.2 Target Population and Event Definition

Once a formal definition of the default event has been established, it will be possible to proceed with the flagging of its occurrence within a 12-month timeframe from a specific reference date. Nevertheless, prior to initiating the flagging process, it is essential to establish a set of objective criteria for evaluating the default event in a fair and impartial manner. The following rules are proposed:

1. The accounts to be considered for measuring the event of default must have been active and up-to-date on the initial date of the event window.

2. The accounts must be present in all periods of the event window.

To illustrate, if we take December 2021 as a reference point, the accounts that are to be considered are those that are zero days past due and still active on that date. In the event that the event window is defined as a 12-month period, the selected accounts must be present in all periods from December 2021 to December 2022.

Figure 10-15 provides an illustrative example of state transitions in terms of days past due. It should be noted that at the outset, the entire target population is considered to be current, with no days past due. It is also noteworthy that at the same point in time, a percentage of the accounts have already reached different levels of days past due. It is evident that to obtain an impartial assessment of delinquency, it is essential that all accounts commence with identical conditions. This is because the objective is not to measure the default event in and of itself, but rather to ascertain the occurrence of the event within a specified timeframe. In light of this reasoning, it is evident that as the composition of the target population shifts at the end of the observation period, the true default occurrence rate is revealed. This rate is calculated based on a population that was current at t_0 and spans a 12-month window.

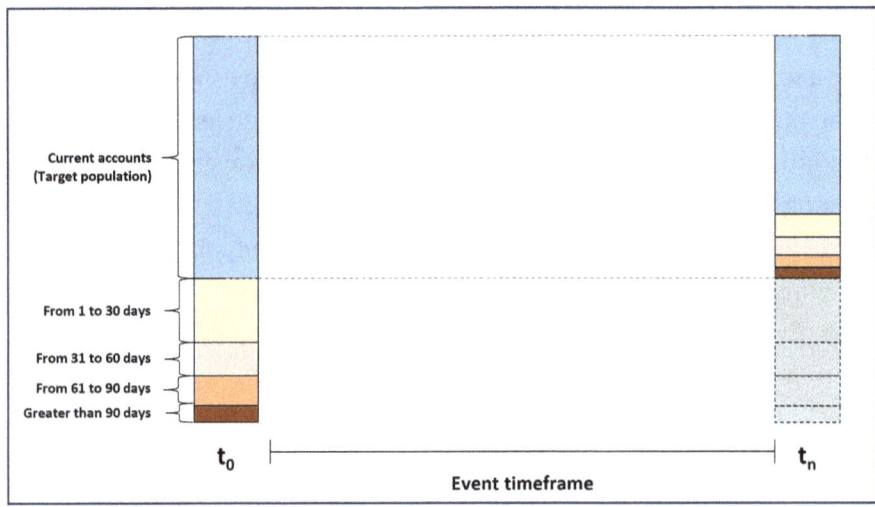

Figure 10-15. Conceptual example of the dynamics of expected transience, in terms of delinquency days, over time.

Figure 10-16 illustrates the phenomenon of information truncation. From this figure, it is important to note that in case 1, the account is present throughout the entirety of the event window, from its inception to its conclusion. Case 2 illustrates an account that was present during the initial period but had its information truncated at some point in its historical record. Case 3 illustrates an exceptional, though plausible, scenario in which an account is present during both the initial and final

CHAPTER 10 THE ABT-BUILDING PROCESS

periods of the analysis, yet its presence is not continuous. Instead, there are intermediate instances where the account's information has been truncated.

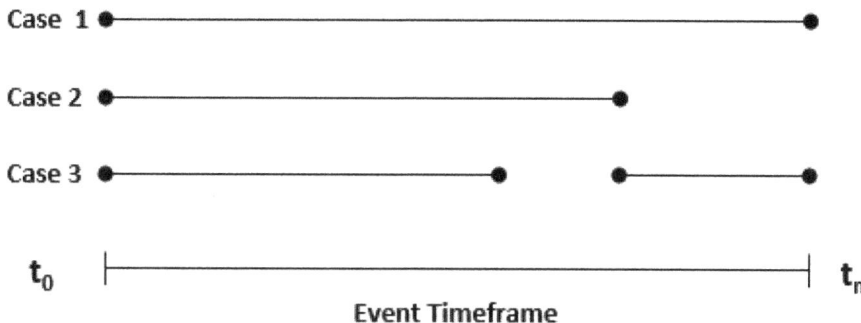

Figure 10-16. *Illustrative cases of information truncation over time.*

Of the three cases presented, only case 1 is deemed to be valid, as it is the sole instance in which the account was present throughout the entirety of the event occurrence window. Consequently, it is the sole case that permits an objective assessment of the default event. It is then imperative, to consider two further rules when selecting the target population, in addition to the elements previously mentioned in Chapter 9. That is, in order for an account to be considered, it must have been active and up to date on the initial date of the event window (reference date), and it must remain present throughout all periods of the event window.

10.7 Stacked ABT

The reference date represents the point of origin for all subsequent calculations. When defining the target as the maximum number of days past due in the next 12 months, it is understood that the subsequent 12 months are to be calculated from the reference date to 12 months later. Similarly, when the average credit card utilization over the past three months is referenced, the three-month period preceding the reference date is being referred to. Therefore, the term *"relative periods"* is used, as all periods are relative to the reference date.

The use of relative periods permits the generalization of a date-specific (DS) chronological sequence, such as May-2020, Jun-2020, Jul-2020, and so forth, to a non-date-specific (NDS) chronological sequence, such as $t+1$, $t+2$, $t+3$, and so forth. It should be noted that in a non-date-specific (NDS) sequence, different points in time can be represented by adding and subtracting periods from the reference time t.

Figure 10-17 illustrates a conceptual representation of a non-date-specific timeline. It should be noted that an NDS timeline functions in a manner analogous to a ruler, enabling straightforward and intuitive navigation from a reference point in time to past and future events. This also facilitates the automation of the ABT construction process, as it allows the process to depend on a single parameter, namely, the reference date. However, the greatest advantage of employing an NDS timeline is the ability to *stack* the ABT.

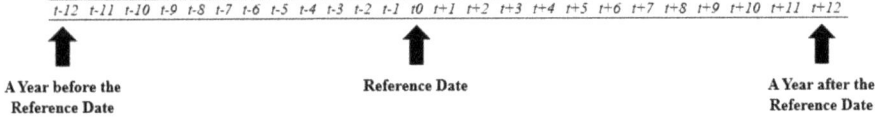

Figure 10-17. Conceptual representation of a non-date-specific timeline.

Introducing variables referring to a relative point in time allows for generalization of the subject's behavior. The last three months of 2020 (t-3, t-2, and t-1) are used, regardless of the actual months. Using a single reference date, such as August 2020 ($t0$), makes relative periods useless, because from August onward, the last three months will always be May, June, and July of 2020.

It is important to note that the objective is to develop a model that is capable of predicting the default event at any given point in time, irrespective of the current date of the year. Accordingly, the model should be trained with a representative sample of all months of the year (a full cycle), which entails having at least 12 reference dates. Thus, when we refer to *"12 reference dates,"* we are not referring to 12 separate ABTs, but rather one ABT with 12 consecutive reference dates, ensuring that our ABT remains a unit.

Figure 10-18 illustrates a conceptual representation of an ABT comprising 12 stacked periods, extending from January 2021 to December 2021. It should be noted that the *t-i* and *t+i* periods are now relative periods, as time *t* is not fixed to a single reference date. In reference to the aforementioned credit card utilization example, it can be stated that the average utilization rate of the last three months is not dependent on any specific month of the year.

The primary objective of the stacked ABT is to utilize time-based variables in a manner that does not compromise the independence of the model from a specific date. Consequently, the incorporation of relative periods for explanatory variables will negate the influence of seasonal variations and cyclical patterns in different months. Therefore, the objective is to guarantee that the model is independent of the calendar month and consistent at any given date. For example, in Section 5.2, we assumed that a credit card utilization level of 65% three months back to the reference date will be significantly correlated with the default event. However, in this example, it is assumed that this variable will be predictive regardless of the specific date on which the 65% utilization occurred.

CHAPTER 10   THE ABT-BUILDING PROCESS

Figure 10-18. Conceptual representation of a stacked ABT with each reference date aligned on top of each other.

CHAPTER 10   THE ABT-BUILDING PROCESS

10.7.1 The Average Period *n*

An understanding of this example is essential for grasping the concept of relative dates and stacked ABTs. This is because the two concepts are inextricably linked; so on the one hand, it is not possible to have relative dates without stacking your ABT, and on the other, relative dates can only make sense in a stacked ABT. However, the fundamental idea behind both concepts is the use of the average period, n. This is because stacking your ABT will average the behavior of all periods in your relative timeline. It is therefore recommended that at least 12 consecutive months be stacked in order to obtain a representative one-year sample.

10.7.2 Handling the Stacked ABT Timeline

Exercise caution when dealing with stacked ABTs, as each stack period adds one month to the total time required to build the ABT. **Figure 10-19** shows a stacked ABT aligned with the corresponding date. This makes the total time interval visible, demonstrating that to construct an ABT comprising 12 months of behavioral data, 12 months to monitor the occurrence of the default event, and another 12 months to allow for the generalization of the average period n, will result in a time interval that goes from January 2020 to December 2022, equating to three years of continuous data.

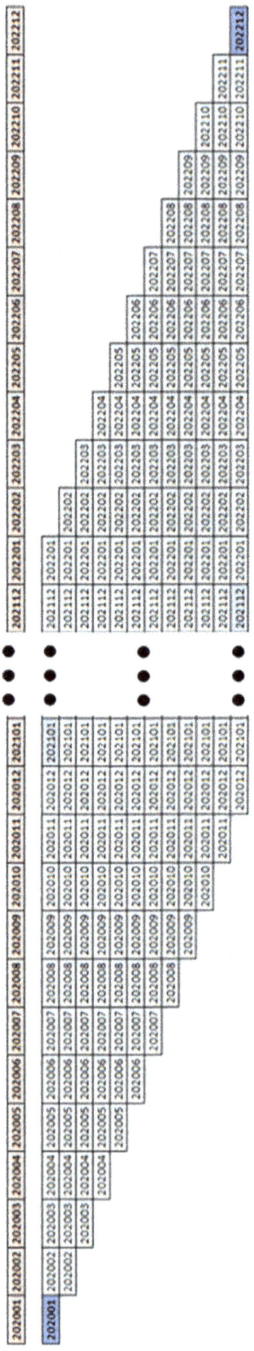

CHAPTER 10   THE ABT-BUILDING PROCESS

Figure 10-19. *Conceptual representation of a stacked ABT, with each reference date aligned with its corresponding chronological date.*

The issue is not the necessity of three years of data to construct a single scorecard model. Rather, the challenge lies in the potential unavailability of information from three consecutive years. To illustrate, consider the following situations:

- Incomplete information, whether at the beginning, in the middle, or at the end of the timeline

- Known inconsistencies, also at the beginning, in the middle, or at the end of the timeline, due to internal or external factors

To exemplify, consider a scenario where a technological shift, such as a transition from one DB system to another, has rendered the disparate data sources essential for ABT construction incompatible. This incompatibility may arise from the accumulation of additional data or the loss of certain variables, resulting in a historical gap. It is also important to consider the possibility of data loss due to negligence. For example, if the platform crashes for a couple of days, there is a risk of losing transactional information.

Inconsistencies are distinct from a lack of information and signify a substantial shift in conduct due to the influence of pivotal and extraordinary circumstances. However, it is essential to distinguish between external and internal factors. Exceptional external factors include events such as wars, pandemics, economic crises, significant shifts in public policy, or changes resulting from the advent of a new government. In contrast, internal factors pertain to modifications within the lender's financial institution, including those related to its collection, debt restructuring, and application policies, among others.

For example, consider the following scenario: today is January 2023, and the objective is to construct a scorecard model using three years of information. However, it is also necessary to exclude certain periods from the training process in order to evaluate the performance of the model with real data before implementing it in a production environment. Therefore, a timeframe between January 2019 and December 2021 is

CHAPTER 10   THE ABT-BUILDING PROCESS

selected, which allows for the scoring of 12 months and the testing of the actual performance of the model on a single, uninfluenced month,[13] namely, January 2022. However, a challenge arises when one recognizes that the global economic disruption caused by the pandemic occurred precisely between March 2020 and March 2021, which falls within the model's designated timeframe. How can this be addressed?

It is not possible to advance the timeframe of our model to March 2021, as the present date is January 2023. This leaves only one year and nine months for training the model, with no time for testing. Conversely, we would be required to advance our training timeframe to March 2018 in order to encompass the requisite three years for our stacked ABT (with no behavioral variables beyond the previous 12 months available). Alternatively, we could simply extend the building timeframe to March 2018; however, this would prompt the question of whether the 2018 data would still be representative of the actual population, given that it is already four years in the past. Nevertheless, there is another option.

A third option would be to utilize the entirety of the dataset, spanning from January 2017 to December 2021, without excluding the information in the middle, in order to stack three years of information, resulting in one year of behavioral information, three years of stack information, and one year of outcome information (five years in total). The rationale for this decision is that although the information between March 2020 and March 2021 will deviate significantly from the information of the other years, this is compensated by the addition of two further years to the stacked ABT. This option becomes even more compelling when we consider that the periodic information in each one of our variables does not belong to a specific date in time, but rather to the average period n. Therefore, it is reasonable to

---

[13] Remember that you can only test your model's actual performance after 12 months have passed by. This is because your model was trained to predict the occurrence of the default event in the next twelve months from the reference date. Therefore, if your reference date is January 2021, then your occurrence window will be from February 2021 to January 2022.

assume that addition of stacked years to our ABT will reduce the impact of the pandemic, while still allowing us to detect behavioral patterns that will enable us to predict the outcome variable. Moreover, if the resulting model demonstrates not only optimal performance during the training phase but also consistent performance during the five-year backtesting period, then it may be inferred that the model is resilient to the effects of the pandemic.

10.8 ABT-Building Recommendations

At this point, you should be able to build your own ABT from start to finish, but there are still some recommendations that you may find very useful.

10.8.1 Build Your Own Data Mart

Thus far, the primary components and operations necessary for the construction of your ABT have been presented as a one-time activity. However, it is important to note that these processes are designed to continue feeding your model once it has been completed and is ready for deployment.

It is important to note that at this stage of the modeling process, the ultimate goal of the model may not be apparent. However, the model is designed to score new data on a recurring basis, which is known as *the scoring process* (for further details, please refer to Chapter 19). The scoring process is comprised of a series of activities that facilitate the construction of a *scoring ABT* and enable the regular scoring of data. Technically speaking, the training and scoring ABTs differ in only two respects: the number of variables to be constructed and the number of records to be included, that is to say, the number of columns and rows. This is due to the fact that variables utilized in scoring ABTs are solely those that were incorporated into the final model. Consequently, scoring ABTs are inherently a subset of the total number of variables employed during the training process. However, at this preliminary stage of the modeling process, it is not yet clear which variables will ultimately be included in the final model.

10.8.1.1 Proactive or Reactive Approach

This issue can be addressed in two ways: proactively or reactively. To pursue a forward-looking approach, design a data mart with the necessary tables and layout. This can serve as input for the 3,132 variables shown in our previous examples. The data mart must be developed and the recursive ETLs designed, constructed, and tested. A backward-looking approach means waiting until the final model is done before looking at the minimum number of tables and columns needed to feed it. This also means building the corresponding data marts and ETLs.

It is this author's recommendation that the forward-looking approach be adopted, as the alternative offers no long-run benefit. I will now elaborate further. It is ill-advised to wait until the final model is finished. This is because there is the theoretical possibility of working in parallel, continuing to develop the model while simultaneously developing the data mart and its corresponding recurring ETLs.

Furthermore, from the perspective of the programmer, there is no distinction between developing a process for loading a table with 100 columns and one billion records and a process for loading a table with 10 columns and 1,000 records. The difficulty in programming the process does not lie in the volume of information, but rather in the complexity of the extraction rules and business logic, as well as the grouping operations and the computation of the base metrics that give way to the actual explanatory variables that feed the model.

Also consider that the construction of a data mart capable of supporting the development of an unlimited number of variables represents a significant investment that will enable the creation of not just one, but an unlimited number of models in the medium and long term. The 3,132 variables that we have discussed are not exclusive to a single type of credit product. They can be used in PD models for credit cards, retail loans, auto loans, mortgages, and even for SME loans. This is because they all measure the same underlying concept.

It should be noted that the continuous variable integration cycle (CVIC) already incorporates cross-referencing of data from products with disparate account numbers through the account-customer bridge. Consequently, if a master account snapshot table is available, it is theoretically possible to construct ABTs for different products by moving across different target populations, that is, different records, and applying different business rules.

Moreover, the 3,132 hypothetical variables are not exclusive to PD models either; they may also be utilized in a multitude of banking-related models, including, but not limited to, those pertaining to collections, LGD, fraud, and propensity, just to name a few.

In conclusion, if one has already initiated the construction of a model, it is recommended to create an *ad hoc* data mart that will serve as the primary input for the current model-building project and as the *foundation* for the construction of numerous subsequent models. It would therefore be prudent to consider referring to the data mart in question as the *"foundation mart."*

10.8.1.2 Recommended Set of Tables for Your Foundation Mart

Table 10-10 presents a selected list of tables pertinent to the foundation mart. The list starts with six snapshot tables that contain periodic balance information for the products of credit card, retail loans, and current or checking accounts. It should be noted that two versions of each product are available: one at the account level, where no grouping operations have been performed, and one at the customer level, where all account information has been summarized at the customer level. This aggregation includes all balances, payments, charges, and other relevant data, resulting in a hypothetical customer-level account. It is recommended that a counter of the total number of distinct accounts be added to the customer-level tables in order to facilitate the tracking of multiaccount customers.

CHAPTER 10 THE ABT-BUILDING PROCESS

Table 10-10. *Selected list of tables for your foundation mart.*

No.	Table	Type	Periodicity	Level	Sublevel	Category	Content
1	FM_SNP_STMT_CRD_ACC	Snapshot	Monthly	Account		Credit Card	Balance Statements
2	FM_SNP_STMT_LON_ACC	Snapshot	Monthly	Account		Retail Loan	Balance Statements
3	FM_SNP_STMT_CUR_ACC	Snapshot	Monthly	Account		Current/Checking	Balance Statements
4	FM_SNP_STMT_CRD_CUS	Snapshot	Monthly	Customer		Credit Card	Balance Statements
5	FM_SNP_STMT_LON_CUS	Snapshot	Monthly	Customer		Retail Loan	Balance Statements
6	FM_SNP_STMT_CUR_CUS	Snapshot	Monthly	Customer		Current/Checking	Balance Statements
7	FM_SNP_MET_STMT_CRD_ACC	Snapshot	Monthly	Account		Credit Card	Derived Metrics
8	FM_SNP_MET_STMT_LON_ACC	Snapshot	Monthly	Account		Retail Loan	Derived Metrics
9	FM_SNP_MET_STMT_CRD_CUS	Snapshot	Monthly	Customer		Credit Card	Derived Metrics
10	FM_SNP_MET_STMT_LON_CUS	Snapshot	Monthly	Customer		Retail Loan	Derived Metrics
11	FM_SNP_MET_STMT_CUR_CUS	Snapshot	Monthly	Customer		Current/Checking	Derived Metrics
12	FM_SNP_MET_COLL_CRD_ACC	Snapshot	Monthly	Account	Collection	Credit Card	Derived Metrics
13	FM_SNP_MET_COLL_CRD_CUS	Snapshot	Monthly	Customer	Collection	Credit Card	Derived Metrics
14	FM_TRX_BHV_USE_CRD_ACC	Transactional	Monthly	Account	Expense type	Credit Card	Derived Metrics
15	FM_TRX_BHV_USE_CRD_CUS	Transactional	Monthly	Customer	Expense type	Credit Card	Derived Metrics
16	FM_TRX_BHV_USE_CUR_ACC	Transactional	Monthly	Account	Expense type	Current/Checking	Derived Metrics
17	FM_TRX_BHV_USE_CUR_CUS	Transactional	Monthly	Customer	Expense type	Current/Checking	Derived Metrics
18	FM_DIM_CUSTOMER	Dimension	N/A	Customer		Customer	Characteristics
19	FM_DIM_CREDITCARD	Dimension	N/A	Card		Credit Card	Characteristics
20	FM_DIM_ACCOUNT_CRD	Dimension	N/A	Account		Credit Card	Characteristics
21	FM_BRD_ACCOUNT_ID_X_ACCOUNT_RK	Bridge	N/A	Account_ID-Account_Rk		Synthetic Keys	Relationship
22	FM_BRD_CUSTOMER_ID_X_CUSTOMER_RK	Bridge	N/A	Customer_ID-Customer_Rk		Synthetic Keys	Relationship
23	FM_BRD_CUSTOMER_X_ACCOUNT_CRD	Bridge	N/A	Customer-Account		Credit Card	Relationship
24	FM_BRD_CUSTOMER_X_ACCOUNT_LON	Bridge	N/A	Customer-Account		Retail Loan	Relationship
25	FM_BRD_CUSTOMER_X_ACCOUNT_CUR	Bridge	N/A	Customer-Account		Current/Checking	Relationship
26	FM_BRD_CARD_X_ACCOUNT_CRD	Bridge	N/A	Card-Account		Card	Relationship

CHAPTER 10 THE ABT-BUILDING PROCESS

The next block (tables 7 to 13) corresponds to the tables containing all the metrics that will give way to the final explanatory variables, such as ratios, variations, increment and decrement flags, and of the sort.

The third block (tables 14 to 17) corresponds to the transaction base tables (TBTs) generated from the corresponding transactional tables. These TBTs are grouped at the month, account, and expense type level, as well as at the month, customer, and expense type level, respectively. It should be noted that transactions may be recorded from credit card operations, as well as from current and checking accounts, among others.

The fourth block (tables 18 to 20) comprises dimension tables that store the attributes pertaining to customers, the physical credit card (plastic), and credit card accounts. It should be noted that the aforementioned dimension tables can be created for any financial product other than the credit card.

The fifth and final block (tables 21 to 26) comprises a series of bridge tables that facilitate cross-referencing of customer and account levels across different products. It is noteworthy that tables 21 and 22 serve as bridges, storing the relationship between the natural business keys (Account ID and Customer ID) and the synthetic keys (Account_Rk and Customer_Rk) for the account and customer numbers, respectively.

It is crucial to highlight that the enumeration of tables presented in **Table 10-10** is merely illustrative and does not represent the sole configuration and/or set of tables within your foundation mart.

10.8.2 Represent Your ABT-Building Process As a Diagram

Rather than conceptualizing the ABT-building process as a series of lines of code, it may be advantageous to view it as a production line. Representing the building process as a *production line* is a convenient approach that allows the code to be depicted as a box-and-line diagram. This representation facilitates the identification of critical points in the process and provides a more comprehensive understanding of the process itself.

CHAPTER 10   THE ABT-BUILDING PROCESS

Figure 10-20 illustrates an example of a data processing diagram created in a spreadsheet. The following features of this diagram are worthy of note:

1. Each box represents a table in your process and displays the more relevant information of your table, that is, the total number of records and the total amount, if available.

2. Tracks the number of records remaining from step to step.

3. Gives you a brief description of the operation that takes place between tables.

4. It shows you the sequence of steps required within each particular branch in a logical order.

5. Branches represent different processes and are identified by Roman numbers.

6. Tables have a logical and systematic nomenclature.

CHAPTER 10   THE ABT-BUILDING PROCESS

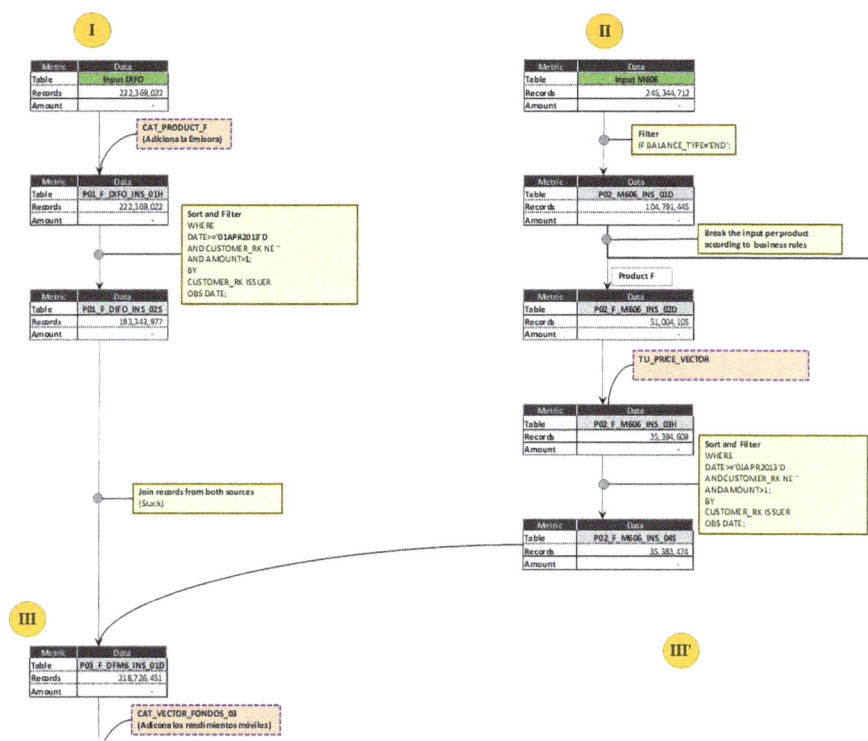

Figure 10-20. Extract from a data processing diagram.

It should be noted that the nomenclature of each table conveys a greater degree of information than is initially apparent. To illustrate, **Figure 10-21** shows the meaning of each one of the elements that comprise a single table name, where

- **P01:** Represents the process number
- **F:** Represents the product's code
- **DIFO:** Represents the specific code of the input
- **INS:** Represents the product's type

259

- **01:** Represents the step number of the corresponding process
- **H:** Represents the type of operation that gave rise to the current table

where

- H represents a hash DATA step
- S represents a sorting operation
- D represent a simple DATA step
- Q represents a SQL query operation (not shown in the previous diagram)

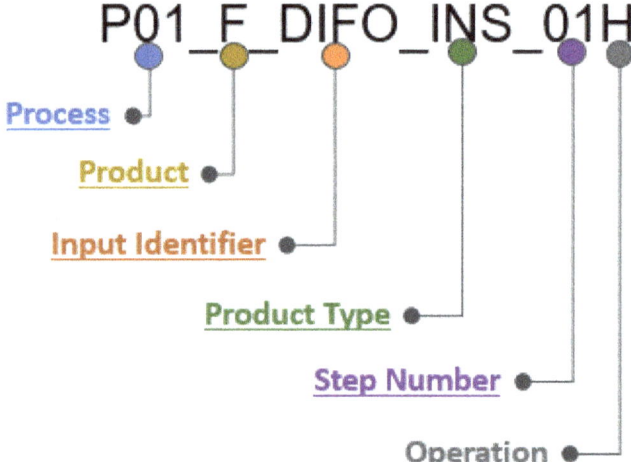

Figure 10-21. Nomenclature of intermediate tables.

The simplicity of this nomenclature makes each table's name easy to understand and helps grasp the overall process. To illustrate, the designation *"P02_F_M606_INS_03H"* indicates that the table is the result of step 3 in process 2, contains product F, type INS, was sourced from M606, and was generated through a hash DATA step.

The ABT-building process diagram can be constructed in a manner that reflects the specific intricacies of the process in question. **Figure 10-22** presents a magnified representation of the elements depicted in **Figure 10-20**. This zoomed-out view depicts over 50 temporary tables (please notice that the fact that the texts in the illustration are illegible is irrelevant for this example), exemplifying the extensive utilization of data processing. This represents merely a portion of the complete diagram. Furthermore, a schematic representation of the process can facilitate the organization of the overall process, the identification of dependencies between processes, and the visualization of inputs and outputs. All of which correspond to a well-known concept, namely, the critical path analysis of the process in question.

CHAPTER 10 THE ABT-BUILDING PROCESS

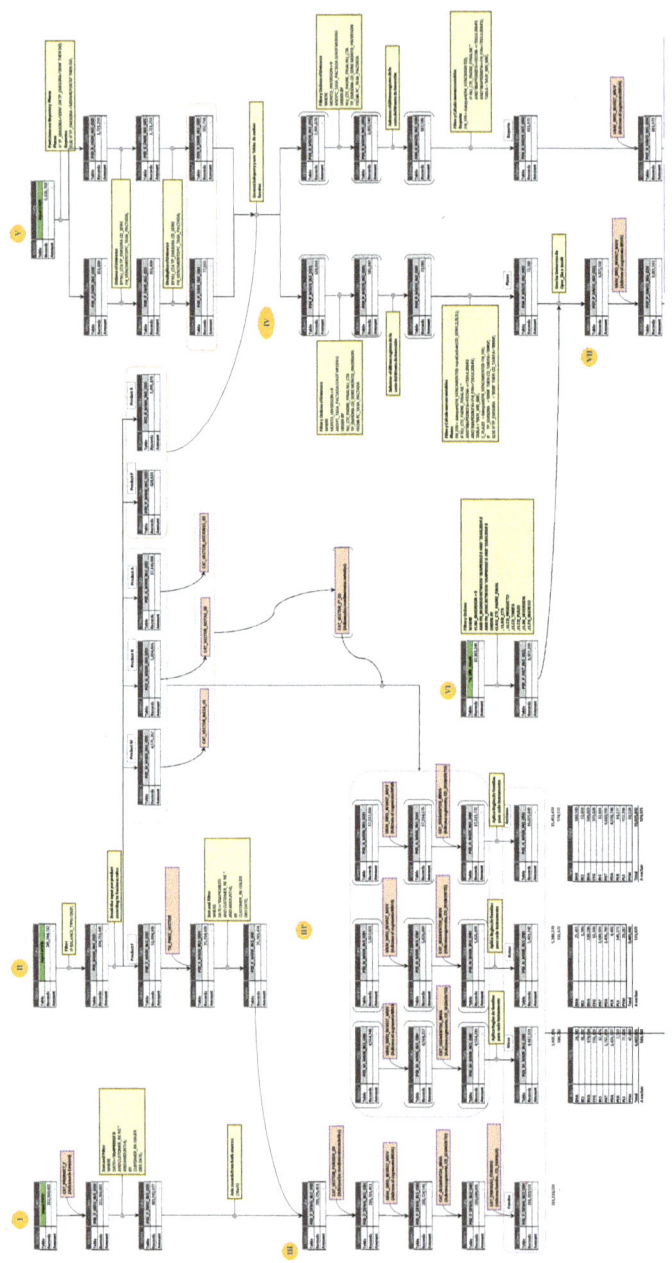

Figure 10-22. Magnified section of a data processing diagram.

CHAPTER 10   THE ABT-BUILDING PROCESS

Figure 10-23 presents a condensed and streamlined representation of the comprehensive data processing diagram. It is notable that in this iteration, the interdependencies and interconnections between the various processes depicted in the diagram become apparent. For example, it is evident that process VIII represents a critical juncture in the overall process, as seven distinct processes must be successfully completed prior to the execution of process VIII. Furthermore, processes I, II, V, VI, XIV, and XV can be executed in parallel, as they lack any dependencies. In terms of inputs and outputs, the entire process depends on only 12 inputs, indicated by green squares, and produces a total of 7 outputs, indicated by purple squares.

However, a diagram such as the one presented in **Figure 10-23** serves not only an illustrative purpose but also that of *automation*, which represents the ultimate objective of all processes. In order to automate a process, it is first necessary to have perfect control over all of its components. As can be observed, as we move from a high-resolution diagram (**Figure 10-20**) to a low-resolution diagram (**Figure 10-23**), we are now able to see the overall *picture*, so to speak. As a result, it is now possible to proceed and link together numerous lines of code as a single block. Furthermore, as the diagram is examined in greater detail, it becomes feasible to control and manage larger and larger blocks.

The construction and automation of a process of this nature may initially appear to be a formidable undertaking. However, with perseverance and dedication, the rewards will become evident over time.

CHAPTER 10 THE ABT-BUILDING PROCESS

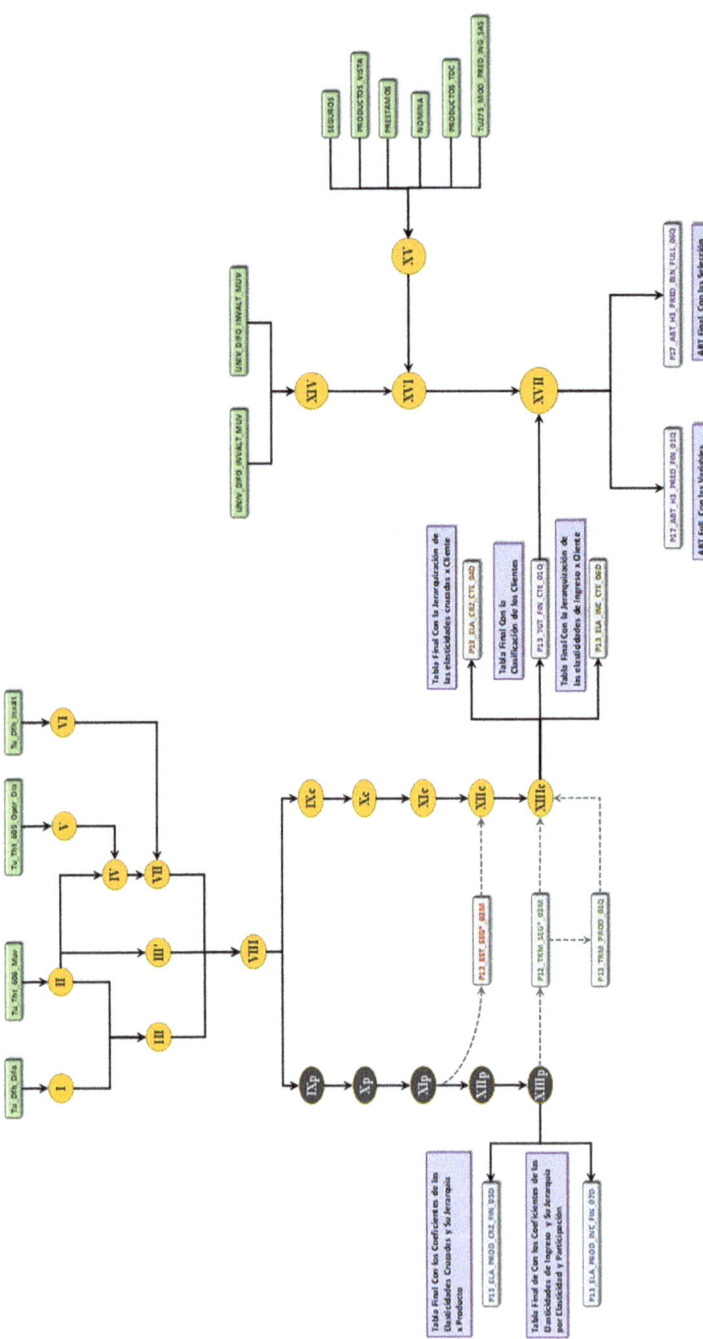

Figure 10-23. Compact representation of the entire data processing diagram.

10.9 Summary

Chapter 10 marks a pivotal moment in our journey, as it is here that all our previous efforts materialize into a tangible, structured entity: the *analytical base table* (ABT). This chapter has demonstrated how the independent pieces of data we have worked on—cleaning, transforming, and engineering—fall into their rightful places to form a cohesive unit called the ABT. It is through this process that we bring structure and purpose to raw data, setting the stage for meaningful analysis.

A key lesson from this chapter is the importance of having a deep understanding of the *use and structure of our data sources*, which significantly facilitates the development of effective integration processes. By recognizing the nuances of each data source, we can design integration strategies tailored to the specific needs of our analysis. However, it is vital to note that variables should *not be constructed directly from the original data sources*. Instead, the recommendation is to create *intermediate structures*, such as a *Data Mart*, that serve as inputs for variable construction. This approach ensures flexibility, scalability, and convenience throughout the process.

Another key takeaway is the emphasis on *safety and reliability* in program design. Our programs must not only achieve their intended objectives but also do so in a way that avoids common pitfalls in data integration, such as mismatched data types, duplicate records, or incomplete joins. To this end, this chapter has introduced us to two of the best methods for efficiently and safely adding new variables to our ABT: Oracle's PL/SQL cursor and SAS's hash DATA step.

The chapter also highlights that the *ABT construction is a continuous process*, evolving as new data becomes available. Therefore, the data integration process should be designed to *continuously add new variables* to the ABT without significantly altering the original programming. To that end, this chapter proposes a continuous variable integration cycle (CVIC) that ensures that the ABT remains relevant and effective over time.

CHAPTER 10 THE ABT-BUILDING PROCESS

Additionally, it was mentioned that the *stacking of the ABT* is what gives meaning to the concepts of *absolute and relative periodicities*, which are essential for capturing general behaviors in the data. Without stacking, it would be impossible to generalize our predictions, rendering models useless for predicting future outcomes.

The *definition of the target variable* was highlighted as a particularly complex task. Determining the event of interest, such as a default, is not as straightforward as it may initially appear. The *roll rate analysis* was introduced as a valuable tool for defining the default event, illustrating the intricacies involved in this process.

Finally, the chapter provided *additional recommendations* to streamline the construction of the ABT. These tips, ranging from data mart feature engineering to visualization techniques, are designed to help you in the design, planning, and construction of both, the training and the scoring ABTs.

In summary, the ABT is more than just a table—it is the culmination of careful planning, integration, and design. It represents the unity of all our efforts, bringing together disparate pieces of data into a single, coherent structure that serves as the foundation for the construction of our predictive model.

Key takeaways:

1. **ABT Integration:** The ABT is where all previous efforts materialize into a unified and structured table.

2. **Data Sources:** Knowledge of data sources is crucial for developing effective integration processes.

3. **Intermediate Structures:** Variables should not be constructed directly from raw data but from intermediate structures (e.g., Data Marts) to ensure flexibility and efficiency.

4. **Program Design:** Programs must be designed to achieve their goals safely and avoid common integration errors.

5. **Continuous Variable Integration:** Data integration processes should accommodate the continuous addition of new variables to the ABT without requiring significant reprogramming.

6. **Stacked ABT:** Stacking the ABT is essential for understanding absolute and relative periodicities and capturing general behaviors.

7. **Event Definition:** Defining the target variable (e.g., a default event) involves considerable complexity, with tools like roll rate analysis aiding in this process.

The culmination of Chapter 10 marks the end of our journey through the complexities of data integration, providing us with the necessary input to fuel our analytical tools and ultimately develop our final scorecard model. However, before proceeding with the construction of the model, it is essential to first familiarize ourselves with the primary tool that will be used from this point onward: *SAS® Enterprise Miner*.

CHAPTER 11

A Brief Introduction to the Use of SAS® Enterprise Miner™

"What is steel, compared to the hand that wields it?"

—Thulsa Doom, Conan the Barbarian (1982)

With the analytical base table (ABT) fully constructed, we now turn our attention to the tools that will enable us to transform this data into actionable insights. This chapter provides an introduction to SAS® Enterprise Miner™ (EM) 15.2, a powerful platform for building predictive models.

It should be noted at the outset that this explanation of how to use SAS® Enterprise Miner™ is not intended to be a comprehensive guide, but rather to present the fundamental concepts required to understand the subsequent chapters of this book.

For those seeking a more in-depth understanding of EM, we recommend *Predictive Modeling with SAS® Enterprise Miner™: Practical Solutions for Business Applications*, 2nd edition, by Kattamuri S. Sarma [59], as well as the SAS training manual *Applied Analytics Using SAS® Enterprise Miner™* [24], both of which are excellent resources.

CHAPTER 11   A BRIEF INTRODUCTION TO THE USE OF SAS® ENTERPRISE MINERTM

By the end of this chapter, you will have a foundational understanding of the key features and functionalities of SAS® Enterprise Miner™, preparing you for its application in the development of our final scorecard model.

11.1 Getting Started with SAS® Enterprise Miner™

SAS® Enterprise Miner™ (EM) is an all-purpose tool based on the construction of process flow diagrams, through the use of an intuitive and friendly interface. **Figure 11-1** shows you the five basic elements that you must become familiar with before starting using EM.

1. The Exploration panel
2. The Properties panel
3. The Quick Help panel
4. The Nodes panel
5. The diagram or canvas panel

CHAPTER 11   A BRIEF INTRODUCTION TO THE USE OF SAS® ENTERPRISE MINER™

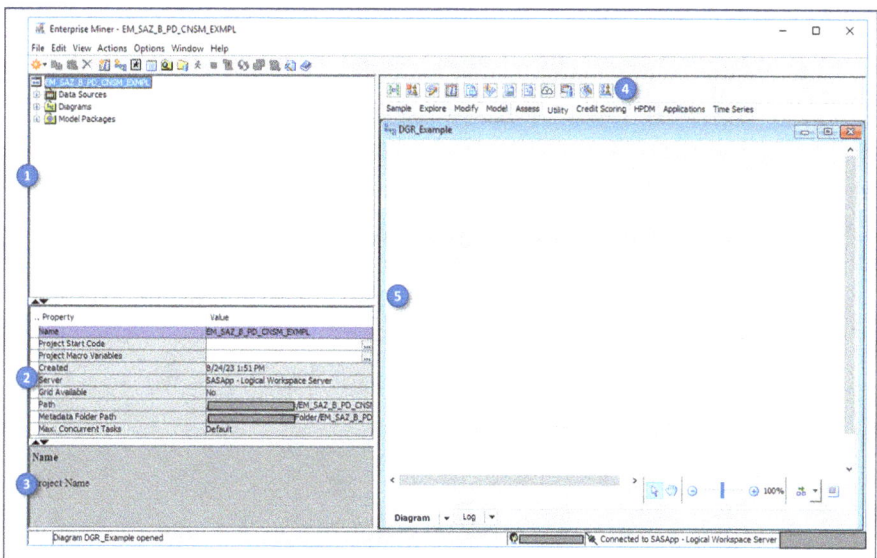

Figure 11-1. *EM workspace. (1) The Exploration panel.*
(2) The Properties panel. (3) The Quick Help panel. (4) The Nodes
panel. (5) The diagram or canvas panel.

11.1.1 The Exploration Panel

The Exploration panel is comprised of four principal components:

1. **The EM's Project**: The panel serves to identify the EM project, which can then be selected to access all of its associated properties.

2. **Data Source**: This is the location where one would typically proceed to incorporate new data sources into the project, thus ensuring their consistent accessibility.

3. **Diagrams**: This is the location where one can create diagrams; however, it is noteworthy that no diagrams are initially available when one first establishes an EM project. Therefore, the diagram depicted in **Figure 11-1** was generated subsequent to the creation of the EM project and was not present previously.

4. **Model Packages**: This section is only applicable when a model package is created for the final model (see Section 19.1.6).

11.1.2 The Properties Panel

Unlike other software where the properties of an object can be accessed by right-clicking on the object, in EM, the properties are always displayed in the Properties panel. This can be a bit confusing at first, but you will get used to it over time.

11.1.3 The Quick Help Panel

As the name suggests, the Quick Help panel provides a quick help on the functionality, options, and other important information when selecting a single property in the Properties panel.

11.1.4 The Nodes Panel

The Nodes panel is the most prominent panel in EM, as it contains all of the data processing and mining algorithms. However, as will be demonstrated in the following chapters, only a small proportion of these are employed during the construction of the model.

11.1.5 The Diagram or Canvas Panel

The diagram panel, or the canvas, as I like to call it, represents the actual workspace and is the location where the design and construction of the process flow are undertaken.

11.1.6 Startup Code

A distinctive attribute of EM is the capacity to execute an initial code each time a project is opened. This is a particularly useful feature as it allows the loading of libraries, macros and macro-variables, thus ensuring their availability throughout the session, eliminating the need to define them each time a code node is executed.

Of particular interest are a set of utility macros that facilitate the manipulation of EM metadata. The capacity to modify EM metadata is a highly convenient feature, as it enables the rejection or approval of candidate variables using the user's own variable selection methods. This is similar to the functionality of native nodes in EM. Furthermore, the capacity to alter EM metadata enables the automation of the variable selection procedure, thereby facilitating the execution of a customized process on any diagram, with any set of variables, and any dataset.

Figure 11-2 illustrates the Project Start Code window and delineates the steps that must be undertaken to access it. The initial step is to select the relevant project from the Exploration panel. Subsequently, select the three-dot icon representing the Project Start Code option in the Properties panel, which will prompt the opening of the Project Start Code window. Once the window is open, the code can be pasted into it. However, it is imperative that the "Run Now" button is clicked before the window is closed. It is recommended that these instructions be followed each time new code is to be added to an EM project.

CHAPTER 11   A BRIEF INTRODUCTION TO THE USE OF SAS® ENTERPRISE MINERTM

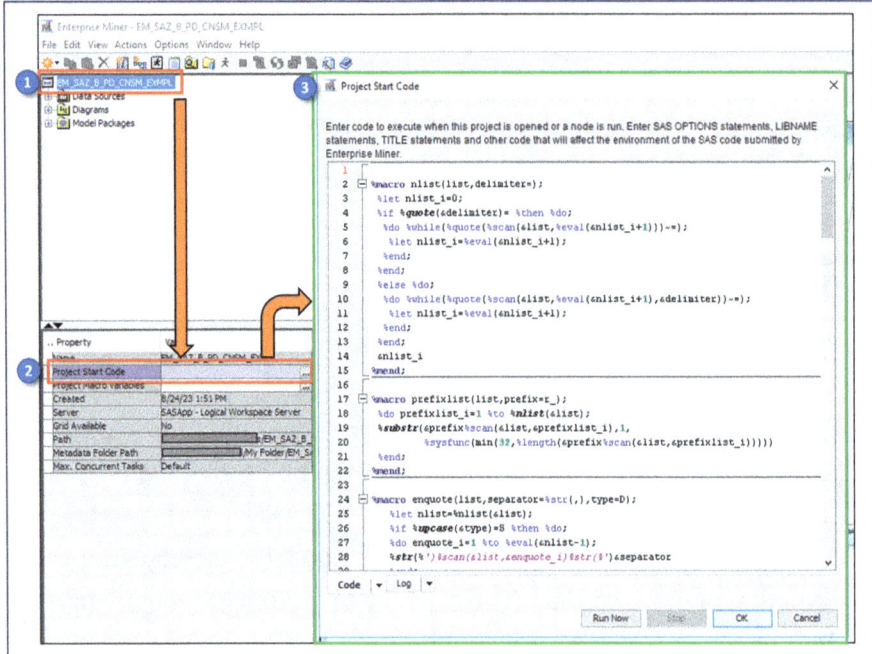

Figure 11-2. Project Start Code option.

11.1.7 The Utility Macros[1]

The utility macros presented in this book are part of the SAS® Advanced Predictive Modeling Using SAS® Enterprise Miner™ 6.1[2] training course [23].

11.1.7.1 NLIST

The %NLIST macro (Listing 11-1) counts the number of tokens in a list. The list is passed to the macro as a positional macro parameter.

---

[1] The utility macros can be downloaded from the accompany material of this book, from https://github.com/saulzapiain/ConceptualVariableDesign.
File: Listing 11- Utility Macros for EM Start Code (complete).sas.
[2] Although this is a discontinued version of EM, the utility macros remain current at any subsequent version of EM.

CHAPTER 11 A BRIEF INTRODUCTION TO THE USE OF SAS® ENTERPRISE MINERTM

Characteristic	Description
Positional parameters	LIST: a list of tokens separated by a delimiter
Keyword parameters	DELIMITER = is used to specify specific characters as delimiters
Dependencies	None

Listing 11-1. NLIST code and example.

(A)
```
%macro nlist(list,delimiter=);
%let nlist_i=0;
%if %quote(&delimiter)= %then %do;
%do %while(%quote(%scan(&list,%eval(&nlist_i+1)))~=);
 %let nlist_i=%eval(&nlist_i+1);
%end;
%end;
%else %do;
%do %while(%quote(%scan(&list,%eval(&nlist_i+1),
&delimiter))~=);
 %let nlist_i=%eval(&nlist_i+1);
%end;
%end;
&nlist_i
%mend;
```
(B)
```
%put %NLIST(a b c);
The value 3 is place in the log.
```

The value **3** is place in the log.

CHAPTER 11 A BRIEF INTRODUCTION TO THE USE OF SAS® ENTERPRISE MINERTM

11.1.7.2 PREFIXLIST

The %PREFIXLIST macro (Listing 11-2) adds the specified prefix to a list of tokens.

Characteristic	Description
Positional parameters	LIST: a list of tokens separated by a delimiter
Keyword parameters	PREFIX = is used to specify the desired prefix
Dependencies	%NLIST macro

Listing 11-2. PREFIXLIST code and example.

(A)
```
%macro prefixlist(list,prefix=r_);
 %do prefixlist_i=1 %to %nlist(&list);
 %substr(&prefix%scan(&list,&prefixlist_i),1,
    %sysfunc(min(32,%length(&prefix%scan(&list,&prefix
    list_i)))))
 %end;
%mend;
```
(B)
```
%put %PREFIXLIST(a b c, prefix=my_);
The value my_a my_b my_c is placed in the log.
```

11.1.7.3 ENQUOTE

The %ENQUOTE macro (Listing 11-3) places quotation marks around items in a list.

CHAPTER 11   A BRIEF INTRODUCTION TO THE USE OF SAS® ENTERPRISE MINERTM

Characteristic	Description
Positional parameters	LIST: a list of tokens separated by a delimiter
Keyword parameters	SEPARATOR= specifies the separator character placed between the elements of the output list. The default is a comma TYPE=specifies the type of quotation marks to use. S is single, and D is double. The default is D
Dependencies	%NLIST macro

Listing 11-3. ENQUOTE code and example.

(A)
```
%macro enquote(list,separator=%str(,),type=D);
   %let nlist=%nlist(&list);
   %if %upcase(&type)=S %then %do;
   %do enquote_i=1 %to %eval(&nlist-1);
   %str(%')%scan(&list,&enquote_i)%str(%')&separator
   %end;
   %str(%')%scan(&list,&nlist)%str(%')
   %end;
   %else %do;
   %do enquotc_i=1 %to %eval(&nlist-1);
   "%scan(&list,&enquote_i)"&separator
   %end;
   "%scan(&list,&nlist)"
   %end;
%mend;
```
(B)
```
%put %ENQUOTE(a b c);
The values "a", "b", "c" are placed in the log.
```

11.1.7.4 REMOVE

The %REMOVE macro (Listing 11-4) removes all occurrences of elements in one list of tokens from another list of tokens.

Characteristic	Description
Positional parameters	LIST: a list of tokens separated by a space to be removed from the FROM= list
Keyword parameters	FROM= is a list of tokens from which all occurrences of the tokens specified by LIST are removed
Dependencies	%NLIST macro

Listing 11-4. REMOVE code and example.

(A)
```
%macro remove(list1,from=);
   %do remove_i=1 %to %nlist(&from);
   %if %index(%upcase(&list1),%scan(%upcase(&from),
   &remove_i))=0 %then %do;
   %scan(&from,&remove_i)
   %end;
   %end;
%mend;
```

(B)
```
%put %REMOVE(a b c,from=1 a 3 c e g);
The value 1 3 e g is placed in the log
```

11.1.7.5 RECODE

The %RECODE macro (Listing 11-5) generates SAS code to transform specified input ranges into new numeric values. The macro is useful for binning interval inputs or recoding categorical inputs. The macro takes the input ranges from a SAS dataset. The dataset must contain variables that specify the name of the variable to be transformed, a range start and a range end, and the transformed value. If the value of the variable to be transformed does not fall within a specified range, the average of the transformed values is used. (If you choose, you can specify a variable in the SAS dataset to weighting the calculation of this average.)

Characteristic	Description
Positional parameters	
Keyword parameters	DATA= name of the SAS dataset with the variable transformation ranges and transformed values
	INPUT=name of variable containing the names of the variables to be recoded
	START=name of the variable containing the starting values of the recoding ranges. To specify negative infinity (or no lower range), use the value -CONSTANT('BIG')
	END=name of the variable containing the ending values of the recoding ranges. To specify positive infinity (or no upper range) use the value CONSTANT('BIG')

CHAPTER 11 A BRIEF INTRODUCTION TO THE USE OF SAS® ENTERPRISE MINERTM

Characteristic	Description
	VALUE=name of the variable containing the transformed values of the variable to be recoded
	PREFIX=prefix to be used (in front of the original variable name) for the new transformed version of the original variable. For example, if the original variable's name is X and the prefix is R_, then the new variable containing the transformed values of X is named R_X
	FREQ=name of the variable containing frequencies used during calculation of the means used for no matching input range
	FILE=fileref to the file that is the transformation code created by the macro
	ACTION= specifies whether the FILE= fileref is appended (ACTION=MOD) or overwritten
Dependencies	SORT procedure
	DATA step

Listing 11-5. RECODE code.

```
%macro recode(data=,input=INPUT,start=START,end=END,value=VALUE,
       prefix=r_,freq=1,file=PRINT,action=);
proc sort data=&data;
 by &input &start;
run;
data _null_;
 attrib s length=$200;
 attrib recode length=$32;
```

CHAPTER 11 A BRIEF INTRODUCTION TO THE USE OF SAS® ENTERPRISE MINERTM

```
retain n_tot weighted_sum;
file &file &action;
set &data;
by &input;
recode="&prefix"||substr(&input,1,min(length(&input),
32-%length(&prefix)));
if first.&input then do;
  n_tot=0;
  weighted_sum=0;
  put ;
  put ;
  s="*** Recode input "||strip(&input)||" to &VALUE values;";
  put s;
  put ;
  s="select;";
  put s;
end;
s="when (";
if &start = &end then do;
 s=strip(s)||strip(&input)||" = '"||trimn(&start)||"'";
end;
else do;
 if &start>-constant('BIG') then s=strip(s)||strip(&start)||" <=";
 s=strip(s)||strip(&input);
 if &end<constant('BIG') then s=strip(s)||" < "||strip(&END);
end;
s=strip(s)||") "||strip(recode)||"="||strip(&value)||";";
put @3 s;
n_tot + &freq;
```

281

```
weighted_sum + (&freq*&value);
if last.&input then do;
  s="otherwise "||strip(recode)||"="||strip(weighted_sum/
  n_tot)||";";
  put @3 s;
  s="end;";
  put s;
 end;
run;
%mend;
```

11.1.7.6 METADATA

The %METADATA macro (Listing 11-6) is used to update the metadata of the training data from within the SAS code node. The macro writes code to be placed in the CDELTA_TRAIN file that updates the metadata after the node is run.

Characteristic	Description
Positional parameters	
Keyword parameters	NEWLEVEL= new metadata level for variables satisfying the WHERE= parameter. Values include interval, nominal, and ordinal
	NEWROLE=new metadata role for variables satisfying the WHERE= parameter
	WHERE=used to select variables metadata update. It can be any valid Boolean expression
	APPEND=append to existing CDELTA_TRAIN. Valid values are Y and N
Dependencies	%NLIST macro

Listing 11-6. METADATA code and example.

(A)
```
%macro metadata(newlevel=,newrole=,where=1,append=N);
   %let newlevel = %upcase(&newlevel);
   %let newrole = %upcase(&newrole);
   %if &append=Y %then %let mode=A;
   %else %let mode=O;
   %let filrf=X;
   %let rc=%sysfunc(filename(filrf,&EM_FILE_CDELTA_TRAIN));
   %let fid=%sysfunc(fopen(&filrf,&mode));
   %if &fid > 0 %then %do;
      %let s= if &where then do%str(;);
      %do metadata_i=1 %to %nlist(%bquote(&s),delimiter=%str(
      )) %by 3;
       %let t=%scan(&s,&metadata_i,%str( ))
      %scan(&s,%eval(&metadata_i+1),%str( ))
      %scan(&s,%eval(&metadata_i+2),%str( ));
       %let rc=%sysfunc(fput(&fid,%bquote(&t)));
       %let rc=%sysfunc(fwrite(&fid));
   %end;
   %if &newlevel~= %then %do;
    %let s=   LEVEL=%str(%')&newlevel%str(%')%str(;);
    %let rc=%sysfunc(fput(&fid,&s));
    %let rc=%sysfunc(fwrite(&fid));
   %end;
```

```
  %if &newrole~= %then %do;
   %let s=  ROLE=%str(%')&newrole%str(%')%str(;);
   %let rc=%sysfunc(fput(&fid,&s));
   %let rc=%sysfunc(fwrite(&fid));
  %end;
  %let s= end%str(;);
  %let rc=%sysfunc(fput(&fid,&s));
  %let rc=%sysfunc(fwrite(&fid));
  %let rc=%sysfunc(fclose(&fid));
  %end;
  %else %put %sysfunc(sysmsg());
  %let rc=%sysfunc(filename(filrf));
%mend;
(B)
%METADATA(newlevel=Interval, newrole=Input,where=(name in (X1 X2 X3)));
This assigns variables X1, X2, and X3 to the level Interval and the role Input.
```

11.1.8 Metadata Properties

Once the start code has been added, the next step is to import and configure the metadata properties associated with the ABT. This is achieved by right-clicking on the Data Sources folder icon, located in the top left corner of the Exploration panel, and selecting "Create Data Source" (**Figure 11-3**). Press Next on the emerging window, it should be the Step 1 of 8 (**Figure 11-4**). Select the appropriate ABT from the SAS library and press OK (**Figure 11-5**). Then, proceed to the subsequent window (**Figure 11-6**) by pressing Next. This will advance the process to Step 3 of 8.

CHAPTER 11 A BRIEF INTRODUCTION TO THE USE OF SAS® ENTERPRISE MINERTM

The subsequent window (**Figure 11-7**) presents two options: Basic and Advanced. Should the Advanced option be selected, a new window will open, where you can specify the following:

- Missing Percentage Threshold, 50% as default value
- Reject Vars with Excessive Missing Values, Yes/No
- Class Levels Count Threshold, 20 as default value
- Detect Class Levels, Yes/No
- Reject Levels Count Threshold, 20 as default value
- Reject Vars with Excessive Class Values, Yes/No
- Database Pass-Through, Yes/No

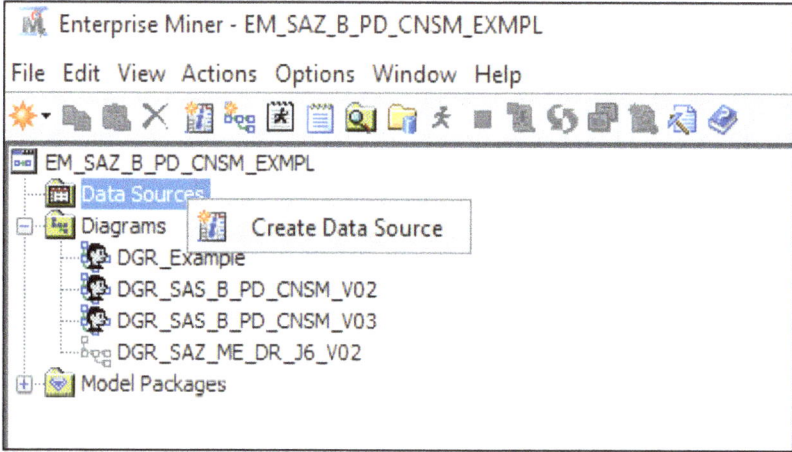

Figure 11-3. How to create a new Data Source in EM.

285

CHAPTER 11 A BRIEF INTRODUCTION TO THE USE OF SAS® ENTERPRISE MINERTM

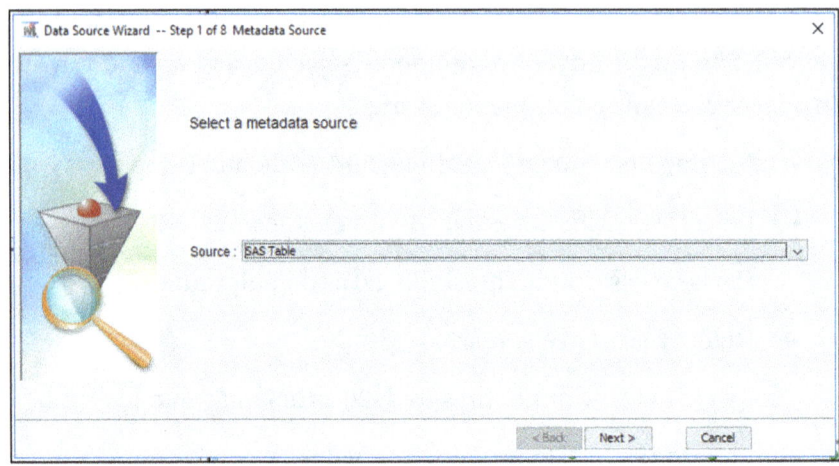

Figure 11-4. Select a metadata source screen.

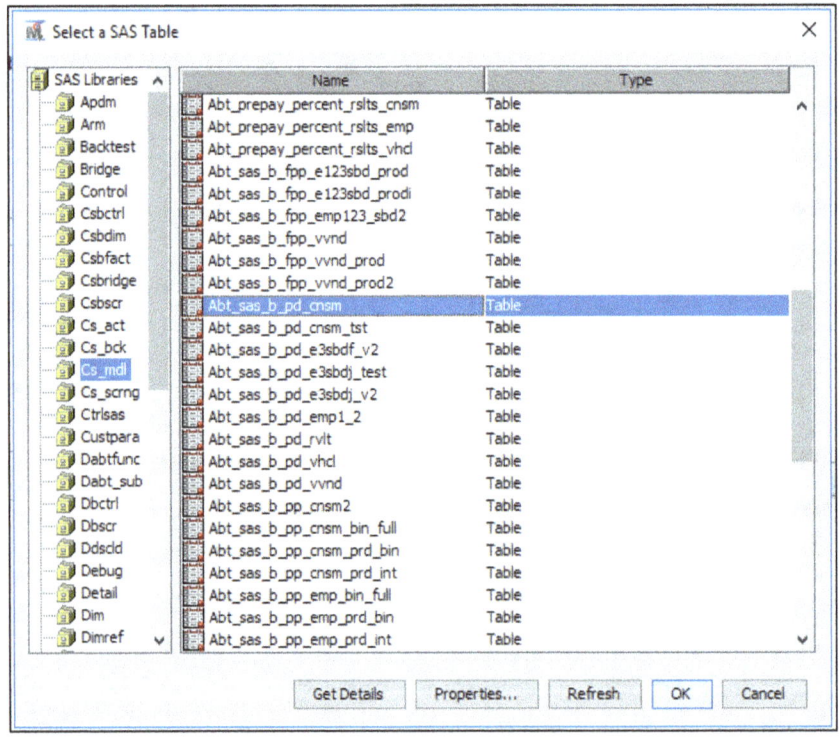

Figure 11-5. How to select your ABT from a SAS library.

CHAPTER 11   A BRIEF INTRODUCTION TO THE USE OF SAS® ENTERPRISE MINERTM

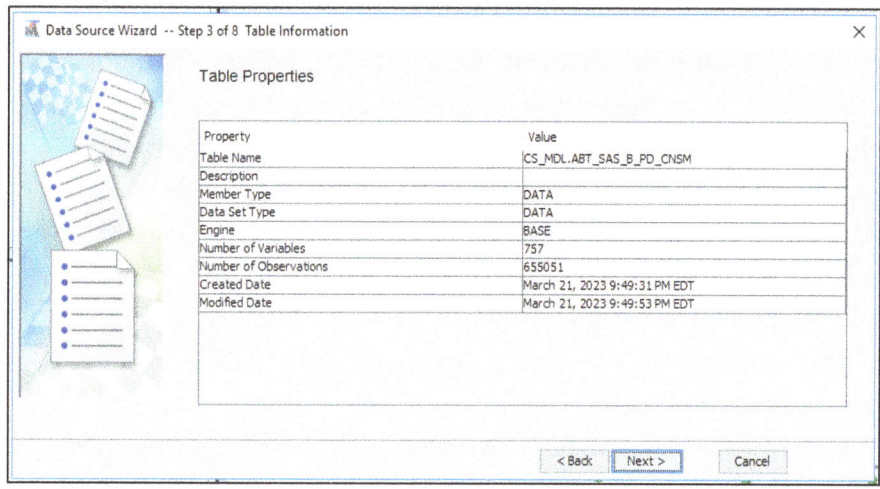

Figure 11-6. *Table Properties screen.*

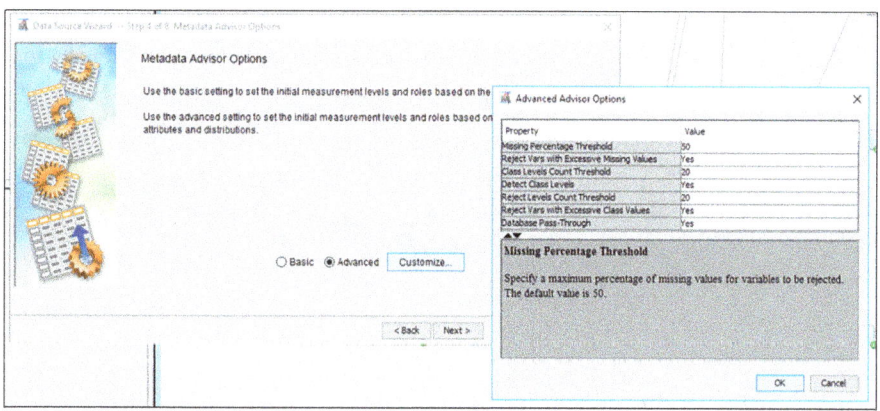

Figure 11-7. *The Metadata Advisor Options screen.*

Upon initial observation, it may appear advantageous to select the advanced options. However, this is not the case. The rationale for avoiding the advanced options is that they impede our capacity to make autonomous decisions. To illustrate, one might wish to utilize a customer's ZIP code as an explanatory variable in a model. This would undoubtedly

287

be a sound decision. However, the sheer number of potential ZIP codes—potentially thousands—makes it challenging to employ this variable in a conventional logistic regression due to its high cardinality. One potential solution is to aggregate the ZIP codes by risk level into a manageable number of groups, such as seven or nine, using the WoE methodology. This prompts the question of why one would wish to reject variables with an excessive number of class values in the first place. The answer is that there is no rationale for the rejection of variables with high cardinality. In light of the aforementioned considerations, we will proceed with the Basic option.

Once the Basic option has been selected, a new screen will appear (**Figure 11-8**), displaying the characteristics of each column within the ABT. In this section, we will focus on two specific properties: the *role* and *level* of our explanatory variables. It is notable that all of the interval variables (those beginning with the letter *I*) have been designated with the role "Classification" and the level "Nominal." This is due to the fact that EM attempts to make an initial estimation regarding the role and level of the columns in an ABT based on their nomenclature. It is unfortunate that the program fails to accurately identify the values for these variables.

CHAPTER 11   A BRIEF INTRODUCTION TO THE USE OF SAS® ENTERPRISE MINERTM

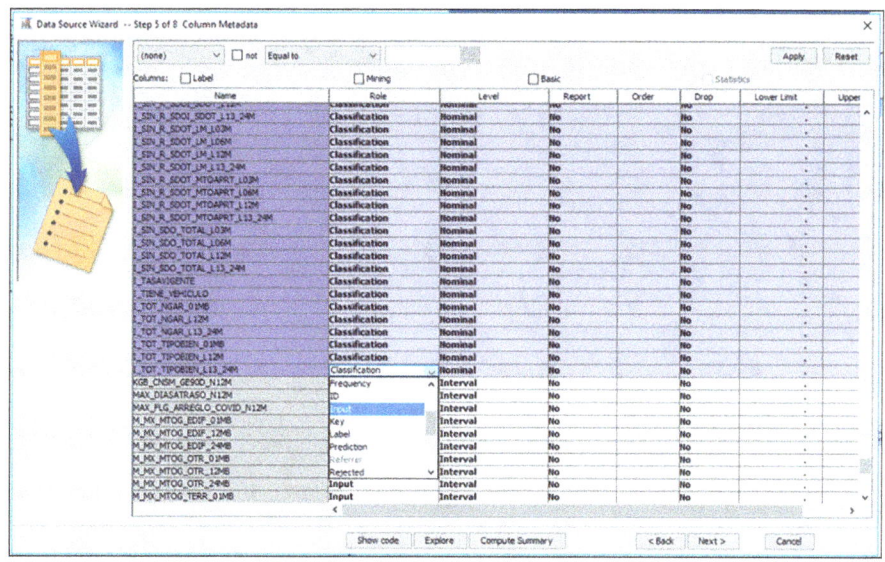

Figure 11-8. *The Column Metadata screen.*

The good news is that implementing a standardized nomenclature effectively resolves the issue. This can be achieved by sorting the variables by name, selecting the block of interval variables (by holding down the Shift key on the keyboard while selecting), and then assigning the correct role and level, which are "Input" and "Interval." As illustrated in **Figure 11-9**, the properties of the interval variables have been corrected, and the binary target, KGB_CNSM_GE90D_N12M, has been defined. Additionally, the support variables, which were previously utilized to generate the final derived variables, have been excluded from further consideration. It can be observed that the supporting variables commence with the letter "M." This is due to the fact that they are of a magnitude nature and thus will not be utilized as explanatory variables (for further details, please refer to Section 4.2).

289

CHAPTER 11 A BRIEF INTRODUCTION TO THE USE OF SAS® ENTERPRISE MINERTM

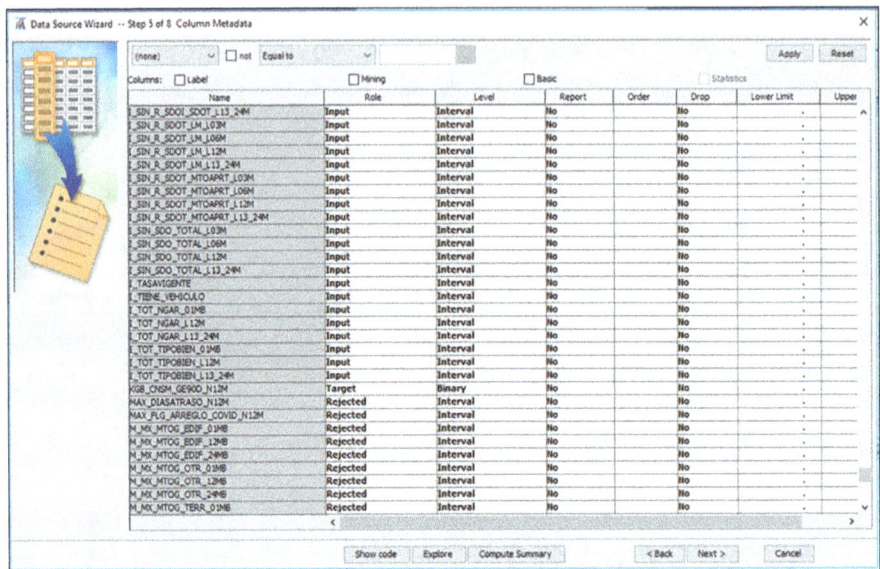

Figure 11-9. Illustrative example of how to correct columns' metadata.

Figure 11-10 demonstrates that nominal variables can be identified and corrected with the same ease as interval variables. However, the underlying reason behind the ease of modifying the metadata of variables is the use of a standardized nomenclature. To illustrate, please direct your attention to **Figure 11-11**. Can you identify the variable whose name contains a typographical error? Was it challenging to locate the aforementioned variable? Or was it rather easy to locate?

CHAPTER 11   A BRIEF INTRODUCTION TO THE USE OF SAS® ENTERPRISE MINERTM

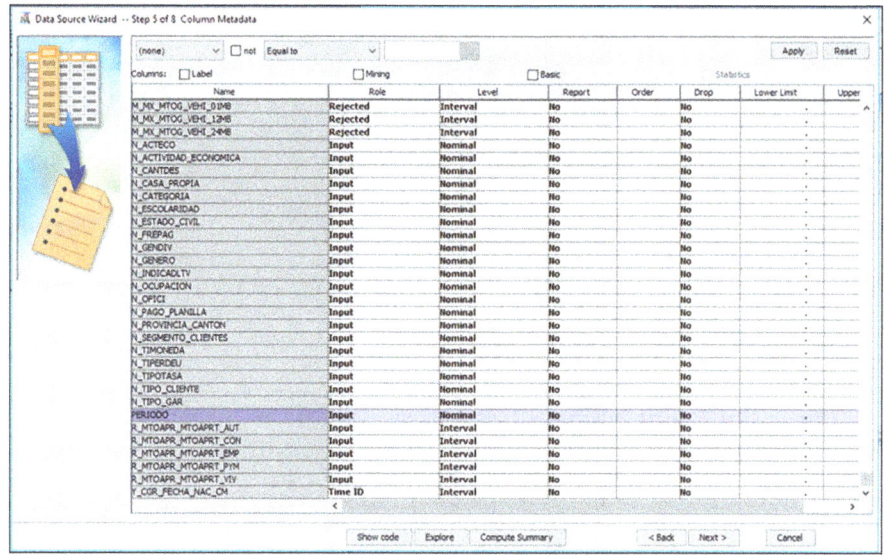

Figure 11-10. *Define nominal variables in the Column Metadata screen.*

CHAPTER 11   A BRIEF INTRODUCTION TO THE USE OF SAS® ENTERPRISE MINERTM

Name
I_R_PGO_SDO_CON_18MB
I_R_PGO_SDO_CON_24MB
I_R_PGO_SDO_EMP_01MB
I_R_PGO_SDO_EMP_03MB
I_R_PGO_SDO_EMP_06MB
I_R_PGO_SDO_EMP_12MB
I_R_PGO_SDO_EMP_18MB
I_R_PGO_SDO_EMP_24MB
I_R_PGO_SDO_PYM_01MB
I_R_PGO_SDO_PYM_03MB
I_R_PGO_SDO_PYM_06MB
I_R_PGO_SDO_PYM_12MB
I_R_PGO_SDO_PYM_18MB
I_R_PGO_SDO_PYM_24MB
I_R_PGO_SDO_VIV_01MB
I_R_PGO_SDO_VIV_03MB
I_R_PGO_SDO_VIV_06MB
I_R_PGO_SDO_VIV_12MB
I_R_PGO_SDO_VIV_18MB
I_R_PGO_SDO_VIV_24MBMB
I_R_SALDES_LM_01MB
I_R_SALDES_LM_02MB
I_R_SALDES_LM_03MB
I_R_SALDES_LM_06MB
I_R_SALDES_LM_08MB
I_R_SALDES_LM_10MB
I_R_SALDES_LM_12MB
I_R_SDODIF_LM_01MB
I_R_SDODIF_LM_02MB
I_R_SDODIF_LM_03MB
I_R_SDODIF_LM_06MB

Figure 11-11. Typo in variable naming.

11.1.8.1 A Standardized Methodology Is a Good Antidote to Operational Risk

It is evident that the ease with which errors can be identified is a consequence of the predictable pattern that characterizes variable names. The adage *"the nail that sticks out gets hammered down first"* can be applied here, in that the name that deviates from the established pattern will be the first to be identified. However, the detection of typographical errors in variable names is not the primary objective of a standardized nomenclature, but to mitigate *operational risk*.

CHAPTER 11   A BRIEF INTRODUCTION TO THE USE OF SAS® ENTERPRISE MINERTM

Operational risk has the potential to bring down entire industries and systems, and it is most often caused by *human error*. The implementation of structured methodologies, best practices, guidelines, and protocols is designed to reduce the occurrence of human error and, consequently, minimize operational risk. Indeed, one of the principal objectives of this book is to assist the reader in reducing operational risk through the utilization of straightforward instruments including equation balances, diagrams, spreadsheets, and a standardized nomenclature. It is important to note, however, that while databases, advanced coding, machine learning, and artificial intelligence can facilitate certain processes, they are not a substitute for a robust *methodology*.

In essence, *a methodology is the way you do things*. However, as with all endeavors in life, there is a correct and an incorrect way to approach a given task. Accordingly, the optimal methodology entails a consistent approach, adhering to a uniform *standard*. This standard is understood to be a model or exemplar established by authority, custom, or general agreement. In this instance, if you are the team leader responsible for the model-building process, then you are the authority.

The key takeaway from this example is that the designation of variables is a crucial yet often overlooked aspect of the modeling process. Despite its seemingly inconsequential nature, this step is fundamental to the success of the modeling endeavor. It is a common misconception among analysts to assume that the modeling process is merely a matter of sophisticated algorithms. Nevertheless, all of the tasks comprising the modeling process are equally important, regardless of the task's scale or complexity.

11.1.9 EM Diagrams

One of the most advantageous features of EM is its capacity to export completed diagrams in XML format, which can be effortlessly integrated into a new project in a matter of seconds. This enables the reuse of all nodes, their configurations, and even the code within each node, thereby ensuring a streamlined and efficient workflow. This is an exceptional feature that has the potential to significantly accelerate the modeling process, particularly following the completion of an initial EM project. Furthermore, EM diagrams are interchangeable, thus enabling the importation of any diagram created by another member of the team, provided that no conflicts exist between the versions of EM utilized.

11.1.9.1 Managing EM Diagrams

Two methods exist for the addition of a new diagram to an EM project. The initial option is to create a new diagram by right-clicking on the Diagrams folder, which is located in the Exploration panel in the top left corner of the screen (as shown in **Figure 11-12**). The second option is to import a diagram from a previously saved diagram in XML format.

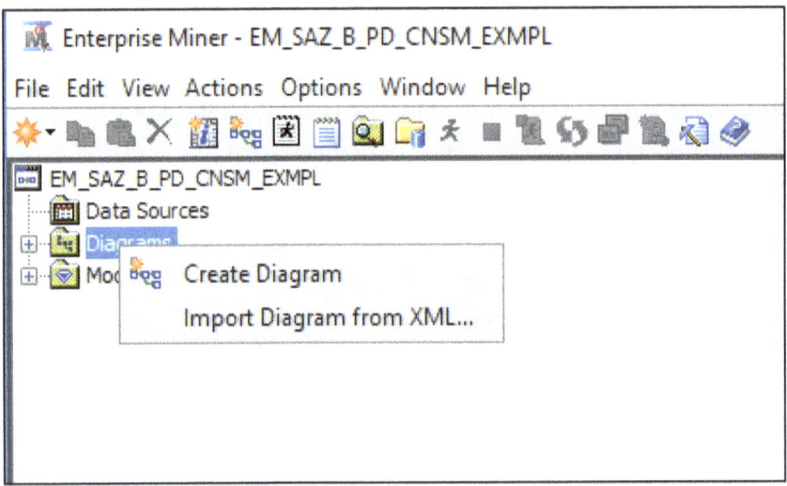

Figure 11-12. How to add a new diagram to your EM project.

The process of exporting an entire diagram is equally straightforward. To do so, simply right-click on the diagram that you wish to save as an XML file and select the desired location for saving (as shown in **Figure 11-13**). It should be noted that the diagram must be open in order to save it.

CHAPTER 11   A BRIEF INTRODUCTION TO THE USE OF SAS® ENTERPRISE MINERTM

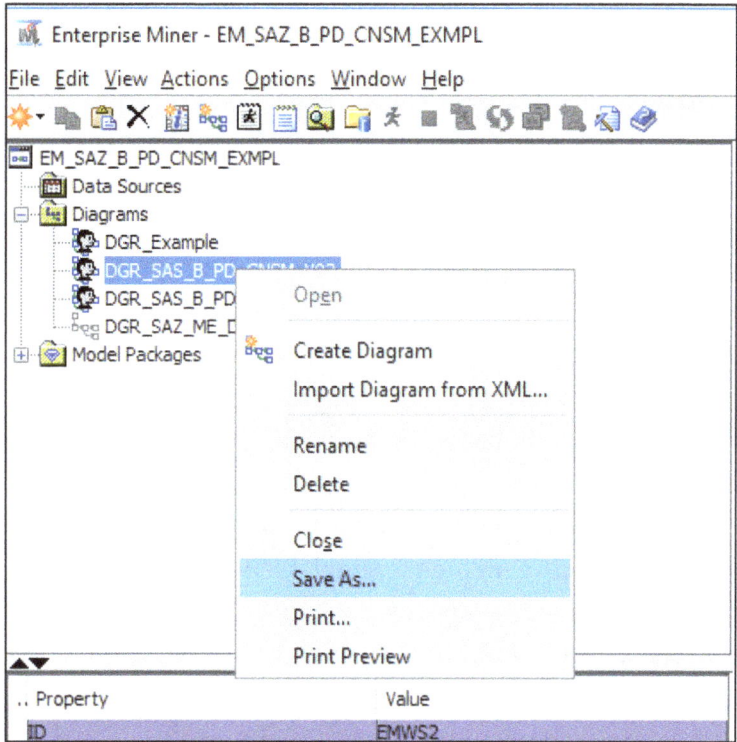

Figure 11-13. How to save a completed diagram in EM as an XML file.

11.1.10 The Code Node in EM

The EM code node (EMCN) [60] is arguably the most powerful and versatile tool in EM. This is due to the fact that the EMCN permits the user to input their own code in a manner that appears to be an inherent part of EM's native functionality. Consequently, in theory, the aforementioned feature has the potential to facilitate the incorporation of an unlimited number of procedures and analytical methods that have been specifically developed by the user. In other words, *"if it doesn't have it, just add it."*

CHAPTER 11 A BRIEF INTRODUCTION TO THE USE OF SAS® ENTERPRISE MINERTM

This is what makes EM such a convenient application, as it enables the user to benefit from the best of both worlds. Consequently, the graphical tool enables the user to observe the model-building process in a systematic and schematic manner. Conversely, users are also able to integrate their own code into the modeling process at any stage, should they so desire. This allows the addition of a vast array of code, offering considerable flexibility.

In contrast, other analytical tools, such as Python, R, or STATA, lack a native graphical representation of the modeling process. Those engaged in modeling who are inclined to code may not perceive this absence of graphical representation as a disadvantage. This is an acceptable perspective; however, it does not negate the fact that coding can become cumbersome, particularly as the complexity of the code increases.

Coding is an effective method for implementing a model-building process. However, coding is often executed in a linear and sequential manner, which may not be optimal. In contrast, EM enables the creation of a graphical representation of the code, facilitating the identification of interdependencies and potential for parallel execution.

Moreover, any individual who has previously constructed an entire model from code can attest to the phenomenon of code becoming so complex that it is challenging to discern the overall structure. Such a graphical representation can assist in the disentangling of the code and in making it more readily comprehensible, even to the coder.

As previously discussed in Section 10.8.2, one effective method for comprehending the data processing of an ABT is to represent it as a diagram, which allows you to

1. Identify critical points
2. See chunks of your code as blocks that perform specific tasks

3. Identify the dependencies between your processes and, consequently, determine which processes should run sequentially and which can run in parallel

4. Understand what your overall process actually does

Similarly, a mining diagram can not only allow you to accomplish all of the above, but it can also allow you to balance and design your experiments, or *variants* (we will discuss this topic in more detail in Chapter 17).

11.1.10.1 Object-Oriented Programming

The EM programming is a type of object-oriented language,[3] but instead of dealing with properties and attributes, the EM programming code deals with elements related to the modeling process.

Table 11-1 presents a selected list of the most commonly used modeling elements in the EM code node. It should be noted that each modeling element is actually a text string with the explicit column names of all variables that fall into a particular role and level, according to their metadata properties (Section 11.1.8).

---

[3] Such as VBA, Java, C++, Python, and PHP, just to mention a few.

CHAPTER 11 A BRIEF INTRODUCTION TO THE USE OF SAS® ENTERPRISE MINER™

Table 11-1. Selected list of the most commonly used modeling elements in the EM code node.

Modeling Element	Summary of Modeling Elements	Elements Returned	Type	Role	Level
EM_TARGET_LEVEL		All targets' levels	Generic	Target	Text
EM_BINARY_TARGET	EM_NUM_BINARY_TARGET	All binary targets	Specific	Target	Binary
EM_ORDINAL_TARGET	EM_NUM_ORDINAL_TARGET	All ordinal targets	Specific	Target	Ordinal
EM_NOMINAL_TARGET	EM_NUM_NOMINAL_TARGET	All nominal targets	Specific	Target	Nominal
EM_INTERVAL_TARGET	EM_NUM_INTERVAL_TARGET	All interval targets	Specific	Target	Interval
EM_BINARY_INPUT	EM_NUM_BINARY_INPUT	All binary inputs	Specific	Input	Binary
EM_ORDINAL_INPUT	EM_NUM_ORDINAL_INPUT	All ordinal inputs	Specific	Input	Ordinal
EM_NOMINAL_INPUT	EM_NUM_NOMINAL_INPUT	All nominal inputs	Specific	Input	Nominal
EM_INTERVAL_INPUT	EM_NUM_INTERVAL_INPUT	All interval inputs	Specific	Input	Interval
EM_BINARY_REJECTED	EM_NUM_BINARY_REJECTED	All binary rejects	Specific	Rejects	Binary
EM_ORDINAL_REJECTED	EM_NUM_ORDINAL_REJECTED	All ordinal rejects	Specific	Rejects	Ordinal
EM_NOMINAL_REJECTED	EM_NUM_NOMINAL_REJECTED	All nominal rejects	Specific	Rejects	Nominal
EM_INTERVAL_REJECTED	EM_NUM_INTERVAL_REJECTED	All interval rejects	Specific	Rejects	Interval
EM_TARGET	EM_NUM_TARGETS	All targets	Generic	Target	Multi-level
EM_INPUT	EM_NUM_VARS	All inputs	Generic	Input	Multi-level
EM_REJECTS	EM_NUM_REJECTS	All rejects	Generic	Rejects	Multi-level
EM_INTERVAL	EM_NUM_INTERVAL	All interval variables	Generic	Multi-role	Interval
EM_CLASS	EM_NUM_CLASS	All class variables	Generic	Multi-role	Nominal

CHAPTER 11   A BRIEF INTRODUCTION TO THE USE OF SAS® ENTERPRISE MINERTM

Table 11-2 presents an example of a list of potential variables with their corresponding role, level, and description. Note that there are six interval variables:

1. I_R_BLNCE_CRDLIM_O0MB
2. I_AVG_R_BLNCE_CRDLIM_L03M
3. I_R_PRCHSES_PYMNTS_02MB
4. I_SIN_R_PRCHSES_PYMNTS_L06M
5. I_R_ATMWD_TOTEXPNS_O0MB
6. I_CNT_CC_TRX_L03M

Three nominal variables:

1. N_CUSTOMER_ADDRSS_ZIPCODE
2. N_EDUCATION_LEVEL
3. N_MARITAL_STATUS

And one binary target: KGB_GT90D_N12M.

Table 11-3 illustrates the values assumed by a series of modeling elements in accordance with the specified set of variables presented in **Table 11-2**. It is notable that each of the modeling elements is capable of identifying specific sets of variables. To illustrate, EM_INPUT enumerates all input variables in a nonexclusive manner, whereas EM_NOMINAL_INPUT and EM_INTERVAL_INPUT are level-specific.

CHAPTER 11 A BRIEF INTRODUCTION TO THE USE OF SAS® ENTERPRISE MINER™

Table 11-2. *Example set of explanatory variables and the target variable, with their role and level.*

Variable	Role	Level	Description
KGB_GT90D_N12M	Target	Binary	Known good and bad, defined as days past due greater than 90 days in the next 12 months
I_R_BLNCE_CRDLIM_00MB	Input	Interval	Ratio of outstanding balance over credit limit in last month
I_AVG_R_BLNCE_CRDLIM_L03M	Input	Interval	Average ratio of outstanding balance over credit limit in the last three months
I_R_PRCHSES_PYMNTS_02MB	Input	Interval	Ratio of purchases over payments, reported two months back
I_SIN_R_PRCHSES_PYMNTS_L06M	Input	Interval	Sum of increments of the ratio of purchases over payments in the last six months
I_R_ATMWD_TOTEXPNS_00MB	Input	Interval	Ratio of ATM withdrawals over total expenses in the last month
I_CNT_CC_TRX_L03M	Input	Interval	Number of credit card transactions in the last three months
N_CUSTOMER_ADDRSS_ZIPCODE	Input	Nominal	Customer's address zip code
N_EDUCATION_LEVEL	Input	Nominal	Customer's education level
N_MARITAL_STATUS	Input	Nominal	Customer's marital status

CHAPTER 11 A BRIEF INTRODUCTION TO THE USE OF SAS® ENTERPRISE MINER™

Table 11-3. *Values taken by the presented macro-variables (modeling elements) using the set of variables presented in Table 11-2.*

Modeling Element	Values Taken by the Macro-Variable
EM_TARGET_LEVEL	Binary
EM_BINARY_TARGET	KGB_GT90D_N12M
EM_NOMINAL_INPUT	N_CUSTOMER_ADDRSS_ZIPCODE N_EDUCATION_LEVEL N_MARITAL_STATUS
EM_INTERVAL_INPUT	I_R_BLNCE_CRDLIM_00MB I_AVG_R_BLNCE_CRDLIM_L03M I_R_PRCHSES_PYMNTS_L03M I_R_PRCHSES_PYMNTS_02MB I_SIN_R_PRCHSES_PYMNTS_L06M I_R_ATMWD_TOTEXPNS_00MB I_CNT_CC_TRX_L03M
EM_TARGET	KGB_GT90D_N12M
EM_INPUT	I_R_BLNCE_CRDLIM_00MB I_AVG_R_BLNCE_CRDLIM_L03M I_R_PRCHSES_PYMNTS_L03M I_R_PRCHSES_PYMNTS_02MB I_SIN_R_PRCHSES_PYMNTS_L06M I_R_ATMWD_TOTEXPNS_00MB I_CNT_CC_TRX_L03M N_CUSTOMER_ADDRSS_ZIPCODE N_EDUCATION_LEVEL N_MARITAL_STATUS

CHAPTER 11 A BRIEF INTRODUCTION TO THE USE OF SAS® ENTERPRISE MINERTM

This example illustrates why the EM code node is such a valuable tool for programming automation—as it allows the generalization of mining codes without the need to specify variable names each time a new model is constructed.

Understanding the Workflow Objects

Most modelers and programmers are familiar with the modeling elements illustrated in **Table 11-1**, but there is another set of objects in EM that not only make reference to modeling elements, but also to their position in the diagram as well. In order to comprehend the workflow objects accessible within the EM code, it is first necessary to gain an understanding of how a line-and-box diagram is represented from a programming perspective.

Figure 11-14 shows a conceptual representation of a line-and-box workflow diagram. Notice that lines represent the direction in which data flows, while boxes on the other hand represent processes. A process can be defined as a transformation stage, where one or more operations, whether simple or complex, take place.

Normally, you would expect to have a one-to-one relationship between the input/output data flows and the process to which they are connected. However, this is not always the case, and sometimes, multiple inputs can feed a single process, or a single process can produce multiple outputs, or both.

For now, we will focus on the one-to-one relationships between the processes and the data flow. From **Figure 11-14**, observe that each process includes an observer representation on top of it. Notice that from the observer's perspective, the same data flow can be seen as both, an output flow, from the producer process, and an input flow, from the receiver process. Knowing when a data flow is input or output then depends entirely on the process you are standing on.

CHAPTER 11 A BRIEF INTRODUCTION TO THE USE OF SAS® ENTERPRISE MINERTM

Figure 11-15 shows a single-process workflow exemplifying the input/output data flow. Notice that the input data in this figure corresponds to the dataset EM_IMPORT_DATA, and the output corresponds to the EM_EXPORT_TRAIN dataset. This is because in the EM code node, each input dataset is represented by the macro-variable &EM_IMPORT_DATA, and any output dataset is represented by the macro-variable &EM_EXPORT_TRAIN.

CHAPTER 11 A BRIEF INTRODUCTION TO THE USE OF SAS® ENTERPRISE MINER™

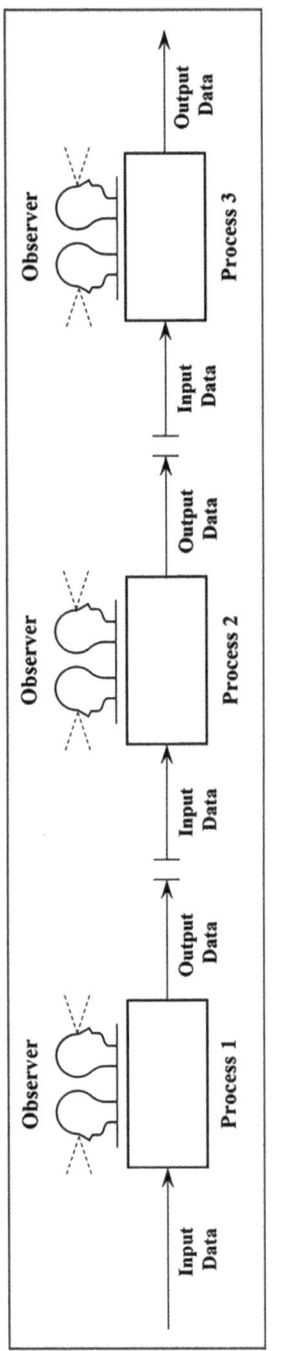

Figure 11-14. Conceptual representation of a line-and-box workflow diagram.

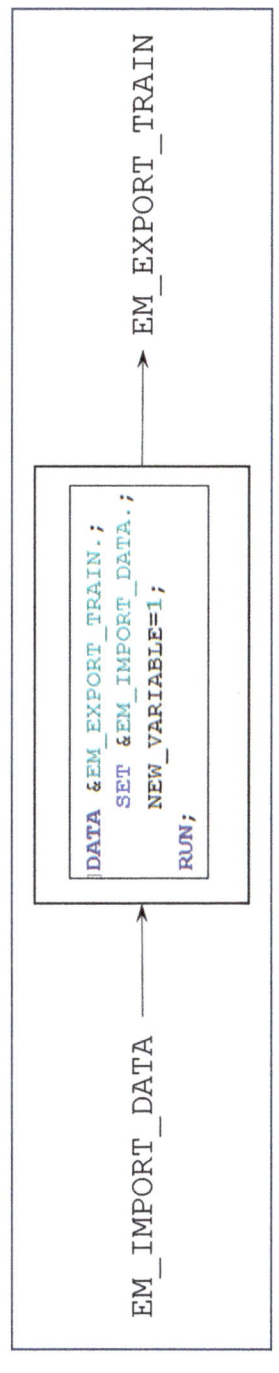

Figure 11-15. A single-process workflow illustrating the concept of import and export data.

CHAPTER 11   A BRIEF INTRODUCTION TO THE USE OF SAS® ENTERPRISE MINERTM

Table 11-4 shows a list of the most commonly used workflow datasets in the EM code node, and note that each input table has its corresponding output, or export counterpart. Also notice that all the elements, in terms of samples, required during the model-building process are present in this list, such as the training, the validation, and the test samples. However, there is one single element that seems to stand out from the rest, and that is the EM_IMPORT_DATA dataset. This is because there is not an EM_IMPORT_TRAIN dataset; instead, there is an EM_IMPORT_DATA dataset. The reason for this apparent inconsistency in nomenclature is because the EM_IMPORT_DATA dataset plays the role of the generic dataset when no partitioning has taken place in the diagram.

Table 11-4. List of the most commonly used workflow datasets in the EM code node.

Workflow input table	Workflow output table	ABT type
EM_IMPORT_DATA	EM_EXPORT_TRAIN	Training
EM_IMPORT_VALIDATE	EM_EXPORT_VALIDATE	Validation
EM_IMPORT_TEST	EM_EXPORT_TEST	Test
EM_IMPORT_SCORE	EM_EXPORT_SCORE	Scoring
EM_IMPORT_TRANSACTION	EM_EXPORT_TRANSACTION	Transaction

To understand this concept, consider the two-node workflow shown in **Figure 11-16**. Notice that the dataset node with the name MDL_ABT_B_PD_TDC_FNL is the starting point of the workflow and that it is located just before the data partition node. In this example, it is clear that the input data, from the point of view of the data partition node, does not match any of the ABT types mentioned in **Table 11-4**, since no partitioning, or any other operation, has yet taken place.

CHAPTER 11   A BRIEF INTRODUCTION TO THE USE OF SAS® ENTERPRISE MINERTM

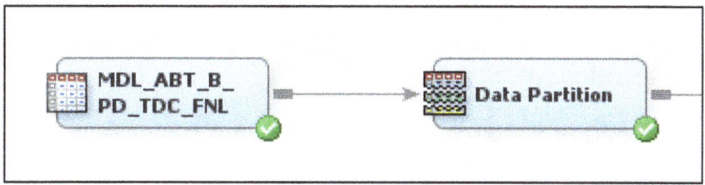

Figure 11-16. *A two-node workflow illustrating the beginning of a mining diagram.*

Therefore, using the generic EM_IMPORT_DATA name to refer to the input dataset is just a clever way to go around the problem of having to deal with two different elements instead of one. From a programmer's point of view, having the same name for the input dataset allows you to use your custom code at any level of your mining diagram, notwithstanding whether or not a partitioning operation has taken place. From this explanation, it should be evident then that the EM_IMPORT_DATA macro-variable always refers to the training partition, from the partition node onward.

11.1.10.2 The Code Editor User Interface

The code editor user interface (UI) is a basic version of the SAS Enterprise Guide (EG) that allows you to enter and edit your SAS code. An interesting feature of the code editor is that it allows you to execute your code in real time without having to close the UI, or you can also execute it by executing the appropriate code node.

Like any other programming interface, the code editor also includes an output and a log pane that allows you to see the results of your real-time executions. According to the SAS Enterprise Miner™ 7.1 Extension Nodes: Developer's Guide [60], the code editor consists of seven components, but for convenience, we will focus on only four of them:

1. The action panel
2. The utilities panel
   a. Macros
   b. Macro-variables
   c. Variables
3. The code editor panel
4. The output panel
   a. Output (list)
   b. Log
   c. Result log

Figure 11-17 shows a screenshot of the code editor window, highlighting its four basic components in blue circles with white numbers. The Developer's Guide provides a comprehensive account of the distinctions between the various code editors located in the action panel and the specific tasks that can be performed by each:

1. Training code
2. Score code
3. Report code

Nevertheless, based on my own experience, the score and report codes are overly specific, and their full functionality can be encompassed by the training code. Consequently, henceforth, our attention will be directed toward the training code, whereby we shall present the reader with the fundamental concepts required to introduce their customized programs into the code node of the EM, thus enabling them to extend their programming ideas from that point onwards.

CHAPTER 11   A BRIEF INTRODUCTION TO THE USE OF SAS® ENTERPRISE MINERTM

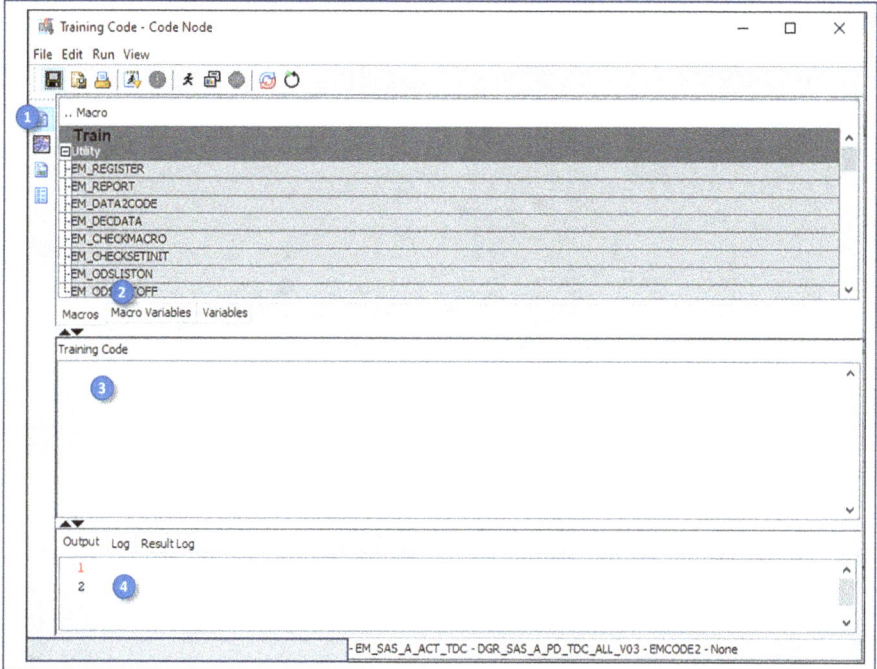

Figure 11-17. *The code editor user interface. (1) The action pane. (2) The utilities pane. (3) The coding pane. (4) The output pane.*

11.1.10.3 Applied Examples of EM Programming

The most effective method for acquiring proficiency in a given task is through direct engagement with the task itself. Accordingly, we will elucidate the utilization of the code editor through the presentation of illustrative, practical examples. **Figure 11-18** illustrates an example of SQL code that queries the number of observations, the number of events, and the proportion of events by period of the dataset of the predecessor node, identified as &EM_IMPORT_DATA. It can be observed that the lower section of the illustration depicts the resulting output of the aforementioned query. Nevertheless, the most significant aspect of **Figure 11-18** is that the preceding SQL code is executed in real time without closing the code

309

editor UI. It is, in fact, possible to execute any code without closing the code editor, provided that no changes are made to the current objects in the mining diagram, e.g., the creation of an EM_EXPORT dataset.

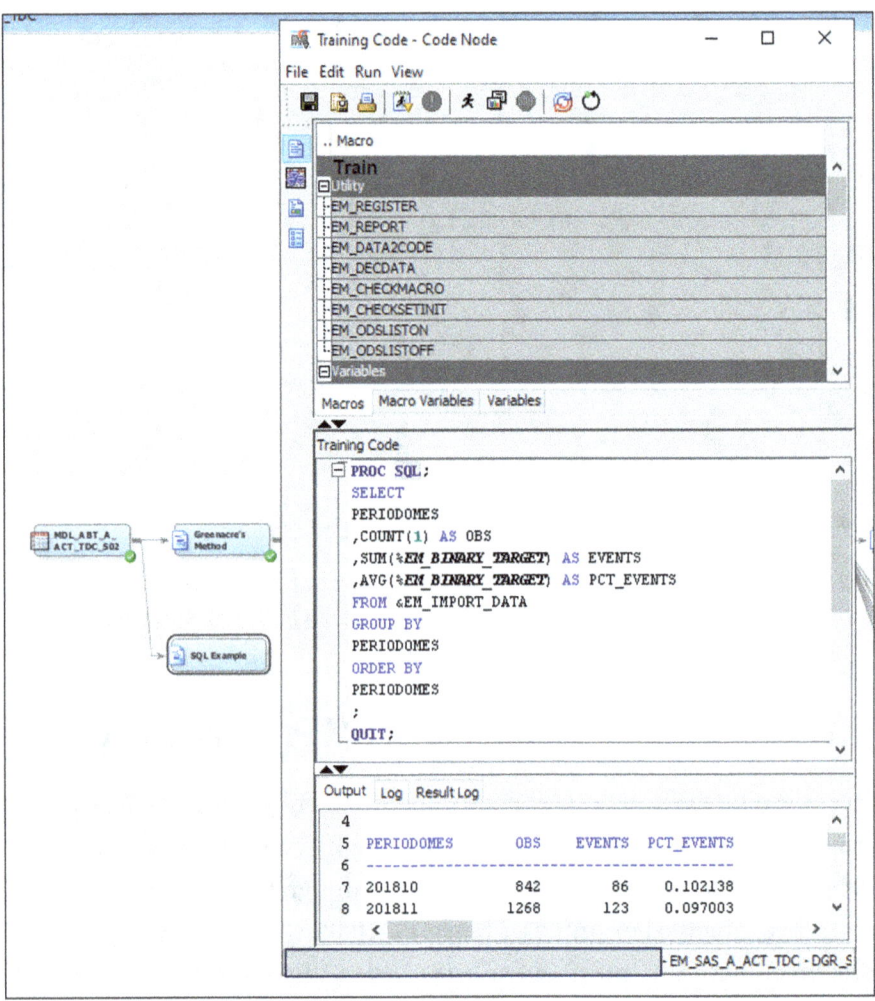

Figure 11-18. *The Training Code section shows an example of an SQL code that queries the number of observations, the number of events, and the proportion of events by period of the predecessor dataset, identified as &EM_IMPORT_DATA. The Output section shows the result obtained after executing the above query.*

CHAPTER 11　A BRIEF INTRODUCTION TO THE USE OF SAS® ENTERPRISE MINERTM

Figure 11-19 illustrates the manner in which any dataset present in the diagram at any given time can be exported with a CREATE TABLE clause, merely by referencing an existing library. Moreover, even in the absence of an assigned library, it is always feasible to incorporate a LIBRARY clause prior to the CREATE TABLE/DATA statement, thereby defining a chosen library. It is also noteworthy that the code editor user interface remains accessible throughout the execution process.

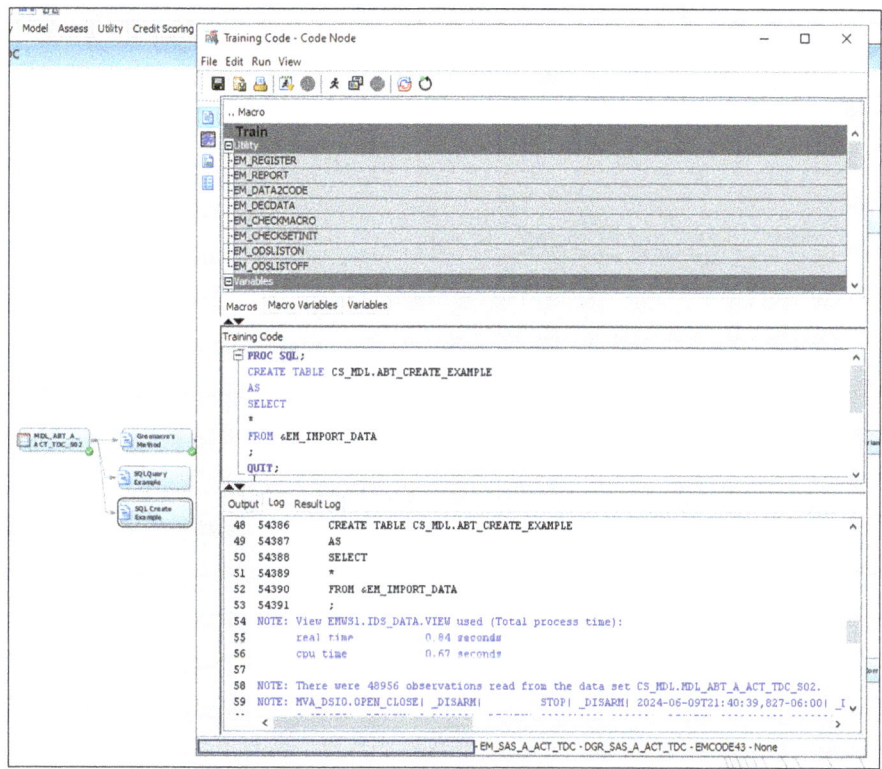

Figure 11-19. Example showing how to export any dataset, at any stage in your diagram using a CREATE TABLE clause.

CHAPTER 11 A BRIEF INTRODUCTION TO THE USE OF SAS® ENTERPRISE MINERTM

Figure 11-20 provides a more illustrative example of the CREATE TABLE clause. In this instance, the code node is directly linked to a scorecard node, thereby enabling the export of not only the original dataset but also any supplementary columns that result from the final scorecard model. It is crucial to recall that EM separates the training and validation partitions into separate datasets. Consequently, in contrast to the preceding example, the UNION ALL clause is essential for consolidating both partitions once more. It is also important to note that, in addition to the UNION ALL clause, a label must be added to identify each partition outside of EM.

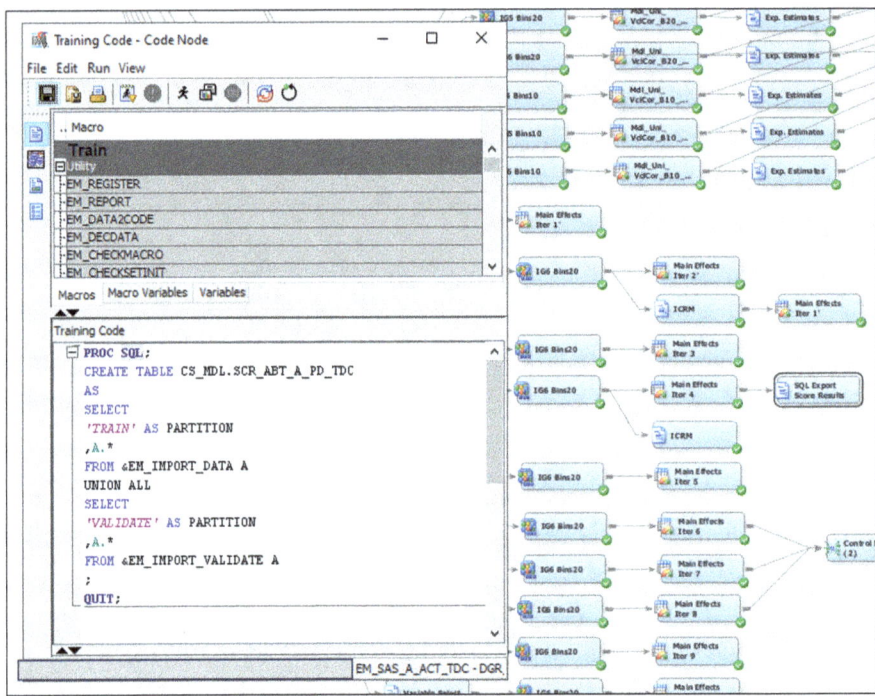

Figure 11-20. *The Training Code section shows an example of an SQL code querying the number of observations, the number of events, and the proportion of events by period of the predecessor dataset, identified as &EM_IMPORT_DATA. The Output section shows the result obtained after the execution of the above query (not shown in this screen).*

CHAPTER 11   A BRIEF INTRODUCTION TO THE USE OF SAS® ENTERPRISE MINER™

The exported dataset comprises the predicted probability of the event, the individual scores of the selected variables, the total score, and additional information. The ability to export these results is a convenient feature that allows users to further explore the results of their model outside of Enterprise Miner in a more programming-friendly interface, such as Enterprise Guide, or even from a database application, such as SQLDeveloper, or Toad. It should be noted that the ability to export the results of a model is not exclusive to the scorecard node. Rather, it is a capability that can be exercised with any EM modeling node, including decision trees, neural networks, gradient boosting, and so forth.

11.2 Data Mining Diagram Design

The construction of a data mining diagram should not be an improvised activity that results in a clumsy and messy series of failed attempts, which may inadvertently lead to the generation of a satisfactory model. A data mining diagram should be a structured, logical, and preplanned series of steps that have a high probability of converging on the construction of an excellent model. It is unfortunate that there is still a misconception that the mining process is a creative process, whereby one can try as many options as possible in a trial-and-error fashion, with no clear path, and without being bound by any rules or predefined methodologies. After all, that is what the creative process is all about, is it not?

Figure 11-21 provides an illustrative example of a data mining diagram for the construction of a scorecard model. It is evident that the diagram has not been created on the basis of improvisation; rather, it has been arranged in a systematic and orderly fashion, following a logical sequence of steps that resembles the structure of a multilayer neural network. It is notable that the diagram is symmetrical; this is not an accidental occurrence but rather a consequence of a balanced number of distinct treatments that contribute to the construction of different models or *variants* (more about variant models in Section 17.1).

CHAPTER 11 A BRIEF INTRODUCTION TO THE USE OF SAS® ENTERPRISE MINERTM

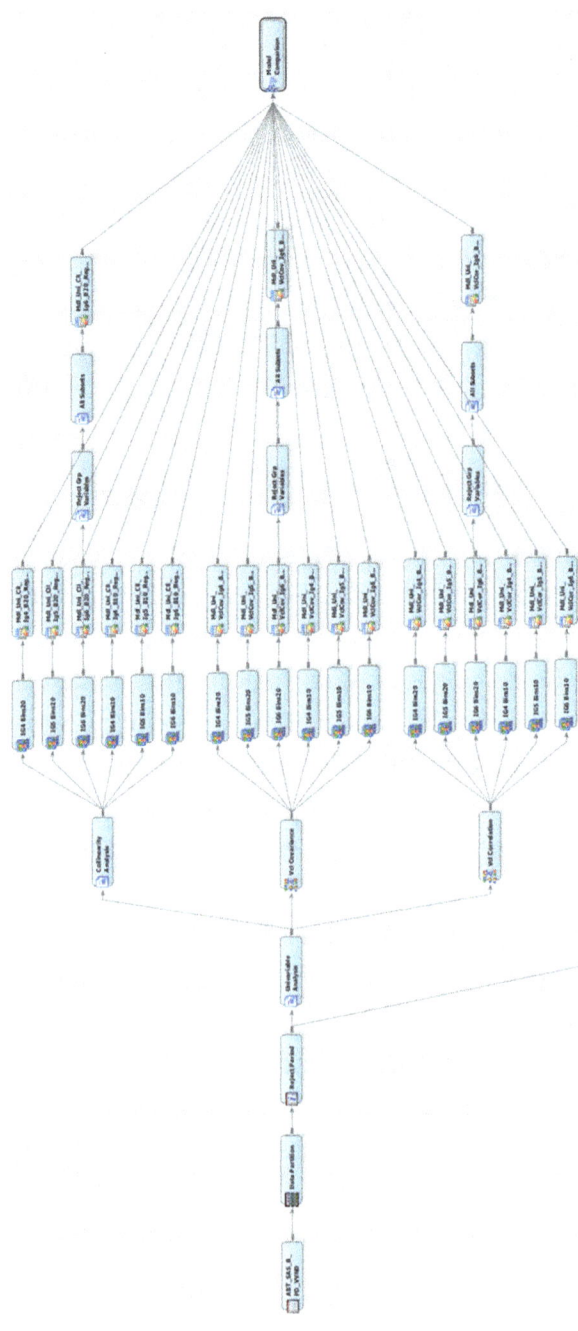

Figure 11-21. Example of an EM diagram to build scorecard models.

CHAPTER 11 A BRIEF INTRODUCTION TO THE USE OF SAS® ENTERPRISE MINERTM

11.2.1 The Six Basic Steps of the Mining Process

Figure 11-22 shows the same diagram as **Figure 11-21**, but **Figure 11-22** highlights the six basic steps of the mining process, or diagram:

1. Data partitioning
2. Univariable selection
3. Redundancy elimination
4. Automatic feature grouping
5. Multivariable selection
6. Models assessment and blending

The initial stage of any mining process[4] is the partitioning of the population into a training (treatment) sample and a validation (control) sample. This allows for *the objective assessment* of the performance of the model using an *unbiased sample* (for a thorough revision of the partitioning process, please see Chapter 12).

The second step is the identification and elimination of *irrelevant variables*, by measuring the individual explanatory power of each variable without any kind of interference from the interaction of other variables.

The third step is the identification and elimination of *redundant variables*, that is, variables that explain the same variability in the dataset.

---

[4] It is important to highlight that this book makes a distinction between the mining process and the statistical modeling process; in that, the statistical modeling process is a subset of the mining process. As a result, from the point of view of the mining process, data partitioning is the initial stage, whereas from the point of view of the statistical modeling, univariable analysis is the initial stage.

CHAPTER 11   A BRIEF INTRODUCTION TO THE USE OF SAS® ENTERPRISE MINERTM

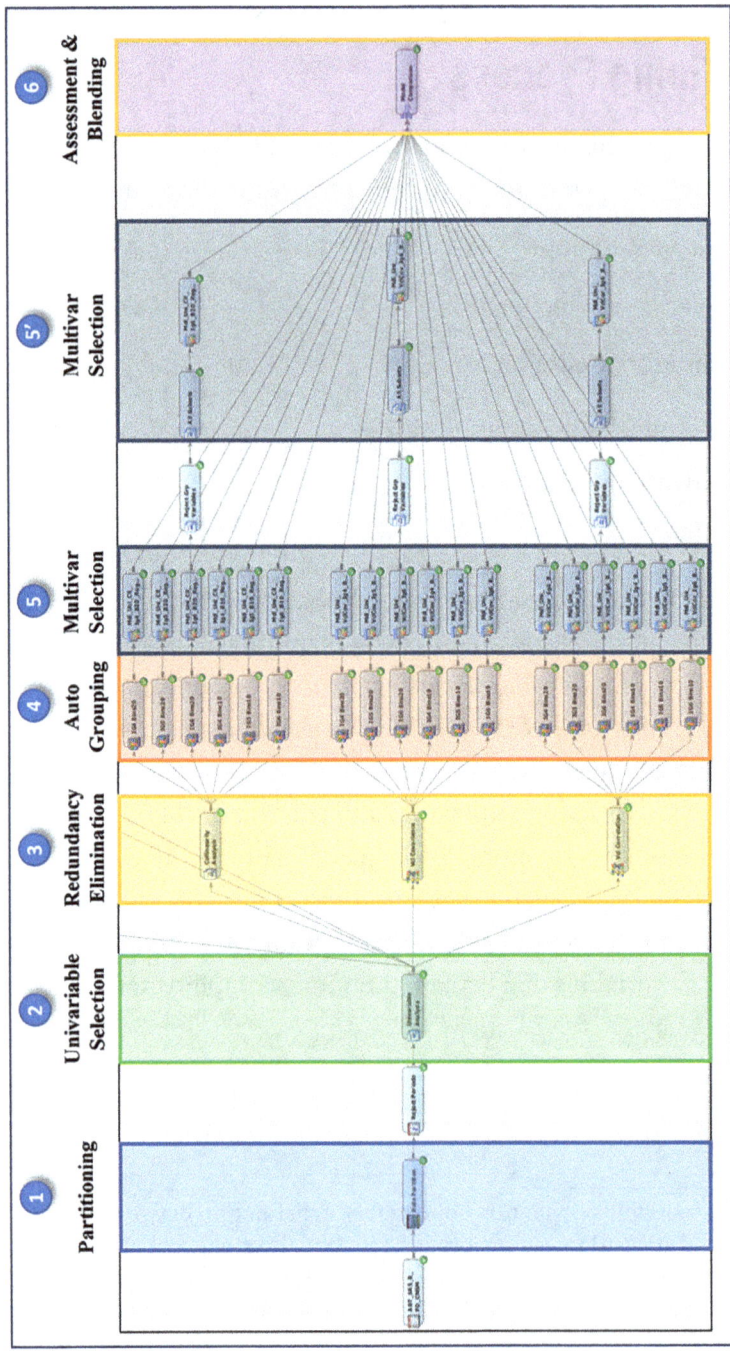

Figure 11-22. The six basic steps of the mining process, highlighted in different colors.

The fourth step represents a crucial phase in the modeling process. This process entails the automatic grouping of explanatory variables into a finite number of bins, which are capable of expressing a linear monotonic relationship between the independent and dependent variables. This step also entails *the standardization of all explanatory variables into a single unit of measurement and scale.* The variables resulting from this step are designated as *WoEs*.

The fifth step is to fit the *first multivariable model* using multivariable selection methods. It should be noted that **Figure 11-22** depicts two distinct multivariable selection procedures, designated as steps 5 and 5' (a detailed discussion on this topic will be presented in Chapter 16).

The sixth step concerns the assessment and combination of all previously fitted models. It is important to note that the objective here is not to select a single *"best model,"* but rather to explore and combine the optimal variables from each model or variant. Accordingly, the terms *"champion model"* and *"challenger model"* are inapplicable in this context.

11.2.2 Keep Improvised Tasks Out of Your Mining Diagram

The objective of this section is not to provide an exhaustive account of the elements depicted in the mining diagram. Instead, it aims to offer a general overview of the fundamental tasks and structural elements that a conventional mining diagram should encompass.

The second objective is to demonstrate how diagrams can be planned and designed in advance, thus eliminating the necessity for improvisation and informality. Nevertheless, supplementary steps and modifications to the ABT are occasionally introduced, which gives rise to the question of whether modifications to the ABT within the mining diagram are permissible.

CHAPTER 11 A BRIEF INTRODUCTION TO THE USE OF SAS® ENTERPRISE MINERTM

The majority of mining software incorporates a suite of data processing tools, many of which appear to be highly beneficial and, in fact, are. It is nevertheless recommended that processes be compartmentalized by isolating each specific set of tasks in its own workspace.

To illustrate, if one is constructing an ABT in Oracle utilizing cursors and PL programming, and subsequently identifies the necessity for alterations to the ABT while engaged in the mining diagram, it is advised that one revert to the original master building process and rebuild the ABT as required.

This action may appear to be excessive, but it is, in fact, the optimal course of action in the long term. To illustrate the argument, consider a scenario in which tasks from disparate workspaces are combined. In such a scenario, one might choose to apply additional filters and incorporate further variables into a mining diagram. Subsequently, a superior-quality scorecard model is successfully developed and the code for the model is extracted for deployment. It should be noted that the ABT code is also required for the scoring process, as the functionality of the model is dependent upon the ABT. The code for the model and the code for the ABT are then provided to an IT department responsible for automating the scoring process—but wait a minute—didn't you leave some part of your code inside of the mining diagram? Were the additional processing steps also extracted for production?

This is a typical operational risk scenario. The decision was made to improvise some tasks and misplaced them in the diagram, causing them to be lost during the extraction process. It is not uncommon for individuals to believe that they will remember a particular item or task at a later time— but this is rarely the case. It is therefore advisable to go back and modify the construction source code rather than improvise within the mining diagram. The adage *"it's better to be safe than sorry"* is applicable here, so instead of relying on memory, rely on safety.

11.3 The Importance of Detail

One of the key objectives of this book is to show readers that the various tasks involved in the construction of a model are inextricably intertwined. This is because the model-building process is the aggregate of multiple elements, each of which is indispensable to the ultimate outcome. For instance, think of the modeling process as a house of cards, wherein each card represents a task delicately balanced on top of the others. Now imagine what would happen if a single card were misplaced.

To illustrate, one might consider the investment of time and attention required to construct an analytical base table (ABT) to feed a model. This process frequently necessitates meticulous examination and verification of each explanatory variable, which may be regarded as a valuable resource. Now, imagine for a moment, a complex mining diagram which employs custom code nodes with sophisticated algorithms that have been developed. Now imagine a model of considerable complexity, comprising an ABT of five million records and over a thousand variables. For the sake of illustration, also assume that each iteration of the mining diagram takes more than 48 hours to produce a result. In addition to this, consider that assessing multiple candidate models (variants) requires an additional day of work. Now imagine a situation where the final model needs to be presented tomorrow morning, and you and your team realize that crucial nominal variables with numeric codes may have been misclassified as interval instead of nominal. Would you consider this to be a tragedy? Can you relate to such a situation?

As illustrated in the preceding example, correctly categorizing the role and level of variables is a crucial but often overlooked aspect of data management. The potential consequences of misclassifying variables, even those initially perceived as trivial, underscore the importance of thorough and precise variable management. However, the crucial point to be derived from this section is not that the definition of the role and level of variables is of paramount importance, but the fact that attention to detail matters, and matters a great deal.

11.4 Summary

In this chapter, we explored the foundational concepts and functionalities of **SAS° Enterprise Miner (EM)**, a powerful tool for predictive modeling. While not exhaustive, this introduction provided the essential knowledge required to effectively navigate the tool and prepare for its application in subsequent chapters. By understanding the basics of EM, you are now equipped to leverage its capabilities in building, evaluating, and refining predictive models.

Key takeaways:

1. **Purpose of SAS° Enterprise Miner**: EM is designed to streamline the process of building predictive models through its user-friendly interface and robust analytical capabilities.

2. **Core Features**: We covered essential features such as data exploration, variable transformations, and model-building nodes.

3. **Focus on Fundamentals**: This chapter emphasized the fundamental concepts necessary for understanding and utilizing EM in the context of this book.

4. **Utility Macros for Automation**: Utility macros play a crucial role in automating repetitive tasks, such as variable selection, inside EM. These tools not only save time but also enhance consistency and efficiency in the modeling process.

5. **The Power of the Code Node**: The **code node** is a critical feature that allows users to extend the capabilities of EM beyond its standard functionalities. By incorporating custom SAS

CHAPTER 11 A BRIEF INTRODUCTION TO THE USE OF SAS® ENTERPRISE MINERTM

code, users can implement virtually any analysis, transformation, or modeling technique they can think of, making EM a highly flexible and customizable tool.

6. **Recommendations**: This chapter has shown how to address operational risk by promoting task compartmentalization—that is, avoiding the contamination of workspaces with improvised tasks. Additionally, the chapter showed us the importance of details by explaining that the various tasks involved in the construction of a model are inextricably intertwined, thereby highlighting the fact that all tasks are important, no matter how seemingly trivial they may seem.

7. **Additional Resources**: For a deeper understanding of EM, readers are encouraged to consult *Predictive Modeling with SAS® Enterprise Miner™* by Kattamuri S. Sarma and the SAS training manual *Applied Analytics Using SAS® Enterprise Miner™*.

With these foundational concepts in mind, you are now ready to move forward and begin applying EM to real-world modeling tasks.

CHAPTER 12

Partitioning

"If you torture your data long enough, they will tell you whatever you want to hear."

—James L. Millis (1993)

"What we see depends mainly on what we look for."

—John Lubbock, The Beauties of Nature and the Wonders of the World We Live In (1892)

In statistical terms, input partitioning is defined as the division of the initial input into two distinct samples: a treatment sample, also referred to as a training partition, and a control sample, also known as a validation partition. Accordingly, the most crucial element to understand in input partitioning is the *representativeness* of the samples/partitions. The term *"representativeness"* is defined as *the homogeneous composition of characteristics between the training and validation partitions.*

CHAPTER 12 PARTITIONING

12.1 Stratified Sampling and Representativeness

Representativeness can be achieved through the use of *stratified sampling*. Stratified sampling permits the definition of an unlimited number of characteristics, ensuring that the proportion of each category remains constant in both resulting samples. In practice, however, incorporating all nominal variables as independent strata is seldom feasible, if ever.

As illustrated in Section 9.3, the replication of 4,320 groups, comprising six nominal variables, represents a significant challenge. However, the primary concern is not the number of groups to be replicated in the training and validation partitions. Instead, it is the number of events present in each group and whether the lack of events is due to a lack of representativeness or a genuine absence of risk, which is highly unlikely.

It is therefore recommended that the number of dimensions included in the sample stratification be limited to a minimum of three dimensions:

1. Space
2. Time
3. Risk

12.1.1 Spatial Dimension

The *spatial dimension* pertains to any variable that is capable of segmenting the population into distinct geographic locations. As previously discussed in Section 6.3, the proximity of individuals to one another exerts a profound influence on their habits and values. This is particularly evident at the community level, such as the neighborhood level, where the population is relatively small and homogeneous, facilitating the formation of shared norms and attitudes. In contrast, at the state level, the population is vast and heterogeneous, rendering it challenging to discern common habits and values.

Consequently, the ZIP code is a superior geolocation variable to the state, as it is more probable that communities with analogous levels of risk will be isolated, thus facilitating the identification of communities with similar habits and values. As a result, utilizing the customer's ZIP code as a stratification variable is an optimal initial approach to stratification, as it ensures geographic representativeness within the samples.

12.1.2 Time Dimension

It should be noted that the analysis employs a stacked ABT, which comprises a series of consecutive time periods that may span several years. The issue is that behavioral patterns do change in relation to the month of the year, and economic cycles change across years. Accordingly, it is essential to ensure temporal representativeness in our samples, ensuring that each month and year in our population is represented equally in our training and validation samples.

12.1.3 Risk Dimension

The risk dimension pertains to the target variable, which is defined as the default event occurring within the next 12 months. Given its nature, the target variable is also a nominal variable with two categories: 0, indicating the absence of the event, and 1, indicating its presence. Consequently, employing the target variable as a stratification variable guarantees that the same proportion of events will be present in both the training and validation partitions.

CHAPTER 12 PARTITIONING

12.2 Stratified Sampling and Overtraining

We can define *overtraining as an overfitting of our model that precludes the generalization of its predictions, making them relevant only in the context of the training data*. To illustrate, imagine a boxer training with his sparring partner in preparation for a fight—could you now tell what would happen if our boxer focused all of his training on developing the tactics and movements necessary to defeat his sparring partner? Would he be able to defeat his real opponent in a real fight?—The answer is that he most likely would not.

The previous example helped to explain the concept of overtraining in a palpable manner, where the boxer is tantamount to *the model*, the sparring partner is tantamount to *the training dataset*, and the real opponent is tantamount to *the validation dataset*. Fortunately, the stratified sampling can reduce the likelihood of *overtraining*. However, it is not the stratification itself that is responsible for this reduction; rather, it is the *statistical representativeness of the sample*. Nevertheless, it is regrettable, that even in the age of ML and AI, many analysts still fail to grasp this concept. They believe that sophisticated algorithms will compensate for their lack of understanding of basic statistical concepts.

It is not uncommon to observe even experienced analysts encountering significant challenges in overcoming severe levels of overtraining (ROC difference >5) in their models. Despite their efforts, they may only achieve marginal improvements, such as a one- or two-point reduction in the ROC index difference (see **Table 17-4** for a suggested list of overtraining levels, based on ROC points difference), between the training and the validation partition, which is insufficient to meaningfully reduce a severe overtraining. At this juncture, it is important to acknowledge that in most cases, the root cause for overtraining is a lack of representativeness in the samples—that is, *the two samples are not properly balanced in terms of their characteristics*.

Stratified sampling can address overtraining in models but is not a guaranteed solution. Rather, the key to preventing overtraining lies in ensuring statistical representativeness. To illustrate, consider using ZIP code as an explanatory variable in a model without stratification of sample by ZIP code. Now, consider that the number of ZIP codes is not only unequally distributed between the two partitions, but they are also noninclusive. It is evident that the model will be unable to generalize a rule for predicting risk levels based on ZIP codes, as the distribution of ZIP codes was not uniform across both samples.

The reason for the increase in the overtraining effect is that the model will be overly optimistic about its ability to predict the default event based on the ZIP code sample that it was given. In reality, however, the model was never trained with the full set of ZIP codes and their corresponding weights; instead, it ended on a biased sample of them. To illustrate, consider the following scenario: You are enrolled in a university course and have been informed by your professor that the final examination will be based on two distinct sets of material, A and B. As a diligent student, you approach the examination with confidence. However, on the examination day, upon reviewing the examination questions, it becomes evident that half of the questions are not aligned with the material that has been studied. Furthermore, you have not previously encountered these topics. In light of the aforementioned information, would you maintain the same level of confidence in your ability to excel on the examination, given the discrepancy between your expectations and the actual content covered in the examination?

It has been demonstrated that ZIP code can be utilized as both a stratification and an explanatory variable. However, this does not imply that variables employed for stratification should always be used as explanatory variables. This suggests that the stratification of a sample should always be based on a geographic variable, irrespective of whether the same variable is used as an explanatory variable.

12.3 Training, Validation, and Test Partitions

The use of three partitions, namely, training, validation, and test, is a common practice during the development of data mining models. However, this is not a mandatory requirement and depends entirely on the statistical methods employed. For instance, the utilization of three partitions is justified when the validation partition is involved in the training process, particularly in instances where overtraining is to be avoided, as is the case for methods such as

1. Validation error
2. Validation misclassification
3. Cross-validation error
4. Cross-validation misclassification

The methodology presented in this book does not employ any of the aforementioned methods and does not utilize validation partitioning as an *auxiliary* tool for training the model. It can thus be guaranteed that the validation partitioning used in this methodology is unbiased and that the use of a third partitioning for testing is unnecessary.

12.4 Data Proportion Between Training and Validation

There is a certain degree of controversy surrounding the proportion that should be employed for training and validation purposes, which has led to the emergence of certain superstitious beliefs. These beliefs include the notion that there is something inherently special about the proportions 80%/20%, 70%/30%, or 60%/40%. Nevertheless, it is evident that the ratio between training and validation is inconsequential, provided that the samples are statistically representative. Accordingly, it is acceptable to utilize a 50/50 partition between training and validation.

Furthermore, a 50/50 split presents a greater challenge in terms of validation due to the size and representativeness of the sample, in comparison to a smaller validation sample of 20% or 30%. This is because smaller validation samples may lack some of the characteristics that are present only in the training partition. As a result, the performance of the model may be overestimated and may not be a true characterization of the model's performance under real-world conditions. However, when should a proportion greater than 50% be used for training?

The sole circumstance in which it is advisable to augment the training proportion beyond 50% is when working with limited samples, where ensuring representativeness is challenging. Therefore, increasing the training sample is not a matter of preference; rather, it is a matter of necessity.

12.5 Applied Example of a Stratified Sample Analysis

Figure 12-1 illustrates a percentile analysis of an ABT comprising 655,051 records. The ABT has been divided into two partitions, one for training and one for validation, in a 50/50 ratio, with a total of 327,085 and 327,966 records, respectively (cells C9 and D9).

CHAPTER 12   PARTITIONING

The two stratum variables, period and ZIP code, were aggregated for both partitions, and the third stratum, the target, was used as a summary variable. **Listing 12-1**[1] illustrates the queries employed to extract and aggregate the results from each partition.[2]

Listing 12-1. SAS PROC SQL example to analyze the distribution of the stratified sample. (A) SQL used to extract the training sample. (B) SQL used to extract the validation sample.

(A)
```
 1  PROC SQL;
 2    CREATE TABLE &EM_LIB..SAMPLE_PROPORTION_TRAIN
 3    AS
 4    SELECT
 5      PERIOD
 6      ,ZIP_CODE
 7      ,COUNT(1) AS OBS
 8      ,AVG(%EM_BINARY_TARGET) AS P_EVENT
 9      ,SUM(%EM_BINARY_TARGET) AS EVENTS
10    FROM &EM_IMPORT_DATA
11    GROUP BY
12      1
13      ,2
14    ;
15  QUIT;
```

(B)
```
17  PROC SQL;
18    CREATE TABLE &EM_LIB..SAMPLE_PROPORTION_VALID
19    AS
20    SELECT
21      PERIOD
22      ,ZIP_CODE
23      ,COUNT(1) AS OBS
24      ,AVG(%EM_BINARY_TARGET) AS P_EVENT
25      ,SUM(%EM_BINARY_TARGET) AS EVENTS
26    FROM &EM_IMPORT_VALIDATE
27    GROUP BY
28      1
29      ,2
30    ;
31  QUIT;
```

It should be noted that three metrics are calculated for each grouping level: the total number of observations, designated as "OBS"; the total number of events, designated as "EVENTS"; and the proportion of events, designated as "P_EVENT". The lower portion of **Figure 12-1** depicts an excerpt from the final report, presented in an Excel spreadsheet. The report contains the following information:

- **PERIOD:** Stratification variable used as a grouping variable

---

[1] The code is available at https://github.com/saulzapiain/Conceptual VariableDesign. File: Listing <InternalRef RefID="PC1">12-1 Stratified Sample Analysis.sas.

[2] Notice that the extraction code is using EM macros and macro-variables to make reference to the output library, the target variable, and each corresponding partition table, where &EM_IMPORT_DATA corresponds to the training partition and &EM_IMPORT_VALIDATE corresponds to the validation partition.

- **ZIP_CODE:** Stratification variable used as a grouping variable
- **Train Observations:** Total number of records per grouping level for the train partition
- **Validation Observations:** Total number of records per grouping level for the validation partition
- **Train % Event:** Proportion of events per grouping level for the train partition
- **Validation % Event:** Proportion of events per grouping level for the validation partition
- **Train Events:** Frequency of events per grouping level for the train partition
- **Validation Events:** Frequency of events per grouping level for the validation partition
- **Abs. Diff.:** Absolute difference between the Train Events and the Validation Events
- **Abs. Prop. Deviation:** The absolute deviation between the Train % Event and the Validation % Event
- **Abs. Freq. Deviation:** The absolute deviation between the Train Events and the Validation Events

CHAPTER 12 PARTITIONING

	A	B	C	D	E	F	G	H	I	J	K
1			Train Observations	Validation Observations	Train % Event	Validation % Event	Train Events	Validation Events	Abs. Diff.	Abs Prop. Deviation	Abs Freq. Deviation
2		Groups	2,916	2,916	2,916	2,916	2,916	2,916	2,916	2,916	2,916
3		Mean+CI	117.1	117.4	2.9%	1.7%	2.50	2.29	1.00	19.8%	20.1%
4		Mean	112.2	112.5	2.8%	1.7%	2.40	2.17	0.97	18.8%	19.1%
5		Mean -CI	107.3	107.6	2.7%	1.6%	2.29	2.06	0.95	17.8%	18.1%
6		CI95%	4.8941	4.9048	0.1%	0.1%	0.10	0.12	0.03	1.0%	1.0%
7		Std.Dev.	134.8406	135.1343	3.2%	2.2%	2.84	3.20	0.73	27.7%	28.3%
8		Obs. Sum.	327,085	327,966	82	49	6,987	6,334	2,837	548	557
9		Max.	1060	1061	33.3%	18.2%	32	34	2	113.3%	100.0%
10		99%	863.85	864	16.7%	9.1%	16	16	2	102.0%	100.0%
11		95%	343	342.25	8.3%	5.5%	7	8	2	91.8%	100.0%
12		90%	241.5	243	5.7%	4.4%	5	5.5	2	50.0%	50.0%
13		80%	142	143	4.0%	3.2%	3	4	2	46.7%	50.0%
14		75%	127	127	3.4%	2.8%	3	3	2	33.3%	33.3%
15		50%	80	81	2.1%	1.1%	2	1	1	0.3%	0.0%
16		25%	38	38	1.1%	0.0%	1	0	0	0.0%	0.0%
17		10%	16	16	0.0%	0.0%	0	0	0	0.0%	0.0%
18		5%	8	8	0.0%	0.0%	0	0	0	0.0%	0.0%
19		Min.	3	3	0.0%	0.0%	0	0	0	0	0
22	PERIOD	ZIP_CODE	Train Observations	Validation Observations	Train % Event	Validation % Event	Train Events	Validation Events	Abs. Diff.	Abs Prop. Deviation	Abs Freq. Deviation
23	202004	5-7	15	16	0.1333	0.0625	2	1	1	113.33%	100.00%
24	201907	7-4	21	22	0.0952	0.0455	2	1	1	109.52%	100.00%
25	201806	5-7	22	23	0.0909	0.0435	2	1	1	109.09%	100.00%
26	201801	7-5	22	23	0.0909	0.0435	2	1	1	109.09%	100.00%
27	201809	7-5	24	25	0.0833	0.0400	2	1	1	108.33%	100.00%
28	201909	1-12	30	31	0.0667	0.0323	2	1	1	106.67%	100.00%
29	201907	2-8	39	40	0.0513	0.0250	2	1	1	105.13%	100.00%
30	201908	2-8	41	42	0.0488	0.0238	2	1	1	104.88%	100.00%
31	202010	6-5	41	42	0.0488	0.0238	2	1	1	104.88%	100.00%
32	201904	6-5	41	42	0.0488	0.0238	2	1	1	104.88%	100.00%
33	201810	5-8	58	59	0.0345	0.0169	2	1	1	103.45%	100.00%
34	202008	1-7	58	59	0.0345	0.0169	2	1	1	103.45%	100.00%
35	201802	5-8	58	59	0.0345	0.0169	2	1	1	103.45%	100.00%
36	201801	2-9	65	66	0.0308	0.0152	2	1	1	103.08%	100.00%
37	201902	2-9	67	68	0.0299	0.0147	2	1	1	102.99%	100.00%
38	201907	1-6	77	78	0.0260	0.0128	2	1	1	102.60%	100.00%
39	202005	6-9	80	81	0.0250	0.0123	2	1	1	102.50%	100.00%
40	202010	4-5	81	82	0.0247	0.0122	2	1	1	102.47%	100.00%
41	202001	6-3	82	83	0.0244	0.0120	2	1	1	102.44%	100.00%

Figure 12-1. Example of a distribution analysis for the resulting stratified sample.

It is important to notice that the grouping levels in this report are sorted by the Abs. Prop. Deviation, in descending order.

The top portion of **Figure 12-1** shows the 5th, 10th, 25th, 50th, 75th, 80th, 90th, 95th, and 99th percentiles, the mean, the 95% confidence intervals (CI), the standard deviation, the minimum and maximum value for each metric, and the total number of grouping levels, which is 2,916 levels.

Figure 12-2 presents a line chart comparing the percentiles of the total number of observations between the training and validation partitions (**Figure 12-2A**) and the event frequency between the training and validation partitions (**Figure 12-2B**). It is notable that in both plots, the total number of observations and event frequency exhibit substantial overlap between the training and validation percentiles. This indicates that the stratified sampling has effectively distributed both observations and events across the grouping levels of the strata in a proportional manner.

CHAPTER 12　PARTITIONING

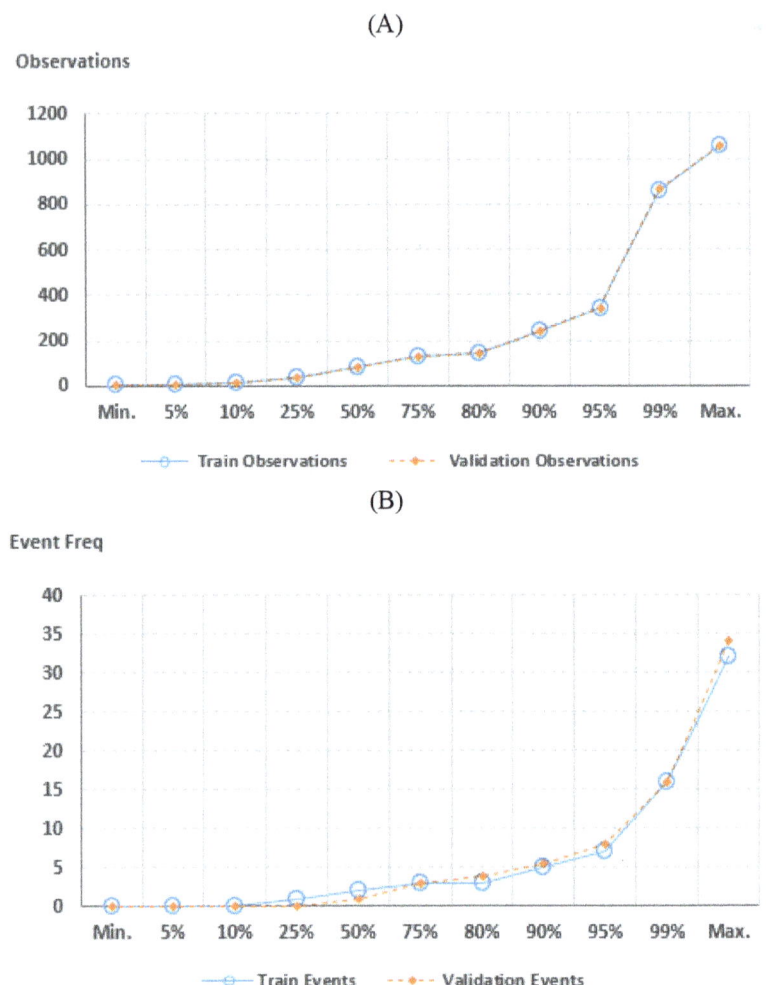

Figure 12-2. *Line graphs comparing percentile cutoffs between the training and the validation partitions. (A) Percentiles of the total number of observations per stratification level. (B) Percentiles of the frequency of events per grouping level.*

Figure 12-3 also presents a line chart comparing the percentiles between the training and validation partitions. In this case, the metric is the proportion of events per grouping level (columns E and F in the spreadsheet). It is notable that the lines do not completely overlap and in fact appear to diverge as we proceed to higher percentiles. This is due to the fact that the stratification sampling algorithm strives to distribute the number of records and events in a proportional manner between the training and validation partitions. However, when a small odd number of events is present, the algorithm exhibits a significant deviation from this objective.

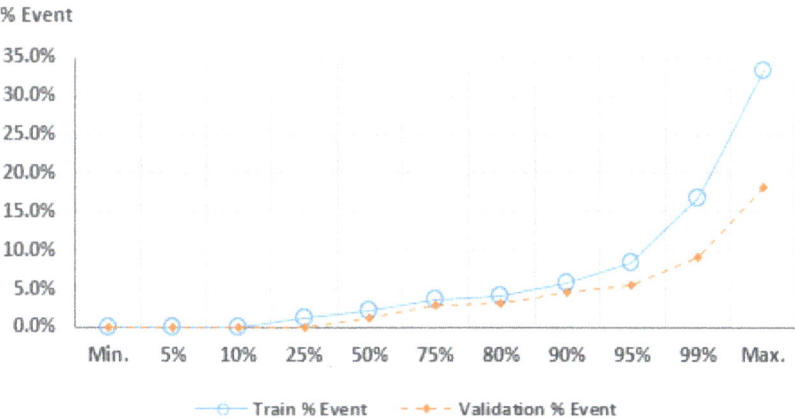

Figure 12-3. *Line chart comparing percentile cutoffs between the training and validation partitions for the proportion of events per grouping level.*

An illustrative example can be found in the lower portion of **Figure 12-1**, spanning rows 23 to 41. As can be observed in this summary, each stratification level comprises a mere three events (columns G and H), which must be distributed between the training and validation partitions. The stratification sampling algorithm consistently prioritizes the training partition over the validation partition, allocating two of the three events to the former and only one to the latter. This results in a significant percentage deviation, leading to considerable discrepancies between the proportion of events in the training and validation samples across all strata when a small odd number of events is present.

CHAPTER 12 PARTITIONING

It is unfortunate that there is no way to circumvent this issue without resorting to mathematical artifices. However, the extent to which this affects the model is dependent on a number of factors.

For example, although there may not appear to be an adequate number of events to perfectly stratify the sample by the three dimensions of time period, ZIP code, and risk, it is important to consider that such a situation is unlikely to arise. Furthermore, a minor lack of representativeness in some grouping levels does not necessarily outweigh the benefits of a stratified sampling approach.

Table 12-1 presents a comparison between the initial population risk level and the risk level of the final stratified samples, training set, and validation set. It is notable that the absolute difference between the partitions and the initial population is nearly identical, with a mere tenth of a percentage point discrepancy.

Table 12-1. Comparison of the risk level of the initial population and the risk level of the final stratified samples, training, and validation.

	% Events	Difference
Train	2.14%	0.102%
Validation	1.93%	−0.103%
Population	2.034%	

However, given the total number of observations in the training partition (327,085), the frequency distribution by percentile, and the absolute difference between the number of events in the training partition

CHAPTER 12 PARTITIONING

and the number of events in the validation partition (approximately one event), it would be advisable to accept the difference and continue building the model.[3]

12.6 Summary

Partitioning is a fundamental concept in statistical analysis, grounded in the principle that the effect of a treatment cannot be properly measured without an unbiased control population for comparison. However, selecting the training (*treatment*) and validation (*control*) samples is far from straightforward. As discussed in this chapter, the cornerstone of proper partitioning lies in statistical representativeness, defined as *the homogeneous composition of characteristics between the training and validation partitions.*

This understanding highlights that the specific proportions of these partitions (e.g., 50/50, 80/20) are inconsequential, except in cases of severe data scarcity. Furthermore, with a clearer grasp of the concepts of *"treatment"* and *"control,"* it becomes evident that three partitions (training, validation, and test) are only necessary when the validation partition is explicitly used to monitor and control overtraining. In all other scenarios, two partitions—whether training and validation or training and test—are sufficient for modeling purposes.

Key takeaways:

1. **Partitioning as a Statistical Foundation:**
   Partitioning is grounded on the principle that you cannot measure the effect of a treatment without an unbiased control population for comparison.

---

[3] As expected, that there were no overtraining issues during the construction of the model, proving that the miniscule difference was indeed negligible.

2. **Statistical Representativeness:** The key to proper partitioning is ensuring statistical representativeness, defined as the homogeneous composition of characteristics between the training and validation partitions.

3. **Proportions Are Inconsequential:** The specific proportions of the partitions (e.g., 50/50, 80/20) are inconsequential unless there is a severe lack of information.

4. **When Three Partitions Are Necessary:** Three partitions (training, validation, and test) are only necessary when the validation partition is used to monitor and control overtraining.

In the next chapter, we will begin our statistical analysis by introducing a preselection method known as *univariable analysis*. Building on the foundation of partitioning, the goal will be to reduce the number of input variables by discarding those that lack a strong association with the target variable. Additionally, Chapter 13 will provide a detailed explanation of the accompanying code and demonstrate how to implement it using the SAS code node in SAS® Enterprise Miner.

CHAPTER 13

Univariable Analysis

"Divide each difficulty into as many parts as is feasible and necessary to resolve it."

—René Descartes

The initial stage of the statistical modeling process is the univariable analysis[1] [32]. The objective of this analysis is to reduce the total number of input variables in the model by eliminating those predictors that do not demonstrate a strong association with the target variable at a reasonable level of significance.

The analysis is conducted by fitting an independent logistic regression for each candidate input variable prior to the implementation of any additional selection processes or transformations. This is a fundamental aspect of the univariable analysis, as it allows for the assessment of whether the input variables possess the capacity to predict the outcome without the influence of partial associations or mathematical transformations. The objective of the univariable analysis is to ascertain the independent significance level of each input variable.

The hypothesis testing is relatively straightforward: does the univariable model significantly predict the outcome in comparison to

---

[1] It is important to note that the partitioning process is the first stage in the mining process, or diagram, whereas the univariable analysis is the first stage of the statistical modeling process.

the baseline model? In the affirmative case, the predictor is accepted and advances to the subsequent step in the workflow. In the event of a negative response, the predictor is rejected. However, it is important to define what constitutes a significant predictor in this context.

According to Hosmer and Lemeshow [32], a *p*-value <0.25 is a good selection criterion for the univariable test. They mention that the reason for using a 0.25 level as a cutoff point is based on the empirical observation that a more conventional level, such as 0.05, frequently fails to identify variables with partial associations in multivariable models. Conversely, adopting a more relaxed threshold may inadvertently include variables of uncertain importance, impeding rather than facilitating the model-building process.

It is therefore evident that the ultimate objective of univariable analysis is not to significantly reduce the number of input variables. Instead, its fundamental objective is to ensure the safe elimination of those variables that are clearly *statistical irrelevant.*

13.1 The Univariable Code: A Step-by-Step Explanation[2]

It is imperative to acknowledge that the contemporary predicament does not stem from an insufficiency of data for model construction; rather, it is characterized by an overwhelming abundance of information, compounded by a lack of effective strategies for its utilization. In Section 5.3, we estimated that we could theoretically construct 3,132 explanatory

---

[2] The complete code can be downloaded from https://github.com/saulzapiain/ConceptualVariableDesign. File: Listing 13- Univariable Analysis (complete).sas.

CHAPTER 13 UNIVARIABLE ANALYSIS

variables, which is equivalent to fitting 3,132 logistic regressions. It is evident that the utilization of an automated process will be instrumental in accomplishing this task.

The univariable code that I have developed over the past few years is capable of assisting analysts in implementing the univariable analysis, irrespective of the number and type of variables that they wish to test. The code is structured around ten principal steps:

1. Set execution parameters
2. Log rerouting
3. Parameter initialization
4. Get reference category for nominal variables
5. Create intermediate storing tables
6. Start main loop and data validation
    a. Assemble class statement (when needed)
    b. Check and reject variables that does not comply data integrity constraints
7. Run regression
8. Estimate significance and save results
9. Loop to the next variable in the ABT (go to step 6)
10. Save final results and reject nonsignificant variables

13.1.1 Setting Execution Parameters

The first step before executing the program is to set the initial parameters, which are

- Log output path
- The cutoff point for your analysis

CHAPTER 13 UNIVARIABLE ANALYSIS

- EM output library
- EM node ID
- User ID
- Select whether you want to generate logs and outputs in EM or to reroute them
- Time initialization parameters

As illustrated by **Listing 13-1**, the prerequisite parameters must be defined prior to executing the univariable code. Line 1 delineates the output path of the logs, a step that is only applicable in the event that the generation of logs in EM is not desired. The second line defines the statistical cutoff level of the analysis, indicating that any variable equal to or greater than the specified cutoff will be rejected. Lines 3 to 5 define the EM and SAS macro-variables. &EM_LIB refers to the OS-level directory of the EM diagram, &EM_NODEID refers to the EM code node ID, which is used for identification purposes, and &SYSUSERID is the user ID of the person executing the analysis. Line 6 enables or disables the generation of EM logs and outputs, with the parameter GenerateOutput set to N, enabling the redirection of all logs and outputs of an execution to a specified DIR location. Lines 10 through 16 define a series of macro-variables that are used to estimate the elapsed execution time and to name and identify the logs and output from the execution.

CHAPTER 13　UNIVARIABLE ANALYSIS

Listing 13-1. Initialization parameters.

```
1   %LET DIR=/warehouse/Models/Output_Logs;  /*If GENERATEOUTPUT=N then you must define the output path*/
2   %LET CUTOFF=0.25;                        /*Recommended significance for univariable analysis*/
3   %LET LIB=&EM_LIB.;                       /*Default Diagram library*/
4   %LET MDL=&EM_NODEID.;                    /*Use the node ID as identifier for output tables*/
5   %LET USER=&SYSUSERID.;                   /*User ID*/
6   %LET GENERATEOUTPUT=N;                   /*N: Don't generate EM output, Y: Generate EM output*/
7   /*-----------------------------------------------------------------
8       --- Timestamp format for log and output
9   -----------------------------------------------------------------*/
10  %LET DATESTART     = %SYSFUNC(DATE());
11  %LET DATESTAMP     = %SYSFUNC(SUM(&DATESTART.),YYMMDDN8.);
12  %LET TIMESTART     = %SYSFUNC(TIME());
13  %LET TIMESTAMP     = %STR(%SYSFUNC(SUM(&TIMESTART.),TIME8.0));
14  %LET TIMESTAMP2    = %SYSFUNC(TRANWRD(&TIMESTAMP.,:,));
15  %LET TIMESTAMP3    = %SYSFUNC(COMPRESS(&TIMESTAMP2.));
16  %LET DATETIMESTAMP = &DATESTAMP._&TIMESTAMP3;
```

Prior to execution, the univariable macro necessitates the incorporation of several supplementary parameters (Listing 13-2). The initial parameter is the designation of the dataset name (line 316), which is defined by the EM macro-variable &EM_IMPORT_DATA. The target variable (line 317) is defined by the EM macro %EM_BINARY_TARGET, and the list of all variables is defined by the EM macros %EM_BINARY_INPUT, %EM_INTERVAL_INPUT, and %EM_NOMINAL_INPUT. The &LIB macro-variable refers to the EM library, and the next two parameters, partition_col and partition_val, are only utilized when the code is employed outside of EM. The subsequent four parameters pertain to the redirection of EM results. In the event that the &GenerateOutput macro-variable is set to "N", it will be necessary to define the log_printto and the output_printto locations. It is noteworthy that both of these macro-variables have all their parameters previously defined by the macro-variables declared above.[3]

---

[3] It is important to note that the &MDL macro-variable refers to the &EM_NODEID macro-variable. This enables the execution of univariable code in multiple nodes within a single diagram. This process prevents the potential for data loss or the overwriting of results.

Listing 13-2. Univariable macro parameters.

```
314  %M_UNIVARIABLE
315  (
316  DSNAME=&EM_IMPORT_DATA.
317  ,DEPENDENT=%EM_BINARY_TARGET
318  ,COVAR= %EM_BINARY_INPUT %EM_INTERVAL_INPUT %EM_NOMINAL_INPUT
319  ,NOMINAL=%EM_NOMINAL_INPUT
320  ,LIB=&LIB.
321  ,PARTITION_COL=
322  ,PARTITION_VAL=
323  ,MDL=&MDL.
324  ,GENERATEOUTPUT=&GENERATEOUTPUT.
325  ,LOG_PRINTTO=&DIR./&DATETIMESTAMP._&USER._UA_LOG_&MDL..LOG
326  ,OUTPUT_PRINTTO=&DIR./&DATETIMESTAMP._&USER._UA_OUTPUT_&MDL..OUT
327  )
328  ;
```

13.1.2 Log Rerouting

It is a well-known issue among users of EM that the software has a tendency to truncate the logs and outputs of nodes once these files reach a certain size. This feature can be annoying at times, necessitating the interruption of the review process to locate the remaining results elsewhere. To circumvent this issue, the univariable code offers the capability to redirect the output of the EM node to a designated location, thus enabling the user to retain a complete version of their logs and outputs. Additionally, the option enables the storage of results generated by identical nodes across multiple executions, with each execution identified by a unique date and timestamp.

Listing 13-3, lines 36 to 54, provides an illustrative example of the code that corresponds to the rerouting of the output of the EM node. It is notable that the PROC PRINTTO is employed for this objective, where a new output path is defined for the output as &output_printto, and the log path is defined by &log_printto. It is important to note that the options ODS RESULTS OFF, ODS OUTPUT CLOSE, and ODS HTML CLOSE are utilized to disable the output of the EM node.

Listing 13-3. Log rerouting.

```
36      %LET TIMESTART = %SYSFUNC(DATETIME());
37      /*-----------------------------------------------
38              --- Start opt sas Parameters
39      -----------------------------------------------
40      DM LOG "CLEAR" CONTINUE;/*clear current log*/
41      DM 'ODSRESULTS' CLEAR;/*Clear current results*/
42      %IF %UPCASE(&GENERATEOUTPUT.)=N %THEN
43          %DO;
44              PROC PRINTTO
45                  FILE="&OUTPUT_PRINTTO."
46                  LOG="&LOG_PRINTTO."
47                  NEW;
48              RUN;
49
50              ODS RESULTS OFF;
51              ODS OUTPUT CLOSE;
52              ODS HTML CLOSE;
53              /*ODS LISTING CLOSE;*//*Only in SAS Base
54          %END;
```

13.1.3 Parameter Initialization

It is a good practice to initialize the main macro-variables before starting the main execution.

Listing 13-4 illustrates the principal macro-variables that are initiated prior to the commencement of the analytical process. Line 58 illustrates the macro-variable &dsname is used to refer to the dataset name or ABT name. Line 59, &dependent, refers to the dependent variables, or target. Line 60, &covar, refers to the list of covariates. Line 61, &Nominal, refers to the list of nominal variables. Line 62, &lib, refers to the EM output library. Lines 63 and 64 pertain to the variable that contains the partition identifier and the partition value for the partition that is to be selected. It should be noted that the application of these lines is reserved for scenarios where the univariable code is executed in SAS Base or Enterprise Guide (EG). Line 65, meanwhile, is pertinent to the total number of variables that are to be processed. Line 66 serves as a supplementary counter for rejected variables. Lines 67 and 68 are macro-variables that are used to store information about the nominal variables; therefore, they must be initialized prior to use.

Listing 13-4. Parameter initialization.

```
55    /*------------------------------------------
56            --- Initialize macro variables
57    ------------------------------------------
58    %LET VDATA        =&DSNAME.;
59    %LET VDEPENDENT   =&DEPENDENT.;
60    %LET VCOVAR       =&COVAR.;
61    %LET VDUMMY       =&NOMINAL.;
62    %LET VLIB         =&LIB.;
63    %LET VPART_COL    =&PARTITION_COL.;
64    %LET VPART_VAL    =&PARTITION_VAL.;
65    %LET N            =%SYSFUNC(COUNTW(&VCOVAR.));
66    %LET N_REJECT     =;
67    %LET REFERENCES   =;
68    %LET REF          =;
```

13.1.4 Obtaining the Reference Category for Nominal Variables

The univariable code employs the PROC LOGISTIC [10] function, which facilitates the construction of models incorporating both interval and nominal variables. In the case of nominal variables, the creation of dummy variables is unnecessary, as the PROC LOGISTIC is capable of handling this task through the use of the CLASS statement. **Listing 13-5** illustrates the application of the CLASS statement in the LOGISTIC procedure. In this example, the dependent variable is designated as "DEFAULT", the independent variable is identified as N_BRANCH, and this

CHAPTER 13 UNIVARIABLE ANALYSIS

refers to the customer's application office. It is notable that the nominal variable "N_BRANCH" is presented in the CLASS statement, yet it also serves as a reference category for the analysis, which in this case is defined by "(PARAM=REF REF="377")".

Listing 13-5. Example of the use of the CLASS statement in the PROC LOGISTIC for nominal variables.

```
1  PROC LOGISTIC
2      DATA=&EM_IMPORT_DATA
3      NAMELEN=200
4      ;
5      CLASS N_BRANCH   (PARAM=REF REF="377");
6      MODEL  DEFAULT (EVENT='1') =  N_BRANCH
7           /
8      ITPRINT
9      TECH=NEWTON
10     MAXITER=1000
11     ;
12     ODS OUTPUT
13          FITSTATISTICS=N_BRANCH_FS;
14 RUN;
```

The univariable code has the capacity to manage nominal variables; nevertheless, in the event of the REF option being utilized, it is first necessary to define a reference category for the analysis. **Listing 13-6** illustrates the segment of the code that is responsible for selecting and extracting the reference category for each nominal variable, which is based on the frequency of each category.

Listing 13-6. Get reference category for nominal variables.

```
/*------------------------------------------------
          --- Get reference category for nominal variables
   ------------------------------------------------
   %LET K=1;
   %DO %WHILE (%SCAN(&VDUMMY.,&K.,' ') NE );
        %LET NOM_VAR=%SCAN(&VDUMMY.,&K.,' ');
        PROC SQL THREADS  OUTOBS=1 NOPRINT;
            SELECT
               &NOM_VAR.
              ,COUNT(1) AS FREQ
            INTO
              :REF
              ,:X
            FROM   &VDATA.
            WHERE &NOM_VAR. IS NOT NULL
            GROUP BY
              &NOM_VAR.
            ORDER BY
              CALCULATED FREQ DESC
            ;
        QUIT;
        %LET REFERENCES=%STR(&REFERENCES.|%QCMPRES(&REF.));
        %LET K=%EVAL(&K.+1);
   %END;

   %LET VREF=&REFERENCES.;
```

It is important to note that the macro-variable &vdummy is a text string comprising all the nominal variables that are iterated in order to extract the category with the highest frequency. Subsequently, the reference category is stored in the macro-variable &REF and concatenated as a vector in the macro-variable &REFERNCES. The process is repeated until all nominal variables have been processed.

It is also noteworthy that the REF option in the CLASS statement can also be defined as LAST or FIRST. In this instance, the reference category is selected on the basis of its order level.

13.1.5 Creating Temporary Storage Tables

In the event of working with a multitude of variables, it is advisable to have a repository in which to store the resulting data in a systematic manner. To this end, the univariable code generates an empty table with the requisite layout.

Listing 13-7 exemplifies the SQL statement employed for the creation of a table that will function as a repository for the results of each regression analysis. It should be noted that only three columns are required: Variable, which stores the name of the variable; G, which stores the absolute difference between the intercept-only model (baseline model) and the intercept and covariates model; and the p_val, which stores the estimated significance level from G.

Listing 13-7. Creating temporary storage tables.

```
95      /*------------------------------------
96              --- Create Results Tables
97      ------------------------------------
98      PROC SQL; CREATE TABLE UNIVARIABLE_MODELS(
99      VARIABLE    CHAR(200)
100     ,G                   NUM
101     ,P_VAL               NUM
102     )
103     ;
104     QUIT;
```

13.1.6 Starting the Main Loop and Data Validation

The essential elements required to commence the iterative assessment of all variables have now been assembled. Nevertheless, for an automated system to function as intended, it is imperative to identify and address the potential scenarios that could lead to its failure.

CHAPTER 13 UNIVARIABLE ANALYSIS

Dealing with a multitude of variables is challenging, as some are unusable or unreliable. The objective is to determine an effective method for checking and validating thousands of variables in a straightforward and expedient manner, while maintaining reliability. The conventional methodology entails conducting a distribution analysis, constructing histograms and percentiles for all covariates. However, this approach is inapplicable when dealing with a vast number of variables and offers minimal contribution to the model-building process. In light of the aforementioned, it is evident that if a reliable statistical test has already demonstrated that the correlation between a covariate and the logit of the target variable is insignificant, there is no reason for including it in a chart.

While statistical tests are effective for relatively clean data, they may not be suitable for variables with an unacceptable proportion of nulls, interval variables filled with zero values, or nominal variables with a single category (unary variables). It is therefore essential to implement a preliminary data cleansing procedure prior to conducting a regression analysis, in order to address the most significant and pervasive issues present in the dataset.

Listing 13-8 is comprised of four sections.[4] The initial section represents the commencement of the primary loop, which will iterate over all variables and is situated at line 120. The second section determines whether the current variable is nominal. In the event of a positive identification, a CLASS statement is assembled for utilization during the execution of the PROC LOGISTIC. The third section determines the existence of the CLASS statement, and based on this determination, it selects one of two distinct SQL SELECT clauses. The fourth and final section performs a query on the current variable. In the case of interval variables, it also performs the following checks:

---

[4] Notice that the %metadata macro is invoked without arguments before starting the main loop; this will reset the metadata in case a previous execution had failed.

CHAPTER 13 UNIVARIABLE ANALYSIS

- The null proportion
- The zero proportion
- The total number of distinct values

Listing 13-8. Main loop and data validation.

```
116     /*----------------------------------------------------------------
117            --- Start Main Loop
118     ----------------------------------------------------------------*/
119     %METADATA();
120     %DO I=1 %TO &N. %BY 1;
121         %LET VAL = %SCAN(&VCOVAR.,&I.);
122     /*----------------------------------------------------------------
123            --- Check if the class statement applies
124     ----------------------------------------------------------------*/
125         %LET VCLASS=;
126         %IF %LENGTH(&VDUMMY.)>0 %THEN
127             %DO;
128                 %LET ND=%SYSFUNC(COUNTW(&VDUMMY.));
129                 %DO I1=1 %TO &ND. %BY 1;
130                     %LET CL = %SCAN(&VDUMMY.,&I1.,' ');
131                     %LET RF = %QSCAN(&VREF.,&I1.,'|');
132                     %IF &CL.=&VAL. %THEN
133                         %LET VCLASS=%STR(CLASS &CL. (PARAM=REF REF="%QCMPRES(&RF.)"););
134                 %END;
135             %END;
136     /*----------------------------------------------------------------
137            --- Select the Validation SQL statement based on the type of variable
138     ----------------------------------------------------------------*/
139         %IF %LENGTH(&VCLASS.)>0 %THEN
140             %DO;
141                 %LET SQL=0;
142             %END;
143         %ELSE
144             %DO;
145                 %LET SQL=%STR(SUM(CASE WHEN &VAL. =0 THEN 1 ELSE 0 END)/&NUMROWS.);
146             %END;
147     /*----------------------------------------------------------------
148            --- Check if the variable has enough information to proceed
149     ----------------------------------------------------------------*/
150         PROC SQL NOPRINT;
151             SELECT
152                SUM(CASE WHEN &VAL. IS NULL THEN 1 ELSE 0 END)/&NUMROWS. AS NULLS
153                ,&SQL. AS ZEROS
154                ,COUNT(DISTINCT &VAL. ) AS VALUES
155             INTO
156                :PNULL
157                ,:PZERO
158                ,:NCATEGORIES
159             FROM &VDATA.
160             ;
161         QUIT;
162
163         %IF (&PNULL.<0.8 AND &PZERO.<0.8 AND &NCATEGORIES.>1)  %THEN
```

CHAPTER 13 UNIVARIABLE ANALYSIS

The distinction between nominal and interval variables lies in the applicability of the zero proportion, a concept that is only applicable to interval variables.

The determination of whether a variable is eligible to proceed to the regression analysis is ultimately determined by a control point (on line 163). The selection criteria are as follows:

- The proportion of null values should be less than 80%.
- The proportion of zeros should be also less than 80%.
- The total number of distinct values should be greater than 1.

This set of conditions is convenient and saves time, as it eliminates the need to debug the code each time a variable with no meaningful value causes the univariable program to fail.

13.1.7 Regression

Once the variables representing the trash have been removed, the regression analysis may then be initiated. This will be achieved through the utilization of the PROC LOGISTIC.[5] **Listing 13-9** provides an illustration of the structure of the PROC LOGISTIC, which is located between lines 176 and 188. The data source is defined in line 177, and the model to be run is defined in line 182 as "model &vdependent (EVENT='1') = &val". It is noteworthy that certain options are enabled, including itprint, tech=newton, and maxiter=1000, from line 184 to line 186. It is also noteworthy that lines 180 and 181 are optional and refer to the partition that is to be used and the CLASS statement, respectively. Line 188 invokes

---

[5] The PROC LOGISTIC is one of the most complex procedures in the SAS repertoire, so delving in its use is beyond the scope of this book. For more information, I recommend you look for the official documentation in support.sas.com.

the Output Delivery System (ODS), which provides access to all available results generated by the PROC LOGIST during its execution. In this instance, the objective is to export the `fitstatistics` results. It is of paramount importance to recognize that the Output Delivery System represents the principal mechanism for automating analytical procedures within the SAS environment.

Listing 13-9. PROC LOGISTIC.

```
163         %IF (&PNULL.<0.8 AND &PZERO.<0.8 AND &NCATEGORIES.>1)   %THEN
164             %DO;
165             /*----------------------------------------------------------------
166                 --- Check for partition
167             ----------------------------------------------------------------*/
168             %LET VPART=;
169             %IF %LENGTH(&VPART_COL.)>0 %THEN
170                 %LET VPART=%STR(WHERE &VPART_COL. = "&VPART_VAL." ;);
171             /*----------------------------------------------------------------
172                 --- Run the Logistic Regression
173             ----------------------------------------------------------------*/
174             TITLE1 "Run Logistic Regression for covariate &VAL. " FONT='arial/bo' JUSTIFY=LEFT;
175
176             PROC LOGISTIC
177                 DATA=&VDATA.
178                 NAMELEN=200
179                 ;
180                 &VPART.
181                 &VCLASS.
182                 MODEL  &VDEPENDENT. (EVENT='1') =  &VAL.
183                 /
184                 ITPRINT
185                 TECH=NEWTON
186                 MAXITER=1000
187                 ;
188                 ODS OUTPUT
189                     FITSTATISTICS=&VAL._FS;
190             RUN;
```

13.1.8 Estimating Significance and Saving Results

The exported goodness-of-fit statistic enables the calculation of the *G* statistic defined as

$$G = -2 \cdot \ln\left[\frac{\text{likelihood without the variable}}{\text{likelihood with the variable}}\right]$$

Equation 13-1. The G statistic.

CHAPTER 13 UNIVARIABLE ANALYSIS

G^6 follows a chi-squared distribution, and as the baseline model is always compared to the univariable model, then the degrees of freedom are always equal to 1. The *p*-value is calculated using SAS's probchi function (Listing 13-10) and is defined as "1-probchi(abs(interceptonly-interceptandcovariates),1,0)". The result is stored in the previously created Univariable_Models table. However, a physical copy is created and replaced at each iteration to ensure the preservation of the results of the last iteration in the event of a program crash.

Listing 13-10. Estimate the significance of the univariable model and save the results.

```
199     PROC SQL NOPRINT;
200         INSERT INTO UNIVARIABLE_MODELS
201         SELECT
202             "&VAL." AS VARIABLE
203             ,ABS(INTERCEPTONLY-INTERCEPTANDCOVARIATES) AS G
204             ,1-PROBCHI(ABS(INTERCEPTONLY-INTERCEPTANDCOVARIATES),1,0) AS P_VAL
205         FROM
206             &VAL._FS
207         WHERE MONOTONIC()=3
208         ;
209     QUIT;
210     /*-----------------------------------------------------------------
211         --- Save Results Permanently and Replace
212     -------------------------------------------------------------------
213     PROC SQL;
214         CREATE TABLE &VLIB..UNIVARIABLE_MODELS_&MDL.
215         AS
216         SELECT
217             *
218         FROM
219             UNIVARIABLE_MODELS;
220     QUIT;
```

---

[6] Notice that SAS's fitstatistics return the −2 log-likelihood result for the baseline and the univariable model, hence that it is not necessary to multiply the result by 2.

13.1.9 Loop to the Next Variable in the ABT

The process is repeated until all variables have been analyzed. It is imperative to acknowledge that any variables failing to meet the stipulated data integrity constraints will be rejected by the %metadata macro (**Listing 13-11**). Consequently, these variables will not be incorporated into the final table.

Listing 13-11. Reject variables that do not satisfy data integrity constraints.

```
222        %ELSE
223            %DO;
224               %LET VARLIST=%ENQUOTE(&VAL.,SEPARATOR=);
225               %METADATA(NEWROLE=REJECTED,APPEND=Y,WHERE=(NAME IN (&VARLIST.) AND UPCASE(ROLE)='INPUT'));
226            %END;
```

13.1.10 Save Final Results and Reject Nonsignificant Variables

Upon termination of the loop, the concluding results are stored in the EM library (**Listing 13-12**, lines 231 through 238) and two reports are generated (lines 240 through 256) enumerating the set of variables to be incorporated in the subsequent stage of the process and those to be rejected (Figure 13-1 shows the results of the univariable analysis exported to an Excel spreadsheet). The final step involves the rejection of all variables that fail to meet the 0.25 significance cutoff, a process that is executed using the %metadata macro. It is important to note that, in contrast to the conventional approach of providing the %metadata macro with a list of all variables to be rejected, we instead input them individually, in an iterative manner. This approach is adopted due to the prior utilization of the %metadata macro in an iterative fashion to reject the noncompliant variables from the data validation process using the APPEND=Y option. Consequently, the rejection of variables in blocks is not a viable option; instead, the %metadata must be input individually, in the same manner, for optimal functionality.

Listing 13-12. Save the final results, print the variables to be included in the next step, and print the variables to be rejected. Finally, modify the metadata and reject any variables that do not meet the specified cutoff.

```sas
228
229     /*-------------------------------------------------------------------
230         --- Save Results Permanently
                                                                           --*/
231     PROC SQL;
232     CREATE TABLE &VLIB..UNIVARIABLE_MODELS_&MDL.
233     AS
234     SELECT
235         *
236     FROM
237         UNIVARIABLE_MODELS;
238     QUIT;
239
240     TITLE1 "Variables to be included in the 1st Multivariable model" FONT='arial/bo' JUSTIFY=LEFT;
241
242     PROC SQL;
243         SELECT * FROM &LIB..UNIVARIABLE_MODELS_&MDL.
244         WHERE P_VAL<&CUTOFF.
245         ORDER BY P_VAL
246         ;
247     QUIT;
248
249     TITLE1 "Variables to be rejected" FONT='arial/bo' JUSTIFY=LEFT;
250
251     PROC SQL;
252         SELECT * FROM &LIB..UNIVARIABLE_MODELS_&MDL.
253         WHERE P_VAL>=&CUTOFF.
254         ORDER BY P_VAL
255         ;
256     QUIT;
257     /*-------------------------------------------------------------------
258         --- Query the significant variables
259                                                                          --*/
260     TITLE1 "Final Selection" FONT='arial/bo' JUSTIFY=LEFT;
261
262     PROC SQL;
263         SELECT
264         VARIABLE INTO :SIGNIFICANT SEPARATED BY " "
265         FROM &LIB..UNIVARIABLE_MODELS_&MDL.
266         WHERE P_VAL<&CUTOFF.;
267     QUIT;
268     /*-------------------------------------------------------------------
269         --- Reset metadata with significant variables at a 0.25 level
270                                                                          --*/
271     DATA UNIVARIABLE_MODELS;
272         SET &LIB..UNIVARIABLE_MODELS_&MDL. (WHERE=(P_VAL>=&CUTOFF.));
273         COUNTER=_N_;
274     RUN;
275
276     PROC SQL; SELECT COUNT(1) INTO :N_REJECT FROM UNIVARIABLE_MODELS; QUIT;
277
278     %IF &N_REJECT.>0 %THEN
279         %DO;
280             %DO I=1 %TO &N_REJECT. %BY 1;
281                 PROC SQL; SELECT VARIABLE INTO :NOSIGNIFICANT FROM UNIVARIABLE_MODELS WHERE COUNTER=&I.; QUIT;
282                 %LET VARLIST=%ENQUOTE(&NOSIGNIFICANT., SEPARATOR=);
283                 %METADATA(NEWROLE=REJECTED,APPEND=Y,WHERE=(NAME IN (&VARLIST.) AND UPCASE(ROLE)='INPUT'));
284             %END;
285         %END;
```

CHAPTER 13 UNIVARIABLE ANALYSIS

Variable	G	p_val
I_R_PGOT_CNSMT_02MB	0.000	0.99988
I_DC_R_SDODIF_LM_01MB	0.000	0.99988
I_IN_R_SDODIF_LM_01MB	0.000	0.99988
I_R_PGOT_SDODIF_10MB	0.000	0.99077
I_DC_R_SDODIF_SDOT_12MB	0.000	0.99073
I_IN_R_SDODIF_SDOT_12MB	0.000	0.99073
I_AV_R_CNSMT_SDCOM_L03M	0.000	0.98803
I_IN_R_PGOT_SDOT_SM_06MB	0.000	0.98278
I_DC_R_PGOT_SDOT_SM_06MB	0.000	0.98278
I_DC_R_SDODIF_LM_12MB	0.001	0.97311
I_IN_R_SDODIF_LM_12MB	0.001	0.97311
I_R_PGOT_SDODIF_08MB	0.003	0.95861
I_SDC_R_SDODIF_SDOT_L06M	0.003	0.95662
I_DC_R_SDOI_SDOT_02MB	0.003	0.95645
I_R_PGOT_CNSMT_08MB	0.004	0.94996
I_IN_R_CNSMT_LM_12MB	0.004	0.94988
I_IN_R_CNSMT_SDCOM_03MB	0.004	0.94946
I_R_SDOI_PGOT_10MB	0.004	0.94842
I_SIN_R_CNSMT_SDCOM_L12M	0.005	0.94486
I_R_PGOT_SDODIF_03MB	0.005	0.94287
I_IN_R_PGOT_SDODIF_02MB	0.007	0.93387
I_DC_R_PGOT_SDODIF_02MB	0.007	0.93387
I_AV_R_PGOT_SDODIF_L06M	0.007	0.93176
I_R_PGOT_SDODIF_06MB	0.008	0.93013
I_R_CNSMT_SDCOM_06MB	0.008	0.92786
I_R_PGO_SDO_CON_24MB	0.008	0.92771
I_R_PGOT_CNSMT_06MB	0.009	0.92418
I_IN_R_SDODIF_SDOT_02MB	0.010	0.92078
I_DC_R_SDODIF_SDOT_02MB	0.010	0.92078
I_R_CNSMT_SDCOM_01MB	0.011	0.91665
I_DC_R_SALDES_LM_06MB	0.013	0.90946

Figure 13-1. Example of results generated by univariable analysis for an ABT with more than 900 variables.

CHAPTER 13 UNIVARIABLE ANALYSIS

13.2 Summary

The univariable analysis serves as a critical first step in statistical modeling, providing a systematic approach to preselecting input variables. By examining the relationship between each input variable and the target variable individually, we can efficiently identify and eliminate *irrelevant* predictors—those with no meaningful association to the target variable. This process ensures that the dataset is streamlined for subsequent modeling steps, focusing only on variables that contribute value to the analysis.

Chapter 13 also underscores the importance of including a practical data cleaning, to avoid the most common issues, e.g., excessive null or zero proportion, and/or unary variables, before conducting the univariable analysis—or any subsequent regression analysis. Without this step, statistical tests may fail or produce unreliable results, complicating the variable selection process.

Finally, the section provided a detailed walkthrough of how to implement the univariable analysis in SAS® Enterprise Miner, ensuring that readers can apply these techniques in real-world scenarios.

Key takeaways:

1. **Purpose of Univariable Analysis:**

    - To preselect input variables by identifying and eliminating *irrelevant* predictors—those with no meaningful association to the target variable.

    - This step reduces the dimensionality of the dataset, simplifying subsequent modeling efforts.

2. **Importance of Data Cleaning:**

    - A thorough data cleaning process is crucial before conducting univariable or regression analysis.

- Ensures accurate statistical results, prevents failed executions, and streamlines the variable selection process.

3. **Significance Thresholds:**

    - The recommended threshold by Hosmer and Lemeshow [32] is a *p*-value <0.25, which allows for the inclusion of important variables with partial associations, while eliminating variables of uncertain importance.

It should be acknowledged that while the univariable analysis eliminates *irrelevant* predictors, it does not address *redundancy*—which corresponds to variables that are highly correlated with one another and provide overlapping information. Redundancy will be addressed in the next section, which focuses on the collinearity analysis. By combining these steps, we can prepare a clean and efficient dataset for multivariable modeling.

CHAPTER 14

Collinearity Analysis

"In the end, there can be only one."

—Juan Sánchez-Villalobos Ramírez, *Highlander (1986)*

Following the elimination of variables with a questionable level of significance through univariable analysis, the next step is to reduce *redundancy* in the variables. This is also known as *collinearity* or *multicollinearity*. Collinearity, within the statistical context, denotes the existence of variables that rival for the explanation of a shared amount of variability in the data. The process of reducing collinearity in variables is also known as *dimensional reduction*.

14.1 Variance Inflation Factor (VIF)

Dimensional reduction is most frequently associated with the implementation of the *principal component analysis* (PCA); nevertheless, the quintessential analysis for measuring collinearity is the *variance inflation factor* (VIF). The VIF is a measure of the strength of a linear relationship between a predictor and the other predictors in a model. It is calculated by fitting a regression model for each predictor using the remaining predictors in the model as explanatory variables and calculating their individual fit with respect to each subset of the remaining variables.

CHAPTER 14   COLLINEARITY ANALYSIS

The VIF is defined as follows:

$$\text{VIF}_j = \frac{1}{1-R_j^2}$$

Equation 14-1. Variance inflation factor (VIF).

Where R_j^2 is the unadjusted coefficient of determination for regressing the *j*th independent variable on the remaining ones.

As demonstrated in **Figure 14-1**, a line chart illustrates the relationship between R^2 and the VIF, indicating that R^2 converges to 1 as the VIF approaches infinity. The range for a VIF cutoff selection recommended in this book is highlighted within a green square, corresponding to an R^2 between 0 and 0.75 and to a VIF between 1 and 4. There does not appear to be a specific VIF cutoff that can be applied in order to effectively reduce the collinearity effect without hindering important interactions in the multivariable model. However, my empirical observations suggest that a VIF of 3 can be used to reduce the number of conflicting variables without eliminating the interaction effects between covariates in the multivariable model.

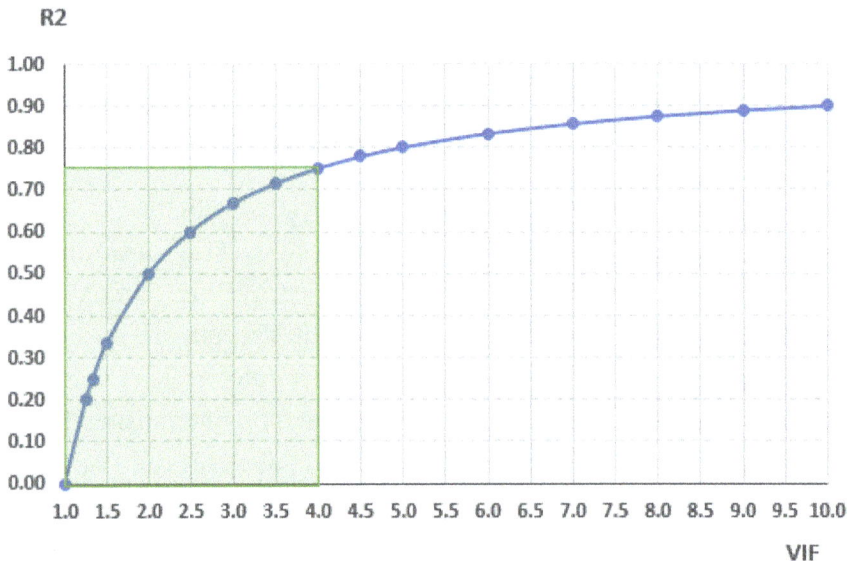

Figure 14-1. *Line chart of the relationship between R^2 and the VIF.*

In the subsequent section, it will be demonstrated that the VIF is not the sole instrument employed to mitigate multicollinearity; rather, it is employed in conjunction with the *proportion of variation* attributed to each predictor in its corresponding principal component.

14.2 Optimization Issues

The issue of collinearity is more numerical in nature than statistical or conceptual. It is imperative to acknowledge that all of the algorithms employed during the model-building process rely on the utilization of optimization methods and automated iterative subroutines. In the context of optimization methods, the issue manifests when two variables possess an equivalent numerical weight with respect to the minimization function, such as the sum of squared errors. This results in a state where the regressors attempt to cancel each other out, leading to coefficients with similar magnitudes but reversed signs.

To illustrate this point, consider an equation such as that shown in **Equation 14-2**. Assuming that the independent variable $x_{2,i}$ is highly correlated with the independent variable $x_{3,i}$, let us say with a $R^2 \approx 0.9$. This implies that $x_{2,i}$ and $x_{3,i}$ possess analogous values at each observation point i. Additionally, it suggests that the contribution of their respective coefficients to the minimization of the objective function will be numerically similar.

$$S = \sum_{i=1}^{n}\left(y_i - \left(\beta_0 + \beta_1 x_{1,i} + \beta_2 x_{2,i} + \beta_3 x_{3,i}\right)\right)^2$$

Equation 14-2. Error function x_1, x_2, and x_3 linear model.

Let us now consider a scenario in which two distinct regressions are fitted, with each regressor being considered individually.

$$S = \sum_{i=1}^{n}\left(y_i - \left(\beta_0 + \beta_1 x_{1,i} + \beta_2 x_{2,i}\right)\right)^2$$

Equation 14-3. Error function for the linear model of x_1 and x_2.

CHAPTER 14 COLLINEARITY ANALYSIS

$$S = \sum_{i=1}^{n}\left(y_i - \left(\beta_0 + \beta_1 x_{1,i} + \beta_3 x_{3,i}\right)\right)^2$$

Equation 14-4. Error function for the linear model of x_1 and x_3.

The result of fitting both models independently would be a β_2 and β_3 with similar magnitudes and both exhibiting the same sign, that is to say $\beta_2 \approx \beta_3$. However, what would be the result if both regressors were to be used simultaneously in the same model, as in **Equation 14-2**?

Given that **Equation 14-2** represents a linear summation of factors, and since it is known that $\beta_2 \approx \beta_3$ in their individual regressions, it can be deduced that $\beta_2 x_{2,i}$ and $\beta_3 x_{3,i}$ cannot be added together, as this would be equivalent to adding the same amount twice in the same equation. Nevertheless, it is possible to fit **Equation 14-2** using ordinary least squares (OLS), given that optimization algorithms are capable of dealing with such conundrums. However, the outcome of this process is that the optimization algorithm will simply invert the sign of the coefficients in order to cancel them out, thus avoiding any perturbation in the equilibrium of the objective function.

It is interesting to note that, in general, fitting a model like the one in **Equation 14-2** will yield an R^2 that is marginally higher than that obtained from fitting the individual models in **Equations 14-3** and **14-4**. This phenomenon can be attributed to the fact that, despite the model in **Equation 14-2** incorporates an additional regressor compared to **Equations 14-3** and **14-4**, the additional regressor, $x_{3,i}$, accounts for a variability explained by $x_{2,i}$, to a similar extent, thus contributing only a negligible amount of the additional variability explained by the model.

The following illustration will demonstrate the issue concerning optimization, employing Example 73.2, Aerobic Fitness Prediction, from support.sas.com [14]. The illustration involves the application of three distinct models to predict the oxygen consumption rate. The estimation

CHAPTER 14 COLLINEARITY ANALYSIS

of oxygen consumption is derived by calculating the difference between postactivity and source oxygen concentrations. The dependent variable is the oxygen concentration postactivity, while the independent variables are simple measurements such as run time, age, weight, run pulse, max pulse, and rest pulse. The three regression models executed with the PROC REG are illustrated in **Listing 14-1**.[1] The first model (model A) incorporates RunTime, Age, Weight, and RunPulse as predictors; the second model (model B) includes RunTime, Age, Weight, MaxPulse, and RestPulse; and the third model (model C) employs both RunPulse and MaxPulse.

Listing 14-1. Example of the collinearity effect using Example 73.2, Aerobic Fitness Prediction, from support.sas.com. A is the regression model without using the MaxPulse variable. B is the regression model without using the RunPulse variable. C is the regression model with no exclusions.

```
proc reg data=fitness;
A:  model Oxygen=RunTime Age Weight RunPulse RestPulse / tol vif collin;
B:  model Oxygen=RunTime Age Weight MaxPulse RestPulse / tol vif collin;
C:  model Oxygen=RunTime Age Weight RunPulse MaxPulse RestPulse / tol vif collin;
run;
```

---

[1] This example is available at https://github.com/saulzapiain/Conceptual VariableDesign. File: Listing 14-1 Collinearity Example with PROC REG .sas.

CHAPTER 14 COLLINEARITY ANALYSIS

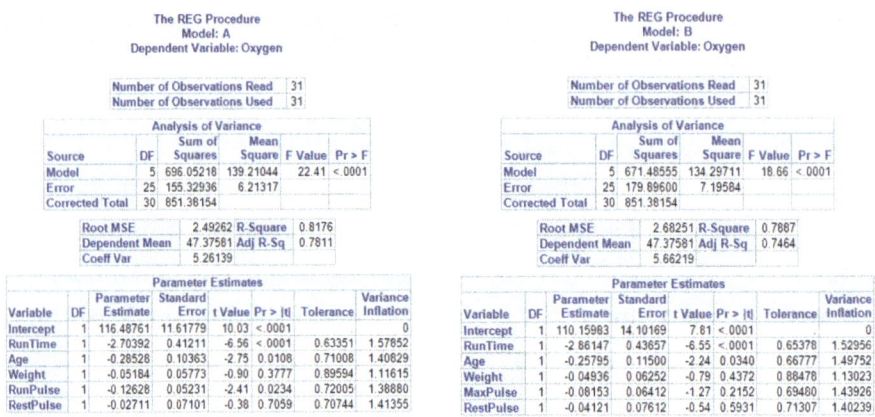

Figure 14-2. Results generated by the PROC REG for models A and B.

Figure 14-2 presents the results of models A and B, respectively. It is noteworthy that both RunPulse and MaxPulse demonstrate the same sign and that they also possess the fourth largest absolute effect relative to their respective models.

Model	Variable	df	Estimate	Std. error	t-value	p-value	Tolerance	VIF
A	RunPulse	1	−0.12628	0.05231	−2.41	0.0234	0.72005	1.3888
B	MaxPulse	1	−0.08153	0.06412	−1.27	0.2152	0.6948	1.43926

But now let us observe the effect of combining both RunPulse and MaxPulse within the same model.

The REG Procedure
Model: C
Dependent Variable: Oxygen

Number of Observations Read	31
Number of Observations Used	31

Analysis of Variance

Source	DF	Sum of Squares	Mean Square	F Value	Pr > F
Model	6	722.54361	120.42393	22.43	<.0001
Error	24	128.83794	5.36825		
Corrected Total	30	851.38154			

Root MSE	2.31695	R-Square	0.8487
Dependent Mean	47.37581	Adj R-Sq	0.8108
Coeff Var	4.89057		

Parameter Estimates

Variable	DF	Parameter Estimate	Standard Error	t Value	Pr > |t|	Tolerance	Variance Inflation
Intercept	1	102.93448	12.40326	8.30	<.0001	.	0
RunTime	1	-2.62865	0.38456	-6.84	<.0001	0.62859	1.59087
Age	1	-0.22697	0.09984	-2.27	0.0322	0.66101	1.51284
Weight	1	-0.07418	0.05459	-1.36	0.1869	0.86555	1.15533
RunPulse	1	-0.36963	0.11985	-3.08	0.0051	0.11852	8.43727
MaxPulse	1	0.30322	0.13650	2.22	0.0360	0.11437	8.74385
RestPulse	1	-0.02153	0.06605	-0.33	0.7473	0.70642	1.41559

Figure 14-3. Results generated by the PROC REG for model C.

Model	Variable	df	Estimate	Std. error	t-value	p-value	Tolerance	VIF
C	RunPulse	1	-0.36963	0.11985	-3.08	0.0051	0.11852	8.43727
C	MaxPulse	1	0.30322	0.1365	2.22	0.036	0.11437	8.74385

Figure 14-3 shows the results of fitting a linear model with both regressors. The two coefficients are similar, but MaxPulse is now positive. Think for a moment on the impact of having both predictors in the same

model from a conceptual point of view. In models A and B, both RunPulse and MaxPulse have been shown to decrease oxygen concentration levels (hence the negative sign), meaning that the higher your run or max pulse, the higher your oxygen consumption. This phenomenon is intuitive from a physiological standpoint, as a higher pulse rate corresponds to a higher number of heartbeats per second, which in turn leads to an increased oxygen consumption. Consequently, both running and maximum pulse can only exhibit a negative correlation with oxygen concentrations. However, MaxPulse is positive when both variables are present, indicating an oxygen production. This is a nonsensical result and a physiological impossibility, given the established fact that humans do not produce oxygen.

This example illustrates the substantial impact of collinearity on the interpretation and understanding of a real phenomenon. Furthermore, it is demonstrated that optimization algorithms invariably yield a numerical approximation for the regressor coefficients.[2] However, it is important to note that successful minimization of the objective function does not guarantee that the regressor coefficients will retain a meaningful interpretation, as they are a product of a *numerical artifice.*

Moreover, a collinearity issue can be regarded as an *over-specified* problem from an optimization standpoint. This implies that, in addition to a single nonsensical solution with regard to interpretability, an infinite number of such solutions may also exist. This explains why the coefficients obtained from different training samples[3] will vary widely, and also why they present large standard errors, as reported in the literature[4,5].

---

[2] Although optimization algorithms may always reach a numerical approximation, most statistical software is capable of identifying perfect collinearity, warning us about the presence of some variables that are a linear combination of the remaining ones. In such cases, a VIF cannot be estimated and neither a regression coefficient.

[3] See Ridge Regression by Schaefer 1986 [12].

[4] Applied Logistic Regression by Hosmer and Lemeshow [32].

[5] Discovering Statistics Using SPSS by Field [20].

14.3 Collinearity Diagnostics

The following examination will now be conducted on the collinearity diagnostics reported by the PROC REG [11] using the "COLLIN" option. The collinearity diagnostics provide the eigenvalues of the scaled, uncentered cross-products matrix, the conditional index, and the variance proportion for each predictor. Each of these eigenvalues corresponds to a principal component, thus rendering the collinearity diagnostics analogous to a PCA.

The focus will now be directed toward the variance proportion highlighted in each model's collinearity diagnostics. It can be observed that the variance proportion for both models, A (Figure 14-4) and B (Figure 14-5), shows the same pattern, where RunTime appears to be dominant over component 4, or the fourth dimension, with a maximum value of 0.744 for A and a maximum value of 0.733 for B, while Age appears to be dominant over component 6, or the sixth dimension, with a value of 0.477 for A and a value of 0.478 for B. Weight seems to be dominant over component 3, with values of 0.372 for A and 0.333 for B, and over component 5, the fifth dimension, with a value of 0.454 for A and 0.541 for B, while RunPulse also seems to be dominant over component 6 with a value of 0.772 for A and of 0.809 for B. It is noteworthy that although Age and RunPulse seemed to be dominant over component 6, there did not appear to be any significant collinearity issues, as their corresponding VIF was considerably low (<1.5). Finally, an analysis of RestPulse reveals its dominance over component 2, with a value of 0.385 for A and of 0.397 for B.

The following section will examine the combined effect of both variables, RunPulse and MaxPulse, when present simultaneously (Figure 14-6). The variance proportion of the other variables does not vary significantly in either magnitude or component location. However, it is evident that both RunPulse and MaxPulse are competing to dominate component 7, or the seventh dimension, with values of 0.913 and 0.984,

CHAPTER 14 COLLINEARITY ANALYSIS

respectively. This phenomenon is distinctly different from the previously observed harmless competition between Age and RunPulse, or between Age and MaxPulse. In these earlier instances, the VIF was less than 1.5. In contrast, the current VIF is greater than 8, indicating a significantly stronger relationship between the variables.

			Collinearity Diagnostics					
		Condition	Proportion of Variation					
Number	Eigenvalue	Index	Intercept	RunTime	Age	Weight	RunPulse	RestPulse
1	5.9527	1.00	0.000	0.000	0.000	0.000	0.000	0.000
2	0.0187	17.86	0.003	0.023	0.164	0.010	0.000	**0.385**
3	0.0142	20.50	0.001	0.101	0.146	**0.372**	0.007	0.025
4	0.0088	26.00	0.011	**0.744**	0.042	0.066	0.007	0.273
5	0.0047	35.65	0.026	0.027	0.171	**0.454**	0.265	0.303
6	0.0010	77.55	0.959	0.105	**0.477**	0.097	**0.722**	0.013

Figure 14-4. Results generated by the PROC REG for model A.

			Collinearity Diagnostics					
		Condition	Proportion of Variation					
Number	Eigenvalue	Index	Intercept	RunTime	Age	Weight	MaxPulse	RestPulse
1	5.9529	1.00	0.000	0.000	0.000	0.000	0.000	0.000
2	0.0187	17.85	0.002	0.028	0.145	0.011	0.000	**0.397**
3	0.0144	20.30	0.001	0.109	0.145	**0.333**	0.007	0.018
4	0.0089	25.90	0.008	**0.733**	0.040	0.084	0.006	0.277
5	0.0044	36.74	0.028	0.061	0.191	**0.541**	0.177	0.298
6	0.0008	88.97	0.961	0.070	**0.478**	0.031	**0.809**	0.008

Figure 14-5. Results generated by the PROC REG for model B.

			Collinearity Diagnostics						
		Condition	Proportion of Variation						
Number	Eigenvalue	Index	Intercept	RunTime	Age	Weight	RunPulse	MaxPulse	RestPulse
1	6.9499	1.00	0.000	0.000	0.000	0.000	0.000	0.000	0.000
2	0.0187	19.29	0.002	0.025	0.146	0.010	0.000	0.000	**0.391**
3	0.0150	21.50	0.001	0.129	0.150	**0.236**	0.001	0.001	0.028
4	0.0091	27.62	0.006	**0.609**	0.032	0.183	0.001	0.001	0.190
5	0.0061	33.83	0.001	0.125	0.113	**0.444**	0.015	0.008	0.365
6	0.0010	82.64	0.800	0.097	**0.497**	0.103	0.069	0.006	0.020
7	0.0002	196.79	0.190	0.015	0.062	0.023	**0.913**	**0.984**	0.006

Figure 14-6. Results generated by the PROC REG for model C.

14.4 Available Methods for Dealing with Collinearity

It has been demonstrated that collinearity can have a detrimental effect on model performance, impacting the magnitude, directionality, and interpretability of regressor coefficients. This interference can impede the optimization process, which is a crucial step in machine learning algorithms. Furthermore, collinearity also affects multivariable selection methods, rendering the automated selection process arbitrary. The question therefore arises as to how collinearity can be effectively dealt with.

A plethora of literature references are available that discuss various measures and methodologies for addressing and reducing collinearity. However, none of these appear to address the problem in a manner that would be useful for our purposes. The most well-known methods for dealing with collinearity are the PCA [20, 59, 62] and the ridge method [12]. However, the PCA method is problematic in that it necessitates the replacement of the original covariates with their linear combinations, which is impractical from a scorecard perspective. This is due to the requirement to interpret the model in terms of the original predictors, which is essential for business understanding. It is evident that the allocation of points to distinct ranks of the principal component function provides no insight into the behavioral patterns that lead to the default event.

In contrast, the ridge method, as outlined in the extant literature (*Regression Analysis by Example* [12]), elucidates that the estimation of the bias parameter k in the ridge regression is subjective and that there is no consensus on the most appropriate method. Furthermore, it is acknowledged that the parameter k of ridge regression can be significantly impacted by the presence of outliers in the data, thereby complicating its utilization as an alternative to address collinearity.

Other authors advocate the utilization of the correlation matrix to visually examine and identify the pairs of variables that are highly correlated. However, the employment of the correlation matrix is not a feasible solution for two fundamental reasons. Primarily, the application of the correlation matrix may only be suitable when dealing with modest academic models, say with less than 20 variables. It is imperative to bear in mind that a model adhering to the methodology outlined in this book would possess thousands of variables, rendering the use of the correlation matrix entirely impractical. Secondly, it is important to note that the correlation analysis is not a multivariable analysis; it is a bivariate analysis. Therefore, it cannot be considered a substitute for the VIF analysis.

A superior option would be to utilize the VIF analysis in conjunction with the proportion of variation explained by each predictor. This would facilitate the identification of the most conflicting variables in terms of the VIF analysis, as well as the competing variable that is causing the inflation in the first place. Consequently, a choice could be made between them based on the amount of variation explained by each variable. The implementation of such a method would offer numerous advantages, including the capacity to reduce collinearity in an iterative manner. Furthermore, it would provide an objective criterion for resolving conflicts between highly correlated covariates—I call this method *The Iterative Multicollinearity Reduction Method*.

14.5 Iterative Collinearity Reduction Method (ICRM)[6]

The iterative collinearity reduction method, or ICRM, for short, is based on six steps:

---

[6] The complete code can be downloaded from https://github.com/saulzapiain/ConceptualVariableDesign. File: Listing 14-(2 to 7) ICRM (complete).sas.

1. Initial collinearity diagnostics.

2. Elimination of perfect collinearity issues.

3. Check for collinearity issues and start main loop accordingly.

4. Identify the maximum VIF variable.

5. Identify competing variables and select the variable with the higher proportion of variance.

6. Reject the variable with the lowest proportion of variance and check again for collinearity issues.

7. If there are no issues, stop the procedure, otherwise go to step 4.

14.5.1 Initial Collinearity Diagnostics

Listing 14-2 shows a two-part code that corresponds to the initial collinearity diagnostics section. The first part corresponds to the creation of a catalog containing all interval variables. The catalog is created by first creating a catalog containing all the columns in the ABT, via the PROC CONTENTS, and then keeping only the interval variables listed by the %EM_INTERVAL_INPUT[7] macro, which can be found from line 77 to line 109. The second part is the actual execution of the PROC REG, which is located from line 113 to line 124. It should be noted that the Output Delivery System (ODS) is employed (line 120 to line 123) to export the table PARAMETERSESTIMATES, which is used to identify and reject the variables with perfect collinearity.

---

[7] The EM_INTRERVAL_INPUT macro is used as one of the macro's parameter to list all interval variables declared in the ABT's metadata. For more information, see the end of this section.

CHAPTER 14 COLLINEARITY ANALYSIS

Listing 14-2. Variables' catalog creation and initial collinearity diagnostics.

```
74
75              --- Create table with all columns names
76
77      PROC CONTENTS
78              DATA=&DS.
79              OUT=VARIABLES_NAMES;
80      RUN;
81      /*-----------------------------------------------------------
82              --- Create table to store interval variables name
83      -----------------------------------------------------------
84      PROC SQL;
85          CREATE TABLE NAMES(NAME CHAR(200));
86      QUIT;
87      /*-----------------------------------------------------------
88              --- Execute loop to populate the NAMES table
89      -----------------------------------------------------------
90      %LET ND=%SYSFUNC(COUNTW(&INPUTS.));
91      %LET QUOTE_INPUTS=;
92      %DO I1=1 %TO &ND. %BY 1;
93          %LET VAR = %SCAN(&INPUTS.,&I1.,' ');
94          PROC SQL;
95              INSERT INTO NAMES VALUES("&VAR.");
96          QUIT;
97      %END;
98      /*-----------------------------------------------------------
99              --- Create catalog considering only the interval inputs
100     -----------------------------------------------------------
101     PROC SQL;
102         CREATE TABLE VARIABLES_INPUTS
103         AS
104         SELECT
105         DISTINCT NAME
106         FROM VARIABLES_NAMES
107         WHERE NAME IN (SELECT NAME FROM NAMES)
108         ;
109     QUIT;
110     /*-----------------------------------------------------------
111             --- Execute initial regression to check for perfect linear combinations
112     -----------------------------------------------------------
113     PROC REG
114         DATA=&DS.
115         PLOTS(MAXPOINTS=NONE)
116         ;
117         &VPART.
118         MODEL &TARGET.=&INPUTS./COLLIN VIF TOL;
119
120         ODS OUTPUT
121             COLLINDIAG=CD_REG
122             PARAMETERESTIMATES=PE_REG
123         ;
124     RUN;
```

CHAPTER 14 COLLINEARITY ANALYSIS

14.5.2 Eliminate Perfect Collinearity Issues

The presence of variables exhibiting perfect collinearity can be discerned from the PARAMETERSESTIMATES report, as their coefficients are unable to be estimated and thus will be equal to zero. **Listing 14-3** illustrates the manner in which variables exhibiting perfect collinearity are physically deleted from the variable catalog via the SQL DELETE clause with the INPUTS macro-variable subsequently updated to reflect the remaining variables. The aforementioned sections can be located between lines 127 and 138. The subsequent step involves reestimating the collinearity diagnostics, which are now free of perfect collinearity. The PROC REG is then executed once more, and the PARAMETERSESTIMATES report is exported.

Listing 14-3. Delete variables with perfect collinearity and reestimate the collinearity diagnostics.

```
124   /*-------------------------------------------------------------------
125         --- Delete variables that are linear function of other variables
126   -------------------------------------------------------------------*/
127   PROC SQL;
128       DELETE FROM VARIABLES_INPUTS
129       WHERE COMPRESS(NAME) IN (SELECT COMPRESS(VARIABLE) FROM PE_REG WHERE ESTIMATE=0 OR VARIANCEINFLATION IS NULL);
130
131       SELECT
132       COMPRESS(NAME)
133       INTO
134       :NEW_INPUTS SEPARATED BY " "
135       FROM VARIABLES_INPUTS
136       ;
137   QUIT;
138   %LET INPUTS=&NEW_INPUTS.;
139   /*-------------------------------------------------------------------
140         --- Rerun regression free of linear functions and create collinearity statistics
141   -------------------------------------------------------------------*/
142   PROC REG
143             DATA=&DS
144             PLOTS(MAXPOINTS=NONE)
145             ;
146       &VPART
147       MODEL &TARGET=&INPUTS/COLLIN VIF TOL;
148   RUN;
149   ODS OUTPUT
150       COLLINDIAG=CD_REG
151       PARAMETERESTIMATES=PE_REG
152       ;
```

Figure 14-7 illustrates the manner in which variables exhibiting perfect collinearity are reported by the REG procedure. It is noteworthy that the two notes in lines 857 and 858 address the issue of parameter estimates not being unique. This outcome is anticipated in the context of an *over-specified problem*. The second note pertains to the fact that certain estimates have been set to zero, given that their respective variables constitute a perfect linear combination of the remaining variables within the model.

CHAPTER 14 COLLINEARITY ANALYSIS

Figure 14-7. Example of how variables with perfect collinearity are reported by the REG procedure.

14.5.3 Check for Collinearity Issues and Start Main Loop Accordingly

The PARAMETERSESTIMATES table is saved as PE_REG and queried to quantify the number of conflicting variables according to the VIF cutoff defined in the initial parameters as selection criteria. The selection of the initial number of conflicting variables can be found in **Listing 14-4** from line 158 to line 166. In the event that no conflicting variables are identified, all inputs are designated as "*collin_free_inputs,*" and the procedure is terminated. Conversely, if a conflict is detected, the main loop is initialized, and the ICRM progresses.

Listing 14-4. Quantify conflicting variables according to the VIF cutoff criteria defined in the initial parameters and start the main loop accordingly.

```
155     /*-----------------------------------------------------------
156             --- Check if the present model has collinearity issues
157     -----------------------------------------------------------
158     TITLE 'Initial Number of conflicts';
159     PROC SQL;
160         SELECT
161         COUNT(DISTINCT VARIABLE) INTO :CONFLICTS
162         FROM PE_REG
163         WHERE
164             VARIANCEINFLATION>=&VIF.
165         ;
166     QUIT;
167
168     %LET N=0;
169
170     %IF &CONFLICTS.=0 %THEN
171         %LET COLLIN_FREE_INPUTS=&INPUTS.;
172     /*-----------------------------------------------------------
173             --- If that so, initiate main loop
174     -----------------------------------------------------------
175     %LET K=0;
176     %DO %WHILE(&CONFLICTS.>0);
```

CHAPTER 14 COLLINEARITY ANALYSIS

14.5.4 Identifying the Maximum VIF Variable

Subsequent to the identification of variables manifesting collinearity issues, the ICRM process is initiated with the selection of the variables exhibiting the highest VIF values. The query employed for this purpose is illustrated in **Listing 14-5**, which can be found between lines 180 and 192. The subsequent step is to ascertain the component that is dominated by the conflicting variables or, in other words, the component in which the conflicting variables account for the largest proportion of variance. The relevant section of the code can be found between lines 196 and 204. Finally, the proportion of variance is extracted and stored in the macro-variable FIRST_LOADING.

Listing 14-5. Identifying the variable with the maximum VIF, locate the component dominated by it, and its variance proportion.

```
177       /*-------------------------------------------------------------
178           --- Get the most conflicting variavle
179       -------------------------------------------------------------
180       TITLE "The first dominant variable and its VIF";
181       PROC SQL;
182           SELECT
183               COMPRESS(VARIABLE)
184               ,VARIANCEINFLATION
185           INTO
186               :FIRST_DOMINANT
187               ,:FIRST_VIF
188           FROM PE_REG
189           HAVING
190               VARIANCEINFLATION=MAX(VARIANCEINFLATION)
191           ;
192       QUIT;
193       /*-------------------------------------------------------------
194           --- Get the maximum loading for this variable and its component (dimension)
195       -------------------------------------------------------------
196       TITLE "Maximum dimension for the first dominant variable";
197       PROC SQL;
198           SELECT
199               STRIP(PUT(NUMBER,9.)) INTO :MAXDIM
200           FROM CD_REG
201           HAVING
202               &FIRST_DOMINANT.=MAX(&FIRST_DOMINANT.)
203           ;
204       QUIT;
205       TITLE "Loading for the first dominant";
206       PROC SQL;
207           SELECT
208               MAX(&FIRST_DOMINANT.) INTO :FIRST_LOADING
209           FROM CD_REG
210           ;
211       QUIT;
```

CHAPTER 14 COLLINEARITY ANALYSIS

Figure 14-8 illustrates the output generated by the ICRM. The initial result is the total number of variables exhibiting collinearity issues (line 17093), which is 271 variables. On line 17100, the variable with the highest VIF is identified as I_R_BALDIF_BALT_06MB, exhibiting a staggering value of 8,227,331. It is not uncommon to observe such large VIFs, which are attributable to variables exhibiting near-perfect collinearity, corresponding to an R^2 of 0.99999988.

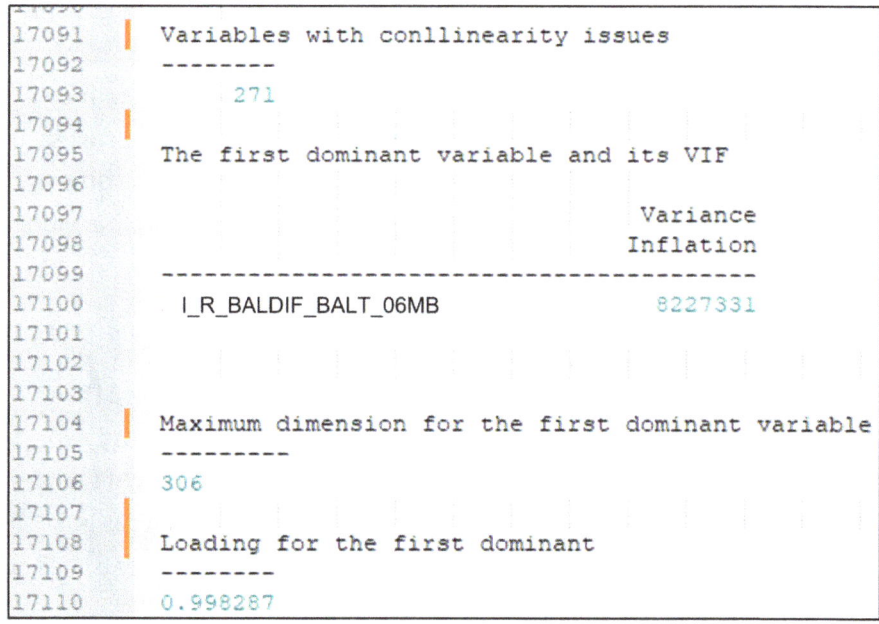

Figure 14-8. *Example of the results reported by the ICRM during the maximum VIF identification.*

The following information pertains to the identification of the component that is predominantly influenced by the variable I_R_BALDIF_BALT_06MB, along with the amount of variation that it accounts for. It is notable that the variable I_R_BALDIF_BALT_06MB exerts a dominant influence on the 306th component, accounting for 0.998287 percent of its variability. This finding indicates that the variability of the 306th component is predominantly explained by the variable I_R_BALDIF_BALT_06MB.

14.5.5 Identifying the Competing Variable and Selecting the Variable with the Higher Proportion of Variance

The next step is to identify the second variable that is competing for the dominance of the component. **Listing 14-6** illustrates the series of steps required to achieve this outcome. The initial step is to transpose the collinearity diagnostics table, which has been designated as CD_REG. The transposition is necessary because the components are arranged in a manner that differs from the conventional presentation of data, namely, as rows rather than columns. This discrepancy complicates the process of identifying the competing variable. It should be noted that the query located between lines 228 and 237 extracts the name of the variable with the largest VIF that is not the first dominant variable.

Subsequent to the identification of the competing variable, the VIF and the proportion of variance from the parameter estimates and the collinearity diagnostics, respectively, can be extracted (lines 241 to 259). The final step is to compare the proportion of variance of the two competing variables. The variable with the smallest proportion of variance is then eliminated from the initial set of variables, and the process is repeated as many times as necessary until no further collinearity issues are identified.

Listing 14-6. Identify the second variable with the largest VIF, that is, competing for component dominance.

```
/*-----------------------------------------------------------------
        --- Transpose the collinearity diagnostics
-----------------------------------------------------------------*/
PROC TRANSPOSE
     DATA=CD_REG
     OUT=CD_REG_T (DROP=_LABEL_)
     PREFIX=COMPONENT
     NAME=INT_VARIABLE
     ;
     VAR
          &INPUTS.
     ;
RUN;
/*-----------------------------------------------------------------
        --- Get the second greatest loading for that component
-----------------------------------------------------------------*/
TITLE "The second dominant variable for the &MAXDIM. component";
PROC SQL;
     SELECT
     COMPRESS(INT_VARIABLE) INTO :SECOND_DOMINANT
     FROM CD_REG_T
     WHERE
          INT_VARIABLE NE "&FIRST_DOMINANT."
     HAVING COMPONENT&MAXDIM.=MAX(COMPONENT&MAXDIM.)
     ;
QUIT;
/*-----------------------------------------------------------------
        --- Get the VIF for the second dominant variable
-----------------------------------------------------------------*/
TITLE "VIF for the second dominant variable";
PROC SQL;
     SELECT
     VARIANCEINFLATION
     INTO
     :SECOND_VIF
     FROM PE_REG
     WHERE  VARIABLE="&SECOND_DOMINANT."
     ;
QUIT;
TITLE "Loading for the second dominant variable for the component &MAXDIM.";
PROC SQL;
     SELECT
     MAX(COMPONENT&MAXDIM.) INTO :SECOND_LOADING
     FROM CD_REG_T
     WHERE
          INT_VARIABLE NE "&FIRST_DOMINANT."
     ;
QUIT;
%IF &FIRST_LOADING.>&SECOND_LOADING. %THEN
     %LET REJECT=&SECOND_DOMINANT.;
%ELSE
     %LET REJECT=&FIRST_DOMINANT.;
```

CHAPTER 14 COLLINEARITY ANALYSIS

Figure 14-9 illustrates the continuation of the report presented in **Figure 14-8**. In the second part of the report, it is determined that the competing variable to I_R_BALDIF_BALT_06MB is the variable I_R_PRCHT_BALT_06MB, with a VIF of 8,227,261 and a variance proportion of 0.998276. In this case, the variable I_R_BALDIF_BALT_06MB explains a greater proportion of variance, with a value of 0.998287. While the discrepancy is slight, it is nevertheless sufficient to justify the selection of one variable over the other.

```
17112        The second dominant variable for the 306        component
17113
17114
17115        ---------------------------
17116          I_R_PRCHT_BALT_06MB
17117
17118
17119
17120        VIF for the second dominant variable
17121
17122            Variance
17123            Inflation
17124        ------------
17125             8227261
17126
17127
17128        Loading for the second dominant variable for the component 306
17129        ---------
17130          0.998276
```

Figure 14-9. Second dominant variable identified for component 306.

It is noteworthy that in this particular instance, the variable with the highest VIF, designated as I_R_BALDIF_BALT_06MB, is also the variable with the greatest proportion of variance. This renders it the selected variable, although it should be noted that this will not always be the case. **Figure 14-10** illustrates a scenario in which the competing variable is the one with the largest proportion of variance, leading to the rejection of the initially identified dominant variable.

CHAPTER 14   COLLINEARITY ANALYSIS

```
36599     Variables with conllinearity issues
36600
36601
36602     --------
36603         266
36604
36605
36606     The first dominant variable and its VIF
36607
36608                                    Variance
36609                                    Inflation
36610     ----------------------------------------
36611     I_AV_R_PRCHT_BALT_L03M             917943
36612
36613
36614     Maximum dimension for the first dominant variable
36615
36616     ---------
36617     303
36618
36619
36620     Loading for the first dominant
36621
36622
36623     --------
36624     0.997114
36625
36626
36627     The second dominant variable for the 303       component
36628
36629
36630     -----------------------------
36631     I_AV_R_BALDIF_BALT_L03M
36632
36633
36634     VIF for the second dominant variable
36635
36636       Variance
36637       Inflation
36638     -----------
36639        909900
36640
36641
36642     Loading for the second dominant variable for the component 303
36643
36644
36645     --------
36646     0.997265
```

Figure 14-10. Example where the competing variable has a greater proportion of variance than the first dominant variable identified.

Another important observation from this example is the fact that I_AV_R_PRCHT_BALT_LO3M and I_AV_R_BALDIF_BALT_LO3M are competing to explain the same variation in the model. In conceptual terms, PRCHT_BALT represents the amount spent (purchases) on the credit card during the previous month, relative to the total balance reported at the end of the billing period. BALDIF_BALT, on the other hand, denotes the credit card's balance difference, defined as the balance amount of the reference month minus the balance amount of the previous month, relative to the total balance reported at the end of the billing period. It is evident that the amount spent during a given month will tend to be approximately equal to the balance difference between the previous month's statement and the actual balance. It is therefore unsurprising that the two variables are in conflict with one another in their ability to explain the same amount of variability in the model.

14.5.6 Delete the Rejected Variable and Recheck for Collinearity Issues

The final step in the process is to physically remove the rejected variable from the variable name catalog, replace the previous set of variables with the new one, rerun the collinearity diagnostics, and check for any remaining collinearity issues. This sequence of actions is illustrated in **Listing 14-7**. From lines 268 to 269, it can be observed that two DELETE clauses are in effect: one that deletes the rejected variable and a second one that checks for the absence of perfect collinearity. The previous set of variables is replaced by executing the code present from lines 272 to 278, which is executed within the same PROC SQL. Subsequently, the PROC REG is executed once more, from line 283 to line 294, and finally, the number of conflicts is quantified once more to ascertain whether the DO WHILE condition has been satisfied.

Listing 14-7. Eliminate the variable with the smaller proportion of variance, run the collinearity diagnostics again, and check to see if collinearity issues still exist.

```sas
        /*------------------------------------------------------------
                --- Get rid off the conflicting variable
        ------------------------------------------------------------*/
        PROC SQL;
            DELETE FROM VARIABLES_INPUTS WHERE COMPRESS(NAME)=COMPRESS("&REJECT.");
            DELETE FROM VARIABLES_INPUTS
                WHERE COMPRESS(NAME) IN (SELECT COMPRESS(VARIABLE) FROM PE_REG WHERE ESTIMATE=0);

            SELECT
                COMPRESS(NAME)
                INTO
                :NEW_INPUTS SEPARATED BY " "
                FROM VARIABLES_INPUTS
            ;
        QUIT;
        %LET INPUTS=&NEW_INPUTS.;
        /*------------------------------------------------------------
                --- Run the linear regression and generate the collinearity diagnostics
        ------------------------------------------------------------*/
        PROC REG
            DATA=&DS.
            PLOTS(MAXPOINTS=NONE)
            ;
            &VPART.
            MODEL &TARGET.=&INPUTS./COLLIN VIF TOL;

            ODS OUTPUT
                COLLINDIAG=CD_REG
                PARAMETERESTIMATES=PE_REG
            ;
        RUN;
        /*------------------------------------------------------------
                --- Check if collinearity is still present
        ------------------------------------------------------------*/
        TITLE 'VARIABLES WITH CONLLINEARITY ISSUES';
        PROC SQL;
            SELECT
                COUNT(DISTINCT VARIABLE) INTO :CONFLICTS
                FROM PE_REG
                WHERE
                    VARIANCEINFLATION>=&VIF.
            ;
        QUIT;

        TITLE;

        %LET N=%EVAL(&N.+1);
        %LET K=%EVAL(&K.+1);
    %END;
```

14.6 Applied Example

Figure 14-11 depicts the initial set of variables exhibiting collinearity issues at a VIF cutoff value of <4. It is notable that the variable I_FLG_GUARANTOR_F (presence of female guarantor) exhibits the highest VIF value of 4.124.

CHAPTER 14   COLLINEARITY ANALYSIS

First Stage, cutoff value of VIF < 4

Variable	DF	Parameter	Srd. Error	t-Value	p-Value	Tolerance	VIF	
Intercept	1	(0.550)	0.352	(1.560)	0.119	.	.	
I_DEC_CPI_N01	1	(0.001)	0.007	(0.140)	0.887	0.450	2.224	
I_DEC_CPI_N04	1	(0.001)	0.006	(0.250)	0.802	0.690	1.450	
I_DEC_CPI_N08	1	0.007	0.005	1.310	0.191	0.739	1.353	
I_DEC_CPI_N09	1	(0.001)	0.006	(0.130)	0.898	0.617	1.622	
I_DEC_CPI_N10	1	(0.000)	0.006	(0.070)	0.946	0.546	1.832	
I_DEC_CPI_N12	1	(0.001)	0.005	(0.180)	0.857	0.764	1.309	
I_AGE	1	(0.001)	0.000	(4.190)	<.0001	0.510	1.961	
I_MAX_GRNTR_AGE	1	(0.000)	0.000	(0.620)	0.534	0.693	1.443	
I_FLG_CODEBTOR	1	0.000	0.007	0.050	0.960	0.503	1.988	
I_FLG_CODEBTOR_F	1	0.001	0.007	0.200	0.843	0.531	1.884	
I_FLG_GUARANTOR_M	1	(0.010)	0.009	(1.070)	0.283	0.246	4.073	Maximum
I_FLG_GUARANTOR_F	1	(0.013)	0.009	(1.370)	0.170	0.243	4.124	Collinearity identified
I_FLG_GUARANTOR_S	1	0.004	0.005	0.680	0.495	0.863	1.159	
I_FLG_PERSONAL_REF	1	0.001	0.005	0.170	0.863	0.865	1.155	
I_FRQ_GUARANTORS	1	0.010	0.005	1.910	0.056	0.	4.124	
I_FRQ_COMMERCIAL_REF	1	(0.006)	0.002	(3.090)	0.002	0.	1.159	
I_CPI_N06	1	0.000	0.003	0.070	0.944	0.		
I_CPI_N10	1	0.005	0.004	1.420	0.157	0.		
I_MONTHS_WORKING	1	(0.000)	0.000	(1.190)	0.234	0.961	1.041	
I_CREDIT_AMOUNT	1	(0.000)	0.000	(1.770)	0.077	0.438	2.282	
I_NUMDEPENDENT	1	(0.000)	0.002	(0.060)	0.952	0.949	1.053	
I_NUMCHILDREN	1	(0.000)	0.000	(0.170)	0.868	0.998	1.002	
I_TERMS_DAYS	1	(0.000)	0.000	(1.760)	0.078	0.264	3.792	
I_R_INSTALLMENT_LOAN	1	(0.157)	0.227	(0.690)	0.488	0.313	3.200	
I_R_INSTALLMENT_INCOME	1	(0.000)	0.001	(0.310)	0.758	0.777	1.287	
I_R_NETINCOME_INCOME	1	(0.000)	0.001	(0.020)	0.981	0.772	1.295	
I_R_APPAGE_MAXGRNTAGE	1	0.007	0.007	0.900	0.368	0.554	1.804	
I_SUM_DEC_CPI_L03M	1	0.003	0.003	0.820	0.411	0.517	1.934	
I_INTERESTRATE	1	0.001	0.001	1.960	0.050	0.482	2.076	

Figure 14-11. Initial set of variables with collinearity issues. VIF cutoff value <4.

Figure 14-12, on the other hand, presents the collinearity diagnostics corresponding to the set of variables shown in **Figure 14-11**. On the left side, an excerpt of the comprehensive diagnostic is observable, which will include a column for each variable. It is notable that the variable I_FLG_GUARANTOR_F attains its maximum proportion of variance on the 25th component, with a value of 0.68. This value conflicts with the variable I_FLG_GUARANTOR_M (presence of male guarantor), which has a variance proportion of 0.64 on the same component. The upper right section of the figure presents a bar chart which serves to illustrate the issue. It is evident

CHAPTER 14 COLLINEARITY ANALYSIS

from the figure that both I_FLG_GUARANTOR_F and I_FLG_GUARANTOR_M compete for the hegemony of the 25th component, as evidenced by bars of almost the same height.

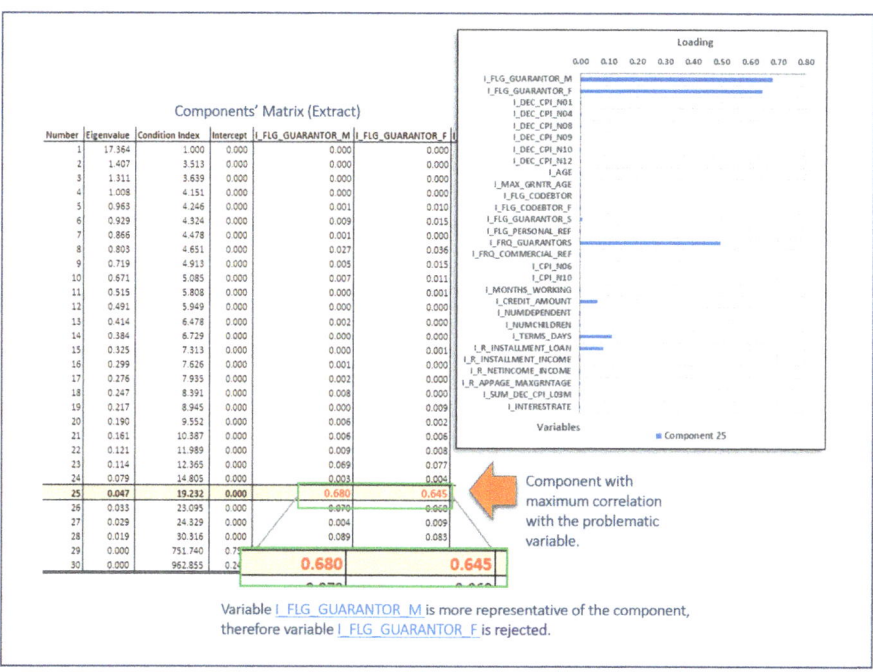

Figure 14-12. Comparison of the proportion of variance between the conflicting variables.

In accordance with the previously delineated methodology, the solution to this issue is to eliminate the variable I_FLG_GUARANTOR_M, since it explains less variance than I_FLG_GUARANTOR_F. As demonstrated in **Figure 14-13**, the elimination of I_FLG_GUARANTOR_M results in a reduction of collinearity. Notice that the collinearity issue has been resolved at a VIF cutoff value <4, and I_FLG_GUARANTOR_F now has a VIF of only 1.077.

CHAPTER 14 COLLINEARITY ANALYSIS

Variable	DF	Parameter	Srd. Error	t-Value	p-Value	Tolerance	VIF
Intercept	1	(0.564)	0.352	(1.600)	0.109	.	.
I_DEC_CPI_N01	1	(0.001)	0.007	(0.150)	0.884	0.450	2.224
I_DEC_CPI_N04	1	(0.001)	0.006	(0.250)	0.800	0.690	1.450
I_DEC_CPI_N08	1	0.007	0.005	1.290	0.198	0.739	1.352
I_DEC_CPI_N09	1	(0.001)	0.006	(0.130)	0.899	0.617	1.622
I_DEC_CPI_N10	1	(0.000)	0.006	(0.070)	0.944	0.546	1.832
I_DEC_CPI_N12	1	(0.001)	0.005	(0.170)	0.865	0.764	1.309
I_AGE	1	(0.001)	0.000	(4.190)	<.0001	0.510	1.961
I_MAX_GRNTR_AGE	1	(0.000)	0.000	(0.640)	0.524	0.693	1.443
I_FLG_CODEBTOR	1	0.000	0.007	0.010	0.996	0.504	1.986
I_FLG_CODEBTOR_F	1	0.002	0.007	0.250	0.805	0.531	1.882
I_FLG_GUARANTOR_M	**1**	**0.001**	**0.005**	**0.200**	**0.840**	**0.929**	**1.077**
I_FLG_GUARANTOR_S	1	0.004	0.005	0.680	0.499	0.863	1.159
I_FLG_PERSONAL_REF	1	0.002	0.005	0.290	0.772	0.872	1.147
I_FRQ_GUARANTORS	1	0.006	0.004	1.440	0.151	0.622	1.609
I_FRQ_COMMERCIAL_REF	1	(0.006)	0.002	(3.120)	0.002	0.894	1.118
I_CPI_N06	1	0.000	0.003	0.070	0.948	0.419	2.386
I_CPI_N10	1	0.006	0.004	1.430	0.152	0.336	2.978
I_MONTHS_WORKING	1	(0.000)	0.000	(1.200)	0.229	0.961	1.041
I_CREDIT_AMOUNT	1	(0.000)	0.000	(1.830)	0.067	0.439	2.278
I_NUMDEPENDENT	1	(0.000)	0.002	(0.020)	0.980	0.950	1.053
I_NUMCHILDREN	1	(0.000)	0.000	(0.160)	0.872	0.998	1.002
I_TERMS_DAYS	1	(0.000)	0.000	(1.690)	0.092	0.265	3.779
I_R_INSTALLMENT_LOAN	1	(0.148)	0.227	(0.650)	0.515	0.313	3.197
I_R_INSTALLMENT_INCOME	1	(0.000)	0.001	(0.300)	0.761	0.777	1.287
I_R_NETINCOME_INCOME	1	(0.000)	0.001	(0.030)	0.975	0.772	1.295
I_R_APPAGE_MAXGRNTAGE	1	0.007	0.007	0.920	0.356	0.554	1.804
I_SUM_DEC_CPI_L03M	1	0.003	0.003	0.820	0.415	0.517	1.934
I_INTERESTRATE	1	0.001	0.001	1.940	0.052	0.482	2.076

Figure 14-13. *Final result after variable elimination.*

It is interesting to note from the bar chart presented in **Figure 14-12** that the I_FRQ_GUARANTORS variable is also representative of the 25th component. This is due to the fact that I_FRQ_GUARANTORS is associated with the *"guarantors"* variable, but instead of identifying whether the guarantor is male (M) or female (F), it quantifies the total number of guarantors. In this sense, the 25th component can be considered as the guarantor's component.

14.7 EM Variable Clustering Node (Varclus)

EM also has its own collinearity reduction procedure called variable clustering, or varclus for short, and it is located on the Explore tab. The varclus procedure is based on a PCA, and according to the SAS help center [15], the algorithm behind the varclus node is both divisive and iterative. By default, the varclus node starts with all variables in a single cluster, and it then repeats the following steps:

1. A cluster is chosen for splitting. Depending on the options specified, the selected cluster has either the smallest percentage of variation explained by its cluster component (using the **Variation Proportion** property) or the largest eigenvalue that is associated with the second principal component (using the **Maximum Eigenvalue** property).

2. The chosen cluster is split into two clusters by finding the first two principal components, performing an orthoblique rotation (raw quartimax rotation on the eigenvectors; Harris and Kaiser, 1964), and assigning each variable to the rotated component with which it has the higher squared correlation.

3. Variables are iteratively reassigned to clusters to try to maximize the variance accounted for by the cluster components. You can require the reassignment algorithms to maintain a hierarchical structure for the clusters (using the Keep Hierarchies property).

CHAPTER 14 COLLINEARITY ANALYSIS

The variable clustering node algorithm stops splitting when either

- The maximum number of clusters as specified by the **Maximum Clusters** property is reached
- Each cluster satisfies the stopping criteria specified in the **Variation Proportion** property and/or the **Maximum Eigenvalue** properties

The varclus node is a good alternative to the ICRM; however, the idea is not to use one or the other, but to use both of them as different treatments simultaneously in a parallel fashion.

The most important aspects of the varclus node can be summarized as follows:

1. The varclus node is not computationally efficient, and it is recommended to be used with less than 100 variables and less than 100,000 rows.

   Consideration: Given the probability of these limits being exceeded, it is advisable to set the **Suppress Sampling Warning** property to yes.

2. The varclus node can also handle nominal variables, but you should be aware that it does so by creating a dummy variable for each category in your nominal variable.

   Consideration: It is imperative to bear in mind that the primary objective of this chapter pertains to the management of collinearity, a phenomenon that is intrinsic to interval variables. To maintain consistency with the methodology outlined in this book, it is essential to refrain from utilizing the varclus node for the purpose of addressing nominal variables.

CHAPTER 14   COLLINEARITY ANALYSIS

3. The PCA of the varclus node allows you to choose between the correlation matrix and the covariance matrix. If the covariance matrix is chosen, variables with a large variance will have a greater effect on the cluster components.

   Consideration: It is acknowledged that determining the most effective method is not possible, as both are data-dependent. Therefore, the most prudent approach is to utilize both methods concurrently in a parallel manner.

4. The varclus node allows you to specify three different stopping criteria:

   a) **Maximum Clusters:** Specifies the maximum number of clusters desired. Valid values are integers greater than or equal to 1.

   b) **Maximum Eigenvalue:** Specifies the cutoff eigenvalue for the second eigenvalue of each cluster. Valid values are real numbers greater than or equal to 1.

   c) **Variation Proportion:** Specifies the minimum variation proportion at which the cluster must be split. The **Variation Proportion** property accepts real numbers between 0 and 1.0.

   Consideration: It is not possible to ascertain which criteria, and at which level, will yield the best result,[8] since the clustering procedure is data-

---

[8] The meaning of "best result" will be explained in detail in Sections 16.1 and 17.1.2.4, but for the time being, the best result refers to the number of output variables obtained from the execution of the varclus procedure. For instance, it is expected that there will be a reduction of around 70% on your input variables after the varclus execution.

dependent. The most effective approach in this case is to initially determine a series of test values for each criterion. This may enable the screening of the solution space, and subsequent testing of each value against its corresponding criterion, while fixing the other two to identify the optimal values for each. To mitigate the intricacy of these trials, it is recommended to focus on the **Maximum Clusters** and the **Variation Proportion** criteria, while leaving the **Maximum Eigenvalue** criterion unaltered. In the event of identifying disparate values that yield equivalent outcomes, it may be advantageous to incorporate additional varclus nodes into the EM diagram, each with distinct parameters. This approach will result in parallel branches, facilitating the exploration, evaluation, and monitoring of the outcomes obtained from each varclus node.

5. The varclus node allows you to select the output variables that you want to export from the clusters; you can choose to replace the original variables with their principal component, or you can choose to select only the most representative variable of each principal component. The varclus node estimates the variance explained by each variable within its corresponding component and therefore selects the best variable based on the $1-R^2$, where the lower the value, the better.

Consideration: As previously discussed, a key characteristic of a scorecard model is its capacity to predict the default event in straightforward terms from its original predictive variables. Consequently, the replacement of the original variables with their principal component is seldom, if ever, employed.

14.7.1 ICRM or Varclus?

It is acknowledged that both the varclus procedure and the ICRM possess distinct advantages and disadvantages. Furthermore, it is recognized that both are data-dependent and have the capacity to regulate the number of output variables through the refinement of their respective stopping criteria. The recommendation is therefore not to choose one over the other, but rather to utilize both concurrently in parallel branches of the EM diagram. **Table 14-1** summarizes some of the key differences between the ICRM and the varclus procedures.

Table 14-1. ICRM and varclus comparison.

ICRM	Varclus
Uses the VIF as stop criterion	Uses the Maximum Clusters, the Maximum Eigenvalue, and the Variation Proportion as stopping criteria
The stopping criterion concept is clearer and can be directly related to the R^2 of each variable with respect to the remaining variables in the model. In addition, there is empirical evidence that a VIF value between 3 and 4 can significantly reduce collinearity, but without eliminating important effects for the multivariable model	There are multiple stopping criteria, none of which have a clear threshold or cutoff point Multiple combinations are possible, making it very difficult to select the best value for each criterion
Makes it easier to choose a cutoff value	Makes it harder to choose a cutoff value
Can control model's size by selecting different VIF values	Can control model's size by selecting different combinations of values for the Maximum Clusters, Maximum Eigenvalue, and the Variation Proportion
Model overfitting may depend on this procedure	Model overfitting may depend on this procedure

14.8 Summary

In this chapter, it was shown that collinearity, caused by high correlations among predictors, leads to redundancy, inflated errors, and instability in regression models. The *variance inflation factor* (VIF) was highlighted as a key diagnostic tool for identifying multicollinearity, with high VIF values

CHAPTER 14 COLLINEARITY ANALYSIS

signaling problematic predictors. It was also demonstrated that collinearity creates *optimization challenges*, making parameter estimates unreliable and model convergence difficult. Of particular importance was the fact that collinearity issues tend to cause a phenomenon called *coefficient inversion*.

Several alternative methods to address collinearity were reviewed, including the *principal component analysis* (PCA), the *ridge regression*, and the *correlation matrix*. However, it was emphasized that these approaches have notable limitations, such as a reduced interpretability (PCA), subjectivity (ridge regression), or impracticality (correlation matrix), which make them unviable options for dealing with collinearity issues, according to the methodology outlined here.

To address these shortcomings, the *iterative collinearity reduction method* (ICRM) was proposed as a systematic approach that combines diagnostics like VIF with iterative removal of redundant predictors, ensuring interpretability and robustness. Additionally, the *varclus node* in Enterprise Miner (EM) was introduced as a native tool that automates variable clustering and collinearity reduction, particularly useful for large datasets.

Key takeaways:

- **Collinearity Detection:** Tools like VIF and PCA diagnostics are essential for identifying multicollinearity.

- **Optimization Challenges:** Collinearity inflates errors and destabilizes parameter estimates, causing a phenomenon called coefficient inversion.

- **Limitations of Alternatives:** The PCA, the ridge regression, and the correlation matrix have interpretability and practicality constraints.

- **ICRM Method:** The ICRM method balances interpretability and efficiency by iteratively removing redundant predictors.

- **Enterprise Miner Integration:** The varclus node is an alternative to the ICRM, since it also effectively automates the collinearity reduction for large-scale data.

- **Branch Your Mining Flow:** There are inherent trade-offs for the varclus and the ICRM, since both of them are data-dependent. Thus, the recommendation is to branch your diagram, creating two *variants* (see Chapter 17) of your model.

In the next chapter, we delve into one of the most critical and indispensable stages in the creation of scorecard models: the calculation of the weight of evidence (WoE). This process is not only rigorous but also demands careful preparation, which highlights the importance of the preceding steps. The progression from the univariable analysis to the collinearity analysis, and finally to the WoE, is deliberate and essential. The univariable analysis eliminates *irrelevancy*, while the collinearity analysis addresses *redundancy*, and together they ensure that the inputs are optimized and ready for the demanding WoE calculation process.

CHAPTER 15

Weight of Evidence

"The greatest challenge facing mankind is the challenge of distinguishing reality from fantasy, truth from propaganda."

—Michael Crichton (2003)

In the initial stages of undertaking a regression analysis, the focus is on simple linear regression, which involves the observation of a straightforward relationship between two continuous variables. One of these variables serves as the independent explanatory variable, while the other is designated as the dependent variable, which is intended to be predicted within a model of the form $y=a+m \cdot x$. Subsequently, the multivariable model is employed, where the independent variable, y, is a function of n explanatory variables, x_i, also known as regressors. The subsequent step involves incorporating nominal variables into the model. Nominal variables can be integrated into a regression model if it is transformed into n-1 binary variables, where n signifies the total number of categories within the nominal variable. The next step considers the range and distribution of interval variables, in the search of the relationship that exists between the continuous variables and the dependent variable. In this regard, techniques such as marginal value

CHAPTER 15 WEIGHT OF EVIDENCE

analysis, scaling, and variable transformation are employed. In summary, the most important factors to consider when building a regression model are as follows:

1. The inclusion of nominal variables through the use of dummy variables

2. The use of robust techniques for the control of outliers, or the identification and elimination of outliers

3. Scaling and/or standardizing the units of the interval variables to stabilize the model

4. Identifying the relationship that exists between the continuous variables and the dependent variable, in order to later look for the appropriate transformation that guarantees the premise of linearity between the dependent and the independent variables, which in the case of logistic regression corresponds to the relationship between the regressor and the logit

What is remarkable, however, is not the fact that the WoE method renders all these activities obsolete, but the fact that the majority of modelers remain unaware of its existence throughout their careers.

15.1 WoE Definition and Calculation

The most effective method of comprehending the weight of evidence (WoE) concept is through the utilization of a case study. **Table 15-1** presents a hypothetical illustration of the WoE calculation for the credit card (cc) utilization (balance amount/credit limit). In this example, the cc utilization has been grouped into ten different groups, where the columns "Min. utilization" and "Max. utilization" refer to the lower and upper limits of the cc utilization.

CHAPTER 15 WEIGHT OF EVIDENCE

Table 15-1. Fictitious example of a WoE calculation for the credit card utilization.

Groups	Min. utilization	Max. utilization	Obs	% Obs	Events	No events	Dist. events	Dist. no events	WOE	DNE-DE	IV
1	<3.2%	3.2%	5,011	9.9%	600	4,411	5.7%	11.0%	0.650	0.053	0.034
2	3.2%	6.0%	5,237	10.4%	720	4,517	6.9%	11.3%	0.492	0.044	0.022
3	6.0%	7.3%	3,899	7.7%	655	3,244	6.3%	8.1%	0.255	0.018	0.005
4	7.3%	8.1%	4,785	9.5%	800	3,985	7.7%	9.9%	0.261	0.023	0.006
5	8.1%	11.5%	4,059	8.0%	820	3,239	7.9%	8.1%	0.029	0.002	0.000
6	11.5%	12.2%	7,466	14.8%	1,700	5,766	16.3%	14.4%	−0.123	−0.019	0.002
7	12.2%	19.5%	4,893	9.7%	1,200	3,693	11.5%	9.2%	−0.221	−0.023	0.005
8	19.5%	41.2%	5,916	11.7%	1,600	4,316	15.3%	10.8%	−0.353	−0.046	0.016
9	41.2%	68.6%	4,525	9.0%	1,550	2,975	14.8%	7.4%	−0.693	−0.074	0.051
10	68.6%	>68.6%	4,703	9.3%	1,800	2,903	17.2%	7.2%	−0.867	−0.100	0.087
			50,494	100.0%	10,438	40,056	100.0%	100.0%			0.228

CHAPTER 15 WEIGHT OF EVIDENCE

From this example, the first and last groups are defined as open intervals and can be defined as follows:

$$A(x)^{ll} = \begin{bmatrix} a_1^{ll} \\ a_2^{ll} \\ \vdots \\ a_m^{ll} \end{bmatrix}$$

Equation 15-1. Vector of cutoff lower limits.

$$A(x)^{ul} = \begin{bmatrix} a_1^{ul} \\ a_2^{ul} \\ \vdots \\ a_m^{ul} \end{bmatrix}$$

Equation 15-2. Vector of cutoff upper limits.

Where $A(x)^{ll}$ refers to the vector of lower limits and $A(x)^{ul}$ refers to the vector of upper limits, both as a function of the x interval variable.

The columns labeled "Obs" and "% Obs" present the total number of cases falling into each utilization group, as well as their vertical proportion, or distribution. The "Events" and "No events" columns present the number of defaults and no defaults within each group, respectively, and can be defined as follows:

$$n_i^E = \sum_{j=1}^{N} \left[if\ a_i^{ll} \leq x_j < a_i^{ul}\ then\ y_j\ else\ 0 \right]$$

Equation 15-3. Number of events.

And

$$n_i^{NE} = \sum_{j=1}^{N} \left[if\ a_i^{ll} \leq x_j < a_i^{ul}\ and\ y_j = 0\ then\ 1\ else\ 0 \right]$$

Equation 15-4. Number of no events.

Where n_i^E and n_i^{NE} are the number of events and no events for the *ith* group, *j* is the *jth* case or observation, and *N* is the total number of cases or observations.

The "Dist. events" and "Dist. no events" correspond to the proportion or distribution of defaults and no defaults across the utilization groups and can be defined as follows:

$$d_i^E = \frac{n_i^E}{N^E}$$

Equation 15-5. Distribution of events.

And

$$d_i^{NE} = \frac{n_i^{NE}}{N^{NE}}$$

Equation 15-6. Distribution of no events.

Where d_i^E and d_i^{NE} are the proportion of events and no events of the *ith* group and N^E and N^{NE} are the total number of events and no events in the sample.

The WoE is then defined as follows:

$$WoE_i = \ln\left[\frac{d_i^{NE}}{d_i^E}\right]$$

Equation 15-7. WoE.

Which is the natural logarithm of the distribution of no events, over the distribution of events for the *ith* group. It is worth noting that the WoE is equivalent to the definition of the *-logit* of the relative risk.

$$WoE \approx -logit_i = \ln\left[\frac{\pi(0)}{\pi(1)}\right]$$

Equation 15-8. WoE as logit.

CHAPTER 15 WEIGHT OF EVIDENCE

Where $\pi(0)/\pi(1)$ refers to the probability of no event over the probability of event. From here, its's derived that the WoE is actually

$$WoE_i = -ln[OR_i]$$

Equation 15-9. WoE as odds ratio.

Where OR refers to the odds ratio, and therefore, the WoE can be seen as a measure of the natural logarithm of the inverse of the relative risk of the *ith* group.

The next column, "DNE-DE," is defined as follows:

$$dd_i = d_i^{NE} - d_i^{E}$$

Equation 15-10. Difference between the distributions of no events and events.

Which is the difference between the no events distribution and the events distribution of the *ith* group.

The distribution difference, dd_i, is used as a mathematical trick to calculate the *information value,* IV, which is defined as follows:

$$IV_i = WoE_i \cdot dd_i$$

Equation 15-11. Information value.

It is important to note that, due to the properties of the natural logarithm, the WoE will be negative for fractional values (<1) and positive for values greater than one unit. Consequently, the product between the *WoE* and the *dd* will always be positive, thus enabling the summation of the individual *IV*s across groups.

CHAPTER 15 WEIGHT OF EVIDENCE

The total sum of the *IV*s provides a measure of the discriminative power of the specific grouping solution for a given variable. The IV is thus a function of the cutoff points of the variables, defined as $A(x)^{ll}$ and $A(x)^{ul}$, and the total number of groups, which can be expressed as follows:

$$IV_i \left(A(x)^{ll}, A(x)^{ul}, m \right)$$

Equation 15-12. IV as a function of the cutoff points and the number of groups.

The *IV* does not have a proper scale, but the rules of thumb presented in **Table 15-2** [17] can be used as a reference to the discriminatory power of the variable in question.

Table 15-2. Rules of thumb for the information value (IV) [17].

Information value	Variable predictiveness
Less than 0.02	Not useful for prediction
0.02 to 0.1	Weak predictive power
0.1 to 0.3	Medium predictive power
0.3 to 0.5	Strong predictive power
>0.5	Suspicious predictive power

As an alternative to the IV, the Gini index can be utilized to measure the overall discriminatory power of a given variable, defined as follows:

$$Gini\ Index = \left(1 - \frac{2 * \sum_{i=2}^{m} \left(n_i^E * \sum_{j=1}^{i-1} n_j^{NE} \right) + \sum_{k=1}^{m} \left(n_k^E * n_k^{NE} \right)}{N^E * N^{NE}} \right) * 100$$

Equation 15-13. The Gini index.

The Gini index may also be defined as a measure of inequality in the distribution of an event, where a higher value indicates greater inequality. In my opinion, the Gini index is preferable to the IV in that it has a well-founded theoretical basis and a defined scale between 0 and 100, which facilitates its interpretation.

15.2 Advantages of Using the WoE

The use of the WoE as a variable transformation offers considerable advantages to the modeling process, which collectively outweigh the benefits of other available transformation types. This approach can effectively address the most critical and prevalent issues encountered during the development of a regression model, including the following:

1. **Outliers**: The elimination of outliers is facilitated by the grouping of interval variables.

2. **Standardization**: The natural logarithm of the distribution of no events over the distribution of events standardizes the units of all variables, leaving all variables in *logit* form.

3. **Linearization**: Monotonic linearization of the interval variable can be achieved by trial-and-error selection of different cutoff points and groupings.

4. **Nominal Variables**: Nominal variables can also be handled by grouping different categories according to their risk level. This reduces the cardinality level of the nominal input, resulting in a single ordinal variable, which also eliminates the use of dummy variables.

5. **Missing Values**: The presence of missing values can be addressed by aggregating them into a separate group and assigning them the corresponding WoE according to their risk level.[1]

The following discussion will examine each of these points in detail to assess its importance and to compare it with alternative methods.

15.2.1 Controlling for the Presence of Outliers

In Section 4.1, an overview of the challenges associated with addressing outliers in the context of overall modeling processes was provided. The issue with outliers is that they deviate significantly from a genuine behavioral trend, thereby complicating the process of model generalization.

The techniques that address the presence of outliers can be classified into two distinct schools of thought. The first of these approaches involves the detection and close examination of outliers, employing techniques such as [41]

- Studentized residuals
- RSTUDENT residuals
- Cook's D
- DFFITS
- DFBETAS
- Haidi's influence measure

---

[1] It is worth mentioning that the lack of information is also information. This seemingly contradictory statement means that the absence of information could be a symptom of an underlying root cause, which is manifested as a different risk level for the population with the absence of information.

The second approach tends to measure the influence of the *outliers*, distinguishing between outliers and *leverage points* using robust techniques (Rousseeuw and Leroy [57]). Unfortunately, both techniques are not feasible from a scorecard perspective.

The primary issue with the initial approach is that it is a passive technique that recommends visual inspection, the utilization of nonlinear models, and even polynomial transformations and interactions. However, these are not viable options for scorecard models, particularly when dealing with thousands of variables. With regard to robust techniques, it would appear that there is no equivalent version of the robust regression for the logistic regression. Consequently, the use of robust techniques is also not a viable option.

However, the presence of outliers does not pose a significant concern when constructing scorecard models for two primary reasons. Firstly, the impact of outliers is mitigated by employing ratios instead of absolute values (see Section 4.2). Secondly, the process of binning interval variables effectively eliminates the presence of outliers. It is noteworthy that the binning technique is not a novel concept within the field of data mining. However, the novel aspect of this technique is the replacement of the original interval variable values that fall into each binning group with their corresponding risk level or WoE.

15.2.2 Standard Units and the Use of Logits

As discussed in Section 4.1.3, the scale and units of regressors have a significant impact on the magnitude and overall stability of the coefficients. The issue was addressed in part through the utilization of recommended ratios, percentages, proportions, and frequencies. However, the employment of the WoE fully rectifies the problem by maintaining consistency in logit units.

The logit is a universal unit for risk measurement, thus obviating the necessity to delve into variable specifics. Furthermore, it facilitates the linearization of the default event and predictor relationship. The utilization of either the logit or the WoE as a unit facilitates the description of default probability as a linear system of points, each delineated in terms of risk and translated into a score. It is evident that no alternative function could possibly accomplish this all at once.

15.2.3 Monotonic Linearization of Interval Variables

It is a fundamental assumption in regression analysis that the relationship between the dependent variable y and the independent variable x should be linear. To verify that the linearity assumption holds, the majority of literature recommends the visual inspection of scatter plots to determine whether or not a mathematical transformation is required. However, it is important to note that this approach is not feasible in practice, particularly when dealing with large numbers of variables or when constructing scorecard models.

The issue with transforming the original variables using functions such as the square root, exponential, or logarithmic functions is the resulting loss of interpretability. It is imperative to recognize that a primary objective of a scorecard model is to elucidate the probability of default in straightforward cause-and-effect terms, while ensuring the global sum of scorecard points is maintained in terms of risk intervals defined in the original units of the predictive variables. To illustrate this, consider a customer between the ages of 25 and 35, who is assigned a value of 30 points, while a customer between the ages of 35 and 50 is assigned a value of 40 points. This interpretation is straightforward, suggesting that as age increases, the probability of default decreases. Now consider an alternative scenario, where the relationship between age and the logit is better represented by the inverse function $1/x$. In this scenario, a customer with an age between 0.04 and 0.028 would receive 30 points, while a customer with an age between 0.028 and 0.02 would receive 40 points. Which of these cases would you consider more favorable in terms of interpretation and clarity?

CHAPTER 15   WEIGHT OF EVIDENCE

The WoE is a method that is designed to achieve monotonic linearization of the predictive variables,[2] obviating the need for mathematical transformations. To illustrate the methodology of linearizing interval variables using the WoE, a bar chart was plotted (utilizing data from **Table 15-1**), displaying the frequency of events per group on the primary axis and a line chart of the WoE as a function of the natural logarithm of the distribution of no events and events on the secondary axis (**Figure 15-1**). It is noteworthy that the WoE demonstrates a monotonic linear relationship with respect to the risk level, exhibiting an IV of 0.228.

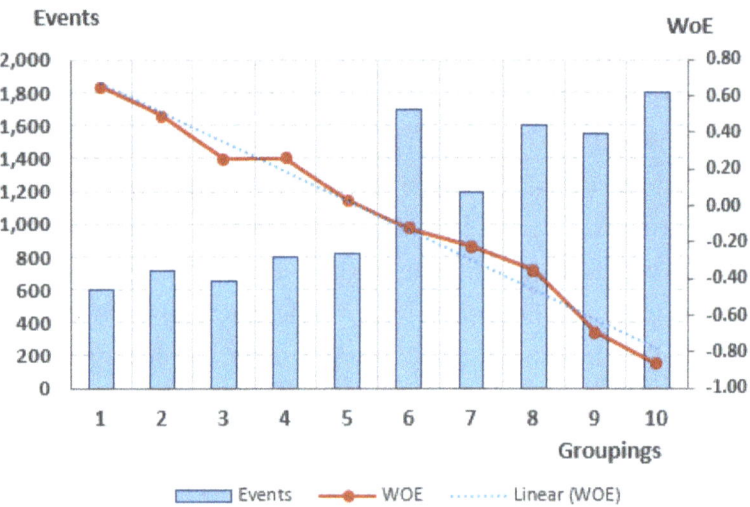

Figure 15-1. Bar chart of the frequency of events per group, on the primary axis. And line chart of the WoE as a function of the natural logarithm of the ratio of the distribution of no events to events, on the secondary axis.

---

[2] It is imperative to acknowledge that while the primary objective of the WoE method is to transform the original variables into monotonic linear functions, this does not imply that the relationship between the independent and dependent variables must be linear. Furthermore, it has been documented that certain relationships may manifest a U-shaped behavior (convex), or an inverted U-shaped behavior (concave).

CHAPTER 15   WEIGHT OF EVIDENCE

Now let us direct our attention to **Table 15-3** and **Figure 15-2**, in which the distribution of events has been modified to facilitate the direct visualization of the WoE function. It is noteworthy that the IV has increased by 31%, reaching a value of 0.298. Furthermore, the application of an additional smoothing technique to the WoE function is demonstrated in **Table 15-4** and **Figure 15-3**. It is notable that the IV has increased to 0.315, indicating a shift from a variable with moderate predictive power to one with strong predictive power, as illustrated in **Table 15-2**.

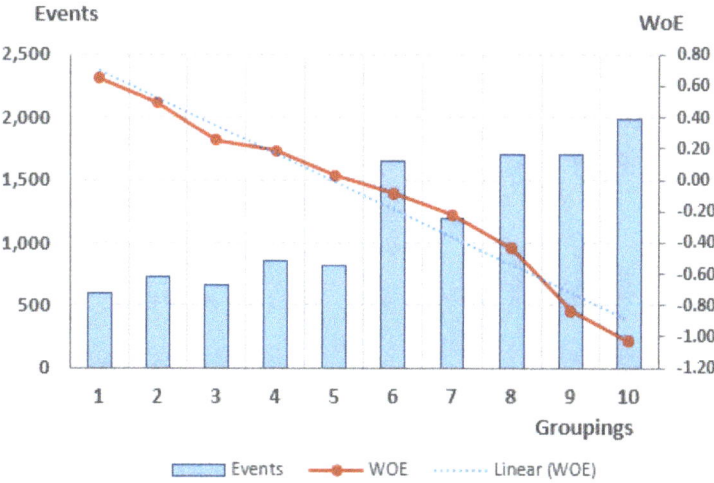

Figure 15-2.   Line chart showing the result after smoothing the WoE curve.

409

CHAPTER 15 WEIGHT OF EVIDENCE

Table 15-3. *Fictitious example where the distribution of events has been manipulated to flatten the WoE curve.*

Bins	Min. utilization	Max. utilization	Obs	% Obs	Events	No events	Dist. events	Dist. no events	WOE	DNE-DE	IV
1	<3.2%	3.2%	5,011	9.9%	600	4,411	5.7%	11.0%	0.650	0.053	0.034
2	3.2%	6.0%	5,237	10.4%	720	4,517	6.9%	11.3%	0.492	0.044	0.022
3	6.0%	7.3%	3,899	7.7%	655	3,244	6.3%	8.1%	0.255	0.018	0.005
4	7.3%	8.1%	4,785	9.5%	850	3,935	8.1%	9.8%	0.188	0.017	0.003
5	8.1%	11.5%	4,059	8.0%	820	3,239	7.9%	8.1%	0.029	0.002	0.000
6	11.5%	12.2%	7,466	14.8%	1,650	5,816	15.8%	14.5%	−0.085	−0.013	0.001
7	12.2%	19.5%	4,893	9.7%	1,200	3,693	11.5%	9.2%	−0.221	−0.023	0.005
8	19.5%	41.2%	5,916	11.7%	1,700	4,216	16.3%	10.5%	−0.437	−0.058	0.025
9	41.2%	68.6%	4,525	9.0%	1,700	2,825	16.3%	7.1%	−0.837	−0.092	0.077
10	68.6%	> 68.6%	4,703	9.3%	1,985	2,718	19.0%	6.8%	−1.031	−0.122	0.126
			50,494	100.0%	10,438	40,056	100.0%	100.0%			0.298

CHAPTER 15 WEIGHT OF EVIDENCE

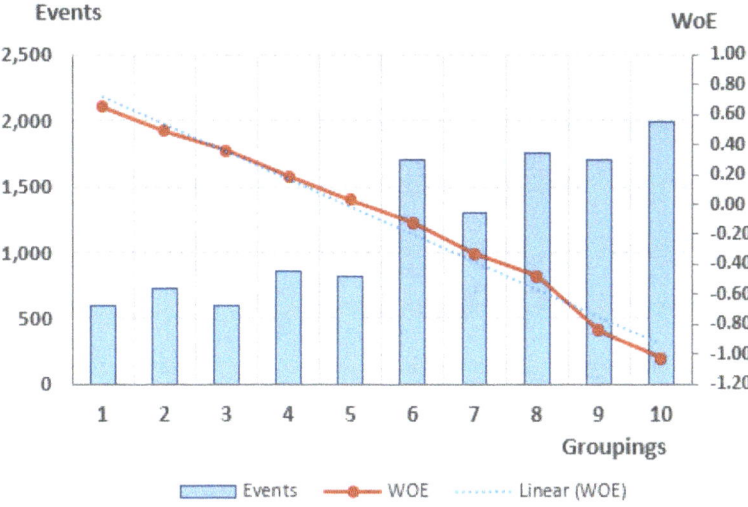

Figure 15-3. *Line chart showing the result after further smoothing of the WoE curve.*

411

CHAPTER 15 WEIGHT OF EVIDENCE

Table 15-4. Fictitious example where the distribution of events has been manipulated to further smooth the WoE curve.

Bins	Min. utilization	Max. utilization	Obs	% Obs	Events	No events	Dist. events	Dist. no events	WOE	DNE-DE	IV
1	<3.2%	3.2%	5,011	9.9%	600	4,411	5.7%	11.0%	0.650	0.053	0.034
2	3.2%	6.0%	5,237	10.4%	720	4,517	6.9%	11.3%	0.492	0.044	0.022
3	6.0%	7.3%	3,899	7.7%	600	3,299	5.7%	8.2%	0.360	0.025	0.009
4	7.3%	8.1%	4,785	9.5%	850	3,935	8.1%	9.8%	0.188	0.017	0.003
5	8.1%	11.5%	4,059	8.0%	820	3,239	7.9%	8.1%	0.029	0.002	0.000
6	11.5%	12.2%	7,466	14.8%	1,700	5,766	16.3%	14.4%	−0.123	−0.019	0.002
7	12.2%	19.5%	4,893	9.7%	1,300	3,593	12.5%	9.0%	−0.328	−0.035	0.011
8	19.5%	41.2%	5,916	11.7%	1,750	4,166	16.8%	10.4%	−0.477	−0.064	0.030
9	41.2%	68.6%	4,525	9.0%	1,700	2,825	16.3%	7.1%	−0.837	−0.092	0.077
10	68.6%	>68.6%	4,703	9.3%	1,985	2,718	19.0%	6.8%	−1.031	−0.122	0.126
			50,494	100.0%	10,438	40,056	100.0%	100.0%			0.315

412

This is the expected dynamic when engaging with the WoE. The objective is to refine the cutoff points[3] in a manner that optimizes the discriminatory power of the original variable while aligning the WoE with a monotonic linear function, consistently adhering to a logical trend.

Consequently, the WoE not only identifies a monotonic linear response and maximizes discriminatory power, but it also enables the determination of whether the observed behavior is logical from a business perspective or can be explained in some way. To illustrate this point, the WoE function is corresponded to the credit card's utilization rate, defined as the amount of pending balance over the credit limit. It is reasonable to hypothesize that an increase in the utilization rate will be accompanied by an increase in the risk of default, which will be evident as a WoE function with a negative slope (it should be noted that the WoE is negative for fractional values). Furthermore, it is anticipated that the WoE curve will exhibit a negative slope and a monotonic decrease as the risk level increases from one grouping to the next.

However, it is important to note that data does not always behave in the anticipated manner. In such cases, it may be necessary to impose a specific form of behavior on the WoE to align it with the expected behavior. To illustrate this phenomenon, one may refer to **Figure 15-4**, which depicts a fictitious example where the empirical data does not align with the conventional principles of logic and commonsense. It is noteworthy that nine out of ten points exhibit a descending straight line, with the exception of the WoE in group 8, which deviates from the expected pattern. It is imperative to bear in mind that the variable under scrutiny is the utilization of the credit card. Consequently, a downward straight line is to be expected, given that the higher the utilization, the higher the risk.

---

[3] It should be noted that in these examples, the distribution of events has been modified for the sole purpose of illustration. Consequently, the cutoff points have been maintained at their original values. It is evident that in a real case, this would not be feasible, as the distribution of events is contingent on the cutoff points.

However, the reason for the deviation of the WoE from the expected pattern remains unclear. This, however, is not a matter of concern, nor is it necessary to inquire further. This is due to the fact that we possess absolute certainty regarding the expected behavior of the WoE curve. In addition, we are not only aware of the expected behavior, but we can also dictate it according to the desired outcome or the behavior we wish to promote.

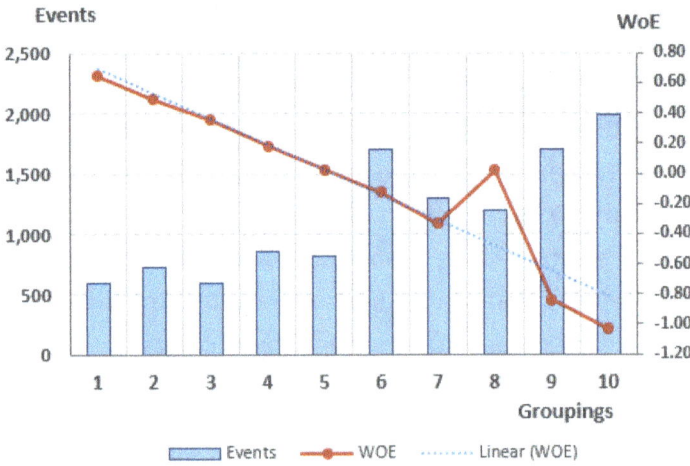

Figure 15-4. *Example of a variable where the empirical data does not reflect the expected behavior.*

This understanding is crucial for comprehending the WoE methodology, which, in contradistinction to alternative transformation techniques, permits not only the direct manipulation of cutoff points and the number of groupings but also the direct manipulation of the WoE value itself. The rationale behind the concept of manipulating the WoE is predicated on the premise that known logical trends cannot be supplanted by a flawed dataset. In other words, *the absence of evidence does not imply the nonexistence of evidence; it merely signifies that the evidence is not currently available.*

15.2.4 Dealing with Nominal Variables

The use of nominal variables in regression analysis is not a novel concept, as the regression analysis and the analysis of variance, more commonly known as *ANOVA* [20, 41, 42], are essentially two manifestations of the same underlying methodology. ANOVA can be conceptualized as an *omnibus* test, whereby the overall significance of the variables is evaluated, rather than the individual significance of each category within the nominal variable. This is analogous to the principle of univariable analysis. The primary objective of ANOVA is to ascertain whether there is a significant discrepancy between the means of multiple populations, or *treatments*, each comprising a distinct number of observations.

It is crucial to elucidate the manner in which the ANOVA test operates, as many analysts tend to overlook the fact that the generation of dummy variables does not result in the creation of n-1 independent binary variables, but in the creation of a single nominal variable whose categories have been conveniently arranged by n-1 columns. Consequently, it is imperative to recognize that dummy variables cannot be treated by multivariable selection methods as mere independent units that can be arbitrarily dropped or selected from the multivariable model. Instead, nominal variables should always be treated as a single unit that happens to be comprised of multiple columns.[4]

This presents a challenge to automated multivariable selection algorithms, as the majority are not programmed to consider the fact that a nominal variable cannot be partially dropped without proper *categorical lumping*. The process of categorizing risk is not arbitrary, but rather is the result of the aggregation of categories with a similar level of risk into

---

[4] It is important to acknowledge that the SAS PROC LOGISTIC is capable of dealing with nominal variables without the necessity to transform them to dummy variables first. See **Listing 13-5** for an applied example of the use of the CLASS statement.

CHAPTER 15   WEIGHT OF EVIDENCE

a total number of groups. This guarantees both a homogeneous level of risk within the members of the same group and a heterogeneous level of risk between the different groups formed. This process is also known as *clustering*.

As previously stated in Section 9.2, categorical clustering can be achieved through a variety of methods, including

1. Greenacre's method [26, 47]
2. The use of decision trees
3. The use of the weight of evidence (WoE)

Both Greenacre's method [26, 47] and the decision trees are effective at reducing the cardinality of a nominal variable to a reasonable number of categories. However, it should be noted that neither approach addresses the issue of having to use dummy variables. In contrast, the WoE is well-suited to handling nominal variables, resulting in the generation of a single new variable that can be utilized alongside interval variables without any discrepancy in units or scale.

Table 15-5 illustrates the application of the WoE methodology to the nominal variable "profession." It is noteworthy that the WoE calculation for a nominal variable is identical to that for an interval variable. Additionally, it is also noteworthy that, in this instance, group 5 represents the missing values or unknown categories that were not available during the WoE training.

CHAPTER 15 WEIGHT OF EVIDENCE

Table 15-5. WoE calculation for profession.

Bins	Profession categories	Obs	% Obs	Events	No events	Dist. events	Dist. no events	WOE	DNE-DE	IV
1	Housewife, other	56,519	17.3%	1,656	54,863	23.7%	17.1%	−0.324	−0.066	0.021
2	Business owner, technician, farmer, blue-collar job	124,534	38.1%	3,030	121,504	43.4%	38.0%	−0.133	−0.054	0.007
3	Health, white-collar job	118,721	36.3%	1,977	116,744	28.3%	36.5%	0.254	0.082	0.021
4	Retired	26,278	8.0%	277	26,001	4.0%	8.1%	0.717	0.042	0.030
5	_MISSING_, _UNKNOWN_	1,033	0.3%	47	986	0.7%	0.3%	−0.781	−0.004	0.003
		327,085	100.0%	6,987	320,098	100.0%	100.0%			0.082

417

It is important to note that the WoE is analogous to Greenacre's method [26, 47] or the decision tree, as the grouping process is identical. However, the distinction lies in the replacement of the resulting groups with their corresponding WoE. At this juncture, it is important to acknowledge that the integration of categorical grouping techniques in conjunction with the WoE makes the use of dummy variables obsolete.

15.2.5 Dealing with Missing Values

As illustrated in the aforementioned example, the WoE is also capable of addressing instances of missing values and unknown categories by assigning them to a distinct group. This feature is highly advantageous for several reasons. Primarily, the final model will invariably be capable of returning a prediction, as even missing values will have a WoE assigned to them. Secondly, it is important to acknowledge that missing information can also be considered as a form of information. This implies that an empirically observed differentiated risk level for a population lacking specific information can be associated with an underlying behavior or characteristic. To illustrate this point, consider a PD retail loan model in which a variable (I_SIN_R_CCBAL_CLM_L06M, **Table 15-6**) representing the sum of increments of the credit card utilisation ratio (pending balance/credit limit) over the preceding six months is employed as an explanatory variable. It is evident that all customers have a retail loan product, since this is the definition of the target population. However, it is less evident that some customers may have both products with the same institution, a retail loan and a credit card, whilst others may not. It is evident that the risk level increases in direct correlation with the number of increments, which is a logical consequence of the fact that the greater the number of consecutive increments in the utilization ratio, the more probable it is that the utilization level will reach a point of no return. It is noteworthy that group 4 (Table 15-6), which contains the missing value category, exhibits the second highest risk level. This observation gives rise to the question of why the absence of information is associated with a different risk level compared to the other groups.

CHAPTER 15 WEIGHT OF EVIDENCE

Table 15-6. Example of WoE interpretation from the group of missing values.

Bins	Intervals	Obs	% Obs	Events	No events	Dist. events	Dist. no events	WOE	DNE-DE	IV
1	I_SIN_R_CCBAL_CLM_L06M<1	18,565	5.7%	120	18,445	1.7%	5.8%	1.210	0.040	0.049
2	1<=I_SIN_R_CCBAL_CLM_L06M<3	32,106	9.8%	323	31,783	4.6%	9.9%	0.764	0.053	0.041
3	3<=I_SIN_R_CCBAL_CLM_L06M	30,769	9.4%	847	29,922	12.1%	9.3%	−0.260	−0.028	0.007
4	_MISSING_	245,645	75.1%	5,697	239,948	81.5%	75.0%	−0.084	−0.066	0.006
		327,085	100.0%	6,987	320,098	100.0%	100.0%			0.102

CHAPTER 15 WEIGHT OF EVIDENCE

Given that the model is at the retail loan level, and therefore the secondary product is the credit card, it can be deduced that the absence of information almost certainly signifies the absence of the credit card product. It is noteworthy that three-quarters of the sample do not possess a credit card, which may indicate a preference for retail loans over credit cards at other financial institutions. Notwithstanding, the question of why customers without credit cards exhibit a heightened default risk remains unaddressed. One hypothesis that may account for this phenomenon is that customers without a credit card are less *committed* to their financial institution and therefore feel less compelled to meet their financial obligations than those who currently have multiple products with their bank—which is also known as *customer loyalty*. It is noteworthy that this observation, though seemingly straightforward, has the potential to elucidate the elevated default risk exhibited by customers devoid of credit cards, as compared to those who possess one. It is also important to acknowledge that this observed risk does not attain the levels of significance that have been observed among individuals with three or more increments in their utilization level over the past six months.

The salient point to be gleaned from this example is that the WoE is sufficiently robust to accommodate missing values and, secondly, that close attention should be paid to missing information, as it is not an error in the data, but rather a feature that may convey significant information about the underlying behaviors and characteristics that drive the risk of default.

15.3 Using Decision Trees to Calculate the WoE

The aforementioned example illustrated a scenario where the total number of groups and their cutoff points were predetermined. However, in practice, it will be necessary to define these parameters from scratch for each explanatory variable in the ABT.

CHAPTER 15 WEIGHT OF EVIDENCE

There are numerous methods that can be employed to determine the number of groups and their cutoff points for the purpose of variable partitioning. The most optimal approach is through the utilization of decision trees. The *Interactive Grouping node* in EM employs *decision trees* to estimate the WoE, a strategy that is advantageous due to their flexibility and robustness. This is due to their capacity to accommodate both interval and nominal variables, in addition to being null-proof.

15.3.1 The Interactive Grouping Node

The Interactive Grouping node (IGN) [13] constitutes a component of the credit scoring nodes suite, situated within the node panel. The IGN is constituted by eight properties, which are

1. Train
2. Interval target options
3. Predefined grouping
4. Interval variable binning options
5. Special code options
6. Grouping options
7. Score
8. Report

15.3.1.1 Tree-Based Grouping Options

As illustrated in **Figure 15-5**, the Interactive Grouping node is located in the **Grouping Options** section, while **Figure 15-6** details the **Tree-Based Grouping Options**. The **Criterion option** enables the user to select either the entropy or *p*-value of the Pearson chi-square statistic.

CHAPTER 15   WEIGHT OF EVIDENCE

The **Missing Values option** permit the user to control how the splitting rules should handle the missing values; the default option isolates the missing values in a separate branch, which is strongly recommended.

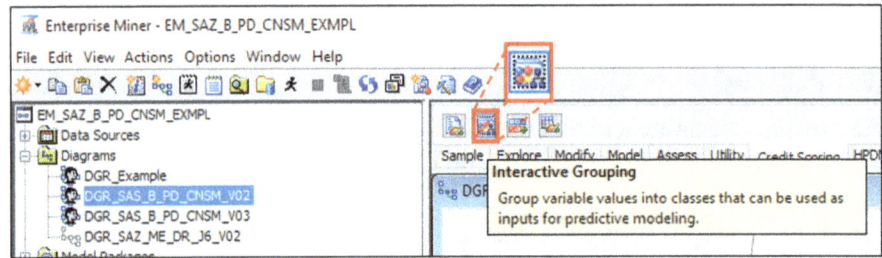

Figure 15-5. *Location of the Interactive Grouping node.*

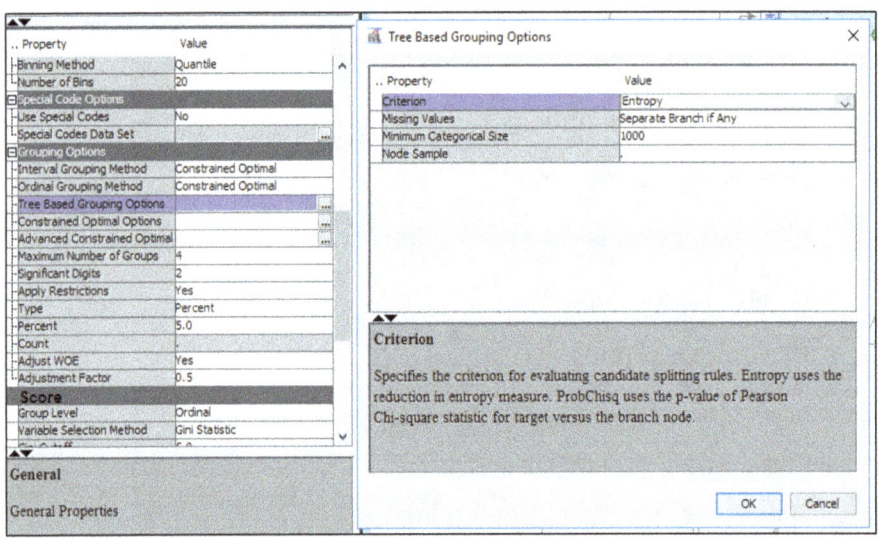

Figure 15-6. *Tree-Based Grouping Options of the Interactive Grouping node.*

422

15.3.1.2 Binning and Grouping Options

Furthermore, within the **Grouping Options** section, the user may select the **Interval Grouping Method**. Four options are available for consideration:

1. Optimal Criterion
2. Quantile
3. Monotonic Event Rate
4. Constrained Optimal

The **Optimal Criterion** employs the Tree-Based Criterion exclusively to identify optimal groupings, often yielding unfavorable outcomes in instances where linear monotonicity is not assured. **Quantile**, on the other hand, creates groups with approximately the same number of observations in each. **Monotonic Event Rate** stipulates that groups must conform to a monotonic event rate. **Constrained Optimal**, on the other hand, creates groups based on predefined constraints.

Prior to initiating the IG procedure, two significant options must be considered: the number of bins, as outlined under the **Interval Variable Binning Options**, and the maximum number of groups, as specified under the **Grouping Options** (see **Figure 15-7**). At first glance, it may appear that the terms *"number of bins"* and *"number of groups"* refer to the same concept. However, this is not the case. In the IG node notation, it is imperative to discern between *"Bins"* and *"Groups."* The former denotes the independent number of slices that can be manipulated to adjust the cutoff points of each group, thereby modifying the WoE. Conversely, groups signify the ultimate risk groups or nodes that are derived from the decision tree.

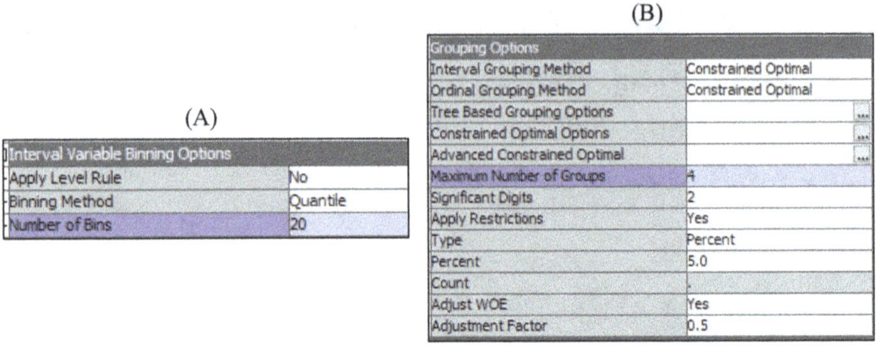

CHAPTER 15   WEIGHT OF EVIDENCE

(A)

Interval Variable Binning Options	
Apply Level Rule	No
Binning Method	Quantile
Number of Bins	20

(B)

Grouping Options	
Interval Grouping Method	Constrained Optimal
Ordinal Grouping Method	Constrained Optimal
Tree Based Grouping Options	
Constrained Optimal Options	
Advanced Constrained Optimal	
Maximum Number of Groups	4
Significant Digits	2
Apply Restrictions	Yes
Type	Percent
Percent	5.0
Count	
Adjust WOE	Yes
Adjustment Factor	0.5

Figure 15-7. *(A) Number of Bins option. (B) Maximum Number of Groups option.*

Figure 15-8 provides an illustrative example of the distinction between bins and groups. As illustrated in the figure, there are five distinct groups,[5] which correspond to

1. _MISSING_
2. I_R_BALDIF_BALT_03MB< -0.05
3. -0.05<= I_R_BALDIF_BALT_03MB< 0.01
4. 0.01<= I_R_BALDIF_BALT_03MB< 0.12
5. 0.12<= I_R_BALDIF_BALT_03MB

---

[5] The variable I_R_BALDIF_BALT_03MB represents the ratio of the balance difference between the previous and the actual period to the total actual balance, as reported three months prior from the reference date.

CHAPTER 15 WEIGHT OF EVIDENCE

Value	Group	Cutoff	Event Count	Non Event Count	Total	Event Rate	WOE
MISSING	1		5776.0	247160.0	252936.0	0.023	-0.06825
I_R_BALDIF_BALT_03MB < -1.5	2	-1.5	17.0	3684.0	3701.0	0.005	1.55397
-1.5 <= I_R_BALDIF_BALT_03MB < -0.44	2	-0.44	16.0	3654.0	3670.0	0.004	1.60641
-0.44 <= I_R_BALDIF_BALT_03MB < -0.18	2	-0.18	25.0	3608.0	3633.0	0.007	1.14746
-0.18 <= I_R_BALDIF_BALT_03MB < -0.09	2	-0.09	43.0	3547.0	3590.0	0.012	0.58808
-0.09 <= I_R_BALDIF_BALT_03MB < -0.05	2	-0.05	66.0	3889.0	3955.0	0.017	0.25163
-0.05 <= I_R_BALDIF_BALT_03MB < -0.03	3	-0.03	85.0	3647.0	3732.0	0.023	-0.06557
-0.03 <= I_R_BALDIF_BALT_03MB < -0.02	3	-0.02	64.0	4225.0	4289.0	0.015	0.36532
-0.02 <= I_R_BALDIF_BALT_03MB < -0.01	3	-0.01	66.0	4128.0	4194.0	0.016	0.31132
-0.01 <= I_R_BALDIF_BALT_03MB < 0	3	0.0	47.0	3073.0	3120.0	0.015	0.35569
0 <= I_R_BALDIF_BALT_03MB < 0.01	3	0.01	72.0	4210.0	4282.0	0.017	0.24398
0.01 <= I_R_BALDIF_BALT_03MB < 0.02	3	0.02	63.0	2227.0	2290.0	0.028	-0.2593
0.02 <= I_R_BALDIF_BALT_03MB < 0.03	4	0.03	293.0	5703.0	5996.0	0.049	-0.856
0.03 <= I_R_BALDIF_BALT_03MB < 0.04	4	0.04	80.0	2108.0	2188.0	0.037	-0.55311
0.04 <= I_R_BALDIF_BALT_03MB < 0.07	4	0.07	90.0	3556.0	3646.0	0.025	-0.14799
0.07 <= I_R_BALDIF_BALT_03MB < 0.12	4	0.12	49.0	3472.0	3521.0	0.014	0.43609
0.12 <= I_R_BALDIF_BALT_03MB < 0.2	5	0.2	29.0	3435.0	3464.0	0.008	0.9499
0.2 <= I_R_BALDIF_BALT_03MB < 0.34	5	0.34	27.0	3748.0	3775.0	0.007	1.10856
0.34 <= I_R_BALDIF_BALT_03MB < 0.56	5	0.56	18.0	3630.0	3648.0	0.005	1.48204
0.56 <= I_R_BALDIF_BALT_03MB < 0.95	5	0.95	19.0	3731.0	3750.0	0.005	1.45542
I_R_BALDIF_BALT_03MB >= 0.95	5		42.0	3663.0	3705.0	0.011	0.64379

Figure 15-8. Illustrative example of the difference between bins and groups in the IG node.

It is important to note that, although there are only five groups, there are 21 bins in total; furthermore, it can be observed that there are five bins in each group, which provides a higher level of resolution regarding the possible cutoff points within each group.

Having a higher level of resolution, in terms of possible cutoff points, is extremely advantageous because it allows us to move the boundary between groups, allowing us to try different cutoff values that could improve the IV. Of paramount importance, however, is the enhancement of the coherence of the WoE in accordance with the logical expected behavior and common sense.

15.3.1.3 Variable Selection Options

The IG node offers a variable selection option (**Figure 15-9**), enabling the user to select either the IV or the Gini index as a selection criterion. As previously stated, the Gini index is preferable to the IV due to its theoretical foundation and the fact that the Gini index has a clearly defined scale between 0 and 100. However, it should be noted that the IG node is not recommended for use as a variable selection node. This is due to the fact that the WoE is a transformation technique and not a multivariable analysis. Furthermore, it is also important to consider that the number of variables has already been reduced by two powerful selection methods—that is, the univariable analysis and the ICRM. The primary objective, therefore, is not to further reduce the number of variables, but to obtain a reasonable transformation capable of capturing a linear and monotonic relationship between the dependent and independent variables.

It is also important to note that reducing the number of variables at this stage of the analysis may be inadvisable due to the fact that the discriminatory power of a single variable does not necessarily guarantee its statistical significance in the presence of the rest of the variables in the multivariable model. For this reason, it is recommended that a relatively low level of discriminatory power, such as a Gini cutoff of 5, be employed

CHAPTER 15 WEIGHT OF EVIDENCE

as a selection criterion. In addition, it is advised that the variables that fall below the selection cutoff be checked to ensure that no important variables[6] are excluded from the multivariable model.

Score	
Group Level	Ordinal
Variable Selection Method	Gini Statistic
Gini Cutoff	5.0
Information Value Cutoff	0.1

Figure 15-9. *Variable selection options.*

15.3.1.4 Creating Grouping Data

A salient feature of the IG node is its capacity to preserve WoE parameters (number of groups and cutoff points) for subsequent refinement in another node, thereby obviating the need to reenter results whenever a different cutoff point is to be tested. As illustrated in **Figure 15-10**, there are specific options that facilitate the preservation of these parameters. The first option is the **Create Grouping Data option**, which is set to "No" by default. It is therefore important to set this option to "Yes" as soon as the IG node is placed in the diagram to avoid losing results by accident.

Report	
Create Grouping Data	Yes
Create Method	Overwrite
Number of Variables	10

Figure 15-10. *Create Grouping Data options.*

The second option permits the user to either overwrite or append the results of the node's execution. The rules of the WoE are stored in the

---

[6] In this context, the term *"important variables"* is used to describe those that are of particular significance from the perspective of a subject-matter knowledge expert.

CHAPTER 15 WEIGHT OF EVIDENCE

diagram workspace, which, in this example, corresponds to the EMWS2. The path is as follows:

/warehouse/EM_SAZ_B_PD_CNSM_EXMPL/Workspaces/EMWS2

As illustrated in **Figure 15-11**(A), the Enterprise Guide (EG) explorer displays the complete path to the EMWS2. It is important to note that the WoE parameters are stored in a SAS dataset and are named after their corresponding IG node (IGN11). Consequently, the WoE dataset is named **ign11_exportgroup.sas7bdat**, as illustrated in **Figure 15-11**(B).

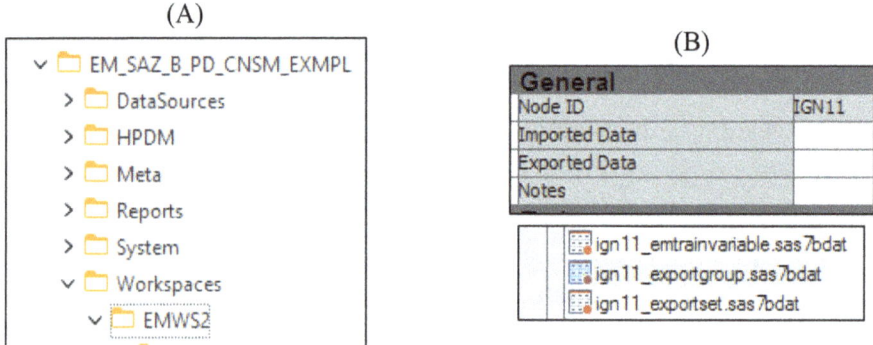

Figure 15-11. *EG file explorer showing the WoE storage path. (A) EG file explorer. (B) IG node name and the resulting dataset with the exported groups.*

Figure 15-12 shows the content of the IG exportgroup that consists of

- **_variable_:** Variable name
- **_split_value_:** Refers to the bin number
- **_level_:** Refers to the variable's level, INTERVAL, or NOMINAL
- **_group_:** Refers to the final risk group
- **binFlag:** Distinguishes interval bins from the rest of information

CHAPTER 15 WEIGHT OF EVIDENCE

- **manualWoe:** Refers to the WoE manually imputed for the corresponding bin

- **calculatedWOE:** Refers to the calculated WoE obtained from the automatic node's calculation, or by manual tunning

	variable	_split_value_	_level_	_group_	binFlag	manualWoe	calculatedWOE
261	I_SIN_R_BALT_CLM_L13_24M	_MISSING_	INTERVAL	5	0		-0.0887
262	I_SIN_R_BALT_CLM_L13_24M	0	INTERVAL	1	1		0.61715
263	I_SIN_R_BALT_CLM_L13_24M	1	INTERVAL	1	1		0.61715
264	I_SIN_R_BALT_CLM_L13_24M	2	INTERVAL	1	1		0.61715
265	I_SIN_R_BALT_CLM_L13_24M	3	INTERVAL	2	1		0.4345
266	I_SIN_R_BALT_CLM_L13_24M	4	INTERVAL	2	1		0.4345
267	I_SIN_R_BALT_CLM_L13_24M	5	INTERVAL	3	1		0.50925
268	I_SIN_R_BALT_CLM_L13_24M	6	INTERVAL	3	1		0.50925
269	I_SIN_R_BALT_CLM_L13_24M	7	INTERVAL	4	1		-0.09026
270	I_SIN_R_BALT_CLM_L13_24M	8	INTERVAL	4	1		-0.09026
271	I_SIN_R_BALT_CLM_L13_24M		INTERVAL	4	1		-0.09026
272	N_ECONOMIC_SECTOR	CNS	NOMINAL	1	0		-0.3378
273	N_ECONOMIC_SECTOR	CRD	NOMINAL	1	0		-0.3378
274	N_ECONOMIC_SECTOR	ELC	NOMINAL	1	0		-0.3378
275	N_ECONOMIC_SECTOR	TUR	NOMINAL	1	0		-0.3378
276	N_ECONOMIC_SECTOR	CMR	NOMINAL	2	0		-0.17098
277	N_ECONOMIC_SECTOR	AGR	NOMINAL	3	0		-0.11035
278	N_ECONOMIC_SECTOR	IND	NOMINAL	3	0		-0.11035
279	N_ECONOMIC_SECTOR	SRV	NOMINAL	3	0		-0.11035
280	N_ECONOMIC_SECTOR	TRN	NOMINAL	3	0		-0.11035
281	N_ECONOMIC_SECTOR	VIV	NOMINAL	3	0		-0.11035
282	N_ECONOMIC_SECTOR	GND	NOMINAL	4	0		0.11928
283	N_ECONOMIC_SECTOR	OTR	NOMINAL	4	0		0.11928
284	N_ECONOMIC_SECTOR	ALM	NOMINAL	5	0		0.94981
285	N_ECONOMIC_SECTOR	GBR	NOMINAL	5	0		0.94981
286	N_ECONOMIC_SECTOR	_UNKNOWN_	NOMINAL	6	0		-0.42534
287	N_ECONOMIC_SECTOR	_MISSING_	NOMINAL	6	0		-0.42534

Figure 15-12. exportgroup contents.

Figures 15-13 and **15-14** show the interactive grouping results for the I_SIN_R_BALT_CLM_L13_24M and the N_ECONOMIC_SECTOR variables, which correspond to the variables shown in the exportgroup contents in **Figure 15-12**. It is noteworthy that the calculated WoE and the number of groups are identical in both the export group and the IG results.

CHAPTER 15 WEIGHT OF EVIDENCE

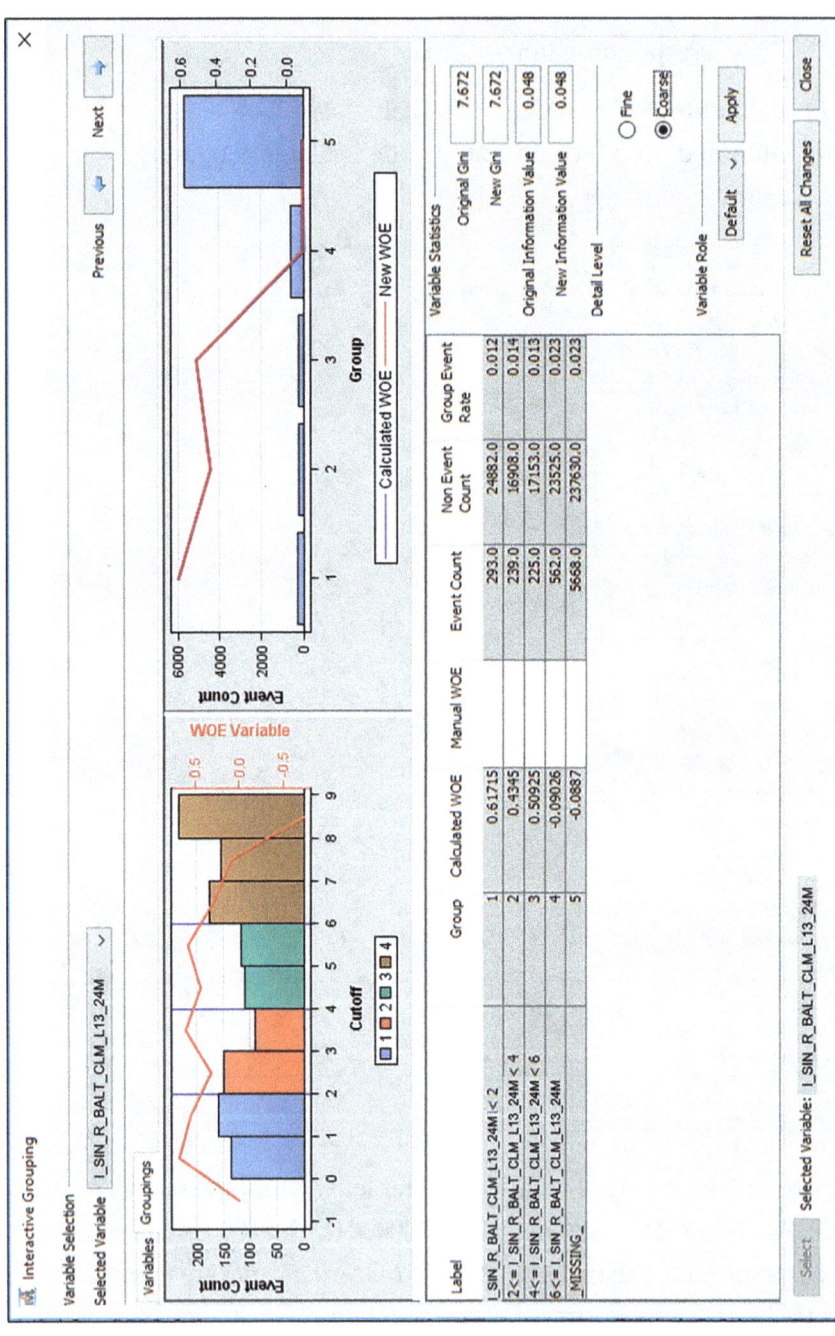

Figure 15-13. Interactive grouping for variable I_SIN_R_BALT_CLM_L13_24M.

CHAPTER 15 WEIGHT OF EVIDENCE

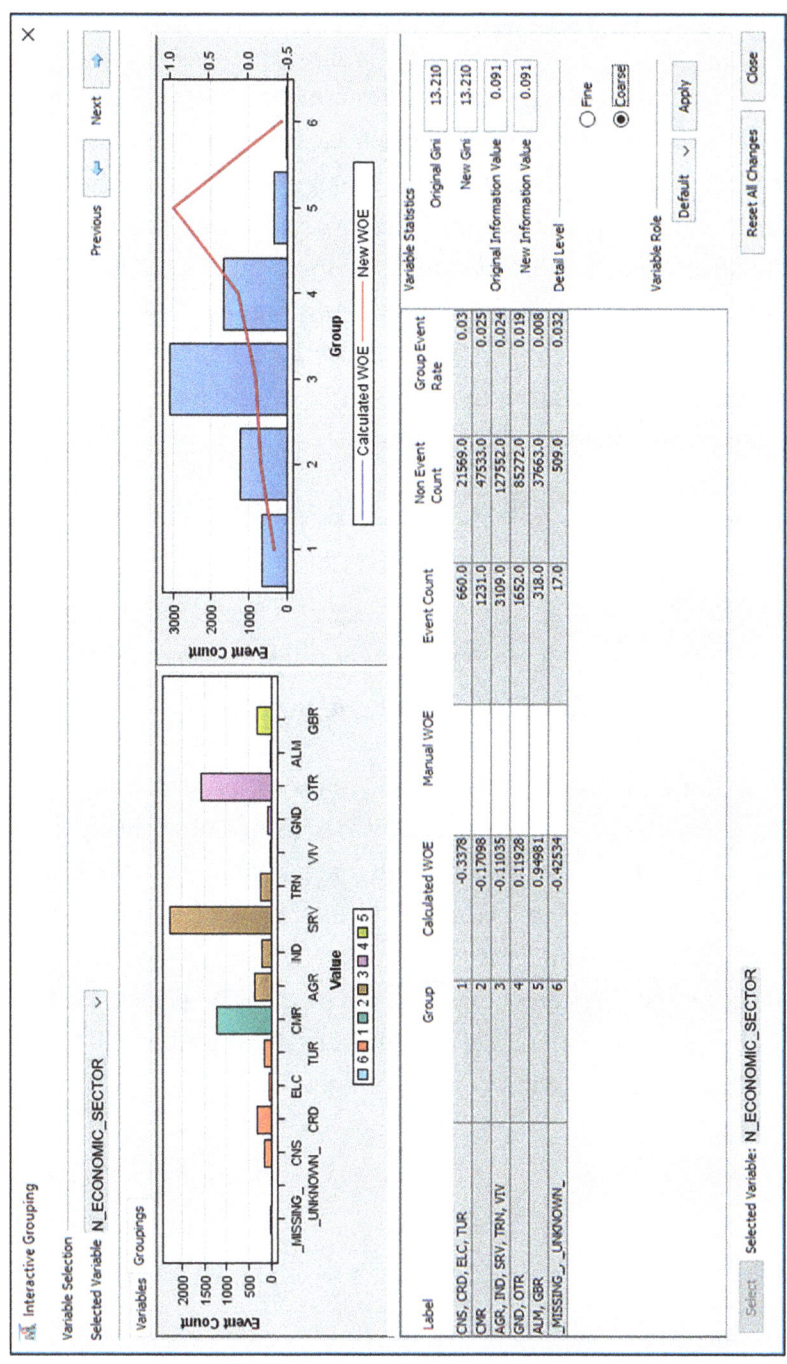

Figure 15-14. Interactive grouping for variable N_ECONOMIC_SECTOR.

15.3.1.5 Predefined Groupings

The primary rationale for exporting the grouping data is to utilize it as a starting point in another IG node, thereby ensuring the retention of previous results. This feature is highly advantageous, as it enables the user to test new cutoff points and make comparisons with previous ones without the need to revert to previous iterations.

Figure 15-15 shows the **Pre-Defined Groupings options**. It is noteworthy that the ign11_exportgroup dataset, previously referenced, is utilized as the predefined grouping data source, with the data source itself being located within the EMWS2 diagram.

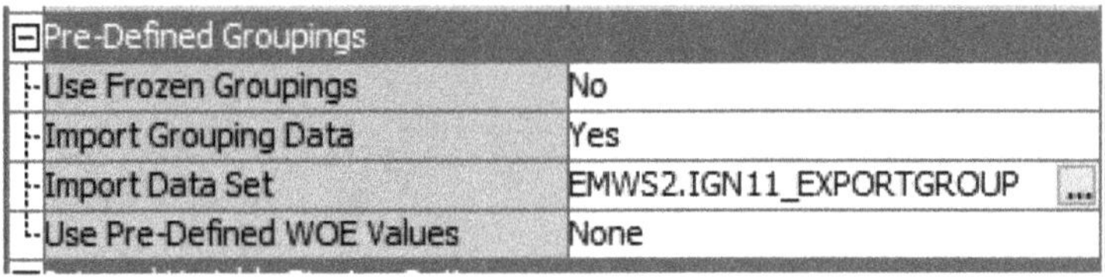

Pre-Defined Groupings	
Use Frozen Groupings	No
Import Grouping Data	Yes
Import Data Set	EMWS2.IGN11_EXPORTGROUP
Use Pre-Defined WOE Values	None

Figure 15-15. Pre-Defined Groupings options.

It is important to note that the EMWS2 library is automatically defined, thereby eliminating the necessity to register a new library for the exportgroup dataset, on the condition that it is situated in the same diagram workspace in which it is to be utilized. However, in the event that exportgroups are situated in other workspaces, it is necessary to manually define the library in the Start code section of the project. **Figure 15-16** illustrates the process of manually defining new libraries that will be accessible in all diagrams within an EM project.

CHAPTER 15 WEIGHT OF EVIDENCE

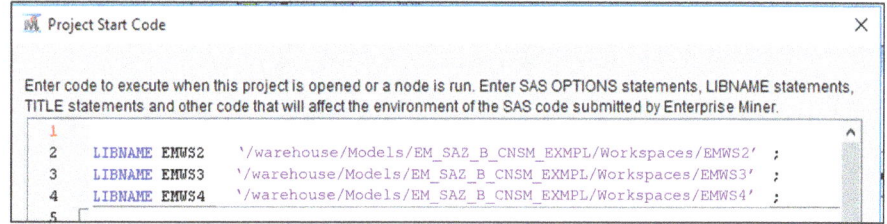

Figure 15-16. Example of manually defined workspaces in the Project Start Code section.

15.4 Setting the Total Number of Groups

In Section 15.3.1.2, an examination was conducted of the binning and grouping options available in the IG node. Among these options is the **Maximum Number of Groups**. As the name suggests, this option determines the maximum number of groups to be formed during the binning and grouping process. This raises the question of why the total number of groups is not also estimated during the binning and grouping process.

My assumption is that this may be attributable to the complexity of the grouping process, given that the binning and grouping process is, in fact, an optimization problem. It is widely acknowledged among those versed in optimization problems that the greater the realism of the problem at hand, the more constraints are required to render the optimal solution meaningful.

For instance, optimizing the total number of groups would result in the objective function becoming a moving target, as it is one thing to collapse existing groups, but quite another to create new groups out of thin air. Fortunately, the IG node is equipped with a comprehensive array of options that permit the configuration of distinct maximum group numbers for each variable, if we so choose.

Figure 15-17 shows the advanced constrained options available in the **Grouping Options**.

433

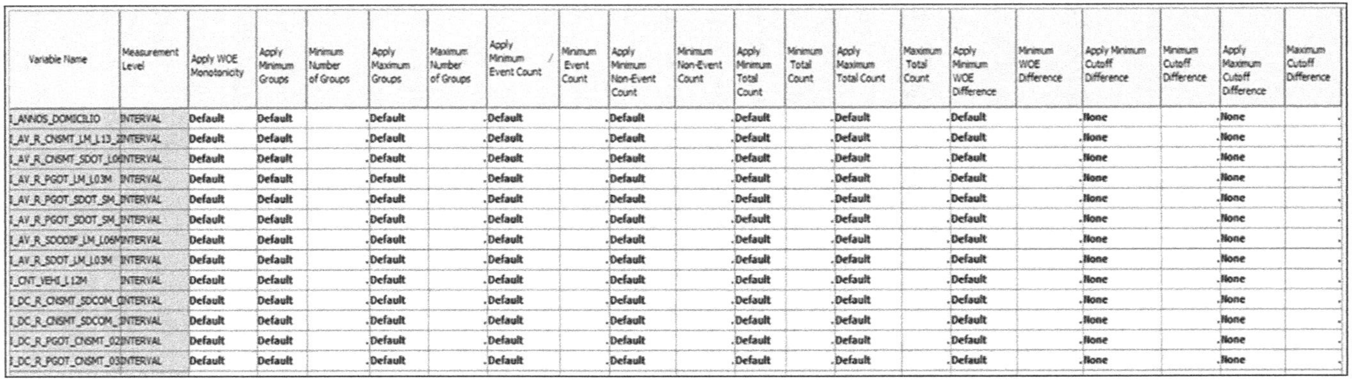

CHAPTER 15   WEIGHT OF EVIDENCE

Figure 15-17. Advanced Constraint Options available in the IG node.

The advanced options that can be defined independently for each variable are

1. Apply WoE Monotonicity: Yes/No
2. Apply Minimum Groups: Default/None/User
3. Minimum number of Groups
4. Apply Maximum Groups: Default/None/User
5. Maximum Number of Groups
6. Apply Minimum Event Count: Default/None/User
7. Minimum Event Count
8. Apply Minimum Non-Event Count: Default/None/User
9. Minimum Non-Event Count
10. Apply Minimum Total Count: Default/None/User
11. Minimum Total Count
12. Apply Maximum Total Count: Default/None/User
13. Maximum Total Count
14. Apply Minimum WoE Difference: Default/Nonc/Uscr
15. Minimum WoE Difference
16. Apply Minimum Cutoff Difference: None/User
17. Minimum Cutoff Difference
18. Apply Maximum Cutoff Difference: None User
19. Maximum Cutoff Difference

CHAPTER 15   WEIGHT OF EVIDENCE

It is evident that the process of establishing parameters for each variable, even for a few dozen of them, is likely to be both time-consuming and laborious. It is therefore recommended that, at the outset of the modeling process, all parameters are fixed for all variables. Rather than establishing distinct parameters for each variable, it would be more efficacious to utilize multiple IG nodes with a distinct set of fixed parameters in each one, thereby regulating the number of *variants*[7] in the diagram. To illustrate, an experiment could be conducted with three distinct maximum numbers of groups, such as 4, 5, and 7, and two disparate minimum WoEs differences. In total, six distinct sets of WoEs (variants) would be generated.

15.4.1 The IV and the Gini Index As a Function of the Number of Groups

It has been established that the maximum number of groups is not a consequence of the grouping process; rather, it is a parameter that must be defined in advance of the grouping process. It is important to note, however, that both the IV and the Gini index are functions of the total number of groups, or bins. This is evidenced by the observation that both the IV and the Gini index are directly proportional to the total number of groups formed from the original variable.

As demonstrated in **Figure 15-18**, there is a direct proportionality between the IV and the Gini index and the total number of groups. It is noteworthy that both the IV and the Gini index exhibit an increase in value as the number of groups increases, reaching a maximum at the 14th group. While this maximum value of both statistics occurs at the 14th group is not a significant detail in this particular example, it is important to note that both the IV and the Gini index are directly proportional to the total

---

[7] We will go in depth about the use of variants in Section 17.1.

CHAPTER 15 WEIGHT OF EVIDENCE

number of groups, at least up to a certain point. This prompts the question of whether it would be advantageous to always use a large number for the maximum number of groups.

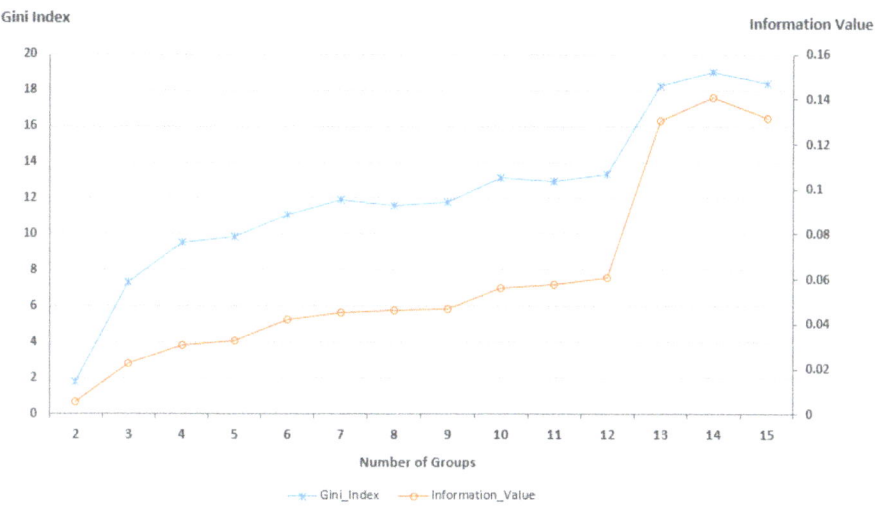

Figure 15-18. Example of the IV and the Gini index as a function of the number of groups.

The answer to this question is negative. While this may enhance the discriminatory power of the variable, it may also increase the complexity of the grouping process, leading to an overfitting of the WoE function. This hinders the ability to obtain a monotonic linear relationship between the independent and target variables, resulting in a nonsensical result.

Figure 15-19 provides an illustrative example of an overfit WoE, wherein the variable of interest is the age of the customer in months. In this particular example, the optimal criterion was selected from both grouping options, and the maximum number of groups was set to 15. As illustrated in the chart on the left side of **Figure 15-19**, the WoE demonstrates a general upward trend, which is consistent with the expected outcome. As customers age, their risk profile tends to decline. However, the general upward trend is obscured by the excessive number of groups, which forces the algorithm to accommodate 12 different groups, causing the WoE to behave erratically.

CHAPTER 15 WEIGHT OF EVIDENCE

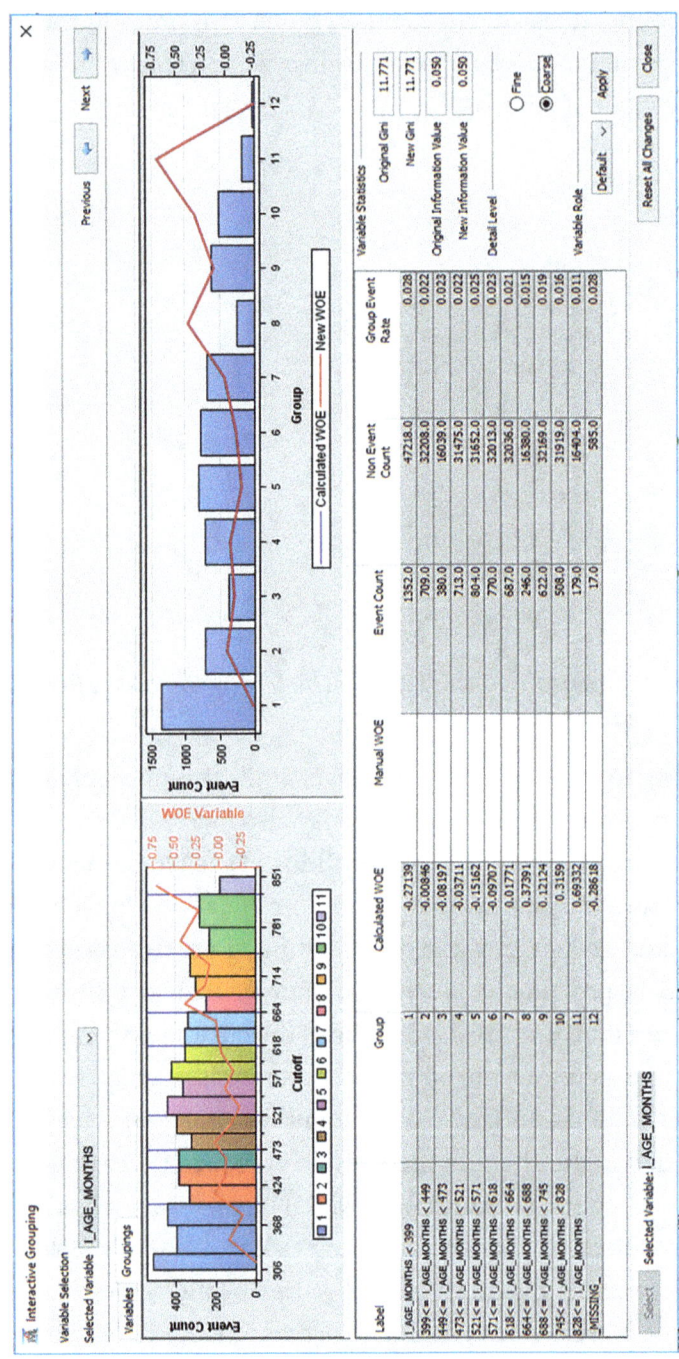

Figure 15-19. Example of an overfitted WoE.

CHAPTER 15 WEIGHT OF EVIDENCE

Figure 15-20 illustrates the grouping of the same variable with a maximum number of groups set to 5, rather than 15, as was the case with the previous example. It is notable that in this instance, the general upward trend is not lost, and the independent variable exhibits a correct monotonic linear relationship with the target variable. It is also noteworthy that despite the reduction in the number of groups from 15 to 5, the Gini index for the 5-group WoE is only slightly smaller (10.43) than that obtained from the 15-group WoE (11.77). This finding suggests that, in the context of WoEs, an increase in the number of groups does not necessarily guarantee a more precise depiction of the independent variable's general trend.

Furthermore, it is crucial to take into account the data-dependent nature of the grouping process when determining the optimal number of groups. As the number of events in the sample increases, one would expect the maximum number of groups to increase as well. However, in my personal experience, I have found that there is no discernible benefit in terms of discriminatory power beyond nine groups. Therefore, it is recommended to keep the maximum number of groups between four and nine.

CHAPTER 15   WEIGHT OF EVIDENCE

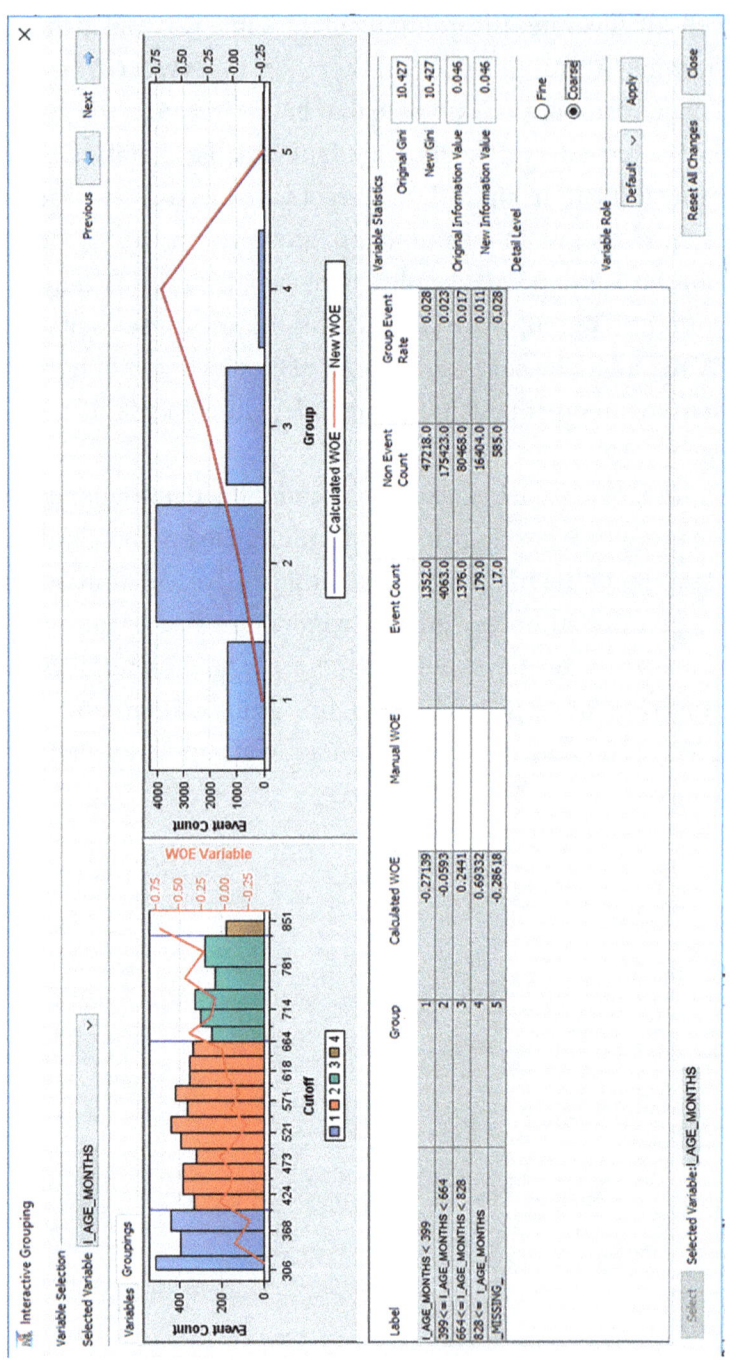

Figure 15-20. Example of a WoE with a proper fit.

15.5 WoE Defined As an Optimization Problem

Despite the fact that this is not a widely utilized approach, it is noteworthy that the WoE for interval variables can also be solved as an optimization problem. In the absence of decision trees, the optimal cutoff points can also be determined by defining the WoE as an optimization problem, with the IV serving as the objective function.

The IV maximization problem can then be defined as follows:
Find

$$A(x)^{ll} = \begin{bmatrix} a_1^{ll} \\ a_2^{ll} \\ \vdots \\ a_m^{ll} \end{bmatrix} \quad A(x)^{ul} = \begin{bmatrix} a_1^{ul} \\ a_2^{ul} \\ \vdots \\ a_m^{ul} \end{bmatrix}$$

that maximize

$$Z = \sum_{i=1}^{m} IV_i$$

subject to

$$a_{i+1}^{ll} = a_i^{ul}$$

$$a_i^{ll} \geq 0$$

$$a_i^{ll} \leq a_i^{ul}$$

Where a_i^{ll} corresponds to the lower limit cutoff point of the *ith* group, while a_i^{ul} corresponds to the upper limit cutoff point of the *ith* group.

Therefore, in theory, it is possible to write a code capable of iteratively finding the optimal cutoff points for each one of your interval predictive variables that maximizes the IV.

15.5.1 WoE Optimization with the OPTMODEL Procedure[8]

The OPTMODEL procedure [12] is a Generalized Algebraic Modeling System (GAMS)[9] specifically designed for optimization models. There are eight basic elements in an OPTMODEL code:

1. Sets
2. Constants
3. Data sources.
4. Variables (parameters)
5. Implicit variables (complex parameters)
6. Constraints
7. The objective function
8. The solver clause
9. Outputs

Listing 15-1 illustrates the initial four elements of an optimization problem. As is evident, the sets are declared as empty arrays (lines 28 and 29), which will be utilized subsequently. The following section delineates the numeric constants (lines 33 through 42), which are

> **INT_I_IM_SUM_09{variable}**: Is the predictive variable that we want to collapse into bins
>
> **INT_TARGET_SS{variable}**: Is the target variable, in this case the default event

---

[8] The complete code can be downloaded from https://github.com/saulzapiain/ConceptualVariableDesign. File: Listing 15- WoE calculation with PROC OPTMODEL.sas.

[9] For more information about GAMS, see SAS documentation.

LL_REF_VALUE{Attributes}: Is the reference for the lower limit values given for the initial iteration of the cutoff values for each bin

UP_REF_VALUE{Attributes}: Is the reference for the upper limit values given for the initial iteration of the cutoff values for each bin

N_Obs=50494: Declare total number of observations

N_E= 10438: Declare total number of events.

N_NE= 40056: Declare total number of no events

Bins=10: Declare number of bins

A_ll_0 =0: Declare minimum variable value

A_UL_M =1209444.444+1: Declare maximum variable value

The subsequent stage is to load the initial sets with the information from the SAS datasets:

- **WOE.INT_I_IM_SUM_09**: Contains only two columns, the variable's values and the binary target
- **INIT_CUTOFFS**: Contains an initial set of cutoff values for the first iteration

Finally, the parameters[10] to be found are defined by the two dynamic sets:

- **var A_ll{1..Bins}**: Lower limits that go from 1 to the total number of bins
- **var A_ul{1..Bins}**: Upper limits that go from 1 to the total number of bins

---

[10] The word "*parameters*" is more commonly used in optimization terminology than the word "*variables.*"

CHAPTER 15 WEIGHT OF EVIDENCE

Listing 15-1. The sets, constants, data sources, and parameters of the optimization problem being modeled.

```
20  PROC OPTMODEL PRINTLEVEL=2;
21      ODS OUTPUT
22              SOLUTIONSUMMARY=WOE.SOLUTIONSUMMARY
23              PROBLEMSUMMARY =WOE.PROBLEMSUMMARY
24      ;
25      /*|----------------------------------------------------------
26      --|Define Sets
27      |----------------------------------------------------------
28      SET <NUMBER> VARIABLE;
29      SET <NUMBER> ATTRIBUTES;
30      /*|----------------------------------------------------------
31      --|Define Numeric Constants
32      |----------------------------------------------------------
33      NUMBER INT_I_IM_SUM_09{VARIABLE};
34      NUMBER INT_TARGET_SS{VARIABLE};
35      NUMBER LL_REF_VALUE{ATTRIBUTES};
36      NUMBER UP_REF_VALUE{ATTRIBUTES};
37      NUMBER N_OBS    =50494;  /*Declare Total Number of Observations*/
38      NUMBER N_E      =10438;  /*Declare total Number of Events*/
39      NUMBER N_NE     =40056;  /*Declare Total Number of No Events*/
40      NUMBER BINS     =10;     /*Declare Number of Bins*/
41      NUMBER A_LL_0   =0;      /*Declare Minimum Variable Value*/
42      NUMBER A_UL_M   =1209444.444+1 ; /*Declare Maximum Variable Value*/
43      /*|----------------------------------------------------------
44      --|Define Data Sources
45      |----------------------------------------------------------
46      READ DATA WOE.INT_I_IM_SUM_09 INTO VARIABLE=[ID] INT_I_IM_SUM_09 INT_TARGET_SS;
47      READ DATA  INIT_CUTOFFS INTO ATTRIBUTES=[ATTRIBUTE] LL_REF_VALUE  UP_REF_VALUE;
48      /*|----------------------------------------------------------
49      --|Define Model's Parameters
50      |----------------------------------------------------------
51      VAR A_LL{1..BINS};
52      VAR A_UL{1..BINS};
```

Listing 15-2 illustrates the subsequent three elements of the optimization problem. The implicit variable (parameter) section (lines 56 to 60) presents the equations that define the number of events (**Equation 15-3**), the number of no events (**Equation 15-4**), and the total number of observations in algebraic notation. The number of events is thus given by

$$n_i^E = \sum_{j=1}^{N}\left[if\ a_i^{ll} \le x_j < a_i^{ul}\ then\ y_j\ else\ 0 \right]$$

Equation 15-14. Conditional sum of events.

CHAPTER 15 WEIGHT OF EVIDENCE

Respectively, the number of no events is given by

$$n_i^{NE} = \sum_{j=1}^{N} \left[if\ a_i^{ll} \le x_j < a_i^{ul}\ and\ y_j = 0\ then\ 1\ else\ 0 \right]$$

Equation 15-15. Conditional sum of no events.

And the total number of observations:

$$n_i^{E} + n_i^{NE}$$

Equation 15-16. Total number of observations.

Lines 64 to 73 define the constraints of our optimization problem and are defined as follows:

> **A_ll[i]>=LL_REF_VALUE[i]:** Lower limits should be greater or equal to the lower limits reference values.
>
> **A_Ul[i]<=UP_REF_VALUE[i]:** Upper limits should be greater or equal to the upper limits reference values.
>
> **A_ll[i+1]=A_Ul[i]:** Upper limits should be equal to the *i*+1 bin of the lower limits.
>
> **A_ll[i]>=0:** All lower limits should be positive numbers.
>
> **A_ll[i]<=A_Ul[i]:** Lower limit bins should be less than upper limit bins.
>
> **A_ll[i]=A_ll_0:** The minimum value should be equal to zero.
>
> **A_UL[i]=A_UL_M:** The upper limit bin should be equal to the maximum value.

CHAPTER 15 WEIGHT OF EVIDENCE

n_NoEvents[i]/N_NE>=0.01: Proportion of no events in each bin should be greater than 1%.

n_Events[i]/N_E>=0.01: Proportion of events in each bin should be greater than 1%.

sum{i in 1..Bins}n_Observations[i]=N_Obs: The sum of all bins should be equal to the total number of observations.

The final section is the objective function, expressed in algebraic notation, which can be defined as

$$\text{Maximize } IV = \sum_{i=1}^{Bins} WoE_i * dd_i$$

Equation 15-17. Simplified version of the objective function for IV.

Where WoE_i and dd_i are defined by **Equations 15-7** and **15-10**, respectively.

It is important to note that the objective function has been expanded due to the fact that the WoE and dd were not previously defined in the implicit variables section.[11]

$$\text{Maximize } IV = \sum_{i=1}^{Bins} \ln\left[\frac{if\ n_i^{NE} = 0\ then\ 1\ else\ n_i^{NE}}{if\ n_i^{E} = 0\ then\ 1\ else\ n_i^{E}}\right] * \left(\frac{n_i^{NE}}{N^{NE}} - \frac{n_i^{E}}{N^{E}}\right)$$

Equation 15-18. Expanded versions of the objective function for IV.

---

[11] The code may be simplified by the addition of the WoE and dd in the implicit variable section as a function of i.

CHAPTER 15 WEIGHT OF EVIDENCE

Listing 15-2. OPTMODEL implicit variables, constraints, and objective function.

```
53   /*|----------------------------------------------------------------|
54   --|Define Complex Parameter's Functions
55   |----------------------------------------------------------------|*/
56   IMPVAR N_EVENTS{i IN 1..BINS}      =SUM{j IN 1..N_OBS}(IF A_LL[i]<=INT_I_IM_SUM_09[j]<A_UL[i]
57                                                           THEN INT_TARGET_SS[j] ELSE 0);
58   IMPVAR N_NOEVENTS{i IN 1..BINS}    =SUM{j IN 1..N_OBS}((IF A_LL[i]<=INT_I_IM_SUM_09[j]<A_UL[i] THEN 1 ELSE 0)
59                                    -(IF A_LL[i]<=INT_I_IM_SUM_09[j]<A_UL[i] THEN INT_TARGET_SS[j] ELSE 0));
60   IMPVAR N_OBSERVATIONS{i IN 1..BINS} =N_EVENTS[i]+N_NOEVENTS[i];
61   /*|----------------------------------------------------------------|
62   --|Define Model's Constraints
63   |----------------------------------------------------------------|*/
64   CONSTRAINT BINS_INITIAL_LOWERLIMIT{i IN 1..BINS-1}:       A_LL[i]>=LL_REF_VALUE[i];
65   CONSTRAINT BINS_INITAL_UPPERLIMIT{i IN 1..BINS-1}:        A_UL[i]<=UP_REF_VALUE[i];
66   CONSTRAINT BINS_EQUALITIY_LAG{i IN 1..BINS-1}:            A_LL[i+1]=A_UL[i];
67   CONSTRAINT BINS_POSITIVE_VALUES{i IN 1..BINS}:            A_LL[i]>=0;
68   CONSTRAINT BINS_UPPERLIMIT_GE_LOWERLIMIT{i IN 1..BINS}:   A_LL[i]<=A_UL[i];
69   CONSTRAINT BINS_FIRST_LOWERLIMIT{i IN 1..1}:              A_LL[i]=A_LL_0;
70   CONSTRAINT BINS_LAST_UPPERLIMIT{i IN BINS..BINS}:         A_UL[i]=A_UL_M;
71   CONSTRAINT NOEVENTS_MIN_DISTRIBUTION{i IN 1..BINS}:       N_NOEVENTS[i]/N_NE>=0.01;
72   CONSTRAINT EVENTS_MIN_DISTRIBUTION{i IN 1..BINS}:         N_EVENTS[i]/N_E>=0.01;
73   CONSTRAINT TOTAL_NUM_OF_OBSERVATIONS:                     SUM{i IN 1..BINS}N_OBSERVATIONS[i]=N_OBS;
74   /*|----------------------------------------------------------------|
75   --|Define The Objective Function
76   |----------------------------------------------------------------|*/
77   MAX Z =SUM{i IN 1..BINS} LOG((IF N_NOEVENTS[i]=0 THEN 1
78                                 ELSE N_NOEVENTS[i])/(IF N_EVENTS[i]=0
79                                 THEN 1 ELSE N_EVENTS[i]))*(N_NOEVENTS[i]/N_NE-N_EVENTS[i]/N_E);
```

Listing 15-3 illustrates the final segment of the optimization program, encompassing the solver syntax, output definitions, and options. Lines 83 through line 100 correspond to the solver engine, which has the following options:

1. **expand:** Expands all equations defined in the program (see **Figure 15-21**)

447

CHAPTER 15 WEIGHT OF EVIDENCE

```
Fix A_ll[1] = 0
Fix A_ul[10] = 1209445.444
Maximize Z=LOG((IF n_NoEvents[1] = 0 THEN 1 ELSE n_NoEvents[1])/IF n_Events[1] = 0 THEN 1 ELSE n_Events[1])*(n_NoEvents[1]/40056 -
n_Events[1]/10438) + LOG((IF n_NoEvents[2] = 0 THEN 1 ELSE n_NoEvents[2])/IF n_Events[2] = 0 THEN 1 ELSE n_Events[2])*(n_NoEvents[2]
/40056 - n_Events[2]/10438) + LOG((IF n_NoEvents[3] = 0 THEN 1 ELSE n_NoEvents[3])/IF n_Events[3] = 0 THEN 1 ELSE n_Events[3])*(
n_NoEvents[3]/40056 - n_Events[3]/10438) + LOG((IF n_NoEvents[4] = 0 THEN 1 ELSE n_NoEvents[4])/IF n_Events[4] = 0 THEN 1 ELSE
n_Events[4])*(n_NoEvents[4]/40056 - n_Events[4]/10438) + LOG((IF n_NoEvents[5] = 0 THEN 1 ELSE n_NoEvents[5])/IF n_Events[5] = 0
THEN 1 ELSE n_Events[5])*(n_NoEvents[5]/40056 - n_Events[5]/10438) + LOG((IF n_NoEvents[6] = 0 THEN 1 ELSE n_NoEvents[6])/IF
n_Events[6] = 0 THEN 1 ELSE n_Events[6])*(n_NoEvents[6]/40056 - n_Events[6]/10438) + LOG((IF n_NoEvents[7] = 0 THEN 1 ELSE
n_NoEvents[7])/IF n_Events[7] = 0 THEN 1 ELSE n_Events[7])*(n_NoEvents[7]/40056 - n_Events[7]/10438) + LOG((IF n_NoEvents[8] = 0
THEN 1 ELSE n_NoEvents[8])/IF n_Events[8] = 0 THEN 1 ELSE n_Events[8])*(n_NoEvents[8]/40056 - n_Events[8]/10438) + LOG((IF
n_NoEvents[9] = 0 THEN 1 ELSE n_NoEvents[9])/IF n_Events[9] = 0 THEN 1 ELSE n_Events[9])*(n_NoEvents[9]/40056 - n_Events[9]/10438)
+ LOG((IF n_NoEvents[10] = 0 THEN 1 ELSE n_NoEvents[10])/IF n_Events[10] = 0 THEN 1 ELSE n_Events[10])*(n_NoEvents[10]/40056 -
n_Events[10]/10438)
Constraint BINS_EQUALITIY_LAG[1]: A_ll[2] - A_ul[1] = 0
Constraint BINS_EQUALITIY_LAG[2]: A_ll[3] - A_ul[2] = 0
Constraint BINS_EQUALITIY_LAG[3]: A_ll[4] - A_ul[3] = 0
Constraint BINS_EQUALITIY_LAG[4]: A_ll[5] - A_ul[4] = 0
Constraint BINS_EQUALITIY_LAG[5]: A_ll[6] - A_ul[5] = 0
Constraint BINS_EQUALITIY_LAG[6]: A_ll[7] - A_ul[6] = 0
Constraint BINS_EQUALITIY_LAG[7]: A_ll[8] - A_ul[7] = 0
Constraint BINS_EQUALITIY_LAG[8]: A_ll[9] - A_ul[8] = 0
Constraint BINS_EQUALITIY_LAG[9]: A_ll[10] - A_ul[9] = 0
Constraint BINS_UPPERLIMIT_GE_LOWERLIMIT[6]: A_ll[6] - A_ul[6] <= 0
Constraint BINS_UPPERLIMIT_GE_LOWERLIMIT[9]: A_ll[9] - A_ul[9] <= 0
Constraint NOEVENTS_MIN_DISTRIBUTION[1]: 0.000024965*n_NoEvents[1] >= 0.01
Constraint NOEVENTS_MIN_DISTRIBUTION[2]: 0.000024965*n_NoEvents[2] >= 0.01
Constraint NOEVENTS_MIN_DISTRIBUTION[3]: 0.000024965*n_NoEvents[3] >= 0.01
Constraint NOEVENTS_MIN_DISTRIBUTION[4]: 0.000024965*n_NoEvents[4] >= 0.01
Constraint NOEVENTS_MIN_DISTRIBUTION[5]: 0.000024965*n_NoEvents[5] >= 0.01
Constraint NOEVENTS_MIN_DISTRIBUTION[6]: 0.000024965*n_NoEvents[6] >= 0.01
Constraint NOEVENTS_MIN_DISTRIBUTION[7]: 0.000024965*n_NoEvents[7] >= 0.01
Constraint NOEVENTS_MIN_DISTRIBUTION[8]: 0.000024965*n_NoEvents[8] >= 0.01
Constraint NOEVENTS_MIN_DISTRIBUTION[9]: 0.000024965*n_NoEvents[9] >= 0.01
Constraint NOEVENTS_MIN_DISTRIBUTION[10]: 0.000024965*n_NoEvents[10] >= 0.01
Constraint EVENTS_MIN_DISTRIBUTION[1]: 0.0000958038*n_Events[1] >= 0.01
Constraint EVENTS_MIN_DISTRIBUTION[2]: 0.0000958038*n_Events[2] >= 0.01
Constraint EVENTS_MIN_DISTRIBUTION[3]: 0.0000958038*n_Events[3] >= 0.01
Constraint EVENTS_MIN_DISTRIBUTION[4]: 0.0000958038*n_Events[4] >= 0.01
Constraint EVENTS_MIN_DISTRIBUTION[5]: 0.0000958038*n_Events[5] >= 0.01
Constraint EVENTS_MIN_DISTRIBUTION[6]: 0.0000958038*n_Events[6] >= 0.01
```

Figure 15-21. Example of the effect of the expand clause in the OPTMODEL output.

2. **Solve:** Invokes the solver engine

3. **Con:** Request the output of constraints

4. **Fix:** Request the output of constant variables

5. **Performance:** Statement that allows the use of performance parameters

6. **Solve with nlp:** Specifies the type of optimization engine to use; in this case is the nonlinear programming engine.

CHAPTER 15 WEIGHT OF EVIDENCE

7. **Printfreq:** Specifies the frequency of the iterations to output to the log
8. **Maxiter:** Specifies the maximum number of iterations before stopping the optimization process
9. **Opttol:** Specifies the convergence precision of the objective function
10. **Feastol:** Specifies the precision of the constraints
11. **Soltype:** Specifies whether to return a local solution, 0, or the best feasible solution in the final report
12. **Hesstype:** Specifies that the full Hessian matrix should be calculated
13. **Maxtime:** Specifies the convergence time limit

The remaining options delineated between lines 95 and 135 stipulate the requested output. It is noteworthy that the syntax in line 122 requests the output of a SAS dataset, which can be utilized to automate the WoE optimization process.

CHAPTER 15   WEIGHT OF EVIDENCE

Listing 15-3. *Different syntax for invoking the solver engine, defining the output variables, and creating an output dataset with the final results.*

```
81     --|Define Solver Parameters
82     |--------------------------------------------------------------------------|*/
83     EXPAND/SOLVE CON OBJ   FIX ;
84     PERFORMANCE NTHREADS=4;
85     SOLVE WITH NLP
86     /
87       PRINTFREQ=1
88       MAXITER=10000
89       OPTTOL=1E-9
90       FEASTOL=1E-6
91       SOLTYPE=1
92       HESSTYPE=FULL
93       MAXTIME=200
94     ;
95     PRINT
96       LL_REF_VALUE
97       UP_REF_VALUE
98       A_LL.SOL
99       A_UL.SOL
100    ;
101    /*|--------------------------------------------------------------------------|
102    --|Define Numeric Outputs
103    |--------------------------------------------------------------------------|*/
104    NUMBER LL_ATTRI{i IN 1..BINS}    =A_LL[i].SOL;
105    NUMBER UL_ATTRI{i IN 1..BINS}    =A_UL[i].SOL;
106    NUMBER TOTAL_OBS{i IN 1..BINS}   =N_OBSERVATIONS[i].SOL;
107    NUMBER EVENT_OBS{i IN 1..BINS}   =N_EVENTS[i].SOL;
108    NUMBER NOEVENT_OBS{i IN 1..BINS}=N_NOEVENTS[i].SOL;
109    NUMBER D_OBS{i IN 1..BINS}       =N_OBSERVATIONS[i].SOL/N_OBS;
110    NUMBER D_EVENTS{i IN 1..BINS}    =N_EVENTS[i].SOL/N_E;
111    NUMBER D_NOEVENTS{i IN 1..BINS}  =N_NOEVENTS[i].SOL/N_NE;
112    NUMBER P_EVENTS{i IN 1..BINS}    =N_EVENTS[i].SOL/N_OBSERVATIONS[i].SOL;
113    NUMBER N_WOE{i IN 1..BINS}       =LOG((IF N_NOEVENTS[i].SOL=0 THEN 1 ELSE N_NOEVENTS[i].SOL)
114                                          /(IF N_EVENTS[i].SOL=0 THEN 1 ELSE N_EVENTS[i].SOL));
115    NUMBER N_DNE_DE{i IN 1..BINS}    = (N_NOEVENTS[i].SOL/N_NE-N_EVENTS[i].SOL/N_E);
116    NUMBER N_IV{i IN 1..BINS}        =LOG((IF N_NOEVENTS[i].SOL=0 THEN 1 ELSE N_NOEVENTS[i].SOL)
117                                          /(IF N_EVENTS[i].SOL=0 THEN 1 ELSE N_EVENTS[i].SOL))
118                                         *(N_NOEVENTS[i].SOL/N_NE-N_EVENTS[i].SOL/N_E);
119    /*|--------------------------------------------------------------------------|
120    --|Create Output Data Set
121    |--------------------------------------------------------------------------|*/
122    CREATE DATA WOE.WOE_RESULTS FROM [I]
123                                 BIN_LL=LL_ATTRI
124                                 BIN_UL=UL_ATTRI
125                                 OBSERVATIONS=TOTAL_OBS
126                                 EVENTS=EVENT_OBS
127                                 NOEVENTS=NOEVENT_OBS
128                                 OBS_DIST=D_OBS
129                                 EVENT_DIST=D_EVENTS
130                                 NOEVENT_DIST=D_NOEVENTS
131                                 EVENTS_PROP=P_EVENTS
132                                 WOE=N_WOE
133                                 DNE_DE=N_DNE_DE
134                                 IV=N_IV
```

15.6 WoEs and Interval Targets

It is evident that a logistic regression model with interval variables will result in a continuous distribution of default probabilities. However, it is not immediately apparent that a logistic regression model that uses WoEs instead of the original interval variables will result in a discrete set of probabilities.

This unanticipated outcome can be attributed to the fact that the WoE transformation converts all explanatory variables into discrete variables, irrespective of whether they were initially interval or nominal variables. This outcome is unsurprising when one considers that each WoE variable is derived from an independent decision tree model.

Figure 15-22 presents a conceptual diagram of the direct relationship between the nodes of the decision tree and the resulting WoE groups of each explanatory variable. This diagram elucidates the rationale behind the finite number of probabilities of default exhibited by scorecard models. This phenomenon occurs because scorecard models do not assign a specific default probability to each individual subject in the ABT; rather, a default probability is assigned to a subpopulation with a particular set of characteristics drawn from the global population.

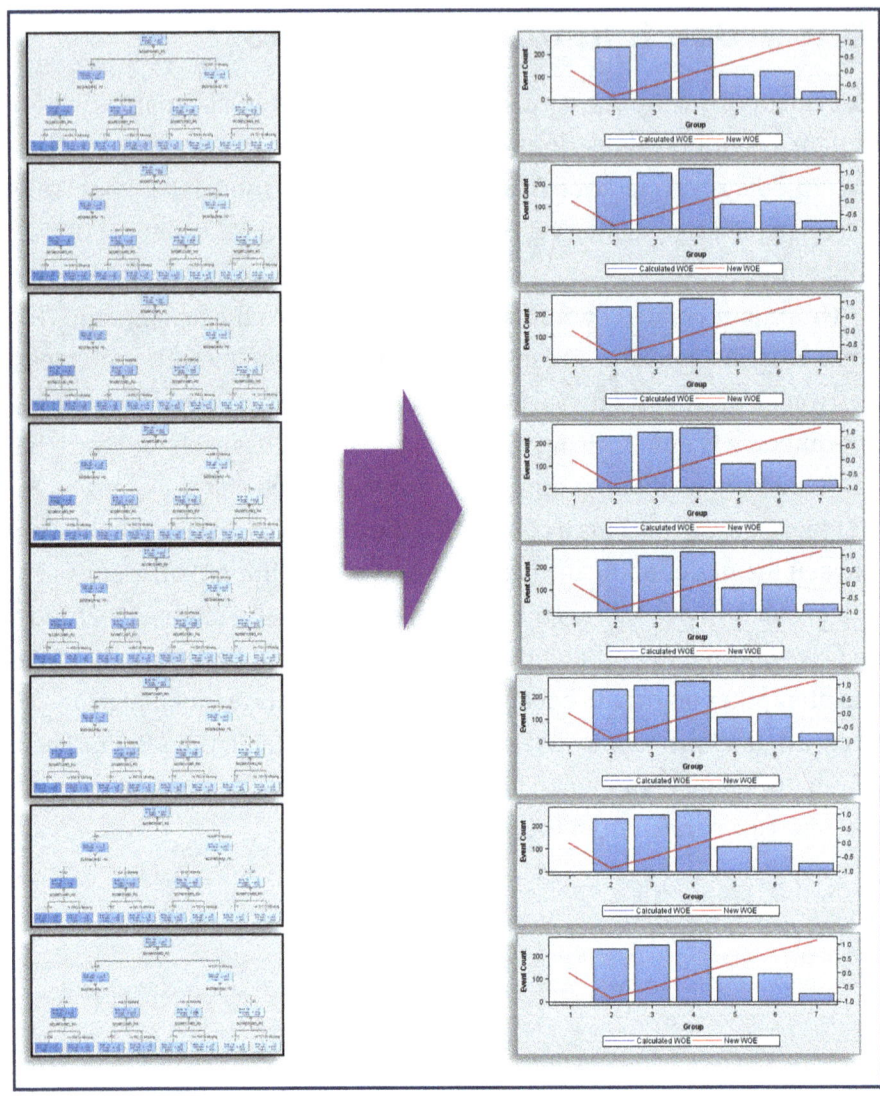

Figure 15-22. Conceptual diagram illustrating the relationship between the nodes of the decision tree and the resulting WoE's groups.

The following expression thus determines the total number of risk subgroups that will result from the scorecard model in question:

$$Risk\ Subgroups = \prod_{i=1}^{N_{var}} Bins(Nodes)_i$$

Equation 15-19. Risk subgroups.

Where $Bins_i$ refers to the total number of bins for the *ith* variable in the specified scorecard model. To illustrate this, consider a scorecard model comprising five variables and the following number of bins:

1. 3 bins
2. 5 bins
3. 4 bins
4. 3 bins
5. 2 bins

This results in a total of 360 risk subgroups, since

$3 \times 5 \times 4 \times 3 \times 2 = 360$

15.6.1 Resolution Issues When Using WoEs for Interval Targets

In the majority of instances, the total number of risk subgroups is not a significant concern, as WoEs are typically employed to predict binary targets. However, it is also possible to employ WoEs in order to predict interval targets[12] [13], such as the default rate (DR). For instance, WoEs can

---

[12] For more information about how to use WoEs for interval target, see the Binary Transformation property in the Interactive Grouping Node documentation, https://documentation.sas.com/doc/en/emref/14.3/p1qzwz7onopjqcn11uc04i18urg7.htm.

CHAPTER 15   WEIGHT OF EVIDENCE

be utilized to construct link functions that predict the DR as a function of macroeconomic information over time. However, in such cases, a higher resolution in terms of risk subgroups may be required.

Figures 15-23 and **15-24** present a line chart of the predicted default rate in comparison to the observed default rate over time. The scorecard model depicted in **Figure 15-23** comprised six explanatory variables and yielded an R^2 value of 0.776. In contrast, the scorecard model depicted in **Figure 15-24** comprises nine explanatory variables, which yielded an R^2 value of 0.882.

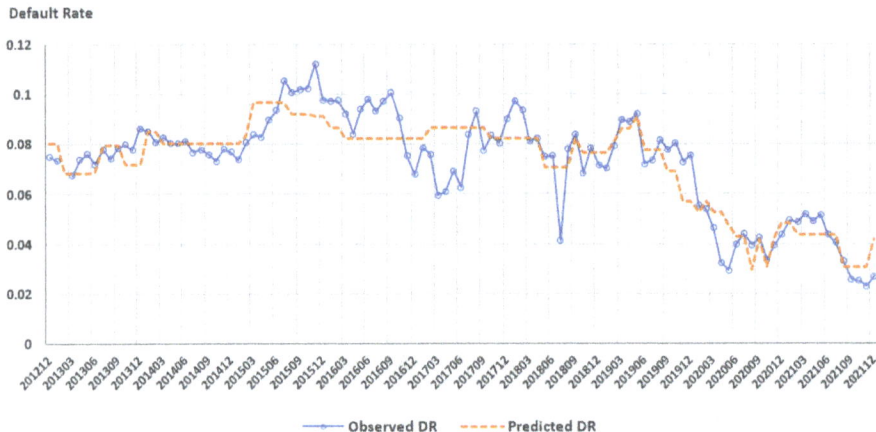

Figure 15-23.   Line chart showing the predicted default rate vs. the observed default rate across time for a six-variable scorecard model.

CHAPTER 15 WEIGHT OF EVIDENCE

$$R^2 = 0.882$$

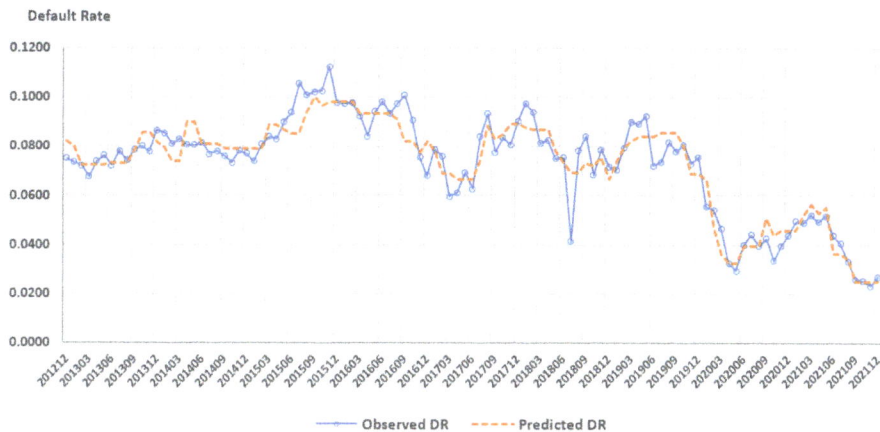

Figure 15-24. Line chart showing the predicted default rate vs. the observed default rate over time for a nine-variable scorecard model.

It is notable that the predicted DR in **Figure 15-23** displays a rectangular wave pattern and remains constant for extended periods (e.g., between April 2016 and January 2017), exhibiting both abrupt drops and increments. In contrast, the predicted DR in **Figure 15-24** demonstrates a smoother continuous trajectory, aligning more closely with the observed DR.

The rectangular wave behavior observed in **Figure 15-23** can be attributed to two factors: the discrete nature of the WoE and the use of bins. When interval variables do not exceed the upper limit of their corresponding bins, the DR value remains constant. This effect can be mitigated by increasing the resolution of the scorecard model, which can be achieved by increasing the number of bins or variables. Consequently, an increase in the number of bins or variables will result in a greater sensitivity of the model to minor fluctuations in the interval variables, thereby enhancing the overall predicted DR over time.

It is evident that the R^2 is directly correlated with the degree of fit exhibited by the scorecard model over time. Nevertheless, it is important to note that the R^2 alone is insufficient for determining the level at which the model is providing a more realistic prediction. In such cases, a visual examination is always a necessary complement.

15.7 Summary

The WoE represents a fundamental aspect of the methodology presented in this book. The information presented in this chapter is particularly comprehensive, especially for those less familiar with the WoE methodology. In light of the aforementioned details, it seems pertinent to present a summary of the most significant aspects of the WoE in eight points:

1. **A Grouping Technique**: The WoE is, in fact, a technique for grouping, or lumping, which can be applied to both interval and nominal variables.

2. **Interval Variables**: In the case of interval variables, the grouping is achieved by establishing a finite number of ranges with open intervals at the extremes of the formed groups, with the objective of obtaining a linear and monotonic distribution of the default risk.

3. **Nominal Variables**: In the case of nominal variables, the categories are grouped together according to their risk level into a finite number of groups. The aim is to minimize the risk difference among the categories in the same group while maximizing the risk difference between the groups formed. Consequently, a linear and monotonic distribution of the default risk is anticipated, occurring in no particular order with respect to the categories within the original variable.

4. **Maximum Number of Groups**: The capacity of the WoE to discriminate between categories is dependent upon the total number of groups, with the maximum number of groups defined by the user. It is nevertheless recommended that the user avoid overfitting the WoE function by setting a large maximum number of groups, as this may impede the grouping algorithm's ability to identify general trends and potentially yield erroneous results.

5. **WoE Function**: The resulting WoE variable is expected to be a discrete function with a linear relationship to the target variable, weighting the risk of default at different levels of the original variable (hence its name).

6. **Expert Subject-Matter Knowledge**: It is not anticipated that the linear and monotonic relationship between the WoE and the target variable will be achieved exclusively through automated methods. Instead, it is imperative to ensure that these methods are consistently guided by expert subject-matter knowledge, logical reasoning, and common sense.

7. **WoE Manipulation**: In this sense, and unlike any other statistical method, the WoE is not entirely data-driven. Therefore, direct manipulations of the WoE for specific bins that contradict well-known logical trends are not only recommended, but expected.

8. **Expected Results**: Consequently, a successful implementation of the WoE method should ensure that all variables

   a. Are numerical in nature
   b. Are a discrete function of the original variables
   c. Have a perfect linear and monotonic relationship with the target variable
   d. Have the same units and scale
   e. Have no outliers or missing values

After applying preselection methods to eliminate *irrelevancy* and *redundancy* and determining *the corresponding discrete functions* for all remaining input variables, we are now ready to construct our first multivariable model. The next chapter will delve into the most effective multivariable selection methods, as highlighted in the latest literature, and provide guidance on their implementation in SAS® Enterprise Miner.

CHAPTER 16

Multivariable Selection Methods

> "...people who run ball clubs...they think in terms of buying players...your goal shouldn't be to buy players; your goal should be to buy wins...."
>
> —Peter Brand, *Moneyball (2011)*

The variable selection process was initiated in Chapter 13, where the univariable procedure was first executed. The univariable procedure was executed in a manner that reduced the number of variables based on statistical significance (p-value <0.25), while also eliminating variables based on data quality (null proportion, zero proportion, and total number of distinct values for nominal variables). In the subsequent Chapter 14, the process of variable selection was advanced by the elimination of redundant variables. These were deemed to be variables that did not significantly contribute to the total variability already explained by their competing variables in the model. This was achieved through the application of the iterative collinearity reduction method, or ICRM. In Chapter 15, we addressed the typical issues of scale, units, linearity, null values, and the inclusion of nominal variables that are inherent in the development of any model. We then demonstrated how these issues can be overcome through the implementation of the WoE transformation.

CHAPTER 16   MULTIVARIABLE SELECTION METHODS

Furthermore, it was demonstrated that the WoE can be employed as an univariable selection method. However, it was advised that this approach should be avoided, as it may potentially disrupt crucial partial associations during the development of the multivariable model. The multivariable selection process is now ready to be initiated. However, it is first necessary to ascertain the number of variables that are anticipated to be included in this phase.

16.1 Variable Survival Rate

Table 16-1 illustrates the results of a variable selection process applied to seven distinct scorecard models, showcasing the number of variables retained at each stage of the process. The seven models correspond to the loan, mortgage, credit card, vehicle, SME1, SME2, and corporate portfolios, with an average initial set of 752 initial variables. **Table 16-2** shows the survival rate of the variables at each step of the modeling process. Finally, **Table 16-3** shows the corresponding variable rejection rate, also for each step of the modeling process.

CHAPTER 16   MULTIVARIABLE SELECTION METHODS

Table 16-1. Number of variables remaining after each variable selection step, for seven selected behavioral PD models for the loan, mortgage, credit card, vehicle, SME1, SME2, and corporate portfolios.

	Loan	Mortgage	Credit card	Vehicle	SME1	SME2	Corporate	Avg.	Min.	Max.
Initial set	735	735	733	735	739	794	794	752	733	794
Data cleaning	552	496	453	552	349	193	191	398	191	552
Univariable	397	381	356	331	235	155	149	286	149	397
Collinearity	103	109	115	102	62	50	35	82	35	115
WoE	18	15	44	45	38	17	21	28	15	45
Multivariable	14	14	29	19	18	14	11	17	11	29

CHAPTER 16 MULTIVARIABLE SELECTION METHODS

Table 16-2. Survival rate after each variable selection step, for seven selected behavioral PD models for the loan, mortgage, credit card, vehicle, SME1, SME2, and corporate portfolios.

	Loan	Mortgage	Credit card	Vehicle	SME1	SME2	Corporate	Avg.	Min.	Max.
Initial set	100.0%	100.0%	100.0%	100.0%	100.0%	100.0%	100.0%	100.0%	100.0%	100.0%
Data cleaning	75.1%	67.5%	61.8%	75.1%	47.2%	24.3%	24.1%	53.6%	24.1%	75.1%
Univariable	54.0%	51.8%	48.6%	45.0%	31.8%	19.5%	18.8%	38.5%	18.8%	54.0%
Collinearity	14.0%	14.8%	15.7%	13.9%	8.4%	6.3%	4.4%	11.1%	4.4%	15.7%
WoE	2.4%	2.0%	6.0%	6.1%	5.1%	2.2%	2.6%	3.8%	2.0%	6.1%
Multivariable	1.9%	1.9%	4.0%	2.6%	2.4%	1.8%	1.4%	2.3%	1.4%	4.0%

CHAPTER 16 MULTIVARIABLE SELECTION METHODS

Table 16-3. *Rejection rate after each variable selection step, for seven selected behavioral PD models for the loan, mortgage, credit card, vehicle, SME1, SME2, and corporate portfolios.*

	Loan	Mortgage	Credit card	Vehicle	SME1	SME2	Corporate	Avg.	Min.	Max.
Initial set	0.0%	0.0%	0.0%	0.0%	0.0%	0.0%	0.0%			
Data cleaning	24.9%	32.5%	38.2%	24.9%	52.8%	75.7%	75.9%	46.4%	24.9%	75.9%
Univariable	28.1%	23.2%	21.4%	40.0%	32.7%	19.7%	22.0%	26.7%	19.7%	40.0%
Collinearity	74.1%	71.4%	67.7%	69.2%	73.6%	67.7%	76.5%	71.5%	67.7%	76.5%
WoE	82.9%	86.2%	61.6%	55.9%	39.3%	65.3%	40.0%	61.6%	39.3%	86.2%
Multivariable	20.4%	6.7%	34.4%	57.7%	52.1%	19.2%	47.6%	34.0%	6.7%	57.7%

463

CHAPTER 16 MULTIVARIABLE SELECTION METHODS

It is important to note that the data cleaning process will result in the elimination of nearly half (46.4%) of the initial set of variables, primarily due to a lack of available information. It is also important to note that the data cleaning process is not a standalone step; rather, it is integrated into the univariable selection process (see Section 13.1.6). For the sake of clarity, the specific contributions have been disaggregated. The univariable procedure results in the rejection of almost 62% of the initial variables. Of these, approximately 46% are rejected as a consequence of data cleaning, while the remaining 15% are excluded due to a lack of statistical significance.

It is noteworthy that the highest rejection rate, from one step to another, is observed in the collinearity analysis, or ICRM, with an average rejection rate of 71.5%. It is also noteworthy that, although the WoE Weight of evidence (WoE) transformation is not a variable selection method in and of itself, it is capable of rejecting 61.6% of the remaining variables from the previous step. As previously stated, the automatic binning procedure (WoE) is not designed to be utilized as an additional variable selection method. However, it is important to acknowledge that the binning process may result in the rejection of numerous variables when the **Constrained Optimal conditions** are not satisfied, a phenomenon that is unrelated to the Gini or IV cutoff. It is important to note that the failure of a variable to satisfy a specific set of binning conditions does not necessarily preclude its selection with a different set of conditions. Indeed, under certain circumstances, the variable may even emerge as a significant component of a multivariable model. However, if a satisfactory number of variables has been defined according to the specified parameters, then a reduction of approximately 62% can be expected.

Figure 16-1 provides a summary of the information presented, illustrating the *survival rate* (**Figure 16-2** shows the complementary *rejection rate*) of the variables at each stage of the model-building process. The chart presents a bar graph, with the mean survival rate displayed at the top of each bar. Error bars, representing the minimum and maximum

CHAPTER 16 MULTIVARIABLE SELECTION METHODS

observed survival rates among the seven models, are also included. It is noteworthy that as the model-building process progresses, the error bars diminish in size, thereby enhancing the precision of the anticipated number of variables in subsequent stages. Consequently, **Figure 16-1** suggests that approximately 4% of the initial variables, or approximately 40 variables on a 1,000-variable basis, will be incorporated into the multivariable selection process. This set of variables will constitute the *first multivariable model.*

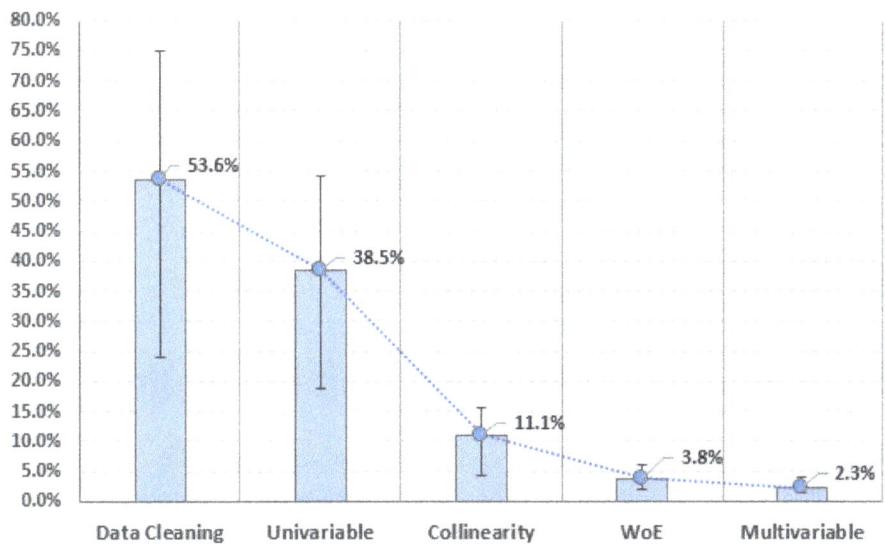

Figure 16-1. *Average variable survival rate per modeling step with confidence intervals at the 95% level of significance.*

CHAPTER 16   MULTIVARIABLE SELECTION METHODS

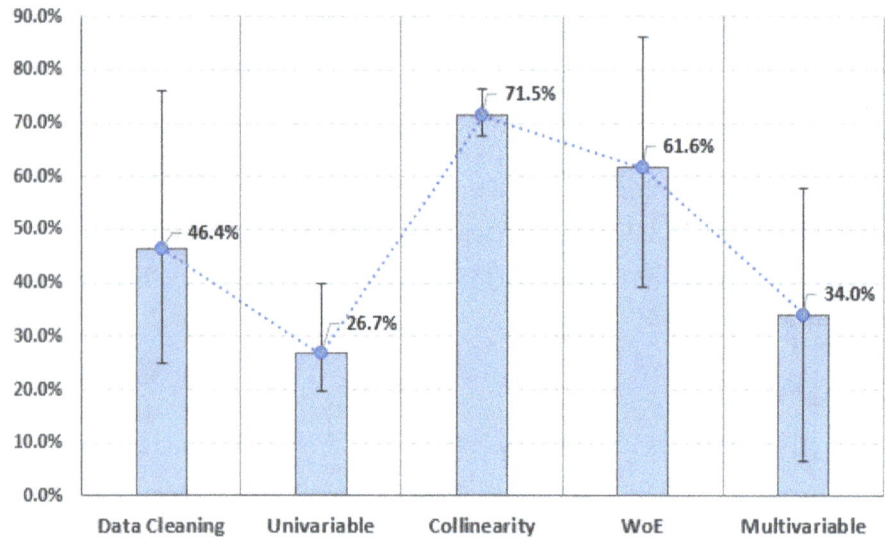

Figure 16-2. *Average variable rejection rate per modeling step, relative to the surviving variables at each step.*

16.2 Multivariable Selection Methods

Multivariable selection methods can be divided into two different strategies [58]. The first strategy involves the sequential selection of variables, either for inclusion or exclusion, based solely on statistical criteria. This objective can be accomplished by employing *forward*, *stepwise*, or *backward* procedures. The second strategy, termed the *all-subsets* method, involves fitting all 2^k (k = variables) possible models and selecting the optimal one by optimizing an information criterion derived from the likelihood function. Examples of such criteria include the AIC (Akaike's information criterion, 1973) and the SBC (Schwarz's Bayesian information criterion, 1978).

16.2.1 Stepwise Procedures

Figure 16-3 shows the multivariable selection methods available in EM, which are backward, forward, and stepwise. According to the SAS® Enterprise Miner™ 14.3 Reference Help, the official definition of each method is as follows:

> **Backward:** By default, this method begins with all candidate effects in the model and then systematically removes effects that are not significantly associated with the target until no other effect in the model meets the Stay Sig. Level, or until the Stop Variable Number that is specified is reached. This method is not recommended when the target is binary or ordinal, and there are many candidate effects or many levels for some classification input variables.[1]
>
> **Forward:** Begins, by default, with no candidate effects in the model and then systematically adds effects that are significantly associated with the target until none of the remaining effects meet the Entry Sig. Level or until the Stop Variable Number criterion is met.
>
> **Stepwise:** As with the forward method, stepwise selection begins, by default, with no candidate effects in the model and then systematically adds effects that are significantly associated with the target. However, after an effect is added to the model, stepwise can remove any effect already in the model that is not significantly associated with the target.

---

[1] This recommendation is based on the SAS® Enterprise Miner™ 14.3 Reference Help and does not coincide with the methodology outlined in this book.

CHAPTER 16   MULTIVARIABLE SELECTION METHODS

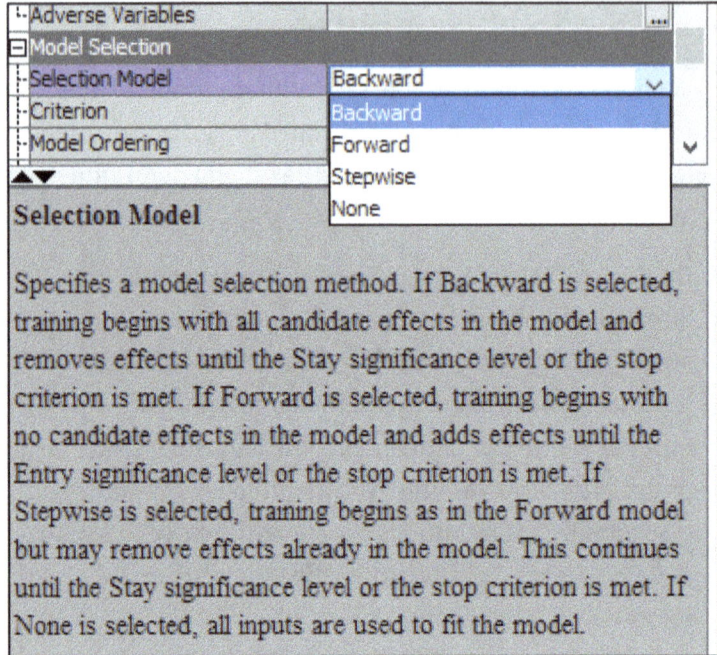

Figure 16-3. Multivariable selection methods available in EM.

According to Royston and Sauerbrei (Multivariable Model-Building [58]), the definitions given by SAS about the stepwise procedures better describe a *backward elimination only*, or BE-Only, and a *forward selection only*, or FS-Only; and the EM stepwise seems to be better described by a pure FS. So, according to Royston and Sauerbrei, the definitions of each method are as follows:

BE-Only: Estimate the full model on $x_1, ..., x_k$. Repeat: as long as least significant term has a *p*-value $\geq \alpha_2$, remove it and reestimate the model.

BE: Estimate the full model on $x_1, ..., x_k$. If the least significant term has a *p*-value $\geq \alpha_2$, remove it and reestimate; otherwise, stop.

Again: if the least significant term has a *p*-value $\geq \alpha_2$, remove it and reestimate; otherwise, stop. Repeat.

- If the most significant excluded term has a p-value $< \alpha_1$, add it and reestimate.
- If the least significant included term has a p-value $\geq \alpha_2$, remove it and reestimate, until neither action is possible.

FS-Only: Estimate the null model. Repeat: as long as the most significant excluded term has a p-value $< \alpha_1$, add it and reestimate.

FS: Estimate null model. If the most significant excluded term has p-value $< \alpha_1$, add it and reestimate; otherwise, stop.

Again, if the most significant excluded term has a p-value $< \alpha_1$, add it and reestimate; otherwise, stop.

Repeat.

- If the least significant included term has a p-value $\geq \alpha_2$, remove it and reestimate.
- If the most significant excluded term has a p-value $< \alpha_1$, add it and reestimate, until neither action is possible.

Where α_1 and α_2 correspond to the including and removing significant levels, known in EM as the entry and stay significant levels.

It is important to note that EM does not incorporate the BE algorithm. However, as highlighted by Royston and Sauerbrei [58], variables excluded by the BE are seldom reentered into the model. Consequently, it is reasonable to expect analogous results with the BE-Only procedure present in EM.

16.2.1.1 Preferable Use of BE over FS

According to Mantel (1970, Why stepdown procedures in variable selection?, Technometrics), BE methods are preferable to FS, especially in the presence of collinearity. This is due to the fact that FS is prone to stop when collinearity is high, without considering better models that include members of a cluster of correlated covariates. As an illustration, we direct our attention to the results presented in **Table 16-4**, which are drawn from Royston and Sauerbrei (2008 [58]).[2] The table compares the results obtained from a Cox survival model with an initial set of 16 covariates after running the full model (all 16 covariates), the BE, and the FS at a 0.05 and at a 0.157 nominal significance level.[3] The results demonstrate that at a nominal significance level of 0.05, the full model and the BE are capable of identifying five and seven significant variables, respectively, in contrast to the FS, which stopped after selecting only two variables from the original set of 16 covariates. However, the important fact to highlight here is that after increasing the nominal value, from 0.05 to 0.157, the FS was still only capable of selecting two significant variables, while the BE results are equal to the all-subsets method (further information on the all-subsets method can be found in the next section). As demonstrated by Royston and Sauerbrei, it can be concluded that BE methods are to be preferred to FS methods on account of their robustness, even in circumstances where nominal significance levels differ.

---

[2] For details about the original data and study, read Krall, J. M., Uthoff, V. A. and Harley, J. B. (1975) A step-up procedure for selecting variables associated with survival, Biometrics 31: 49-57.

[3] The all-subsets method was also evaluated; however, we will not go into detail about this procedure until the next section.

CHAPTER 16 MULTIVARIABLE SELECTION METHODS

Table 16-4. *Myeloma study (65 patients, 48 events). Results of applying variable different multivariable selection methods. * denotes that a variable is significant at the relevant p-level in the full model.*

Variable	Nominal significance level, α							All-subsets
	0.05			0.157				
	Full	BE	FS	Full	BE	FS		AIC
X_1	*	✓	✓	*	✓	✓		✓
X_2			✓			✓		
X_3	*	✓		*	✓			✓
X_4		✓		*	✓			✓
X_5								
X_6		✓		*	✓			✓
X_7	*	✓		*	✓			✓
X_8				*	✓			✓
X_9								
X_{10}								
X_{11}								
X_{12}	*	✓		*	✓			✓
X_{13}	*	✓		*	✓			✓
X_{14}				*				
X_{15}								
X_{16}								

CHAPTER 16 MULTIVARIABLE SELECTION METHODS

It is noteworthy that Royston and Sauerbrei [58] mentioned that the FS and the BE may select the same model under low collinearity conditions. However, it should be noted that, despite the significant reduction in collinearity achieved through the ICRM in our model, the objective is to develop an automated, streamlined, and robust model-building process. Consequently, it is imperative that we refrain from subjecting our variable selection process to a vulnerable method, such as the FS. Consequently, the FS will never be a viable option, as it is incompatible with the objectives of the methodology outlined in this book.

16.2.1.2 Stay Significance Value and Predictor Strength

In the previous example, Royston and Sauerbrei [58] employed a stay/entry significance cutoff of 0.05 and 0.157 with a sample size of less than 100 observations. However, for large training datasets (>100,000 observations), it is important to consider that the standard error, $SE(\hat{\beta})$, decreases with the sample size n. Similarly, the strength of the predictor x is a function of the estimated coefficient, $\hat{\beta}$, and its corresponding variance, as well as the sample size n. Now, given that the Wald test is defined as

$$W = \frac{\hat{\beta}}{SE(\hat{\beta})}$$

Equation 16-1. Wald test.

It can then be reasonably hypothesized that, for substantial training datasets, the nominal significance level can be set to a p-value<0.001 without risking loss of main effects during the multivariable selection process. However, it is worth noting that in practice, it has been necessary to reduce the stay significance level of the BE-Only of EM to even a p-value<0.0001, and yet, a relatively large number of variables (>15) were still being selected by the BE method.

CHAPTER 16   MULTIVARIABLE SELECTION METHODS

16.2.1.3 Model Selection Criterion

In the preceding section, it was demonstrated that the most effective multivariable selection method for use in EM is the backward elimination method, which corresponds to the BE-Only. Additionally, it was established that this method is largely analogous to the BE version recommended by Royston and Sauerbrei [58]. The present section will therefore proceed to identify the optimal *model selection criterion*. **Figure 16-4** shows the different model selection criteria available in EM and their official description, which according to the SAS® Enterprise Miner™ 14.3 Reference Guide [16] are

- **None:** (Default if you did not define a profit or loss matrix with two or more decisions) The last model produced by the effects selection method is chosen as the final model.

- **AIC:** Akaike's information criterion = $n*\ln(SSE/n) + 2p$, where n is the number of cases, SSE is error sum of squares, and p is the number of model parameters. The model with the smallest AIC value is chosen.

- **SBC:** Schwarz's Bayesian criterion = $n*\ln(SSE/n) + p*\ln(n)$. The model with the smallest SBC value is chosen.

- **Validation Error:** Chooses the model that has the smallest error rate for the validation dataset. For logistic regression models, the error is the negative log-likelihood. For linear regression, the error is the error sum of squares (SSE). This option is grayed out if a validation predecessor dataset is not input to the regression node.

- **Validation Misclassification:** Chooses the model that has the smallest misclassification rate for the validation dataset. This option is unavailable and appears dimmed if a validation predecessor dataset is not input to the regression node.

- **Cross-Validation Error:** This criterion chooses the model that has the smallest cross-validation error rate for the training dataset. For logistic regression models, the error is the negative log-likelihood. For linear regression, the error is the error sum of squares.

- **Cross-Validation Misclassification:** This criterion chooses the model that has the smallest cross-validation misclassification rate for the training dataset.

CHAPTER 16   MULTIVARIABLE SELECTION METHODS

Model Selection	
Selection Model	Backward
Criterion	Schwarz Bayesian Criterion (...
Model Ordering	Default
Use Selection Defaults	None
Entry Significance Level	Akaike Information Criterion (AIC)
Stay Significance Level	Schwarz Bayesian Criterion (SBC)
Start Variable Number	Validation Error
Stop Variable Number	Validation Misclassification
Force Candidate Effects	Cross Validation Error
Maximum Number of Steps	Cross Validation Misclassification

Criterion

If you choose the Forward, Backward, or Stepwise effect selection method for Model Selection, then you can specify a selection criterion to be used to select the final model. The criteria are AIC (Akaike's Information Criterion), Default (which corresponds to NONE), SBC (Schwarz's Bayesian Criterion), VERROR (Validation Error), VMISC (Validation Misclassification), XERROR (Cross Validation Error) or XMISC(Cross Validation Misclassification).

Figure 16-4. Model selection criteria available in EM.

As outlined in the EM documentation [16], the selection of the forward, backward, or stepwise effect selection method allows the specification of a selection criterion to be employed in the final model selection process. Once a selection criterion has been specified, as previously described, the node initiates the effect selection process, which generates a set of candidate models corresponding to each step in the process. The selection of effects is based on the levels of significance set for the entry and/or stay criteria. Upon completion of the effect selection process, the candidate model that optimizes the selection criterion is selected as the final model. This outcome indicates that the FS and BE-Only methods executed by EM are not solely based on statistical significance. However, the question remains as to which selection criterion is the most optimal and why.

CHAPTER 16 MULTIVARIABLE SELECTION METHODS

As previously stated, the BE-Only method represents the optimal choice as a multivariable selection algorithm. However, the selection criterion optimization method now encompasses six potential options, or, alternatively, the option of not utilizing any method at all. To simplify the decision-making process, it is important to recall the information presented in Section 12.3. This section demonstrated that there is no rationale for utilizing three partitions during the training phase, provided that the validation partition remains impartial. Consequently, the four selection criteria pertaining to the utilization of the validation partition, as illustrated in **Figure 16-4**, can be disregarded, leaving only three options: the SBC, the AIC, or the absence of criteria.

The decision-making process is dependent on the desired outcome; however, it is essential to have a clear understanding of the purpose and rationale behind the use of selection criteria. For instance, the absence of any criterion in the multivariable selection process will result in a model whose variables have been selected solely on the basis of statistical significance. While there is nothing inherently problematic with this approach, it is important to consider whether there is a rationale for employing a selection criterion at all if the absence of such a criterion will yield an equally optimal model. The principal rationale for employing such a criterion is to reduce the number of parameters in a model, thereby reducing its overall complexity. It is important to note that if the modeling methodology outlined in this book is followed, a vast number of variables will likely be retained following the multivariable selection process. It would therefore be prudent to employ a method capable of reducing the number of covariates that is based on a criterion other than statistical significance.

Equations 16-2 and **16-3** illustrate the SBC and AIC criteria, respectively. It should be noted that both criteria introduce a second term to the -2 log-likelihood function, which is a penalty parameter.

Consequently, both the SBC and the AIC can be considered penalized versions of the -2 log-likelihood function. The distinction between the SBC and the AIC lies in the level of penalization they employ. Specifically, the AIC penalization parameter is defined as $2p$, where p represents the total number of parameters in the model. In contrast, the penalty parameter for SBC is $p \cdot ln[n]$, where p represents the total number of parameters and n denotes the total number of observations in the sample. It is evident that the SBC is a more rigorous criterion than the AIC. The SBC penalty is dependent not only on the number of parameters but also on the size of the training sample, whereas the AIC penalty function is solely dependent on $2p$. Consequently, the SBC criterion will tend to yield smaller models than the AIC criterion. Therefore, the decision to use one or the other will be guided by the objective of either reducing the size of the model or increasing it.

$$SBC = -2LL + p \cdot ln[n]$$

Equation 16-2. Schwarz's Bayesian criterion.

$$AIC = -2LL + 2p$$

Equation 16-3. Akaike's information criterion.

16.2.1.4 EM Stepwise Considerations

Prior to the implementation of the stepwise techniques delineated in EM, it is imperative to acquire a comprehensive understanding of the distinctions between the FS, FS-Only, BE, and BE-Only and to acknowledge that EM does not employ a BE, but rather a BE-Only algorithm. Furthermore, it is imperative to acknowledge that EM's multivariable selection algorithms lack the capacity to effectively manage polychotomous variables that have been previously converted to n-1 dummy variables, as previously discussed in Section 15.2.4. This is due to the fact that dummy variables cannot be treated by multivariable selection methods as if they were independent entities that could be eliminated or selected from the multivariable model at will. Fortunately, this problem does not arise at all when the WoE transformation is applied. Nevertheless, in the absence of this transformation, it is imperative to acknowledge this inherent limitation of multivariable selection.[4]

16.2.2 All-Subsets Procedure

Unlike the stepwise procedures, the all-subsets procedure fits the 2^k possible models, with the optimal model being selected through the optimization of an information criterion. It is important to note that the all-subsets procedure is not included as a native option in the EM scorecard/regression node. However, it is included in the LOGISTIC procedure using the SELECTION=SCORE option. **Listing 16-1** shows the precode (see **Figure 16-5** for details of the expected nodes layout), which is executed prior to the all-subsets procedure. This is necessary to reject the *"GRP"*[5] variables that were generated as a by-product of the IG node (see **Figure 16-6** for details). **Listing 16-2** illustrates the initial part of the all-subsets code implemented in EM. The START (EM_PROPERTY_MININPUTS) and STOP (EM_PROPERTY_MAXINPUTS) parameters specify the number of inputs to be at the start and the end of the process, respectively. The BEST (EM_PROPERTY_MODELSPERLEVEL) option specifies the number of models of each complexity to be fitted. It is worth mentioning that the SCORE option of the LOGISTIC procedure does not fit the 2^k possible

---

[4] Hosmer and Lemeshow recommend the use of the BMDPLR program for the proper treatment of polychotomous variables during the multivariable selection process.

[5] The IG node's output comprises two types of variables: WOE variables and GRP variables. The WOE variables contain the corresponding WoE value for each bin level of the original variables, while the GRP variables only contain the corresponding bin number as an ordinal variable.

CHAPTER 16 MULTIVARIABLE SELECTION METHODS

models *per se*, but instead employs the *branch and bound algorithm*, developed by Furnival and Wilson (Regressions by Leaps and Bounds, 1974 [22]), which is capable of identifying the best subsets without examining all possible subsets. This algorithm has been demonstrated to run even faster than stepwise algorithms for less than 60 inputs, which is a significant advantage in this context. As mentioned in Section 16.1, the average number of covariates expected to enter the multivariable selection process is approximately 40 variables (on a 1,000-variable basis).

Listing 16-1. Reject GRP variables created as a by-product of the IG node.

```
PROC CONTENTS
            DATA=&EM_IMPORT_DATA.
            OUT=VARIABLES_NAMES
            NOPRINT;
RUN;

PROC SQL ;

    SELECT
    DISTINCT NAME
    INTO
    :GROUP_VARIABLES SEPARATED BY " "
    FROM VARIABLES_NAMES
    WHERE NAME LIKE "GRP_%"
    ;
QUIT;

%LET VARLIST=&GROUP_VARIABLES.;

%LET VARLIST=%ENQUOTE(&VARLIST.,SEPARATOR=);
%METADATA(NEWROLE=REJECTED,WHERE=(NAME IN (&VARLIST.) AND UPCASE(ROLE)='INPUT'));
```

Figure 16-5. *Example of the expected nodes layout when using the all-subsets procedure.*

CHAPTER 16 MULTIVARIABLE SELECTION METHODS

I_RISK_LEVEL	GRP_I_RISK_LEVEL	WOE_I_RISK_LEVEL	I_R_CLTV_CM	GRP_I_R_CLTV_CM	WOE_I_R_CLTV_CM
8	2	0.45892	1.4074242002	4	-0.23994
1	1	-2.03799	2.3042307646	5	-0.40976
5	1	-2.03799	0.818985253	2	0.03936
8	2	0.45892	0.6592932162	1	0.26115
8	2	0.45892	0.5533443365	1	0.26115
8	2	0.45892	1.0821670876	3	-0.07517
8	2	0.45892	0.3352815853	1	0.26115
8	2	0.45892	0.3615964581	1	0.26115

Figure 16-6. *Example of the expected IG node output for significant variables (according to the Gini index or the IV), consisting of a GRP, and a WOE Weight of evidence (WoE) variable for each original variable that entered into the IG procedure.*

Listing 16-2. Initial part of the all-subsets code implemented in EM.

```sas
%LET EM_PROPERTY_MODELSPERLEVEL = 10;   /*Number of models to show per level*/
%LET EM_PROPERTY_MININPUTS      = 1;    /*Minimum number of inputs*/
%LET EM_PROPERTY_MAXINPUTS     = MAX;   /*Max number of inputs*/
%LET MDL=&EM_NODEID.;                   /*Use the node ID as identifier for output
%LET LIB=&EM_LIB.;                      /*Default Diagram library*/

/*-------------------------------------------------------------------
         --- Set the EM_REGISTER macro for chart generation
   ----------------------------------------------------------------*/
%EM_REGISTER(KEY=ASSESS,TYPE=DATA);
%EM_REGISTER(KEY=MALLOWS_CQ,TYPE=DATA);

/*-------------------------------------------------------------------
         --- Set start and stop parameters
   ----------------------------------------------------------------*/
** stop=maxinputs value or count(# interval inputs);
%LET STOP=%SCAN(&EM_PROPERTY_MAXINPUTS. %NLIST(%EM_INTERVAL_INPUT),
                %EVAL((%INDEX(%UPCASE(&EM_PROPERTY_MAXINPUTS.),MAX)>0)+1));
** start=min(max(mininput value,1),stop);
%LET START=%SYSFUNC(MIN(%SYSFUNC(MAX(&EM_PROPERTY_MININPUTS.,1)),&STOP.));

ODS LISTING CLOSE;
ODS OUTPUT BESTSUBSETS=WORK.SCORE
           RESPONSEPROFILE=WORK.PROFILE;
/*-------------------------------------------------------------------
         --- Execute logistic regression with the "SCORE" option for all subsets
   ----------------------------------------------------------------*/
PROC LOGISTIC DATA=&EM_IMPORT_DATA. DES NAMELEN=200;
    MODEL %EM_BINARY_TARGET=%EM_INTERVAL_INPUT / SELECTION=SCORE
                                        BEST=&EM_PROPERTY_MODELSPERLEVEL.
                                        START=&START.
                                        STOP=&STOP.;
RUN;
```

16.2.2.1 Score Test Approximation to Mallows' C_q

The central issue is no longer how to identify the optimal subsets without actually implementing all possible models, but rather, how to select the most suitable model from among them. Mallows (1973 [39]) developed a criterion capable of measuring the predictive squared error for all-subsets of linear regression, called C_q.[6] **Equation 16-4** shows the C_q criterion:

$$C_q = \frac{X^2 + \lambda^*}{X^2 / (n-p-1)} + 2(q+1) - n$$

Equation 16-4. Mallows' C_q.

Where

$$X^2 = \sum \left\{ \frac{(y_i - \hat{\pi}_i)^2}{[\hat{\pi}_i(1-\hat{\pi}_i)]} \right\}$$

Equation 16-5. Person's chi-square statistic.

Where X^2 is the Person chi-square statistic for the model with p variables and λ^* is the multivariable Wald test statistic for the hypothesis that the coefficients in the $p - q$ variables not in the model are equal to zero.

Hosmer and Lemeshow [32] obtained an approximation of linear regression's Mallows' C_q for logistic regression. They assumed that the Person's chi-square statistic is equal to its mean, $\therefore X^2 \approx (n - p - 1)$. Next, they assumed that the Wald statistic for the $p - q$ excluded covariates can

---

[6] Other authors prefer the notation C_p instead of C_q; however, we are being consistent with the notation used by Hosmer and Lemeshow (Applied Logistic Regression), where q refers to a subset of variables, whereas p refers to the total number of possible variables.

CHAPTER 16 MULTIVARIABLE SELECTION METHODS

be approximated by the difference between the values of the score test for all p covariates and the score test for q covariates, namely, $\lambda_q^* \approx S_p - S_q$. This leads to the following equations:

$$C_q = \frac{X^2 + \lambda^*}{X^2/(n-p-1)} + 2(q+1) - n \approx \frac{(n-p-1)+(S_p - S_q)}{1} + 2(q+1) - n$$

$$\approx S_p - S_q + 2q - p + 1$$

Equation 16-6. Score test approximation to Mallows' C_q.

Listing 16-3 presents the SQL code in the EM code node that calculates the score test approximation to Mallows' C_q, using the identical formula presented in **Equation 16-6** (line 44), which is described in terms of the SCORE table variables as follows:

$$\texttt{sp-scorechisq+2*numberofvariables-p+1}$$

It should be noted that the subsequent two steps merely sort and number the number of models within each complexity level. As illustrated in **Figure 16-7**, the output produced by **Listing 16-3** reveals that there are 10 best models generated per complexity level, i.e., 10 univariable models, 10 bivariable models, 10 tri-variable models, and so on.

Listing 16-3. *The score test approximation to Mallows'* C_q.

```sas
34  /*----------------------------------------------------
35           --- Estimate Mallows Cq
36  -----------------------------------------------------*/
37  PROC SQL;
38  CREATE TABLE &EM_LIB..MALLOWS_CQ_&MDL.
39  AS
40      SELECT
41      VARIABLESINMODEL AS MDLCOV 'Model Covariates'
42      ,NUMBEROFVARIABLES AS INPUT_COUNT
43      ,SCORECHISQ AS SQ 'Sq'
44      ,SP-SCORECHISQ+2*NUMBEROFVARIABLES-P+1 AS CQ 'Cq'
45      ,NUMBEROFVARIABLES+1 AS CQ_CRITERION LABEL='Cq Criterion'
46      FROM
47      WORK.SCORE A
48      ,(
49      SELECT
50      MAX(SCORECHISQ) AS SP
51      ,MAX(NUMBEROFVARIABLES) AS P
52      FROM WORK.SCORE) B
53      ;
54  QUIT;
55  /*----------------------------------------------------
56           --- Sort by number of inputs and Cq
57  -----------------------------------------------------*/
58  PROC SORT
59          DATA=&EM_LIB..MALLOWS_CQ_&MDL.
60          THREADS
61          SORTSIZE=MAX;
62      BY
63          INPUT_COUNT
64          CQ;
65  RUN;
66  /*----------------------------------------------------
67           --- Models numbered by complexity
68  -----------------------------------------------------*/
69  DATA &EM_USER_MALLOWS_CQ.;
70      RETAIN BEST;
71      SET
72          &EM_LIB..MALLOWS_CQ_&MDL. END=LAST;
73      BY
74          INPUT_COUNT;
75      IF FIRST.INPUT_COUNT THEN
76          BEST=0;
77      BEST+1;
78  RUN;
```

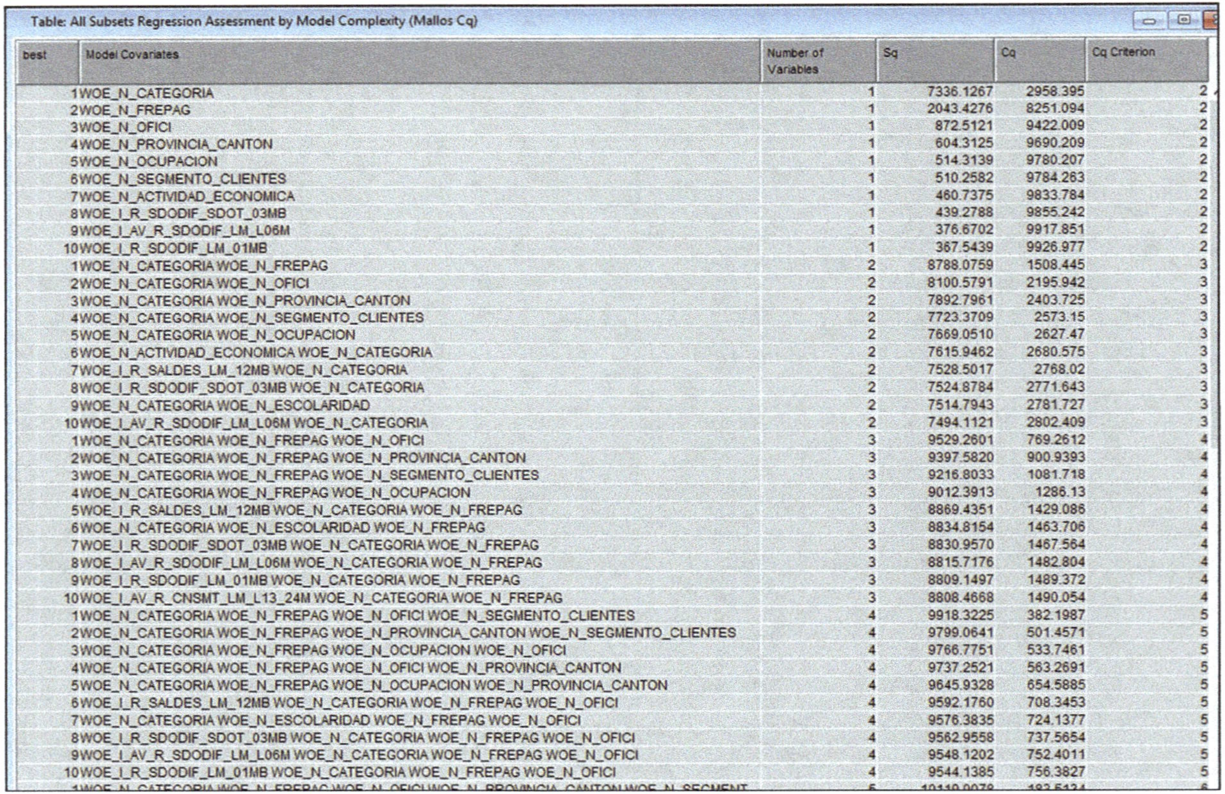

Table: All Subsets Regression Assessment by Model Complexity (Mallos Cq)

best	Model Covariates	Number of Variables	Sq	Cq	Cq Criterion
1	WOE_N_CATEGORIA	1	7336.1267	2958.395	2
2	WOE_N_FREPAG	1	2043.4276	8251.094	2
3	WOE_N_OFICI	1	872.5121	9422.009	2
4	WOE_N_PROVINCIA_CANTON	1	604.3125	9690.209	2
5	WOE_N_OCUPACION	1	514.3139	9780.207	2
6	WOE_N_SEGMENTO_CLIENTES	1	510.2582	9784.263	2
7	WOE_N_ACTIVIDAD_ECONOMICA	1	460.7375	9833.784	2
8	WOE_I_R_SDODIF_SDOT_03MB	1	439.2788	9855.242	2
9	WOE_I_AV_R_SDODIF_LM_L06M	1	376.6702	9917.851	2
10	WOE_I_R_SDODIF_LM_01MB	1	367.5439	9926.977	2
1	WOE_N_CATEGORIA WOE_N_FREPAG	2	8788.0759	1508.445	3
2	WOE_N_CATEGORIA WOE_N_OFICI	2	8100.5791	2195.942	3
3	WOE_N_CATEGORIA WOE_N_PROVINCIA_CANTON	2	7892.7961	2403.725	3
4	WOE_N_CATEGORIA WOE_N_SEGMENTO_CLIENTES	2	7723.3709	2573.15	3
5	WOE_N_CATEGORIA WOE_N_OCUPACION	2	7669.0510	2627.47	3
6	WOE_N_ACTIVIDAD_ECONOMICA WOE_N_CATEGORIA	2	7615.9462	2680.575	3
7	WOE_I_R_SALDES_LM_12MB WOE_N_CATEGORIA	2	7528.5017	2768.02	3
8	WOE_I_R_SDODIF_SDOT_03MB WOE_N_CATEGORIA	2	7524.8784	2771.643	3
9	WOE_N_CATEGORIA WOE_N_ESCOLARIDAD	2	7514.7943	2781.727	3
10	WOE_I_AV_R_SDODIF_LM_L06M WOE_N_CATEGORIA	2	7494.1121	2802.409	3
1	WOE_N_CATEGORIA WOE_N_FREPAG WOE_N_OFICI	3	9529.2601	769.2612	4
2	WOE_N_CATEGORIA WOE_N_FREPAG WOE_N_PROVINCIA_CANTON	3	9397.5820	900.9393	4
3	WOE_N_CATEGORIA WOE_N_FREPAG WOE_N_SEGMENTO_CLIENTES	3	9216.8033	1081.718	4
4	WOE_N_CATEGORIA WOE_N_FREPAG WOE_N_OCUPACION	3	9012.3913	1286.13	4
5	WOE_I_R_SALDES_LM_12MB WOE_N_CATEGORIA WOE_N_FREPAG	3	8869.4351	1429.086	4
6	WOE_N_CATEGORIA WOE_N_ESCOLARIDAD WOE_N_FREPAG	3	8834.8154	1463.706	4
7	WOE_I_R_SDODIF_SDOT_03MB WOE_N_CATEGORIA WOE_N_FREPAG	3	8830.9570	1467.564	4
8	WOE_I_AV_R_SDODIF_LM_L06M WOE_N_CATEGORIA WOE_N_FREPAG	3	8815.7176	1482.804	4
9	WOE_I_R_SDODIF_LM_01MB WOE_N_CATEGORIA WOE_N_FREPAG	3	8809.1497	1489.372	4
10	WOE_I_AV_R_CNSMT_LM_L13_24M WOE_N_CATEGORIA WOE_N_FREPAG	3	8808.4668	1490.054	4
1	WOE_N_CATEGORIA WOE_N_FREPAG WOE_N_OFICI WOE_N_SEGMENTO_CLIENTES	4	9918.3225	382.1987	5
2	WOE_N_CATEGORIA WOE_N_FREPAG WOE_N_PROVINCIA_CANTON WOE_N_SEGMENTO_CLIENTES	4	9799.0641	501.4571	5
3	WOE_N_CATEGORIA WOE_N_FREPAG WOE_N_OCUPACION WOE_N_OFICI	4	9766.7751	533.7461	5
4	WOE_N_CATEGORIA WOE_N_FREPAG WOE_N_OFICI WOE_N_PROVINCIA_CANTON	4	9737.2521	563.2691	5
5	WOE_N_CATEGORIA WOE_N_FREPAG WOE_N_OCUPACION WOE_N_PROVINCIA_CANTON	4	9645.9328	654.5885	5
6	WOE_I_R_SALDES_LM_12MB WOE_N_CATEGORIA WOE_N_FREPAG WOE_N_OFICI	4	9592.1760	708.3453	5
7	WOE_N_CATEGORIA WOE_N_ESCOLARIDAD WOE_N_FREPAG WOE_N_OFICI	4	9576.3835	724.1377	5
8	WOE_I_R_SDODIF_SDOT_03MB WOE_N_CATEGORIA WOE_N_FREPAG WOE_N_OFICI	4	9562.9558	737.5654	5
9	WOE_I_AV_R_SDODIF_LM_L06M WOE_N_CATEGORIA WOE_N_FREPAG WOE_N_OFICI	4	9548.1202	752.4011	5
10	WOE_I_R_SDODIF_LM_01MB WOE_N_CATEGORIA WOE_N_FREPAG WOE_N_OFICI	4	9544.1385	756.3827	5

Figure 16-7. *Example of an output generated by* ***Listing 16-3***.

Hosmer, Jovanovic, and Lemeshow (1989) demonstrated that, under the assumption that the model fit is the correct one, the approximate expected values of X^2 and λ^* are $(n - p - 1)$ and $p - q$, respectively. Substituting these values into **Equation 16-4** yields $C_q = q + 1$ (see column labeled 'Cq Criterion' in **Listing 16-3**, line 45), thereby indicating that all models with a C_q less than or equal to $q + 1$ can be considered as candidates for the best model. In accordance with the logic of all-subsets linear regression programs, the best all-subsets selected from the &EM_LIB..MALLOWS_CQ_&MDL table will be the subset with the smallest C_q. The SQL code that implements this criterion to select the covariates from the best subset is shown in **Listing 16-4**.

Listing 16-4. Selection criterion to select the covariates from the best subset.

```
93  /*------------------------------------------------------------
94              --- Select best model based on Mallows criterion
95  ------------------------------------------------------------
96  PROC SQL;
97    SELECT
98    MDLCOV INTO :SIGNIFICANT
99    FROM &LIB..MALLOWS_CQ_&MDL.
100   WHERE CQ<=CQ_CRITERION
101   AND CQ=(SELECT MIN(CQ) FROM &LIB..MALLOWS_CQ_&MDL. WHERE CQ<=CQ_CRITERION)
102   ;
103 QUIT;
```

16.2.2.2 Graphical Solution to C_q's Optimization Problem

The selection of the optimal subset from the score test approximation can be considered an optimization problem. As illustrated in **Figure 16-8**, a line chart has been employed to demonstrate C_q as a function of the number of variables, with the series representing the top ten subsets for each level of complexity. It is noteworthy that C_q is highest for models with a limited number of variables and undergoes an exponential decrease as the number of variables in the model rises, or as complexity escalates. It is

CHAPTER 16 MULTIVARIABLE SELECTION METHODS

imperative to bear in mind that C_q is actually a measure of the predictive error; thus, the objective is to identify the subset that minimizes the predictive error. **Figure 16-9** provides an expanded view of **Figure 16-8** which presents the graphical solution to the C_q's optimization problem. The dots circled in the large red ellipse represent the four bottom models that are less than or equal to $q + 1$, as presented in **Table 16-5**. It is notable that the minimum is attained by the model with eleven covariates, which is encircled in **Figure 16-9** with a purple double-lined ellipse, that corresponds to the point or model selected by the condition Cq=(select min(Cq) from &lib..Mallows_Cq_&mdl where Cq<=Cq_Criterion) in **Listing 16-4**.

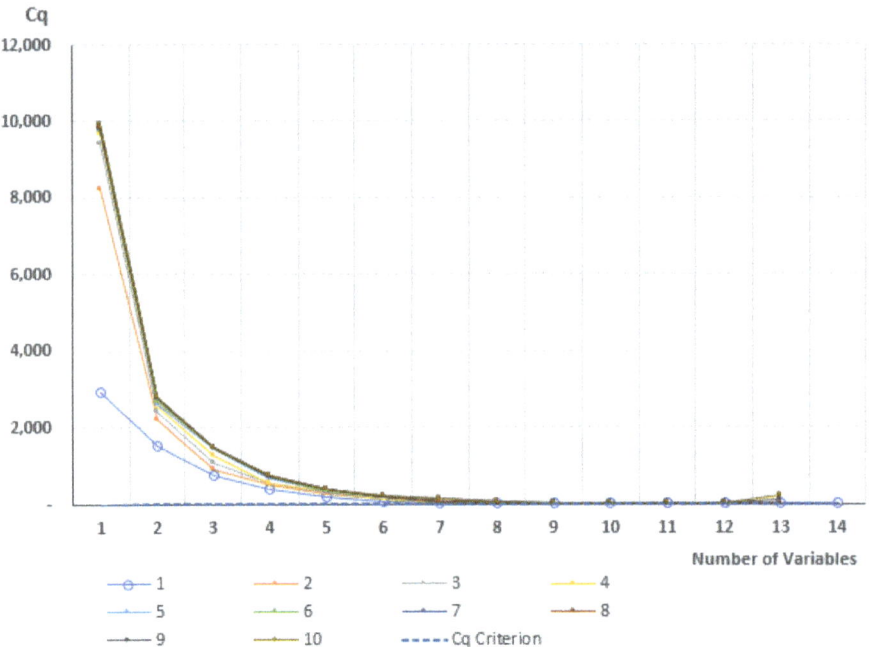

Figure 16-8. *Line chart of C_q as a function of the number of variables, the series representing the top ten subsets for each one of the levels of complexity.*

CHAPTER 16 MULTIVARIABLE SELECTION METHODS

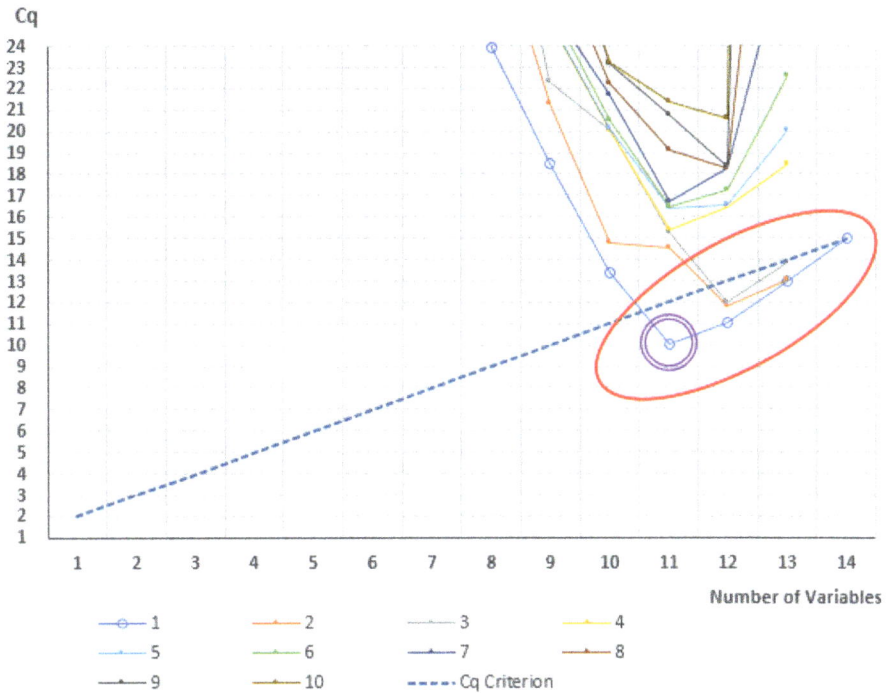

Figure 16-9. *Graphic solution of the C_q optimization problem.*

CHAPTER 16 MULTIVARIABLE SELECTION METHODS

Table 16-5. *Top models at each level of complexity for a 15-covariate model, in descending order by C_q.*

best	Model Covariates	q	Sq	Cq	Cq Criterion
1	WOE_N_CATEGORY	1	7,336.13	2,958.39	2
1	WOE_N_CATEGORY WOE_N_PAYFRQ	2	8,788.08	1,508.45	3
1	WOE_N_CATEGORY WOE_N_PAYFRQ WOE_N_BRANCH	3	9,529.26	769.26	4
1	WOE_N_CATEGORY WOE_N_PAYFRQ WOE_N_BRANCH WOE_N_CUTOMER_SEGMENT	4	9,918.32	382.20	5
1	WOE_N_CATEGORY WOE_N_PAYFRQ WOE_N_BRANCH WOE_N_STATE_DISTRICT WOE_N_CUTOMER_SEGMENT	5	10,119.01	183.51	6
1	WOE_N_CATEGORY WOE_N_PAYFRQ WOE_N_OCCUPATION WOE_N_BRANCH WOE_N_STATE_DISTRICT WOE_N_CUTOMER_SEGMENT	6	10,236.27	68.25	7
1	WOE_I_R_BAL_CLM_12MB WOE_N_CATEGORY WOE_N_PAYFRQ WOE_N_OCCUPATION WOE_N_BRANCH WOE_N_STATE_DISTRICT WOE_N_CUTOMER_SEGMENT	7	10,269.91	36.61	8
1	WOE_I_R_BAL_CLM_12MB WOE_N_ECONOMIC_SECTOR WOE_N_CATEGORY WOE_N_PAYFRQ WOE_N_OCCUPATION WOE_N_BRANCH WOE_N_STATE_DISTRICT WOE_N_CUTOMER_SEGMENT	8	10,284.57	23.95	9
1	WOE_I_AV_R_PRCHT_CLM_L13_24M WOE_I_R_BAL_CLM_12MB WOE_N_ECONOMIC_SECTOR WOE_N_CATEGORY WOE_N_PAYFRQ WOE_N_OCCUPATION WOE_N_BRANCH WOE_N_STATE_DISTRICT WOE_N_CUTOMER_SEGMENT	9	10,291.99	18.53	10
1	WOE_I_AV_R_PRCHT_CLM_L13_24M WOE_I_R_BAL_CLM_12MB WOE_I_R_BALDIF_BALT_03MB WOE_N_ECONOMIC_SECTOR WOE_N_CATEGORY WOE_N_PAYFRQ WOE_N_OCCUPATION WOE_N_BRANCH WOE_N_STATE_DISTRICT WOE_N_CUTOMER_SEGMENT	10	10,299.12	13.41	11
1*	WOE_I_AV_R_PRCHT_CLM_L13_24M WOE_I_R_BAL_CLM_12MB WOE_I_R_BALDIF_BALT_03MB WOE_N_ECONOMIC_SECTOR WOE_N_CATEGORY WOE_N_EDUCATION WOE_N_PAYFRQ WOE_N_OCCUPATION WOE_N_BRANCH WOE_N_STATE_DISTRICT WOE_N_CUTOMER_SEGMENT	11	10,304.49	10.03	12
1	WOE_I_AV_R_PRCHT_CLM_L13_24M WOE_I_R_BAL_CLM_12MB WOE_I_R_BALDIF_BALT_03MB WOE_I_R_BALDIF_BALT_06MB WOE_N_ECONOMIC_SECTOR WOE_N_CATEGORY WOE_N_EDUCATION WOE_N_PAYFRQ WOE_N_OCCUPATION WOE_N_BRANCH WOE_N_STATE_DISTRICT WOE_N_CUTOMER_SEGMENT	12	10,305.48	11.04	13
1	WOE_I_AV_R_PRCHT_CLM_L13_24M WOE_I_AV_R_BALDIF_CLM_L06M WOE_I_R_BAL_CLM_12MB WOE_I_R_BALDIF_BALT_03MB WOE_N_ECONOMIC_SECTOR WOE_N_CATEGORY WOE_N_EDUCATION WOE_N_PAYFRQ WOE_N_OCCUPATION WOE_N_BRANCH WOE_N_STATE_DISTRICT WOE_N_CUTOMER_SEGMENT	13	10,305.52	13.00	14
1	WOE_I_AV_R_PRCHT_CLM_L13_24M WOE_I_AV_R_BALDIF_CLM_L06M WOE_I_R_BAL_CLM_12MB WOE_I_R_BALDIF_BALT_03MB WOE_I_R_BALDIF_BALT_06MB WOE_N_ECONOMIC_SECTOR WOE_N_CATEGORY WOE_N_EDUCATION WOE_N_PAYFRQ WOE_N_OCCUPATION WOE_N_BRANCH WOE_N_STATE_DISTRICT WOE_N_CUTOMER_SEGMENT	14	10,305.52	15.00	15

N_CATEGORY = internal rating category; N_PAYFRQ = payment frequency; N_BRANCH = banking branch; N_CUSTOMER_SEGMENT = customer's internal segment; N_STATE_DISTRICT = concatenated state and district codes; N_OCCUPATION = customer's occupation; N_ECONOMIC_SECTOR = customer's economic sector; I_AV_R_PRCHT_CLM_L13_24M = average ratio of total purchases to credit limit, over the last 24 to 13 months from the reference date; I_R_BALDIF_BALT_06MB = ratio of the balance difference, between the previous and the actual period, to the total actual balance, reported six months back from the reference date; I_R_BALDIF_CLM_06MB = ratio of the balance difference, between the previous and the actual period, to the credit limit, reported six months back from the reference date; I_R_BAL_CLM_12MB = ratio of the actual balance to the credit limit, reported 12 months back from the reference date.

The final part of the all-subsets procedure is illustrated in **Listing 16-5** where a logistic regression with the best subset solution is executed to analyze the results. The best subset variables are then selected using the %metadata macro.

Listing 16-5. Best subset regression and variable selection.

```
105  /*------------------------------------------------------------
106          --- Run Logistic Regression with best subset
107  ------------------------------------------------------------*/
108  PROC LOGISTIC
109       DATA=&EM_IMPORT_DATA.
110       NAMELEN=200
111       ;
112       MODEL  %EM_BINARY_TARGET (EVENT='1')= &SIGNIFICANT.
113       /
114       LINK=GLOGIT
115       CLPARM=WALD
116       TECH=NEWTON
117       ;
118  RUN;
119  ODS OUTPUT
120       TYPE3=&LIB..T3_ALLSUBSETS_&MDL.
121  ;
122
123  /*------------------------------------------------------------
124          --- Applied code in order to set the new metadata|--
125  ------------------------------------------------------------
126  %LET VARLIST= &SIGNIFICANT.;
127
128  %LET VARLIST=%ENQUOTE(&VARLIST.,SEPARATOR=);
129  %METADATA(NEWROLE=REJECTED,WHERE=NAME NOT IN (&VARLIST.) AND UPCASE(ROLE)='INPUT');
```

16.2.3 Stepwise or All-Subsets Procedure?

The all-subsets procedure is predicated on the exploration of all possible subsets, a process which renders it more likely to include the most significant effects from the remaining variables. Consequently, in contradistinction to stepwise methods, which have a propensity to yield smaller models, the all-subsets procedure is likely to produce larger models. However, in accordance with the principle of parsimony, should the objective not be to always strive for simple, and thus smaller, models?

As demonstrated in previous sections, the challenge posed by the methodology outlined in this book is not a shortage of variables, but rather an excess. It is important to note that the model-building process has been described as a linear flow, when in fact it is more accurately represented as a network. It is therefore recommended that the reader revisits **Figure 11-21**. This figure provides an example of an EM diagram using the methodology proposed in this book. It is evident that, in contrast to a single model-building flow, there are 21 flows, comprising 18 scorecard models that employ the BE-Only method[7] and the remaining 3 that utilize the all-subsets method.

In light of the aforementioned, the response to the initial inquiry is that, although the parsimony principle should be adhered to, it is still being implemented in its early stages. Consequently, the recommendation is not to select between the stepwise and the all-subsets approaches, but to employ both concurrently, as illustrated in **Figure 11-21**. This then gives rise to the questions of how many scorecards should be created, how many should utilize the stepwise approach, and how many should employ the all-subsets approach, as well as the rationale behind these decisions.

16.3 Summary

Chapter 16 provides a thorough review of the multivariable selection methods, focusing on two primary approaches: *the stepwise procedures* (with an emphasis on backward elimination) and *the all-subsets method*. These methods are essential for identifying the most important effects in a model while ensuring that variables with weak associations, which may contribute to a better overall model, are not prematurely excluded.

The *backward elimination* (BE) algorithm is highlighted as the preferred stepwise procedure over forward selection. BE avoids convergence to local optima and has a better chance of retaining the main effects of the model. Notably, recent research suggests that using a staying significance level of 0.157 can yield results comparable to the all-subsets method. The *all-subsets method*, in contrast, is the only approach that evaluates all possible variable combinations, making it uniquely capable of capturing all main effects in a model.

---

[7] The scorecards using the BE-Only method are identified because they are directly connected to the IG nodes.

CHAPTER 16   MULTIVARIABLE SELECTION METHODS

A critical takeaway from the chapter is that multivariable selection should not rely solely on statistical significance. Instead, additional selection criteria, such as the *Schwarz Bayesian criterion* (SBC) and *Mallows' C_q*, are introduced to guide model selection. SBC is recommended for obtaining smaller, simpler models, while Mallows' C_q, through a score test approximation, minimizes the predictive error and helps identify the best model.

The chapter concludes with a recommendation to use both stepwise and all-subsets methods concurrently. This aligns with the book's broader methodology of constructing a multibranch diagram, where each final node represents a variant of the model.

Key takeaways:

1. **Backward Elimination (BE):**

    - Preferred over forward selection because it avoids convergence to local optima and better retains the main effects of the model.

    - Using a staying significance level of 0.157 can produce results comparable to the all-subsets method.

2. **All-Subsets Method:**

    - The only method capable of evaluating all possible variable combinations, ensuring that all main effects are included in the model.

    - Provides a comprehensive exploration of variables but is computationally intensive.

3. **Schwarz Bayesian Criterion (SBC):**

    - Recommended for use with the BE-Only method to obtain smaller, more interpretable models.

CHAPTER 16 MULTIVARIABLE SELECTION METHODS

4. **Mallows' Cq:**

    - Introduced as a criterion for the all-subsets method, where it minimizes the predictive error through a graphical optimization approach.

5. **Concurrent Use of Methods:**

    - Both BE and all-subsets methods should be employed simultaneously in the modeling process.

    - This is consistent with the multibranch diagram methodology, where each branch represents a variant of the model.

The next step in the modeling process moves beyond selecting a single optimal model to systematically exploring multiple feasible regions. Chapter 17 focuses on defining these regions through experimental design and fine-tuning their parameters via hyperoptimization, thereby expanding the solution space allowing for the identification of a broader set of main effects.

CHAPTER 17

Experimental Design and Hyperoptimization

"An approximate answer to the right problem is worth a good deal more than an exact answer to an approximate problem."

—John Tukey

The model-building process has thus far been conceptualized as a linear and sequential series of steps. Initially, an ABT was constructed with explanatory and outcome variables. Subsequently, the ABT was partitioned into a training (treatment) and a validation (control) partition. The variable selection process was initiated, beginning with the univariable procedure and subsequently followed by the collinearity reduction method. The original variables were then replaced with their corresponding WoEs, and a multivariable selection procedure was executed, resulting in the initial multivariable model. The objective is to demonstrate that the final multivariable model represents the optimal solution within a feasible region, constrained by a set of predetermined conditions that were defined concurrently with the modeling process. For instance, it is plausible that the subset of variables incorporated into the

CHAPTER 17 EXPERIMENTAL DESIGN AND HYPEROPTIMIZATION

final model would have been different had the varclus correlation method been selected in place of the ICRM. Furthermore, alternative grouping options in the IG node may result in a different selection of variables (due to the influence of the rejection rate on the stringency of the grouping constraints) and may also increase the discriminatory power of some variables over others, simply by selecting a better cutoff point during the grouping process. Consequently, the feasible region within which the optimal model is contained is defined by the specific set of options, whether methods, parameters, or values, selected during the model-building process. It can thus be concluded that there must be a different optimal solution for each possible feasible region. The process of defining the number of feasible regions to explore is known as *experimental design*, while the process of defining the parameters that constrain each feasible region is known as *hyperoptimization*.

17.1 Experimental Design

Each feasible region can be defined as an independent experiment consisting of

- Response
- Conditions, also known as *factors*
- Factors' levels
- Effects
- Balance

Consequently, each experiment corresponds to a specific mining branch or modeling process, as illustrated in **Figure 11-21**. The response is determined by evaluating the final multivariable model at each branch. The factors are defined by the options that are available at each stage of the modeling process, in terms of statistical methods, mathematical

CHAPTER 17   EXPERIMENTAL DESIGN AND HYPEROPTIMIZATION

techniques, selection algorithms, and cutoff values. The factor levels represent the set of potential options or values that a given factor can assume. To illustrate, collinearity represents a factor that affects the response, with three distinct levels depending on the statistical method employed to mitigate it. These include the ICRM, the varclus correlation matrix, and the varclus covariance matrix, in addition to others. Similarly, the effect denotes the anticipated outcome on the response. To elaborate, the significance level of the backward elimination (BE) algorithm and the selection criterion together determine the size and hence the complexity of the optimal multivariable model, which in turn influences the performance of the model. It is then reasonable to assume that the expected effect of modifying the significance level of the BE and the selection criterion factors will affect

- The number of variables
- The size of the coefficients of the regressors
- The overall performance of the model

The final experimental element is balance, which pertains to the total number of model's variants and their specific combination of factors and factor's levels, which can also be termed *treatments*. Thus, each variant is the consequence of implementing a distinct set of treatments, with each treatment comprising a particular combination of factors and factor's levels. An example of experimental balance is illustrated in **Figure 17-1**, where the mining flow is divided by the collinearity reduction method into three blocks of seven different combinations of binning options and multivariable selection methods in a symmetrical manner. This particular example is noteworthy because it offers a tangible illustration of how the concept of balance can be defined as *symmetry*, thereby indicating that a symmetrical mining diagram is indicative of a balanced experimental design.

CHAPTER 17   EXPERIMENTAL DESIGN AND HYPEROPTIMIZATION

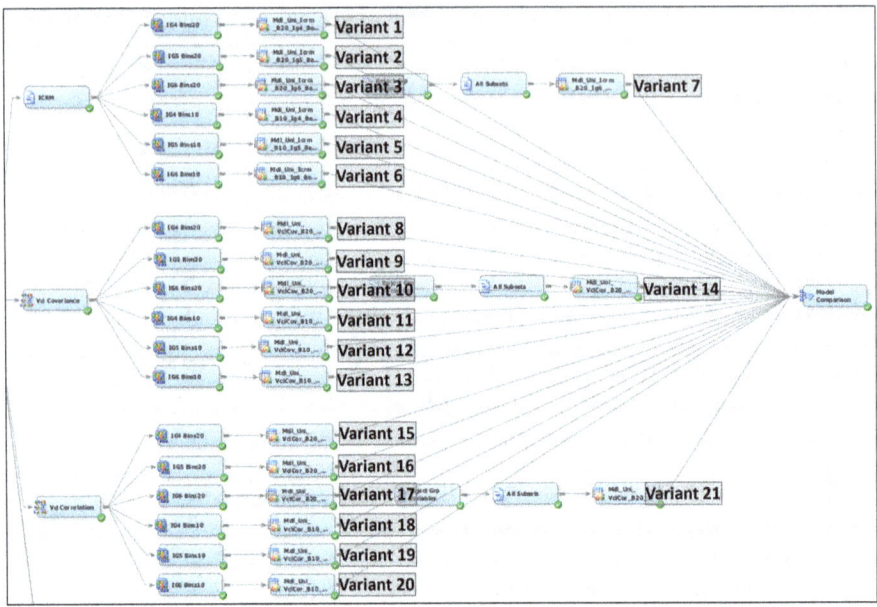

Figure 17-1. Identification of model variants in the mining diagram.

17.1.1 Response (Model's Assessment)

The optimal multivariable model's assessment provides the response, which in turn raises the question of what should be assessed about the model at study. A multitude of factors can be evaluated with regard to the quality of a given model. These include, for instance, performance, fit, and stability, each of which can be further broken down into a number of tests, diagnostics, and metrics. However, based on my professional experience, I have arrived at the conclusion that only three aspects should be considered when evaluating the overall quality of a model:

1. Performance
2. Complexity
3. Overfit

Performance is concerned with the capacity of a model to distinguish between nonevents and events. In the context of credit scoring, this is analogous to the differentiation between *goods* and *bads*. A plethora of metrics can be utilized to evaluate the efficacy of a model. However, a select few have been identified as being particularly robust and prevalent[1] [5], namely, *the area under the ROC curve, the cumulative accuracy ratio* (also known as the *Gini coefficient*), and *the Kolmogorov-Smirnov test statistic*. The term *"complexity"* is used to describe the size of a model in terms of the number of variables it contains. However, it is also used to describe the extent to which each variable contributes to the overall performance of the model. The overfitting phenomenon pertains to the extent to which a model performs similarly on new data as it did on the training data.

17.1.1.1 Area Under the ROC Curve

The receiver operating characteristic (ROC) curve is a statistical curve that plots the true positive rate (TPR) against the false positive rate (FPR). In contrast to the *sensitivity* (TPR) and *specificity* (FPR) metrics, which are evaluated at a single cutoff point, the ROC curve quantifies the ability to discriminate between events and nonevents at multiple cutoffs. This is achieved by integrating the area under the curve of multiple sensitivity points vs. 1-specificity. That is to say, the ratio of true positives to false positives (see **Figure 17-2**). The result is a metric that measures the overall classification accuracy of the model.

---

[1] According to the Basel Committee on Banking Supervision, Working Paper No. 14, asserts that the Gini coefficient and the ROC curve are the most significant performance indices due to their statistical properties.

CHAPTER 17   EXPERIMENTAL DESIGN AND HYPEROPTIMIZATION

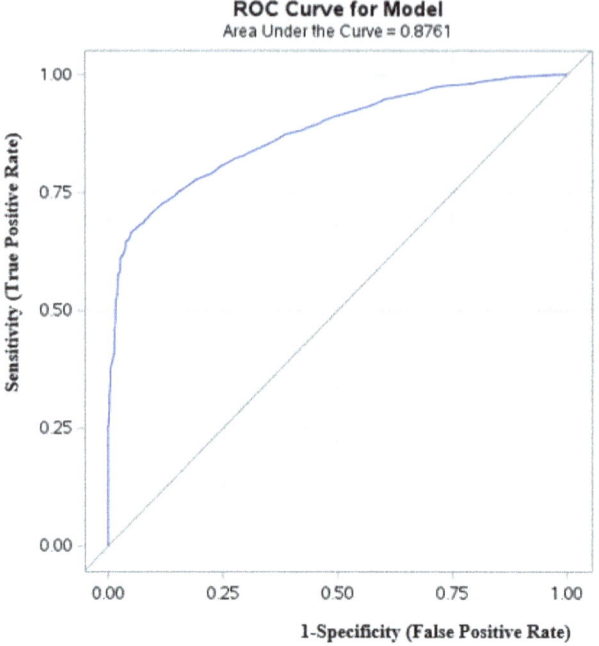

Figure 17-2. Example of the ROC curve plot.

In general, models can be classified in terms of their discriminatory level based on the ROC curve [32] as shown by Table 17-1:

Table 17-1. Suggested model's assessment based on their ROC level.

ROC range	Discrimination level
0.5	Random model
$0.5 < \text{ROC} < 0.7$	Poor or unacceptable
$0.7 \leq \text{ROC} < 0.8$	Fair
$0.8 \leq \text{ROC} < 0.9$	Excellent
$\text{ROC} \geq 0.9$	Outstanding

17.1.1.2 The Cumulative Accuracy Ratio (Gini Coefficient)

The cumulative accuracy ratio, otherwise referred to as the Gini coefficient, is defined as the area located between the *egalitarian* and the *Lorenz curves* (A), divided by the total area under the egalitarian curve (A+B), and is given by

$$Gini = \frac{A}{A+B}$$

Equation 17-1. Gini coefficient described by the Lorenz curve.

Figure 17-3 presents a chart of the cumulative event distribution vs. the cumulative population distribution, ordered by the probability of default (PD) estimated by the final multivariable model. It is noteworthy that the red diagonal line at 45° represents the theoretical perfectly equal distribution of events across the population's distribution, which is also known as the egalitarian curve. The blue line in **Figure 17-3** represents the Lorenz curve, which is the actual distribution of events given by the model. As a result, a perfect model would be obtained when B = 0 and a random model when A = 0. The Gini coefficient is then a measure of the bias in the cumulative distribution of events that favors larger PDs.

CHAPTER 17 EXPERIMENTAL DESIGN AND HYPEROPTIMIZATION

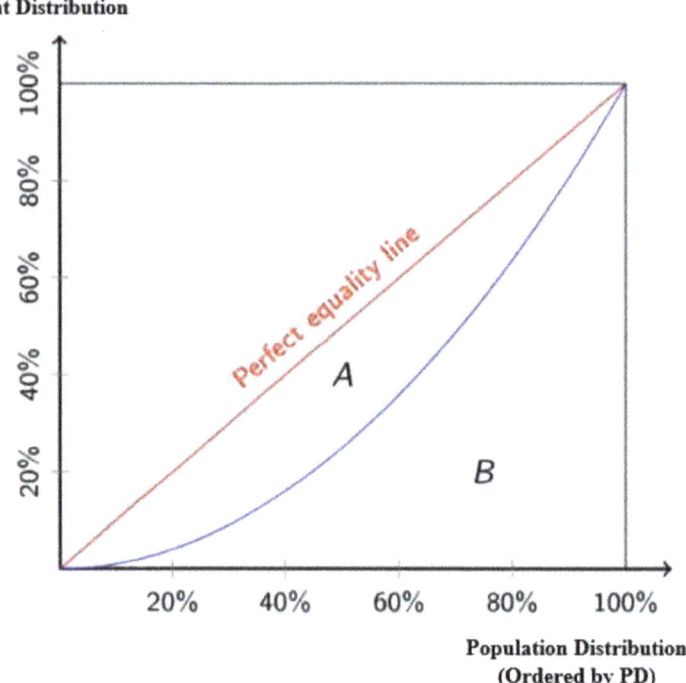

Figure 17-3. Lorenz curve for Gini coefficient calculation.

The following equation is derived from the established mathematical relationship between the ROC curve and the Gini coefficient [18].

$$Gini = 2 \cdot ROC - 1$$

Equation 17-2. Gini coefficient as a function of the ROC curve.

The following general rules can be derived for the classification of models based on the Gini coefficient as shown by Table 17-2:

Table 17-2. *Suggested model's assessment based on their Gini's level.*

Gini range	Discrimination level
0	Random model
0 < Gini < 0.4	Poor or unacceptable
0.4 ≤ Gini < 0.6	Fair
0.6 ≤ Gini < 0.8	Excellent
Gini ≥ 0.8	Outstanding

17.1.1.3 The Kolmogorov-Smirnov Test Statistic

The Kolmogorov-Smirnov (KS) test statistic is a metric that quantifies the maximum distance between the cumulative distributions of nonevents and events, as ordered by the probability of event (PD) derived from the final multivariable model. **Figure 17-4** shows the cumulative distributions of events and nonevents, ordered by PD. The test statistic D is the maximum absolute difference between the two cumulative distribution functions and is given by

$$D = \max_i \left| CD_E(p_i) - CD_{NE}(p_i) \right|$$

Equation 17-3. Kolmogorov-Smirnov test statistic.

Where $CD_E(p_i)$ refers to the cumulative distribution of events as a function of the predicted probability of the *i-th* bin, while the latter, $CD_{NE}(p_i)$, refers to the cumulative distribution of nonevents as a function of the predicted probability of the *i-th* bin.

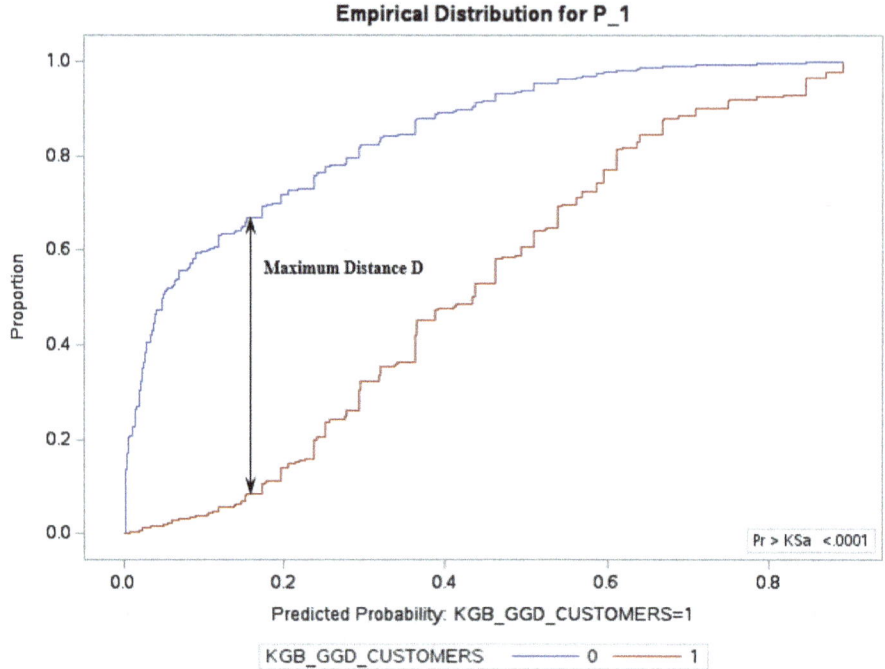

Figure 17-4. *Plot of cumulative distributions for events and nonevents ordered by the predicted probability of default (PD).*

It is important to note that for models with a high discriminatory power, the cumulative distribution function for nonevents exhibits exponential growth at low predicted probabilities. Conversely, the cumulative distribution function for events commences its exponential growth with a delay and progressively accelerates as the nonevent distribution undergoes a gradual decline in speed. In essence, the KS quantifies the disparity in growth rates between nonevent and event distributions. A greater contrast in growth rates corresponds to a higher KS value, indicating a more robust discriminatory power.

The KS can range from 0 to 1.1, yet there appears to be no consensus regarding the discriminatory levels based on the KS value. Nevertheless, based on my empirical observations, a working rule of thumb can be formulated as shown by Table 17-3:

Table 17-3. *Suggested model's assessment based on their KS's level.*

KS range	Discrimination level
0	Random model
0 < KS < 0.30	Poor or unacceptable
0.30 ≤ KS < 0.42	Fair
0.42 ≤ KS < 0.65	Excellent
KS ≥ 0.65	Outstanding

17.1.1.4 Complexity

In accordance with the principle of parsimony, the development of smaller models should be pursued. The number of covariates in the final multivariable model is quantified; however, the total number of covariates does not necessarily provide insight into the quality or individual contribution of each covariate to the overall performance of the model. An effective method for approximating this is the *average performance per variable* (APPV) based on the ROC curve:

$$APPV_{ROC} = \frac{ROC}{df - 1}$$

Equation 17-4. Average ROC performance per variable.

Where *df* refers to the degrees of freedom of the final multivariable model.

Consequently, the complexity comparison between variants is determined by the total number of covariates in the optimal multivariable model obtained by each feasible region and by a measure that determines whether the complexity at hand is justified or not.

17.1.1.5 Overfitting

In Section 12.2, the topic of overfitting, otherwise referred to as overtraining, was discussed. It was determined that the primary cause of overtraining was the lack of statistical representativeness in the training and validation samples. The most efficacious method for rectifying this issue was determined to be the stratification of the sample. It was recommended that three primary stratification dimensions be considered: periodicity, geographic location, and risk level. The implementation of these measures was found to be an effective strategy for mitigating the risk of overfitting the model. However, it is important to acknowledge that complete elimination of overfitting is unattainable. Instead, the objective is to minimize its impact, which gives rise to the question of what constitutes an acceptable level of overfitting.

The resulting discussion will focus on a methodological approach for evaluating overfitting, as delineated in **Table 17-4**. This method is predicated on the analysis of the discrepancy in receiver operating characteristic (ROC) points between the training and the validation partitions. It is imperative to recognize that this guideline is founded on empirical observations. An ROC overfit of less than one is considered excellent, an overfit of between one and two is deemed good, an overfit of between two and three is regarded as acceptable, and an overfit of between three and five may indicate an issue with the automatic binning process.[2] A difference greater than five points indicates a more critical issue that cannot be rectified solely by modifying the binning options. This may indicate that the sampled data is lacking in representativeness.

---

[2] In this case, the recommendation to reduce the overfit is to decrease the maximum number of groups or to increase the grouping constraints.

Table 17-4. *Suggested overfit assessment based on ROC points difference between the training and the validation partitions.*

ROC points difference	Overfit assessment
< 1	Excellent
1–2	Good
2–3	Fair
3–5	Needs correction
> 5	Unacceptable

It is important to note that there is yet another type of overfitting that manifests itself in the distribution of the empirical odds across the scorecard bins. A more detailed examination of this topic will be provided in Chapter 18, during the refinement process of the *multivariable main-effects model*.

17.1.2 Factors and Factor Levels

In order to analyze the contributing and subtracting forces affecting the response, it is first necessary to understand the system at hand (**Figure 1-2**). As previously discussed, the response can be categorized into three distinct components: performance, complexity, and overfitting. According to the balance equation, the forces that positively impact our system are

- Big sample size
- Representativeness of the target variable in the sample
- High predictive power of the input variables
- Independent variation explained by each input variable

CHAPTER 17 EXPERIMENTAL DESIGN AND HYPEROPTIMIZATION

- Linearity between input and output variables
- High discriminatory power and a monotonic response
- Main effects retained after multivariable selection

Likewise, the forces that negatively impact our system are

- Small sample size
- Lack of representativeness of the target variable in the sample
- Irrelevant input variables
- Explained variation shared among the input variables
- Lack of linearity between input and output variables
- Low discriminatory power and lack of monotonicity response
- Loss of main effects after multivariable selection

Therefore, the balance can be simplified into six specific forces:

1. Sample size
2. Statistical representativeness
3. Predictive power
4. Collinearity
5. Variable transformation
6. Multivariable selection

These forces can be regarded as global factors that exert an influence on the system under scrutiny. However, due to the complexity of their structure, it is not possible to exercise complete control over them. Consequently, it is necessary to break them down into their fundamental components in order to ascertain the actual set of specific factors that exert control over our system.

17.1.2.1 Sample Size

The global factors can now be classified into two distinct categories: those that are within the control of the experimenter and those that are not. The size of the sample is contingent upon the availability of data. However, it can be reasonably assumed that an ample sample will be available, as previously mentioned, given that, in the current era, the issue is not the lack of information but rather the overwhelming amount of it.

17.1.2.2 Statistical Representativeness

The issue of statistical representativeness, or the lack thereof in the target variable (or default event within the sample), can be addressed through the implementation of a stratified sampling strategy. As previously discussed in Section 12.2, the representativeness of the target variable is assumed to be fixed and controlled through the partitioning of the initial sample in a stratified manner.

17.1.2.3 Predictive Power

The univariable procedure is employed to remove irrelevant variables, either through data cleaning or via the statistical significance criterion. Given that the data cleaning parameters are fixed (although they can be modified if desired), the significance cutoff value of p-value <0.25 is the sole factor defining variable irrelevancy. It is important to note that the 0.25 cutoff is based on the empirical observation that a lower value is unlikely to identify variables with partial associations in the multivariable model; conversely, a higher value will only promote the inclusion of irrelevant variables. Accordingly, we will consider the presence of input variables with a reasonable predictive power to be covered, and this specific factor can be considered fixed at a p-value <0.25.

17.1.2.4 Collinearity

As previously discussed in Chapter 14, collinearity can be reduced using either the ICRM or the varclus procedure. With regard to the ICRM, the specific factor is the VIF cutoff point. Conversely, within the varclus framework, the global collinearity factor is influenced by multiple specific subfactors, including

- The PCA's matrix type:
  - Correlation
  - Covariance
- The stopping criteria:
  - Maximum clusters
  - Maximum eigenvalue
  - Variation proportion

As demonstrated in Section 14.7, the complexity of the stopping criteria can be reduced to two specific factors: **Maximum Clusters** and **Variation Proportion**. However, it should be noted that both of these factors can take an infinite number of values. **The Maximum Clusters factor** accepts any integer value from 1 to infinity, while **the Variation Proportion factor** accepts any fractional value between 0 and 1. One potential solution to this issue is to fix each specific factor at a time and use different values for the other, with the aim of determining an optimal value for both of them.

However, it has been observed that the varclus procedure frequently selects an excessive number of variables, even when constraints are imposed on **Maximum Clusters** and **Variation Proportion**. This results in the creation of larger, more complex models. Addressing this issue may be achieved by constraining the number of clusters and variation proportion, thereby reducing the total number of variables selected. In my experience,

CHAPTER 17   EXPERIMENTAL DESIGN AND HYPEROPTIMIZATION

I have found that a range of 5 to 7 for the maximum clusters and a value of 70% to 80% for the variation proportion may reduce the number of variables selected to a reasonable number.

With regard to the two possible PCA matrix types, the recommendation is to bifurcate the mining diagram into a correlation and a covariance branch, with each branch assigned to a different subset of models or variants. This process is referred to as *blocking*.

The VIF, on the other hand, can take any value; however, as demonstrated in Section 14.1, empirical observations indicate that a VIF of 3 can reduce the number of conflicting variables while maintaining important interaction effects between covariates in the multivariable model. In light of the aforementioned considerations, the ICRM can be regarded as a fixed specific factor with a VIF value of 3.

As demonstrated in **Figure 17-5**, the global collinearity factor is comprised of its constituent specific factors. It is imperative to acknowledge that the global factor influencing the system is collinearity; nevertheless, collinearity itself is influenced by numerous specific subfactors. The global collinearity factor can be simplified by defining it as a function of the statistical method, which in turn is a function of its corresponding parameters (**Equation 17-5**).

CHAPTER 17   EXPERIMENTAL DESIGN AND HYPEROPTIMIZATION

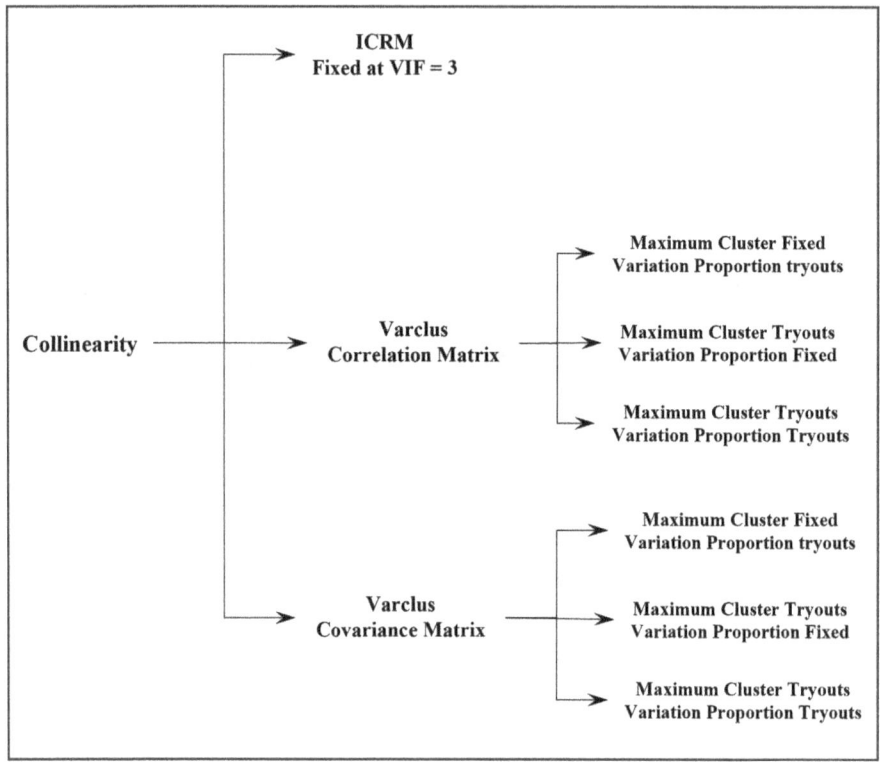

Figure 17-5. The collinearity factor and its subfactors.

$$Collinearity\big(Stat.\ Method\big(Parameters\big)\big)$$

Equation 17-5. Collinearity as a nested function of the statistical method and its corresponding parameters.

17.1.2.5 Variable Transformation

The WoE transformation is the determining factor in the linearity, monotonicity, and discriminatory power of a variable. Consequently, it has been determined that these three factors will be combined into a single global factor, designated as the variable transformation factor.

CHAPTER 17 EXPERIMENTAL DESIGN AND HYPEROPTIMIZATION

The IG node, which is responsible for the WoE calculation, is one of the nodes with the largest number of options and parameters, which are themselves specific factors. In order to reduce the complexity of the IG node, it is recommended that the following options or factors be fixed:

- Interval Variable Binning Options → **Binning Method** = Quantile
- Grouping Options → **Interval Grouping Method** = Constrained Optimal
- Grouping Options → Tree-Based Grouping Options → **Criterion** = Entropy
- Grouping Options → Tree-Based Grouping Options → **Missing Values** = Separate Branch if any
- Grouping Options → Tree-Based Grouping Options → **Minimum Categorical size** = 1000 (assuming a sample of 100,000 observations or greater)
- Grouping Options → Constrained Optimal Options → **Apply Minimum Number of Groups** = Yes
- Grouping Options → Constrained Optimal Options → **Minimum Number of Groups** = 2
- Grouping Options → Constrained Optimal Options → **Apply Minimum Non-Event** = Yes
- Grouping Options → Constrained Optimal Options → **Minimum Non-Event Count** = 500 (assuming a sample of 100,000 observations or greater)
- Grouping Options → Constrained Optimal Options → **Apply Minimum Event** = Yes
- Grouping Options → Constrained Optimal Options → **Minimum Event Count** = 5 (assuming a minimum default level per bin of less than 1%)

CHAPTER 17   EXPERIMENTAL DESIGN AND HYPEROPTIMIZATION

- Grouping Options → Constrained Optimal Options → **Apply Minimum Total Count** = No

- Grouping Options → Constrained Optimal Options → **Apply Minimum WOE Difference** = Yes

- Grouping Options → Constrained Optimal Options → **Minimum WOE Difference** = 0.015

- Score → **Variable Selection Method** = Gini Statistic

- Score → **Gini Cutoff** = 5

The following options are the specific factors and their trial factor levels:

- Interval Variable Binning Options → **Number of Bins** = 10/20

- Grouping Options → **Maximum Number of Groups** = 4/5/6

It is important to note that the fixed option values and the tryout values of the parameters are founded on empirical observations, assuming training samples of a minimum of 100,000 observations and a default level of 1% or greater. It must therefore be recognized that the results obtained depend on the particular characteristics of the dataset under consideration. Nevertheless, these parameters may be utilized as a starting point for the fine-tuning of the automatic binning process.

17.1.2.6 Multivariable Selection

The final global factor affecting the model-building system is the multivariable selection method, which in turn depends on the specific options of the selection method, the significance cutoff point, and the optimization criterion. As previously discussed in Section 16.2.1.1, the BE methods are preferred over the FS methods, because the FS methods are prone to stop prematurely in the presence of high collinearity without

CHAPTER 17 EXPERIMENTAL DESIGN AND HYPEROPTIMIZATION

considering better models that include members of a cluster of correlated covariates.

As indicated in Section 16.2.1.2, for datasets exceeding 100,000 observations, it is advised to adjust the backward elimination only (BE-Only) stay significance level of the EM to a p-value of less than 0.001. Consequently, the number of multivariable selection-specific factors can be reduced by setting the preferred stepwise method to BE-Only and the stay significance level to a p-value less than 0.001. Similarly, the selection criterion can be set to Schwarz's Bayesian criterion (SBC), given that simpler models are always preferable to complex ones.

Conversely, the all-subsets method cannot be dismissed outright, as there is always a possibility that it will identify a better-fitting model that is not considered by the BE (Royston and Sauerbrei, 2009 [58]). This is because the all-subsets method is the only method that examines all possible models. Accordingly, the multivariable selection-specific factor will be divided into two levels, corresponding to the two predefined selection methods: the BE-Only (with the SBC criterion), with a fixed stay significance level of p-value <0.001, and the all-subsets method.

17.1.3 Final Specific Factors and Total Number of Variants

In view of the aforementioned considerations, a list of the main factors having a direct impact on the system can be identified as follows:

1. The collinearity reduction method
   a. Varclus correlation matrix
      i. The maximum cluster, discrete variable with a minimum value of 1
      ii. The variation proportion, interval variable with values within 0 and 1

CHAPTER 17   EXPERIMENTAL DESIGN AND HYPEROPTIMIZATION

    b. Varclus covariance matrix

        i. The maximum cluster, discrete variable with a minimum value of 1

        ii. The variation proportion, interval variable with values within 0 and 1

    c. The ICRM, fixed at VIF = 3

2. The number of bins, preset to 10 and 20 bins
3. The maximum number of groups, preset to 4, 5, and 6
4. The multivariable selection method, preset to

    a. BE-Only, fixed at p-value <0.001, using the SBC criterion

    b. All-subsets

Assuming that an optimal value for the **Maximum Cluster** and the **Variation Proportion** specific factors can be set and that this optimal value is the same for both the correlation and the covariance matrix, then, in theory, a total of 36 *variants* could be possible. **Table 17-5** presents a list of the 36 potential variants that may be generated through the combination of the specific factors.

CHAPTER 17 EXPERIMENTAL DESIGN AND HYPEROPTIMIZATION

Table 17-5. *List of all possible treatments/variants resulting from the combination of root factors.*

Treatment/Variant	Collinearity Reduction Method	Number of Bins	Maximum Number of Groups	Multivariable Selection Method
1	Varclus Correlation Matrix	10	4	BE-Only (SBC), p-value < 0.001
2	Varclus Correlation Matrix	10	5	BE-Only (SBC), p-value < 0.001
3	Varclus Correlation Matrix	10	6	BE-Only (SBC), p-value < 0.001
4	Varclus Correlation Matrix	20	4	BE-Only (SBC), p-value < 0.001
5	Varclus Correlation Matrix	20	5	BE-Only (SBC), p-value < 0.001
6	Varclus Correlation Matrix	20	6	BE-Only (SBC), p-value < 0.001
7	Varclus Correlation Matrix	10	4	All-Subsets
8	Varclus Correlation Matrix	10	5	All-Subsets
9	Varclus Correlation Matrix	10	6	All-Subsets
10	Varclus Correlation Matrix	20	4	All-Subsets
11	Varclus Correlation Matrix	20	5	All-Subsets
12	Varclus Correlation Matrix	20	6	All-Subsets
13	Varclus Covariance Matrix	10	4	BE-Only (SBC), p-value < 0.001
14	Varclus Covariance Matrix	10	5	BE-Only (SBC), p-value < 0.001
15	Varclus Covariance Matrix	10	6	BE-Only (SBC), p-value < 0.001
16	Varclus Covariance Matrix	20	4	BE-Only (SBC), p-value < 0.001
17	Varclus Covariance Matrix	20	5	BE-Only (SBC), p-value < 0.001
18	Varclus Covariance Matrix	20	6	BE-Only (SBC), p-value < 0.001
19	Varclus Covariance Matrix	10	4	All-Subsets
20	Varclus Covariance Matrix	10	5	All-Subsets
21	Varclus Covariance Matrix	10	6	All-Subsets
22	Varclus Covariance Matrix	20	4	All-Subsets
23	Varclus Covariance Matrix	20	5	All-Subsets
24	Varclus Covariance Matrix	20	6	All-Subsets
25	ICRM, VIF = 3	10	4	BE-Only (SBC), p-value < 0.001
26	ICRM, VIF = 3	10	5	BE-Only (SBC), p-value < 0.001
27	ICRM, VIF = 3	10	6	BE-Only (SBC), p-value < 0.001
28	ICRM, VIF = 3	20	4	BE-Only (SBC), p-value < 0.001
29	ICRM, VIF = 3	20	5	BE-Only (SBC), p-value < 0.001
30	ICRM, VIF = 3	20	6	BE-Only (SBC), p-value < 0.001
31	ICRM, VIF = 3	10	4	All-Subsets
32	ICRM, VIF = 3	10	5	All-Subsets
33	ICRM, VIF = 3	10	6	All-Subsets
34	ICRM, VIF = 3	20	4	All-Subsets
35	ICRM, VIF = 3	20	5	All-Subsets
36	ICRM, VIF = 3	20	6	All-Subsets

CHAPTER 17   EXPERIMENTAL DESIGN AND HYPEROPTIMIZATION

It is noteworthy that the collinearity reduction method (varclus correlation, varclus covariance, or ICRM) can naturally divide the 36 variants into three blocks of 12 variants each. It is important to note that each of these blocks can be further subdivided into two blocks of six variants, a process that is facilitated by the multivariable selection method that has been employed. The construction of an EM diagram can then be based on this experimental design. However, it is recommended that the number of variants be reduced from 36 to 21. This can be achieved by retaining only the all-subsets variants that are more likely to yield a more complex model from each of the collinearity blocks.

Table 17-6 presents a simplified list of 21 variants, comprising only those all-subsets variants that are more likely to yield a more complex model. It is noteworthy that the likelihood of a more complex model is indicated by the fact that the all-subsets variants possess the highest number of **Bins** (20) and the highest number of **Maximum Groups** (6).

CHAPTER 17   EXPERIMENTAL DESIGN AND HYPEROPTIMIZATION

Table 17-6. *Simplified list of treatments/variants resulting from the combination of root factors.*

Treatment/Variant	Collinearity Reduction Method	Number of Bins	Maximum Number of Groups	Multivariable Selection Method
1	Varclus Correlation Matrix	10	4	BE-Only (SBC), p-value < 0.001
2	Varclus Correlation Matrix	10	5	BE-Only (SBC), p-value < 0.001
3	Varclus Correlation Matrix	10	6	BE-Only (SBC), p-value < 0.001
4	Varclus Correlation Matrix	20	4	BE-Only (SBC), p-value < 0.001
5	Varclus Correlation Matrix	20	5	BE-Only (SBC), p-value < 0.001
6	Varclus Correlation Matrix	20	6	BE-Only (SBC), p-value < 0.001
7	**Varclus Correlation Matrix**	**20**	**6**	**All-Subsets**
8	Varclus Covariance Matrix	10	4	BE-Only (SBC), p-value < 0.001
9	Varclus Covariance Matrix	10	5	BE-Only (SBC), p-value < 0.001
10	Varclus Covariance Matrix	10	6	BE-Only (SBC), p-value < 0.001
11	Varclus Covariance Matrix	20	4	BE-Only (SBC), p-value < 0.001
12	Varclus Covariance Matrix	20	5	BE-Only (SBC), p-value < 0.001
13	Varclus Covariance Matrix	20	6	BE-Only (SBC), p-value < 0.001
14	**Varclus Covariance Matrix**	**20**	**6**	**All-Subsets**
15	ICRM, VIF = 3	10	4	BE-Only (SBC), p-value < 0.001
16	ICRM, VIF = 3	10	5	BE-Only (SBC), p-value < 0.001
17	ICRM, VIF = 3	10	6	BE-Only (SBC), p-value < 0.001
18	ICRM, VIF = 3	20	4	BE-Only (SBC), p-value < 0.001
19	ICRM, VIF = 3	20	5	BE-Only (SBC), p-value < 0.001
20	ICRM, VIF = 3	20	6	BE-Only (SBC), p-value < 0.001
21	**ICRM, VIF = 3**	**20**	**6**	**All-Subsets**

The primary rationales for reducing the number of all-subsets variants are based on my personal experience, where I have found that the subset solutions of each variant exhibit minimal differences, with a notable overlap in the variables identified as optimal subsets. This outcome is unsurprising in light of the fact that strong effects (p-value <0.001) are almost invariably incorporated within the model (Royston and Sauerbrei [58]) and that the input variables entering the multivariable selection process are themselves a subset of preselected variables (Section 16.1). Therefore, a large overlap of variables among variants is not only unsurprising but expected.

It is evident that a considerable degree of overlap will also be observed among the BE-Only variants. This prompts the question of why the BE-Only should be favored over the all-subsets. The answer to this question can be attributed to the principle of parsimony. This is due to the fact that the all-subsets models will invariably yield larger and more intricate models, given that they assess all potential models. Conversely, the BE-Only, with a significant level of p-value <0.001 and an SBC selection criterion, is highly probable to produce the smallest and less complex models.

It is important to note that the response of the experiment does not consist of a single optimal multivariable model, but rather, it is comprised of all optimal variants. Consequently, our objective is not to obtain the simplest models or to generate the most complex ones. Instead, our aim is to identify the most diverse subset of variables from each variant, thereby enabling us to select a larger number of variables, regardless of their level of complexity.

17.1.4 Considerations and the Curse of Statistics

It is a well-known fact among those versed in statistical analysis that the inherent difficulty of this field lies in the inability to ascertain the significance of an effect, parameter, cutoff value, or other such variable until it is tested. This underscores the crucial role of experimental design in this process. It should be acknowledged that the experimental design presented earlier does not represent the sole possibility for configuration; there are indeed numerous viable alternatives and variations that could have been implemented. However, it is crucial to recognize that the incorporation of an increased number of factors and factor levels will invariably result in a more complex experimental design. The question that naturally follows is that of identifying the optimal number of factors and levels to include. Nevertheless, it is not possible to provide definitive answers to these questions, but there are a number of guidelines that can assist in the process of experimental design.

It is imperative to acknowledge that all statistical techniques are data-dependent, implying that the efficacy of a particular technique may vary depending on the specific problem being addressed. Nevertheless, this does not imply that a completely new approach must be employed each time a model is constructed. Indeed, the unified model-building process presented in this book is the culmination of over two decades of applied theory and empirical observations. Consequently, it encompasses the most critical elements inherent to the development of predictive models. It is therefore strongly recommended that the experimental design presented here be used as a starting point, from which variants can be derived or new features added. However, it is equally essential that the underlying analytical sequence of univariable, collinearity, WoE transformation, and multivariable selection be maintained as the foundation of the experimental design.

17.1.5 Alternative Factors and Factor Levels

The varclus can be regarded as a complex procedure due to the substantial number of available parameters and the presence of both discrete and interval levels. The optimal methodology for addressing the varclus variants is to initially decompose them into a principal component correlation matrix and a principal component covariance matrix, as previously recommended in the experimental design. The recommendation to utilize both the correlation and the covariance matrices is based on the premise that both matrices possess inherent characteristics that may confer a degree of favoritism toward certain variables over others, thereby allowing for the delineation of two distinct subsets of variables. It is important to note that it is rare, if not impossible, for two different statistical techniques to result in completely different subsets of variables. The objective, therefore, is to identify a small but significant number of variables that distinguish one subset of variables from the other.

For instance, in the aforementioned example, the varclus blocks were not further subdivided into different variants of the **Maximum Clusters** or the **Variation Proportion**. As previously stated in Section 17.1.2.4, empirical observations have indicated that a **Maximum Cluster** of 5–7 and a **Variation Proportion** of 70–80% have proven effective in controlling the number of output variables. It is therefore not advisable to attempt every possible combination of these two parameters; instead, it would be more fruitful to fix one of them and then vary the other. As an illustration, one may set the variation proportion to 70% and attempt a maximum cluster of 5 and 7. It is important to note that the addition of two more variants at the varclus level will result in a significant increase in the total number of variants. In the extended version, the total number of variants will increase from 36 to 60, while in the all-subsets simplified version, it will increase from 21 to 45.

One might be inclined to augment the number of levels in the **Maximum Number of Groups**, given that both the IV and the Gini index are functions of the total number of groups (see **Figure 15-18**), as previously discussed in Section 15.4.1. However, should this course of action be pursued, it would be prudent to consider the potential for increased risk of overfitting the model and the possibility of obtaining nonsensical solutions (**Figure 15-19**). Nevertheless, it is possible to increase the difference between levels without risking overfitting the model by considering the **Maximum Number of Groups** to be 4, 5, 7, or 9. In order to circumvent the risk of overfitting while preserving a substantial **Maximum Number of Groups**, such as 12, it is advisable to employ the constrained optimal options instead of the optimal criterion. However, it is imperative to acknowledge that a substantial number of variables may be rejected if the constraints are not met. Furthermore, it may be tempting to add more levels to the number of bins, given that the number of bins increases the resolution, by including a larger range of possible cutoff points for group formation. However, it is crucial to note that this approach may also lead to a more complex optimization process, potentially resulting in a less stable solution. In any case, should an increase in the

CHAPTER 17 EXPERIMENTAL DESIGN AND HYPEROPTIMIZATION

number of levels be desired, one may consider using 10, 20, 30, or 40 bins, or alternatively, 10, 20, and immediately proceed to 40 bins in order to circumvent the addition of another block of variants. Moreover, alternative values may be considered for the remaining **Constrained Optimal Options**. However, it is recommended to maintain the values utilized in Section 17.1.2.5, with the exception of the **Minimum Non-Event Count** and the **Event Count**, which should be adjusted in accordance with the size of the dataset and the average event proportion.

An additional factor that can be incorporated into the experimental design is the **Minimum WOE Difference**. The default value of the **Minimum WOE Difference** is 0.01; however, alternative values may be considered, such as 0.01, 0.015, 0.02, and 0.025. The **Minimum WOE Difference** widens the search for an optimal cutoff point by increasing the difference in event proportions between each risk group, thereby increasing the likelihood of identifying truly distinct subpopulations in terms of their risk level. Prior to the introduction of additional variants for the **Minimum WOE Difference**, two considerations must be taken into account. Firstly, it is important to note that the WoE is expressed in logarithmic units; therefore, caution should be exercised to ensure that trial values are not extended beyond their optimal range. Secondly, the larger the WoE difference, the more variables will be rejected by the IG algorithm, potentially to the point of rejecting them all. It is therefore important to note that constraining the **Minimum WOE Difference** to infeasible values may result in an error being displayed in the scorecard node due to missing inputs. It is crucial to acknowledge that the maximum feasible cutoff point for the **Minimum WOE Difference** is dependent upon the specific characteristics of the dataset in question. In some instances, it may be possible to set a relatively high value for the **Minimum WOE Difference** without significantly reducing the number of output variables, thereby potentially facilitating the identification of variables with the greatest discriminatory power, thus expediting the selection and refinement process. In other instances, however, it may not be feasible to increase the **Minimum WOE Difference** without losing the majority or entirety of the variables. It may therefore be advisable to subject the **Minimum WOE Difference** to a stress test in a single IG node, prior to determining the number of variants or whether it is feasible to add any variants above the **Minimum WOE Difference**.

The essential concept to be derived from this section is that the development of an experimental design does not have to be conducted in a blind manner. Instead, the design will be the result of multiple iterations of trial and error, which will assist in the refinement of the design until the number of factors and factor levels included are definitively determined. Furthermore, even after executing the EM diagram and obtaining the results of all the models comprising the multi-model subset, it may still be advisable to fine-tune the design. This refinement process is a vital step in the overall process, as the experimental design should not be regarded as a fixed and predetermined procedure. Instead, it should be viewed as a dynamic process, subject to change and open to evolution and improvement at any time.

17.1.6 Applied Example

Figure 17-6 illustrates an exemplar of an experimental design developed as a data mining flow. It is important to note that the variable selection process begins with the univariable analysis procedure. Thereafter, the flow is divided into three distinct blocks according to the collinearity reduction method employed. These correspond to

1. Collinearity analysis (ICRM)
2. Vcl (varclus) covariance
3. Vcl (varclus) correlation

CHAPTER 17 EXPERIMENTAL DESIGN AND HYPEROPTIMIZATION

Next, each collinearity method is divided once again into six different blocks in the following variants:

1. Bins 10, IG4
2. Bins 10, IG5
3. Bins 10, IG6
4. Bins 20, IG4
5. Bins 20, IG5
6. Bins 20, IG6

In this context, the term *"Bins"* refers to the **Number of Bins**, while *"IG"* refers to the **Maximum Number of Groups**. The variants of the IG nodes are directly connected to the scorecard and all-subsets nodes.[3] The scorecard nodes execute the BE-Only multivariable selection process, incorporating the SBC criterion, while the all-subsets nodes implement the branch and bound algorithm, which explores the 2^k possible models. It is important to note that the multivariable selection step does not branch the mining flow in the conventional manner. Instead, this specific experimental design employs the strategy delineated in **Table 17-6**, which permits the examination of 21 distinct variants, or treatments. The final results (response) are summarized in the model comparison node.

The experimental design details are as follows:

- Univariable procedure:
  - (FP) Select variables with a p-value <0.25
  - (FP) Proportion of null values <0.8

---

[3] Remember that the all-subsets procedure is comprised of two steps, the reject GRP variables, coming from the IG nodes, and the all-subsets node, where the actual selection routine takes place.

CHAPTER 17   EXPERIMENTAL DESIGN AND HYPEROPTIMIZATION

- ○ (FP) For interval variables, proportion of zeros <0.8
- ○ (FP) For nominal variables, total number of distinct categories >1
- Collinearity analysis, or ICRM:
  - (FP) Execute until all covariates have a VIF ≤3
- Varclus:
  - **(FL) Clustering Source = Correlation/Covariance**
  - (DO) Keep Hierarchies = Yes
  - (DO) Include Class Variables = No
  - (DO)Two Stage Clustering = Auto
  - (TE)Maximum Clusters = 7
  - (DO) Maximum Eigenvalue = None
  - (TE) Variation Proportion = 0.8
  - (FP) Suppress Sampling Warning = Yes
  - (FP) Variable Selection = Best Variables
- Interactive Grouping procedure:
  - (DO) Binning Method = Quantile
  - **(FL) Number of Bins = 10/20**
  - (FP) Interval Grouping Method = Constrained Optimal
  - (FP) Ordinal Grouping Method = Constrained Optimal
  - Tree-Based Grouping Options:
    - ○ (DO) Criterion = Entropy
    - ○ (DO) Missing Values = Separate Branch if Any
    - ○ (TE) Minimum Categorical Size = 1000

CHAPTER 17 EXPERIMENTAL DESIGN AND HYPEROPTIMIZATION

- Constrained Optimal Options:
    - (DO) Apply WoE Monotonicity = Yes
    - (DO) Apply Minimum Groups = Yes
    - (DO) Minimum number of Groups = 2
    - (FP) Apply Minimum Non-Event = Yes
    - (TE) Minimum Non-Event Count = 1000
    - (FP) Apply Minimum Event = Yes
    - (TE) Minimum Event Count = 10
    - (DO) Apply Minimum Total Count = No
    - (FP) Apply Minimum WoE Difference = Yes
    - (TE) Minimum WoE Difference = 0.02
- **Maximum Number of Groups = 4/5/6 (FL)**
- (DO) Significant Digits = 2
- (DO) Apply Restrictions = Yes
- (DO) Type = Percent
- (DO) Percent = 5
- (DO) Adjust WoE = Yes
- (DO) Adjustment Factor = 0.5
- (DO) Group Level = Ordinal
- (FP) Variable Selection Method = Gini Statistic
- (FP) Gini Cutoff = 5
- **Multivariable selection method:**
    - **(FL) Scorecard procedure (regression):**
        - (FP) Selection Model = Backward

- CHAPTER 17 EXPERIMENTAL DESIGN AND HYPEROPTIMIZATION

    - (FP) Criterion = Schwarz Bayesian Criterion (SBC)
    - (FP) Use Selection Defaults = No
    - (DO) Entry Significance Level = 0.05
    - (TE) Stay Significance Level = 0.0001
  - **(FL) All-subsets**
    - (FP) EM_PROPERTY_modelsperlevel = 3
    - (FP) EM_PROPERTY_mininputs = 1
    - (FP) EM_PROPERTY_maxinputs = Max

Where:

- **DO:** Refers to a **default option** of EM nodes, no modification has been made.

- **FL:** Denotes a **factor level** wherein disparate values or options have been employed, thereby yielding disparate variants or treatments.

- **FP:** Denotes a **fixed parameter**. Fixed parameters differ from default options in that their value or option is not derived from the original default value; rather, it is predetermined based on theoretical foundations or previous empirical observations.

- **TE:** Refers to values or options that have been determined through a process of **trial and error**. In contrast to factor levels, trial-and-error values did not result in the generation of additional variants, as their optimal values had been determined prior to the finalization of the experimental design.

CHAPTER 17   EXPERIMENTAL DESIGN AND HYPEROPTIMIZATION

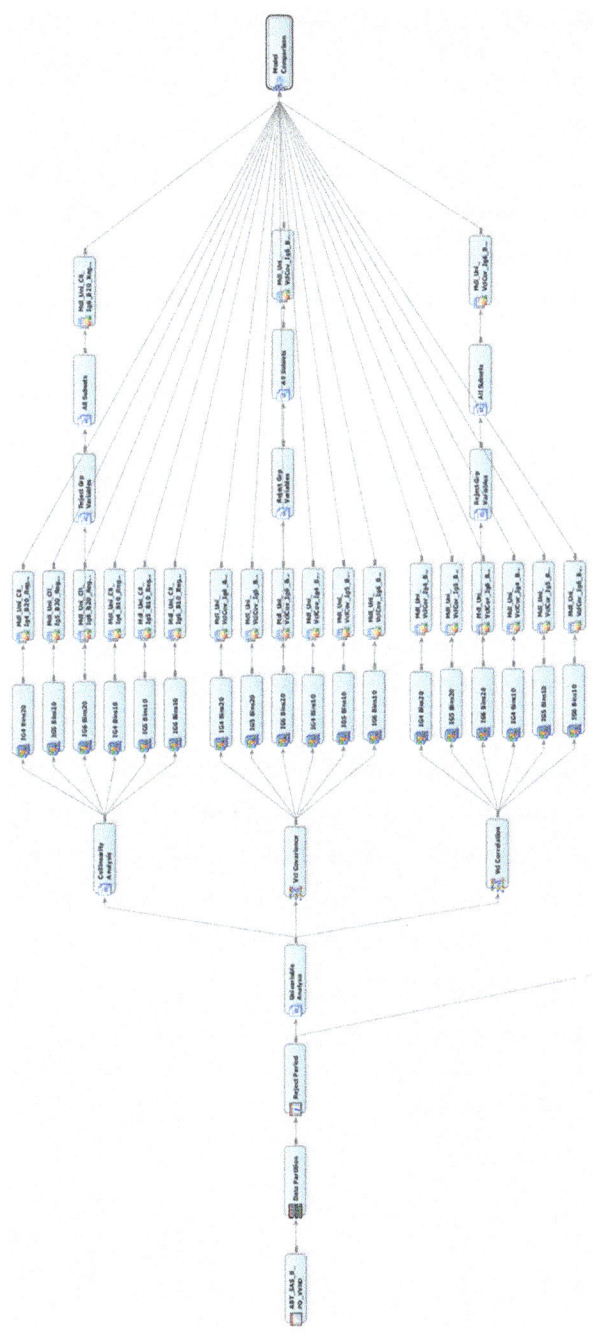

Figure 17-6. Example of an EM diagram arranged as an experimental design.

CHAPTER 17 EXPERIMENTAL DESIGN AND HYPEROPTIMIZATION

17.1.7 Naming Your Variants

Upon the addition of a new node to the mining diagram, EM automatically assigns a name. However, this suggested name is merely the original node's name with a sequential number, which is inadequate for the purpose of properly identifying each element in the diagram in an orderly and logical manner. Consequently, it is imperative that each node is assigned a descriptive yet concise designation that facilitates the intuitive identification of both the element in question and the function it performs.

As illustrated in **Figure 17-6**, an orderly mining diagram representing our experimental design enables the identification of each element by name and logical and sequential structure. However, as we proceed to the final multivariable models at each branch, the precise identification of each model becomes increasingly challenging. This is primarily due to the fact that each model represents a specific treatment or variant with a distinctive combination of factors. It is therefore prudent to have a nomenclature system that can readily distinguish the unique set of factors that gave rise to each model.

Figure 17-7 illustrates a proposed nomenclature structure that identifies each factor that contributed to the creation of the final model. It is important to note that the nomenclature follows the same sequence of events as the modeling process, from the univariable analysis to the multivariable selection method, from left to right. The final model is comprised of the following elements:

- Model:
  - Mdl (only option)
- Univariable Analysis
  - Uni (only option)

CHAPTER 17   EXPERIMENTAL DESIGN AND HYPEROPTIMIZATION

- Collinearity Reduction Method:
  - ICRM: Iterative collinearity reduction method
  - VclCor: Varclus with correlation matrix
  - VclCov: Varclus with covariance matrix
- Number of Bins:
  - B##
- Maximum Number of Groups:
  - IG#
- Multivariable Selection Method:
  - BE_SBC: Backward elimination with Schwarz's Bayesian criterion
  - AllSub_Mcq: All-subsets with Mallows' C_q

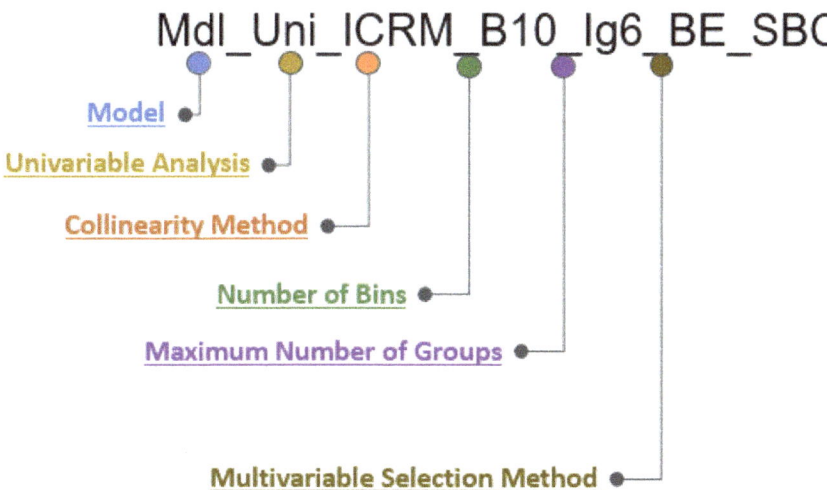

Figure 17-7. Logical structure of the proposed nomenclature for identifying and distinguishing each response model in the multimodel subset.

It should be noted that the proposed nomenclature may be modified as necessary. For instance, the code *"Uni"* may be removed since all models in this experimental design possess the same feature. The rationale behind its inclusion in the nomenclature was to ensure comprehensive representation.

17.2 Hyperoptimization

The specific sets of parameters or factors defined during the experimental design phase, in turn, define multiple feasible regions within which a local optimum model should occur. Subsequently, each modeling process operates as an optimization search function, gradually converging to a local optimum multivariable model.

The concept of hyperoptimization is illustrated in **Figure 17-8**. For the sake of argument, we may assume that the shaded areas represent optimal models that occur within three different feasible regions. The significant variables for variant 1 are v_1, v_2, v_3, v_{11}, v_{12}, v_9, v_{10}, v_8, v_6, and v_4. Similarly, for variant 2, they are v_{20}, v_{17}, v_8, v_{19}, v_{11}, v_{12}, v_9, v_{10}, v_8, v_6, v_4, v_{13}, and v_{16}. And finally, for variant 3, the significant variables are v_{14}, v_{15}, v_7, v_5, v_9, v_8, v_6, v_{13}, v_{16}, and v_4.

From this fictional example, it is important to note that despite each feasible region being defined by a distinct set of parameters, such as different collinearity reduction methods, different numbers of bins, and/or different cutoff points, these regions are not far enough apart to contain exclusively different sets of significant variables.

CHAPTER 17   EXPERIMENTAL DESIGN AND HYPEROPTIMIZATION

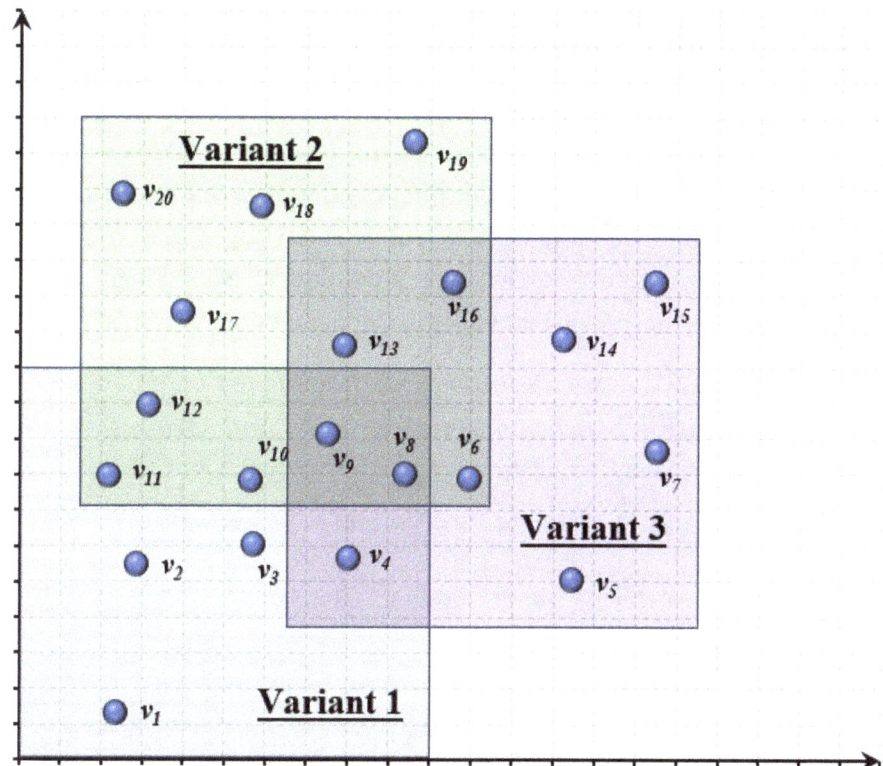

Figure 17-8. *Graphical representation of a hyperoptimization problem.*

This example then helps us to understand that the hyperoptimization process is not intended to produce *exclusive models*, but rather, *equivalent models*. Moreover, it underscores the absence of a perfect model, instead highlighting the existence of a collection of models, the majority of which possess the potential to emerge as *the champion model*.

The term *"champion model"* is therefore a misleading concept that assumes the existence of a single *"champion model,"* when in fact (as will be demonstrated shortly) there is a population of optimal models that are virtually indistinguishable from a statistical performance standpoint. The hyperoptimization process is therefore not intended to assist in the selection of a single *"champion model,"* but rather to facilitate the identification of a more expansive set of optimal variables that would not have been selected in the absence of the use of multiple feasible regions.

As an illustration, **Table 17-7** presents the association matrix corresponding to the feasible regions illustrated in **Figure 17-8**. It is noteworthy that had the three variants not been employed, the 20 predictor variables would not have been identified, all of which are extremely significant (p-value <0.0001) in their respective variants.

Table 17-7. Association matrix showing the relationship between optimal variables and variants.

Variables	Variant$_1$	Variant$_2$	Variant$_3$
v_1	X		
v_2	X		
v_3	X		
v_4	X		X
v_5			X
v_6		X	X
v_7			X
v_8	X	X	X
v_9	X	X	X
v_{10}	X	X	
v_{11}	X	X	
v_{12}	X	X	
v_{13}		X	X
v_{14}			X
v_{15}			X
v_{16}		X	X
v_{17}		X	
v_{18}		X	
v_{19}		X	
v_{20}		X	

We will now examine an applied hyperoptimization example, wherein multiple variants are assessed in terms of their performance, complexity, and degree of overfit.

17.2.1 Assessing the Hyperresponse

We will now proceed to examine the response, or rather, the hyperresponse of our hyperoptimization process, that is comprised of our set of variant models. To this end, two cases will be presented: case I, will present the results of 21 optimal credit card multivariable models, while case II, will present the results of 21 optimal retail loan multivariable models.

17.2.1.1 Case of Study I: Credit Card PD Model

Table 17-8 presents the assessment results of the 21 optimal multivariable models that were generated through the application of our experimental design.[4] It should be noted that the three performance statistics are presented in pairs, with one representing the validation set and the other the training set. The ROC curve is displayed on the left, the Gini coefficient is shown in the middle, and the KS is presented on the right. The discrepancy between the validation and training partitions is presented immediately following each pair of results. The degrees of freedom (*df*) and the mean performance per variable are located at the far right of the table.

---

[4] The performance statistics and the degrees of freedom presented in **Table 17-8** were generated and extracted from the model comparison node. **Figure 17-9** shows the model comparison's options used.

CHAPTER 17 EXPERIMENTAL DESIGN AND HYPEROPTIMIZATION

Table 17-8. *Assessment results from the 21 optimal multivariable models for a credit card PD model.*

No.	Model Description	Valid ROC	Train ROC	Diff ROC	Valid Gini	Train Gini	Diff Gini	Valid KS	Train KS	Diff KS	df	APPV$_{roc}$
1	Mdl_Uni_ICRM_B20_Ig6_AllSub_MCq	**0.816**	**0.818**	**0.002**	**0.631**	**0.637**	**0.006**	**0.478**	**0.478**	**0**	**16**	**0.0510**
2	Mdl_Uni_VclCov_B20_Ig6_AllSub_MCq	0.816	0.82	0.004	0.632	0.64	0.008	0.488	0.479	-0.009	18	0.0453
3	Mdl_Uni_VclCov_B20_Ig6_BE_SBC	0.816	0.82	0.004	0.631	0.64	0.009	0.486	0.475	-0.011	15	0.0544
4	Mdl_Uni_VclCov_B20_Ig6_AllSub_MCq	0.815	0.819	0.004	0.63	0.638	0.008	0.487	0.479	-0.008	14	0.0582
5	**Mdl_Uni_ICRM_B10_Ig6_BE_SBC**	**0.814**	**0.816**	**0.002**	**0.628**	**0.633**	**0.005**	**0.477**	**0.479**	**0.002**	**12**	**0.0678**
6	Mdl_Uni_ICRM_B20_Ig6_BE_SBC	0.814	0.817	0.003	0.628	0.634	0.006	0.479	0.476	-0.003	13	0.0626
7	Mdl_Uni_VclCov_B10_Ig6_BE_SBC	0.814	0.82	0.006	0.629	0.639	0.01	0.487	0.475	-0.012	15	0.0543
8	**Mdl_Uni_VclCor_B20_Ig6_BE_SBC**	**0.814**	**0.818**	**0.004**	**0.628**	**0.637**	**0.009**	**0.484**	**0.475**	**-0.009**	**12**	**0.0678**
9	**Mdl_Uni_VclCor_B10_Ig6_BE_SBC**	**0.814**	**0.818**	**0.004**	**0.628**	**0.637**	**0.009**	**0.484**	**0.476**	**-0.008**	**12**	**0.0678**
10	Mdl_Uni_VclCov_B20_Ig5_BE_SBC	0.814	0.818	0.004	0.628	0.635	0.007	0.48	0.477	-0.003	13	0.0626
11	Mdl_Uni_VclCov_B10_Ig5_BE_SBC	0.814	0.817	0.003	0.627	0.635	0.008	0.481	0.476	-0.005	13	0.0626
12	Mdl_Uni_ICRM_B10_Ig5_BE_SBC	0.812	0.815	0.003	0.624	0.629	0.005	0.472	0.479	0.007	12	0.0677
13	Mdl_Uni_ICRM_B20_Ig5_BE_SBC	0.812	0.816	0.004	0.624	0.631	0.007	0.475	0.478	0.003	13	0.0625
14	Mdl_Uni_VclCor_B10_Ig5_BE_SBC	0.812	0.817	0.005	0.624	0.634	0.01	0.482	0.48	-0.002	12	0.0677
15	Mdl_Uni_VclCor_B20_Ig5_BE_SBC	0.812	0.817	0.005	0.624	0.634	0.01	0.482	0.48	-0.002	12	0.0677
16	Mdl_Uni_ICRM_B20_Ig4_BE_SBC	0.808	0.812	0.004	0.615	0.623	0.008	0.469	0.47	0.001	13	0.0622
17	Mdl_Uni_ICRM_B10_Ig4_BE_SBC	0.807	0.81	0.003	0.614	0.621	0.007	0.47	0.468	-0.002	12	0.0673
18	Mdl_Uni_VclCov_B20_Ig4_BE_SBC	0.807	0.81	0.003	0.614	0.621	0.007	0.47	0.468	-0.002	12	0.0673
19	Mdl_Uni_VclCov_B10_Ig4_BE_SBC	0.807	0.81	0.003	0.614	0.621	0.007	0.47	0.468	-0.002	12	0.0673
20	Mdl_Uni_VclCor_B10_Ig4_BE_SBC	0.807	0.81	0.003	0.614	0.621	0.007	0.47	0.468	-0.002	12	0.0673
21	Mdl_Uni_VclCor_B20_Ig4_BE_SBC	0.807	0.81	0.003	0.614	0.621	0.007	0.47	0.468	-0.002	12	0.0673

CHAPTER 17   EXPERIMENTAL DESIGN AND HYPEROPTIMIZATION

Train	
Variables	
Assessment Reports	
‑Number of Bins	20
‑ROC Chart	Yes
‑Recompute	No
Model Selection	
‑Selection Data	Default
‑Selection Statistic	ROC
‑HP Selection Statistic	Default
‑SAS Viya Selection Statistic	
‑Selection Table	Validation
‑Selection Depth	10

Figure 17-9. *Model comparison options used to generate performance statistics and degrees of freedom from* **Table 17-10***.*

As illustrated in **Table 17-8**, the resulting variants are ordered in descending fashion with respect to the validation ROC. Consequently, the variant exhibiting the largest ROC values is positioned as the initial variant on the list. It is notable that the top three variants exhibit indistinguishable ROC values, as previously discussed, yet demonstrate varying degrees of complexity. Additionally, it is notable that, with regard to the ROC, all models exhibit an exceptional degree of discrimination (ROC >0.8, as illustrated in **Table 17-1**) and that they are not substantially different from one another.

The absence of overfitting in all variants can be inferred from the discrepancy between the validation and training partitions of the performance statistics. This result also corroborates the assertion that the initial stratified sampling methodology, which yielded the training and validation partitions, effectively mitigated the issue of overfitting, leaving minimal residual evidence of it.

In terms of complexity, the variants range from 11 to 17 covariates per model. However, variants 5, 8, and 9 stand out due to their high yield, with an $APPV_{ROC}$ of 0.0678, despite having the lowest number of covariates (11).

It is noteworthy that the top three variants exhibit relatively low $APPV_{ROC}$ values, with a range of 0.0510 to 0.0544. This outcome can be attributed to the fact that the highest validation ROC of 0.816 is not markedly disparate from the ROC values observed in the remaining variants. Consequently, the $APPV_{ROC}$ will be highly susceptible to alterations in the number of covariates within the model.

17.2.1.2 Case of Study II: Retail Loan

The results of case II are analogous to those of case I (**Table 17-10**), as all models exhibit excellent discriminatory power and are not significantly disparate from one another. However, two notable differences emerge between the two cases. Firstly, the range of complexity in case I spanned 12 to 18 covariates, while in case II, the range of covariates extended from 13 to 37. Secondly, while both sets of models (case I and case II) can be regarded as excellent based on their ROC value, which falls within the range of $0.8 \leq ROC \leq 0.9$, the ROC values in case I span from 0.807 to 0.816, while those in case II range from 0.873 to 0.897. It can thus be concluded that the set of models in case II is of a markedly superior quality to those presented in case I.

As will be demonstrated in the following sections, one of the primary factors contributing to the notable discrepancy in performance between variants is the average number of significant variables selected in the final multivariable models. It is noteworthy that in both cases, the BE-Only significance level remained at a p-value of less than 0.0001. However, some models in case II were capable of selecting up to 37 variables, indicating the substantial presence of significant variables throughout the selection process. Furthermore, the cutoff point for the VIF in case I was the recommended value of 3, while in case II, it was reduced to 1.5 to avoid the overwhelming selection of significant variables by the ICRM. Interestingly, the varclus methods did not require any adjustment to reduce the number of variables being selected. Paradoxically, the adjustments made were to increase the number of output variables instead of reducing them.

CHAPTER 17   EXPERIMENTAL DESIGN AND HYPEROPTIMIZATION

Table 17-9 provides a side-by-side comparison of the collinearity parameters used in case I and those employed in case II.

Table 17-9. Comparison of the collinearity parameters used for case I and case II.

Method	Parameter	Case I	Case II
ICRM	VIF	3	1.5
Varclus[5]	Variation proportion	0.8	0.6
	Number of clusters	7	9

It is imperative to emphasize that the observed discrepancies in both cases cannot be attributed to the differences in the parameters of the collinearity reduction techniques employed. Rather, they are a consequence of the distinctive attributes inherent to the data utilized to construct the model, the variables' configuration, and the robustness and consistency of the target's definition. These factors are more closely associated with the conceptual variable design and skillful data manipulation than with the utilization of sophisticated algorithms.

In the next section, a comprehensive list of the variables selected by the 21 variants in both case I and case II will be provided, along with a detailed rationale for the retention of nearly all of these variables.

---

[5] Includes both, the covariance and the correlation matrix variants.

Table 17-10. *Assessment results from the 21 optimal multivariable models for a retail loan PD model.*

No.	Model Description	Valid ROC	Train ROC	Diff ROC	Valid Gini	Train Gini	Diff Gini	Valid KS	Train KS	Diff KS	df	APPV_ROC
1	Mdl_Uni_ICRM_B20_Ig4_BE_SBC	0.897	0.898	0.001	0.794	0.796	0.002	0.655	0.658	0.003	30	0.0309
2	Mdl_Uni_ICRM_B20_Ig6_AllSub_MCQ	0.897	0.898	0.001	0.793	0.795	0.002	0.653	0.656	0.003	37	0.0249
3	Mdl_Uni_ICRM_B20_Ig6_BE_SBC	0.897	0.897	0	0.793	0.795	0.002	0.652	0.656	0.004	27	0.0345
4	Mdl_Uni_ICRM_B10_Ig6_BE_SBC	0.897	0.898	0.001	0.794	0.795	0.001	0.656	0.658	0.002	30	0.0309
5	Mdl_Uni_ICRM_B10_Ig4_BE_SBC	0.896	0.896	0	0.791	0.792	0.001	0.652	0.654	0.002	30	0.0309
6	Mdl_Uni_ICRM_B20_Ig5_BE_SBC	0.896	0.897	0.001	0.792	0.793	0.001	0.652	0.655	0.003	27	0.0345
7	Mdl_Uni_ICRM_B10_Ig5_BE_SBC	0.896	0.897	0.001	0.793	0.794	0.001	0.655	0.657	0.002	30	0.0309
8	Mdl_Uni_VclCor_B20_Ig6_BE_SBC	0.896	0.898	0.002	0.792	0.795	0.003	0.652	0.656	0.004	17	0.0560
9	Mdl_Uni_VclCor_B20_Ig6_AllSub_MCQ	0.896	0.898	0.002	0.792	0.795	0.003	0.652	0.657	0.005	24	0.0390
10	Mdl_Uni_VclCor_B10_Ig6_BE_SBC	0.896	0.897	0.001	0.791	0.795	0.004	0.651	0.656	0.005	19	0.0498
11	Mdl_Uni_VclCor_B20_Ig5_BE_SBC	0.895	0.897	0.002	0.79	0.794	0.004	0.651	0.655	0.004	18	0.0526
12	Mdl_Uni_VclCor_B10_Ig5_BE_SBC	0.895	0.897	0.002	0.79	0.794	0.004	0.65	0.655	0.005	20	0.0471
13	Mdl_Uni_VclCor_B10_Ig4_BE_SBC	0.894	0.896	0.002	0.788	0.792	0.004	0.651	0.656	0.005	22	0.0426
14	Mdl_Uni_VclCor_B20_Ig4_BE_SBC	0.894	0.895	0.001	0.788	0.791	0.003	0.65	0.655	0.005	22	0.0426
15	**Mdl_Uni_VclCov_B10_Ig6_BE_SBC**	**0.876**	**0.878**	**0.002**	**0.752**	**0.755**	**0.003**	**0.595**	**0.599**	**0.004**	**13**	**0.0730**
16	Mdl_Uni_VclCov_B20_Ig6_AllSub_MCQ	0.876	0.878	0.002	0.752	0.756	0.004	0.595	0.599	0.004	14	0.0674
17	**Mdl_Uni_VclCov_B20_Ig6_BE_SBC**	**0.876**	**0.878**	**0.002**	**0.752**	**0.756**	**0.004**	**0.595**	**0.599**	**0.004**	**13**	**0.0730**
18	Mdl_Uni_VclCov_B10_Ig5_BE_SBC	0.875	0.877	0.002	0.75	0.753	0.003	0.593	0.597	0.004	14	0.0673
19	Mdl_Uni_VclCov_B20_Ig5_BE_SBC	0.875	0.877	0.002	0.75	0.753	0.003	0.594	0.597	0.003	14	0.0673
20	**Mdl_Uni_VclCov_B20_Ig4_BE_SBC**	**0.874**	**0.874**	**0**	**0.747**	**0.749**	**0.002**	**0.591**	**0.591**	**0**	**13**	**0.0728**
21	**Mdl_Uni_VclCov_B10_Ig4_BE_SBC**	**0.873**	**0.875**	**0.002**	**0.747**	**0.749**	**0.002**	**0.59**	**0.592**	**0.002**	**13**	**0.0728**

17.2.2 Assessing the Hyperspace

In this section, we will explore the expanded set of variables selected from all the feasible regions, or hyperspace, and provide the rationale that explains which variables should remain for the construction of the main-effects model, and why.

17.2.2.1 Case of Study I: Credit Card PD Model

Table 17-11 presents the association matrix for the regression estimates between the 21 variants and their corresponding 22 covariates. It is important to note that the matrix is ordered in descending sequence according to the frequency of the covariate across the variants. It is noteworthy that the first nine covariates are present in all variants, despite the fact that each variant has its own feasible region. This phenomenon can be explained by two factors. The first reason for this phenomenon is that, although different constraints define different feasible regions, these regions are not necessarily orthogonal to each other. It is therefore to be expected that there will be some overlap between them. As a result, it is not unexpected that regions with an overlap will converge to the same variables. The second reason, which is more closely related to the nature of the covariates than to the feasible regions, can be explained in optimization terms. One may conceptualize a strong covariate as a powerful force capable of encompassing the solution space in a manner analogous to that observed in black holes. Now, consider the possibility that these strong covariates exert a downward force on the surrounding space, creating a vortex-like region. It is postulated that the diameter of the vortex will increase in proportion to the strength of the covariate. Consequently, it is anticipated that no algorithm will be capable of evading the affected region, repeatedly converging on the same solution. It can be reasonably assumed that these nine covariates exert a significant influence on the final multivariable model, and therefore, it is prudent to retain them in the model.

Table 17-11. Association matrix between estimates and variants for case I.

	Occurrence	Mdl_Uni_lCRM_B20_lg4_BE_SBC	Mdl_Uni_lCRM_B20_lg5_BE_SBC	Mdl_Uni_lCRM_B20_lg6_BE_SBC	Mdl_Uni_lCRM_B10_lg4_BE_SBC	Mdl_Uni_lCRM_B10_lg5_BE_SBC	Mdl_Uni_lCRM_B10_lg6_BE_SBC	Mdl_Uni_lCRM_B20_lg6_AllSub_Mcq	Mdl_Uni_VelCov_B20_lg4_BE_SBC	Mdl_Uni_VelCov_B20_lg5_BE_SBC	Mdl_Uni_VelCov_B20_lg6_BE_SBC	Mdl_Uni_VelCov_B10_lg4_BE_SBC	Mdl_Uni_VelCov_B10_lg5_BE_SBC	Mdl_Uni_VelCov_B10_lg6_BE_SBC	Mdl_Uni_VelCov_B20_lg6_AllSub_Mcq	Mdl_Uni_VelCor_B20_lg4_BE_SBC	Mdl_Uni_VelCor_B20_lg5_BE_SBC	Mdl_Uni_VelCor_B20_lg6_BE_SBC	Mdl_Uni_VelCor_B10_lg4_BE_SBC	Mdl_Uni_VelCor_B10_lg5_BE_SBC	Mdl_Uni_VelCor_B10_lg6_BE_SBC	Mdl_Uni_VelCor_B20_lg6_AllSub_Mcq
Intercept	0																					
X_1	21	-4.63	-4.63	-4.62	-4.63	-4.63	-4.62	-4.62	-4.63	-4.62	-4.62	-4.63	-4.62	-4.62	-4.62	-4.63	-4.62	-4.62	-4.63	-4.62	-4.62	-4.62
X_2	21	-0.31	-0.31	-0.3	-0.31	-0.31	-0.3	-0.28	-0.31	-0.3	-0.29	-0.31	-0.24	-0.25	-0.28	-0.31	-0.24	-0.31	-0.31	-0.24	-0.24	-0.28
X_3	21	-0.25	-0.24	-0.25	-0.25	-0.25	-0.25	-0.22	-0.25	-0.24	-0.25	-0.25	-0.25	-0.3	-0.23	-0.25	-0.24	-0.25	-0.25	-0.24	-0.24	-0.21
X_4	21	-0.73	-0.73	-0.72	-0.72	-0.72	-0.72	-0.72	-0.73	-0.71	-0.73	-0.73	-0.71	-0.73	-0.71	-0.78	-0.74	-0.73	-0.78	-0.74	-0.71	-0.75
X_5	21	-0.78	-0.75	-0.74	-0.78	-0.75	-0.74	-0.75	-0.78	-0.74	-0.73	-0.78	-0.35	-0.37	-0.37	-0.31	-0.35	-0.37	-0.31	-0.35	-0.37	-0.37
X_6	21	-0.31	-0.35	-0.37	-0.31	-0.35	-0.37	-0.36	-0.31	-0.78	-0.77	-0.31	-0.78	-0.77	-0.77	-0.77	-0.78	-0.77	-0.77	-0.78	-0.77	-0.77
X_7	21	-0.77	-0.78	-0.77	-0.77	-0.78	-0.77	-0.77	-0.77	-0.43	-0.41	-0.77	-0.42	-0.41	-0.45	-0.77	-0.43	-0.42	-0.45	-0.43	-0.42	-0.44
X_8	21	-0.45	-0.44	-0.43	-0.45	-0.43	-0.43	-0.45	-0.45	-0.34	-0.3	-0.39	-0.34	-0.3	-0.3	-0.39	-0.33	-0.3	-0.39	-0.33	-0.3	-0.29
X_9	21	-1.15	-1.05	-1.04	-1.15	-1.05	-1.04	-1.06	-1.15	-1.06	-1.05	-1.15	-1.06	-1.05	-1.04	-1.15	-1.06	-1.05	-1.07			-1.07
X_{10}	18	-0.86	-0.85	-0.85	-0.85	-0.85	-0.85	-0.84	-0.86	-0.84	-0.82	-0.86			-0.82	-0.86	-0.85	-0.85	-0.84			
X_{11}	10	-0.63	-0.6		-0.64	-0.53	-0.53	-0.55	-0.63	-0.47	-0.46	-0.63	-0.45	-0.47	-0.43	-0.63	-0.54	-0.54	-0.53	-0.54	-0.53	-0.53
X_{12}	8		-0.61					-0.5														
X_{13}	5				-0.47	-0.34	-0.34	-0.28														
X_{14}	4									-0.3	-0.31		-0.36	-0.27								
X_{15}	4										-0.36			-0.3								
X_{16}	3												-0.84	-0.35								
X_{17}	3													-0.82								
X_{18}	3										-0.24		-0.86		-0.42							-0.44
X_{19}	3														-0.16							
X_{20}	2																					
X_{21}	1							-0.21														
X_{22}	1														-0.01							

CHAPTER 17 EXPERIMENTAL DESIGN AND HYPEROPTIMIZATION

Figure 17-10 shows a bar chart of the absolute value of the estimates of the top nine variables. Note that the bars are grouped into sets of three according to the collinearity method that generated them, where ICRM is the iterative collinearity reduction method, VclCov is the varclus covariance matrix, and VclCor is the varclus correlation matrix. In addition, each plot contains seven blocks corresponding to the seven possible treatments:

1. **B20_IG4_BE_SBC:** 20 Bins, 4 Max. Groups, BE-Only, SB Criterion

2. **B20_IG5_BE_SBC:** 20 Bins, 5 Max. Groups, BE-Only, SB Criterion

3. **B20_IG6_BE_SBC:** 20 Bins, 6 Max. Groups, BE-Only, SB Criterion

4. **B10_IG4_BE_SBC:** 10 Bins, 4 Max. Groups, BE-Only, SB Criterion

5. **B10_IG5_BE_SBC:** 10 Bins, 5 Max. Groups, BE-Only, SB Criterion

6. **B10_IG6_BE_SBC:** 10 Bins, 6 Max. Groups, BE-Only, SB Criterion

7. **B20_IG6_AllSub_MCq:** 20 Bins, 4 Max. Groups, All-Subsets, Mallows' C_q

which when combined result in the 21 variants previously defined.

What is interesting to note from this series of charts is the effect that each treatment has on the relative importance of the variables, based on their respective weights. For instance, one would expect that when the same variable is selected by different (but equivalent) models, this one should display a similar numerical value across all variants. However, this is not the case, and markedly differences are reported for almost all combinations of factors (treatments). However, it is noticeable that the only factors that present a slight discrepancy in values are those related to the collinearity method employed. This is the case for variables x_5, x_8, and, to a lesser extent, x_9.

But what is remarkable about this exercise is the fact that it allows us to see a tangible example of what a hyperspace is and the undeniable effect that different hyperparameters (factors) have on the resulting optimal multivariable model.

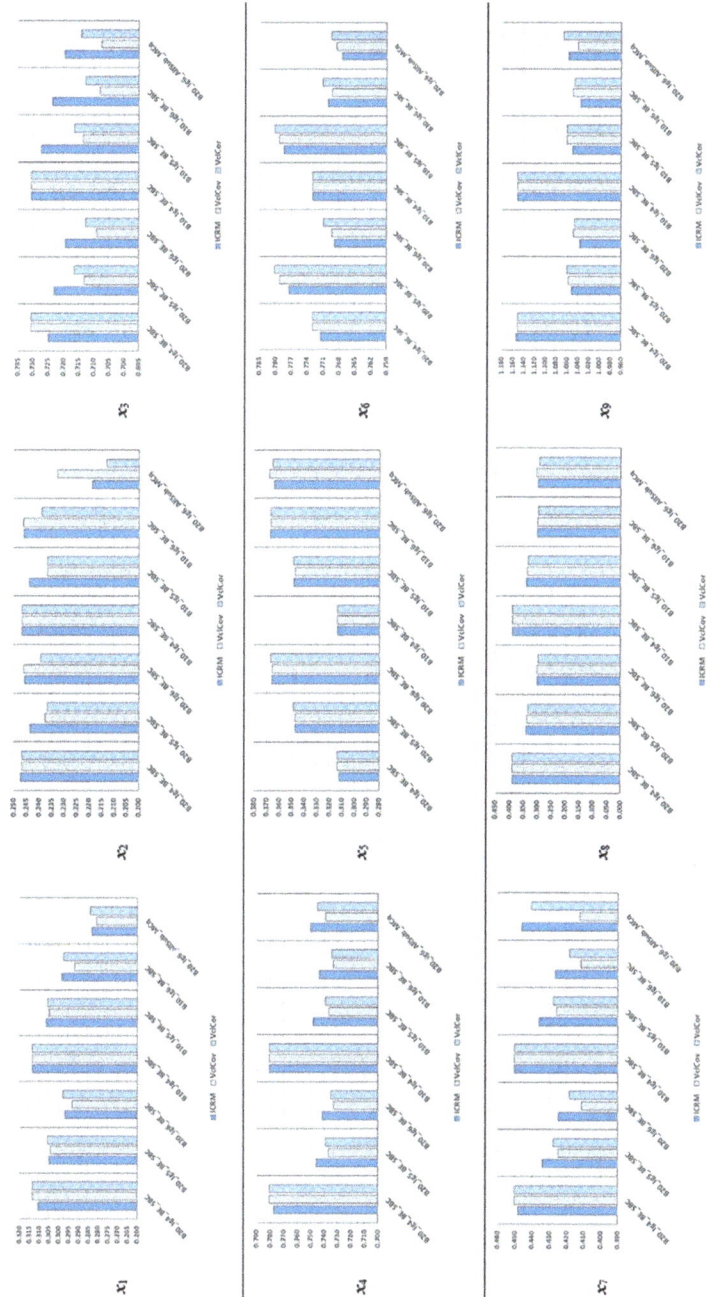

Figure 17-10. *Bar charts for the top nine covariates comparing the estimates (in absolute values) of the 21 variants in blocks of three, according to their collinearity method.*

CHAPTER 17 EXPERIMENTAL DESIGN AND HYPEROPTIMIZATION

We now turn our attention to **Figures 17-11**, **17-12**, and **17-13**, which illustrate the resulting empirical odds of the validation partition for the six BE-Only variants of the ICRM block, the covariance varclus, and the correlation varclus, respectively. As the term implies, the empirical odds plot the empirical or observed odds against the average scorecard points in ascending order. The scorecard points[6] are a linear function of the odds estimated by the model. As a consequence, for a stable model, we would anticipate a descending, smooth, linear relationship between the empirical odds and the average scorecard points. However, as illustrated in **Figure 17-11**, the relationship appears to be distorted as the number of bins and the maximum number of groups increase. It is notable that the top three variants of **Figure 17-11**, defined at 10 bins, demonstrate resilience to the maximum number of groups, maintaining a quasilinear behavior across the increasing levels of average scorecard points. In contrast, the bottom three variants of **Figure 17-11**, defined at 20 bins, exhibit a hook-like behavior at extreme values of the average scorecard points. This suggests that the risk of default is even higher at extreme values of the score distribution, rather than at the lowest, which would be the expected result. As the number of bins or maximum number of groups increases, it becomes evident that the empirical odds undergo a progressive distortion, as illustrated in **Figures 17-12 and 17-13**, which correspond to the variants belonging to the covariance and the correlation varclus, respectively. The only exception is the variant `Mdl_Uni_VclCov_B20_Ig6_BE_SBC`, which unexpectedly does not exhibit the hook-like behavior and maintains the corresponding descending linear trend, despite being defined with the highest number of bins, 20, and the maximum number of groups, 6. However, it can be expected that as these parameters increase, the distortion will be higher, rather than the

---

[6]. For more information about calculate the scorecard points read Scaling Calculation from Siddiqi's Intelligent Credit Scoring.

opposite. This serves as an exemplar of the issue previously discussed in Section 17.1.5, namely, that increasing the resolution, as indicated by the number of bins, will, as previously stated, result in a less stable model, rather than the desired outcome. This prompts the following research question: Are the number of bins and the maximum number of groups the sole determinants of the stable linear trend, or is a combination of these parameters and the presence of specific covariates also a contributing factor?

CHAPTER 17 EXPERIMENTAL DESIGN AND HYPEROPTIMIZATION

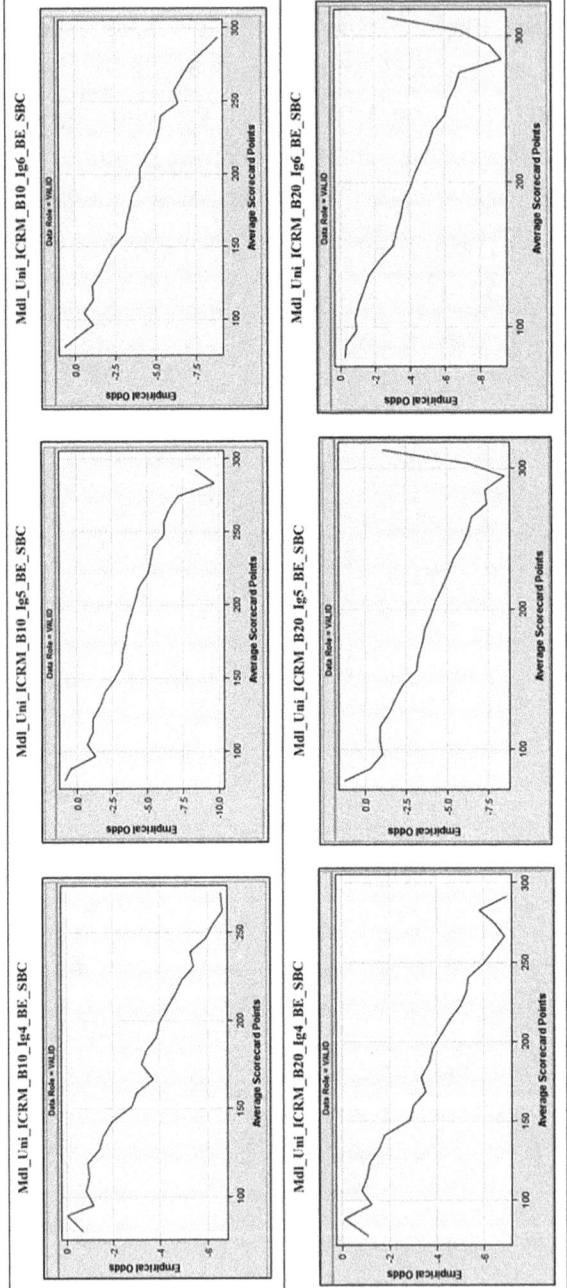

Figure 17-11. *Validation empirical odds plots for the six BE-Only variants for the ICRM block for case I.*

CHAPTER 17 EXPERIMENTAL DESIGN AND HYPEROPTIMIZATION

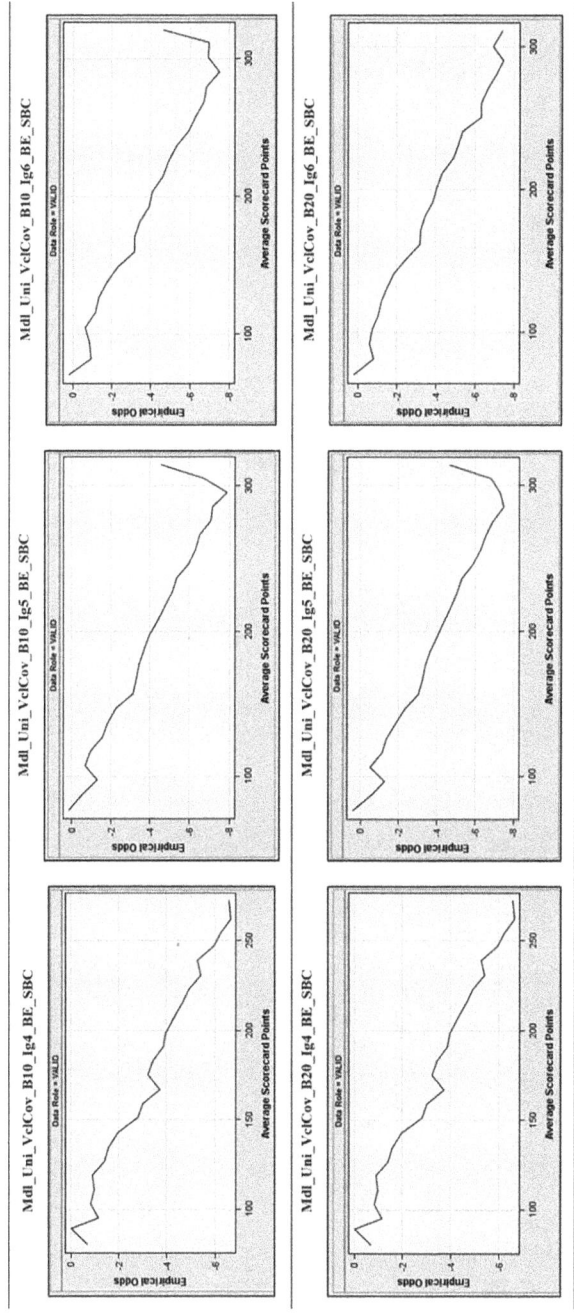

Figure 17-12. *Validation empirical odds plots for the six BE-Only variants for the covariance varclus block, for case I.*

CHAPTER 17 EXPERIMENTAL DESIGN AND HYPEROPTIMIZATION

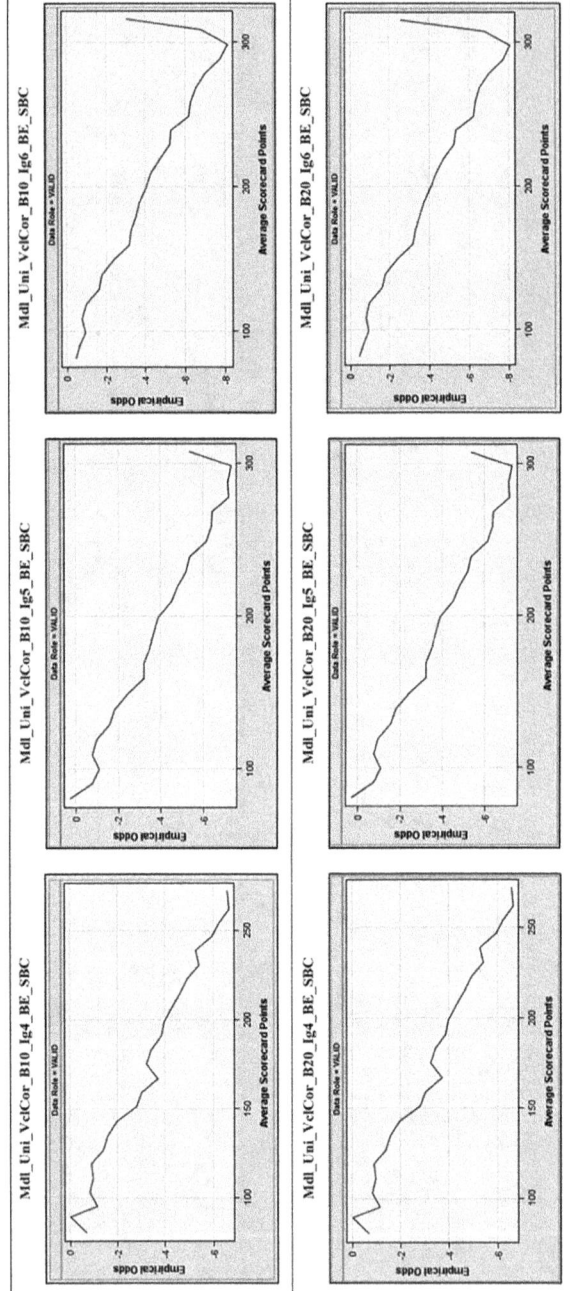

Figure 17-13. Validation empirical odds plots for the six BE-Only variants for the correlation varclus block, for case I.

CHAPTER 17   EXPERIMENTAL DESIGN AND HYPEROPTIMIZATION

We will now investigate the hypothesis that specific covariates may be responsible for the stable linear trend observed in the following nine variants (heading in red font in **Table 17-11**):

1. Mdl_Uni_ICRM_B20_Ig4_BE_SBC
2. Mdl_Uni_ICRM_B10_Ig4_BE_SBC
3. Mdl_Uni_ICRM_B10_Ig5_BE_SBC
4. Mdl_Uni_ICRM_B10_Ig6_BE_SBC
5. Mdl_Uni_VclCov_B20_Ig4_BE_SBC
6. Mdl_Uni_VclCov_B20_Ig6_BE_SBC
7. Mdl_Uni_VclCov_B10_Ig4_BE_SBC
8. Mdl_Uni_VclCor_B20_Ig4_BE_SBC
9. Mdl_Uni_VclCor_B10_Ig4_BE_SBC

In order to achieve this objective, the initial nine common covariates will be dismissed, with the remaining nine, x_{10} to x_{18}, being the primary focus. It is noteworthy that x_{10}, x_{12}, x_{13}, and x_{14} are present in the ICRM block; x_{10}, x_{11}, x_{12}, and x_{15} to x_{18} are present in the covariance varclus block; and finally, only x_{10} and x_{12} are present in the correlation varclus block.

CHAPTER 17 EXPERIMENTAL DESIGN AND HYPEROPTIMIZATION

It can be dismissed that x_{10} and x_{11} are associated with the correction of the linear trend, as both of them are present in both the distorted and the nondistorted plots. However, their prevalence in the variants may justify their inclusion in the final multivariable model, as they are the second and third most common occurrences.

In contrast, x_{12} is present in six out of nine stable variants and only two out of eight unstable variants. This suggests that x_{12} should be included in the final multivariable model, given its strong likelihood of being related to the stabilization of the model. While x_{13} and x_{14} appear to lack a clear correlation with a stabilizing effect, they may be included for the sake of diversity, given their presence only in the ICRM block. The variables x_{15} to x_{18} merit particular attention, as they are all present in the variant Mdl_Uni_VclCov_B20_Ig6_BE_SBC, which was the sole exception to the aforementioned rule. Despite having the highest number of bins and groups, this variant is nevertheless among the nine stable variants. It can be reasonably deduced, therefore, that the combination of variables in the variant Mdl_Uni_VclCov_B20_Ig6_BE_SBC is exceptional in terms of stability. Accordingly, the final multivariable model will also include the covariates from x_{15} to x_{18}.

Having completed an exhaustive examination of all potential variables associated with variant stability, we now direct our attention to the remaining four effects, spanning x_{19} to x_{22}. While it may appear that x_{19} should be excluded from the analysis due to its presence in only three of the 21 variants, this would be an oversight as x_{19} is only present in the all-subsets variants. The rationale behind the inclusion of the all-subsets variants was to augment the total number of effects, given that the all-subsets algorithm is the sole one that examines all possible subsets. Consequently, the multivariable selection algorithm is more likely to select all the main effects required for the model. Therefore, x_{19} will also be included in the final model. The final model will also include the remaining variables x_{20}, x_{21}, and x_{22} (which were selected based on the

Covariance Varclus and ICRM, respectively) due to their contribution to the diversity of the final model.

17.2.2.2 Case of Study II: Retail Loan PD Model

The conclusion of case I is that none of the 22 effects selected during the hyperoptimization process should be discarded. However, it is acknowledged that in certain instances, the available evidence suggests the potential for the rejection of some of the selected effects, a decision that is primarily influenced by a *coefficient inversion*. **Table 17-12** presents the association matrix for case II. In contrast to the findings in case I, this table demonstrates that variables x_{16}, x_{29}, x_{40}, x_{42}, and x_{44} have positive coefficients across all variants. Likewise, variables x_{48}, x_{49}, and x_{50} exhibit positive coefficients, yet, in contrast to the aforementioned group, they are not shared across variants.

CHAPTER 17 EXPERIMENTAL DESIGN AND HYPEROPTIMIZATION

Table 17-12. Association matrix between estimates and variants for case II.

	Mdl_Uni_ICRM_B10_Ig4_BE_SBC	Mdl_Uni_ICRM_B10_Ig5_BE_SBC	Mdl_Uni_ICRM_B10_Ig6_BE_SBC	Mdl_Uni_ICRM_B20_Ig4_BE_SBC	Mdl_Uni_ICRM_B20_Ig5_BE_SBC	Mdl_Uni_ICRM_B20_Ig6_AllSub_MCQ	Mdl_Uni_ICRM_B20_Ig6_BE_SBC	Mdl_Uni_VdCor_B10_Ig4_BE_SBC	Mdl_Uni_VdCor_B10_Ig5_BE_SBC	Mdl_Uni_VdCor_B10_Ig6_BE_SBC	Mdl_Uni_VdCor_B20_Ig4_BE_SBC	Mdl_Uni_VdCor_B20_Ig5_BE_SBC	Mdl_Uni_VdCor_B20_Ig6_AllSub_MCQ	Mdl_Uni_VdCor_B20_Ig6_BE_SBC	Mdl_Uni_VdCov_B10_Ig4_BE_SBC	Mdl_Uni_VdCov_B10_Ig5_BE_SBC	Mdl_Uni_VdCov_B10_Ig6_BE_SBC	Mdl_Uni_VdCov_B20_Ig4_BE_SBC	Mdl_Uni_VdCov_B20_Ig5_BE_SBC	Mdl_Uni_VdCov_B20_Ig6_AllSub_MCQ	Mdl_Uni_VdCov_B20_Ig6_BE_SBC
	30	30	30	30	27	37	27	22	20	19	22	18	24	17	13	14	13	13	14	14	13
x_1	-2.316	-2.319	-2.32	-2.313	2.319	-2.318	2.319	-2.321	-2.32	-2.32	-2.322	-2.32	-2.32	-2.32	2.338	2.336	2.337	2.339	2.337	2.337	-2.337
x_2	-0.473	-0.459	-0.437	-0.45	0.455	-0.418	0.424	-0.32	-0.315	-0.271	-0.312	-0.305	-0.277	-0.273	0.433	0.426	-0.39	0.437	0.424	-0.385	-0.385
x_3	-0.219	-0.215	-0.186	-0.213	0.215	-0.191	0.178	-0.201	-0.212	-0.208	-0.209	-0.215	-0.194	-0.203	0.312	0.322	0.307	0.317	0.327	-0.312	-0.311
x_4	-1.03	-0.74	-0.738	-0.437	0.709	-0.723	0.725	-0.866	-0.864	-0.864	-0.867	-0.864	-0.862	-0.863	0.914	0.914	0.913	0.914	0.914	-0.913	-0.913
x_5	-0.427	-0.47	-0.437	-0.447	0.476	-0.443	0.444	-0.323	-0.361	-0.325	-0.325	-0.359	-0.32	-0.322	0.466	0.504	0.461	0.464	0.501	-0.459	-0.459
x_6	-0.377	-0.394	-0.431	-0.38	0.388	-0.438	0.426	-0.338	-0.348	-0.404	-0.34	-0.349	-0.392	-0.4	0.444	0.443	-0.49	0.443	0.441	-0.487	-0.486
x_7	-0.301	-0.289	-0.299	-0.278	-0.27	-0.274	0.276	-0.139		-0.126	-0.127			-0.547	0.298	0.275	-0.28	0.301	0.276	-0.277	-0.277
x_8	-0.534	-0.551	-0.54	-0.54	0.555	-0.535	-0.54	-0.53	-0.549	-0.546	-0.524	-0.542	-0.552	-0.29	0.508	0.478	0.475	0.507	0.472	-0.47	-0.469
x_9	-0.768	-0.735	-0.744	-0.728	0.721	-0.714	0.729	-0.29	-0.268	-0.269	-0.306	-0.293	-0.312	-1.959							
x_{10}	-0.162	-0.209	-0.205	-0.144	0.187	-0.194	0.184	-1.898	-1.916	-1.923	-1.777	-1.914	-1.959	-0.223	0.795	0.754	0.786	0.789	-0.74	-0.747	-0.748
x_{11}	-0.09	-0.103	-0.091	-0.13	0.115	-0.11	0.106	-0.235	-0.179	-0.211	-0.251	-0.208	-0.206		0.128	0.143	0.147	0.122	0.142	-0.144	-0.144
x_{12}	-0.139	-0.157	-0.17	-0.086	-1.06	-0.169	1.039	-0.143	-0.11		-0.144	-0.104	-0.031			0.085			0.085	0.0038	
x_{13}	-0.195					-1.054		-0.101	-0.072		-0.098		-0.064								
x_{14}	-1.056	-1.016	-1.018	-1.086				-0.636	-0.642	-0.631	-0.646	-0.631	-0.619	-0.624							
x_{15}								0.204	0.2346	0.2366	0.2113	0.3259	0.2614	0.3215	0.397	0.405	0.406	0.357	-0.368	-0.384	-0.385
x_{16}																					
x_{17}															0.516	0.485	-0.48	0.525	-0.486	-0.478	-0.478
x_{18}																					

554

CHAPTER 17 EXPERIMENTAL DESIGN AND HYPEROPTIMIZATION

(continued)

X_{19}	7	-0.526	-0.497	-0.497	-0.5	0.496	-0.486	0.494		-0.286	-0.278	-0.264	-0.297	-0.288	-0.279	-0.283	
X_{20}	7	0.1904	-0.14	-0.141	-0.445	0.151	-0.135	0.133									
X_{21}	7	-0.331	-0.412	-0.344	-0.324	0.397	-0.33	-0.33									
X_{22}	7	-0.291	-0.231	-0.251	-0.257	-0.23	-0.196	0.227									
X_{23}	7	-0.883	-0.886	-0.885	-0.885	0.877	-0.909	0.877									
X_{24}	7	-0.218	-0.192	-0.189	-0.239	0.216	-0.246	0.221									
X_{25}	7	-0.322	-0.373	-0.369	-0.342	0.363	-0.364	0.382									
X_{26}	7									-0.45	-0.437	-0.445	-0.438	-0.434	-0.445	-0.448	0.659
X_{27}	7									-0.33	-0.337	-0.344	-0.34	-0.336	-0.34	-0.341	0.641
X_{28}	7																0.629
X_{29}	7	0.2119	0.2366	0.2741	0.2561	0.244	0.3349	0.276		-0.649	-0.668	-0.662	-0.649	-0.676	-0.664	-0.665	0.647
X_{30}	7									-0.386	-0.366	-0.36	-0.353	-0.349	-0.354	-0.358	0.633
X_{31}	7																0.628
X_{32}	7																0.629
X_{33}	7	-0.445	-0.421	-0.431	-0.456	0.443	-0.475	0.452									
X_{34}	7	-0.508	-0.498	-0.492	-0.502	0.496	-0.491	0.492									
X_{35}	7	-0.389	-0.362	-0.357	-0.319	0.301	-0.289	-0.3									
X_{36}	7	-0.356	-0.36	-0.367	-0.396	0.389	-0.375	0.393									

CHAPTER 17 EXPERIMENTAL DESIGN AND HYPEROPTIMIZATION

Table 17-12. *(continued)*

Model	Count	X37	X38	X39	X40	X41	X42	X43	X44	X45	X46	X47	X48	X49	X50	X51	X52	X53
		7	6	6	5	5	3	3	2	2	1	1	1	1	1	1	1	1
Mdl_Uni_ICRM_B10_I4_BE_SBC		-0.365	-0.223		-0.168	0.386												
Mdl_Uni_ICRM_B10_I5_BE_SBC		-0.352	-0.168	-0.282	-0.178	0.3312												
Mdl_Uni_ICRM_B10_I6_BE_SBC		-0.348	-0.142	-0.234	-0.183	0.3294												
Mdl_Uni_ICRM_B20_I4_BE_SBC		-0.379	-0.186	-0.245	-0.177			0.1258	-0.114									
Mdl_Uni_ICRM_B20_I5_BE_SBC		-0.379	-0.178	-0.316														
Mdl_Uni_ICRM_B20_I6_AllSub_MCQ		-0.39		-0.422	-0.156			0.0894	-0.141	-0.007			0.1722	0.0271	-0.041	0.079	-0.041	
Mdl_Uni_ICRM_B20_I6_BE_SBC		-0.375		-0.163	-0.447													
Mdl_Uni_VdCor_B10_I4_BE_SBC					0.2146			-0.087										
Mdl_Uni_VdCor_B10_I5_BE_SBC					0.2001													
Mdl_Uni_VdCor_B10_I6_BE_SBC					0.1889													
Mdl_Uni_VdCor_B20_I4_BE_SBC					0.1764			-0.093										
Mdl_Uni_VdCor_B20_I5_BE_SBC																		
Mdl_Uni_VdCor_B20_I6_AllSub_MCQ					0.1939			-0.006			-0.111	0.0665						
Mdl_Uni_VdCor_B20_I6_BE_SBC																		
Mdl_Uni_VdCov_B10_I4_BE_SBC																		
Mdl_Uni_VdCov_B10_I5_BE_SBC																		
Mdl_Uni_VdCov_B10_I6_BE_SBC																		
Mdl_Uni_VdCov_B20_I4_BE_SBC																		
Mdl_Uni_VdCov_B20_I5_BE_SBC																		
Mdl_Uni_VdCov_B20_I6_AllSub_MCQ																		
Mdl_Uni_VdCov_B20_I6_BE_SBC																		

CHAPTER 17 EXPERIMENTAL DESIGN AND HYPEROPTIMIZATION

A positive estimate is indicative of a coefficient inversion, which could suggest the presence of collinearity issues with these variables. Consequently, while all effects within each variant constitute the optimal solution of a feasible region, this does not guarantee the coherence of the optimal solution (see Section 14.2). It should be recalled that the WoE transformation is derived from the expression

$$WoE = \ln\left[\frac{d_i^{NE}}{d_i^{E}}\right]$$

Due to its logarithmic properties, positive values are assigned when the nonevent to event ratio is greater than one and negative values are assigned when it is fractional. Consequently, the direction of the risk is inherently associated with the values of the WoE, and thus, it can only be interpreted in a single direction with respect to the original variable. To illustrate this, consider the case of the utilization variable, defined as the outstanding balance in excess of the credit limit. It is reasonable to expect that the WoE values will shift from positive to negative as the utilization level increases. It is therefore vital to note that all WoE variables possess a critical value that differentiates two states: one where the WoE is negative, indicating an increased probability of default as the original variable increases, and one where the WoE is positive, indicating a decreased probability of default as the original variable decreases. To illustrate this concept, consider the logistic regression equation utilized by the scorecard node:

$$(x) = \frac{1}{1 + e^{-g(x)}}$$

Equation 17-6. Logistic equation.

Where $g(x)$ represents the logit and is given by

$$g(x) = a_0 + x_1 a_1 + x_2 a_2 + \cdots + x_n a_n$$

Equation 17-7. Logit equation.

In a conventional logistic regression, the coefficients are responsible for determining both the magnitude and the direction of the covariates.

However, when the covariates in question are WoEs, it is anticipated that the direction is to be determined by the WoE level itself, while the magnitude is to be determined by the regression coefficients. Consequently, within a scorecard model, it is imperative that the regression coefficients are aligned with the direction of the risk derived from the WoE calculation.

A negative coefficient would serve to reduce the probability of default in instances where the WoE values are positive and, conversely, would serve to increase said probability in instances where the WoE values are negative. It is important to note that positive coefficients are not applicable in a logistic regression with WoEs as covariates, as they would reverse the interpretation of the variable regarding the risk direction. To illustrate this point, we return to our utilization example. A positive coefficient would reverse the interpretation of the variable, increasing the probability of default as the utilization decreases—the exact opposite of what is expected.

The primary reason for a positive coefficient slope is the presence of collinearity issues. As previously discussed in Section 14.2, the presence of collinearity issues can result in a phenomenon known as coefficient inversion. This occurs when two variables, which contribute equally to the minimization or maximization of an objective function, are inverted to achieve balance in the objective function. At this juncture in the modeling process, our objective is not to investigate the cause of this collinearity issue. Consequently, it is recommended that these variables be excluded from the *multivariable main-effects model* and that the investigation proceed to the subsequent stage. It is noteworthy that variable x_{12} also exhibits a positive coefficient. However, in contrast to x_{16} and x_{29}, x_{12} exhibits a positive slope in only 2 out of the 12 variants. Consequently, there is insufficient evidence to conclude that its inclusion will invariably result in conflict with the remaining main effects.

Finally, the corresponding empirical odds plots for case II are presented in **Figures 17-14**, **17-15**, and **17-16**. However, no interpretation is provided, thereby enabling readers to practice by drawing their own conclusions regarding the adherence of the different treatments to the expected pattern, or to identify any exceptions, if present.

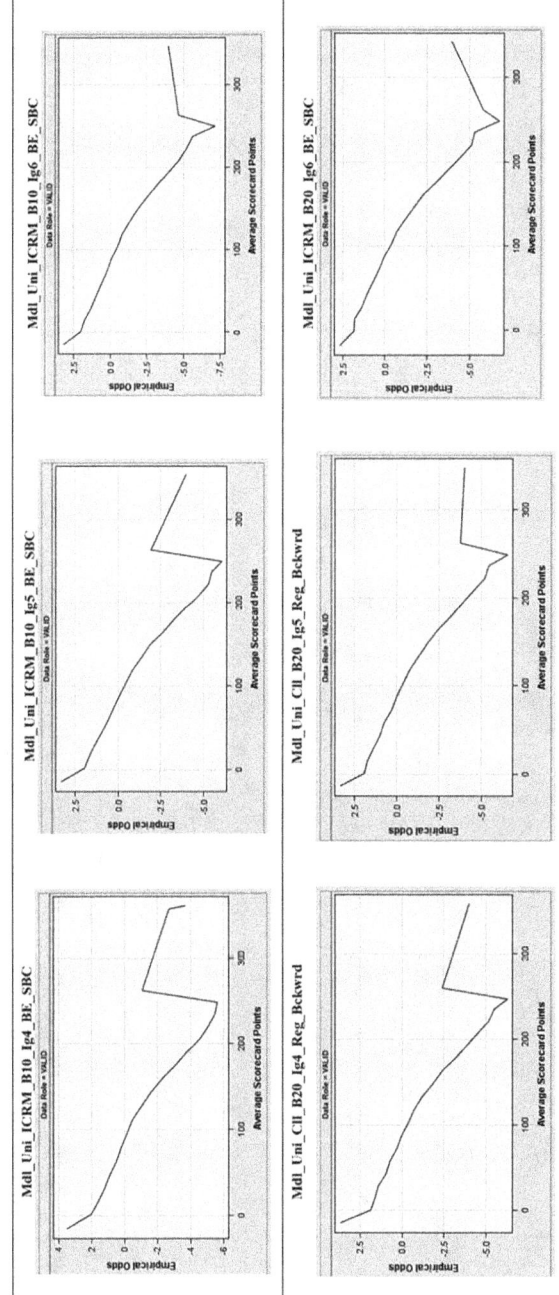

Figure 17-14. Validation empirical odds plots for the six BE-Only variants for the ICRM block, for case II.

CHAPTER 17 EXPERIMENTAL DESIGN AND HYPEROPTIMIZATION

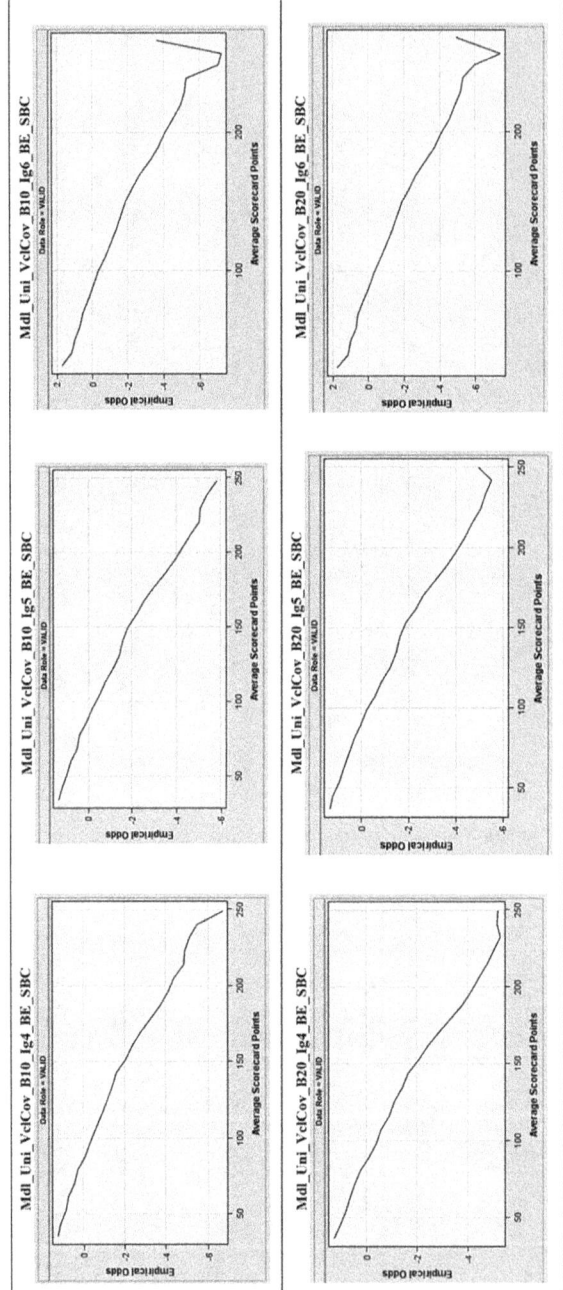

Figure 17-15. Validation empirical odds plots for the six BE-Only variants for the covariance varclus block, for case II.

CHAPTER 17 EXPERIMENTAL DESIGN AND HYPEROPTIMIZATION

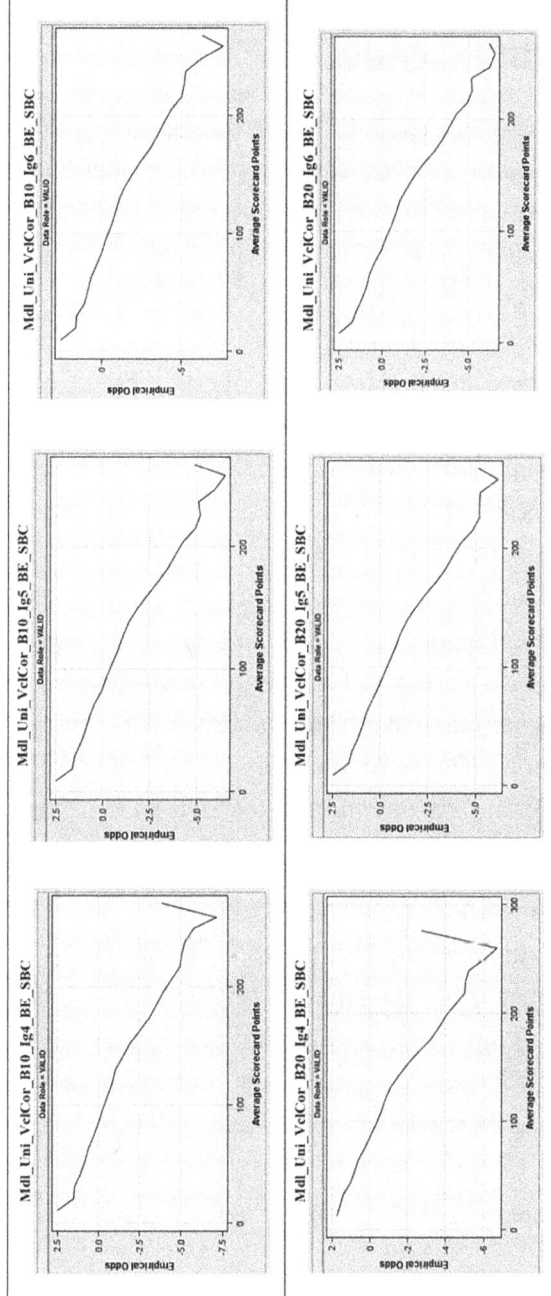

Figure 17-16. Validation empirical odds plots for the six BE-Only variants for the correlation varclus block, for case II.

CHAPTER 17   EXPERIMENTAL DESIGN AND HYPEROPTIMIZATION

17.3 Models Are Not Selected, They Are Constructed

From the methodology outlined in this chapter, it is clear that the notion that one can simply select the most optimal model from a set of competing models and assume that it will emerge as the indisputable victor is fundamentally flawed. Unfortunately, this notion has been the primary focus of the sales pitch employed by numerous data mining software companies for decades. But the fundamental issue lies in the fact that software companies endeavor to generate profit by making exaggerated claims regarding the capabilities of their software, without fully comprehending the intricacies of model construction. A more significant concern is that they have successfully misled numerous generations of analysts over a prolonged period, to the extent that they have influenced the formal data mining (DM) and machine learning (ML) literature, which not only addresses the subject of automated model selection but also advocates for its implementation. This situation has resulted in the emergence of numerous software technicians who exhibit a disproportionate focus on the tools they utilize, relative to their understanding of the underlying rationale for their usage. This issue has already been addressed by Hosmer and Lemeshow [32], who stated that[7]

> *The problem is not the fact that the computer can select such models, but rather that the analyst fails to scrutinize the resulting model carefully, and reports such results as the final, best model. The wide availability and ease with which [computer] methods can be used has undoubtedly reduced*

---

[7] I took the liberty to replace the word "*stepwise*" for "*computer*" in the sentence "*The wide availability and ease with which computer methods*" because the word "*computer*" is more appropriate in this context and does not alter the main idea of the quote.

CHAPTER 17   EXPERIMENTAL DESIGN AND HYPEROPTIMIZATION

some analysts to the role of assisting the computer in model selection rather than the more appropriate alternative. It is only when the analyst understands the strengths, and specially the limitations, of the methods that these methods can serve as useful tools in the model-building process. The analyst, not the computer, is ultimately responsible for the review and evaluation of the model.

It would therefore be erroneous to suggest that one can simply select the optimal model from the list of viable models presented in **Table 17-10**. This is due to the fact that the objective of experimental design and hyperoptimization was never to create a random subset of models from which one can simply select the best one and conclude the model-building process. Instead, the objective of the experimental design and hyperoptimization was to investigate how different constraints on the system can define multiple solution spaces and influence the search for the optimal multivariable model. Furthermore, the assessment of the variants presented in **Table 17-10** revealed that all variants possess excellent discriminatory power. However, the apparent gain in discriminatory power of some variants does not outweigh their gain in complexity.

Consequently, it would be challenging to select a single model over the others, given that they all entail trade-offs. However, it is conceivable that, in a limited number of instances, certain variants may be demonstrably superior or inferior to others. In such instances, it may be tempting to focus exclusively on the most promising variants. However, it should be evident to the reader that all variants, regardless of their performance, are the result of multiple search algorithms converging on an optimal solution. Consequently, the probability that the variables selected in each variant will ultimately be of high importance in the final multivariable model is high. Therefore, it is advised that a single model from the subset of variants not be selected, but rather that all optimal variables (without coefficient inversion) be used as inputs for the construction of the *main-effects model*.

CHAPTER 17 EXPERIMENTAL DESIGN AND HYPEROPTIMIZATION

17.4 Summary

Chapter 17 delves into the critical processes of *experimental design* and *hyperoptimization*, which are essential for extending the solution space in the model-building process. The chapter emphasizes that the goal is not to identify a single best-performing solution but to systematically explore feasible regions and refine parameters to uncover additional main effects that may have been excluded due to earlier methodological constraints. By varying methods, parameters, and constraints, experimental design allows for the structured exploration of alternative feasible regions, while hyperoptimization focuses on fine-tuning the parameters within these regions to expand the solution space.

The chapter demonstrates how these techniques can uncover models that capture more main effects, which might otherwise have been rejected due to overly stringent conditions applied during earlier stages of the modeling process. This approach ensures that the final model is not only robust but also more inclusive of important variables. Chapter 17 underscores the importance of flexibility and adaptability in the modeling process, but at the same time, it provides a framework for a standardized methodology.

Ultimately, the chapter reveals a fundamental flaw in the notion that a *"champion model"* can truly exist. This is proven by the fact that all models are optimal solutions that correspond to a particular feasible region. Consequently, the selection of a single model from a set of variants is arbitrary, and this practice is not endorsed by the methodology delineated in this book.

The following chapter will employ the selected main effects, derived from the hyperoptimization process, as inputs for a new modeling process. The process will commence with the automatic binning process, during which a comprehensive review of the resulting WoE functions for each main effect will be conducted, and any necessary adjustments will be made.

CHAPTER 17 EXPERIMENTAL DESIGN AND HYPEROPTIMIZATION

Key takeaways:

1. **Goal of Experimental Design and Hyperoptimization:**

    - The primary objective is to expand the solution space, allowing for the inclusion of additional main effects that might have been excluded under previous constraints.

2. **Experimental Design:**

    - Provides a systematic framework for exploring multiple feasible regions by varying methods, parameters, and constraints.

    - Helps identify alternative configurations that might lead to models with greater explanatory power.

3. **Hyperoptimization:**

    - Focuses on refining parameters within each feasible region to enhance the discriminatory power of variables and ensure that the model captures more main effects.

4. **Flexibility in Model Building:**

    - By relaxing certain constraints and exploring alternative configurations, these techniques ensure that the modeling process remains adaptable to different scenarios.

5. **Iterative Process:**

    - Although a prebuilt diagram could be used, it is important to highlight the fact that both experimental design and hyperoptimization are of iterative nature, requiring repeated evaluation of feasible regions and hyperparameters, that ultimately will depend on the data at hand.

6. **Emulating a Neural Network (NN):**

    - Unlike other methodologies that present the modeling process as a linear series of steps, the one presented here proposes a network-like configuration, where each branch represents a feasible region being explored. This structure resembles a neural network, where the hidden layer is represented by the IG nodes, which optimize multiple decision trees, which, in turn, generate multiple logistic activation functions that take place in the scorecard node.

Having explored the expansive hyperspace of effects in the previous chapter, *Experimental Design* and *Hyperoptimization*, where we employed advanced techniques to identify and optimize potential predictors, we now focus our attention to refining and consolidating these effects into a robust and interpretable model. While the prior chapter emphasized the technical aspects regarding variants and hyperparameters, the next chapter, *The Main-Effects Model*, shifts back to a conceptual understanding, where the art of modeling comes to life in the manual refinement of the *main-effects model*, which can only be achieved through a combination of common sense, business logic, and statistical rigor.

CHAPTER 18

The Main-Effects Model

"Models should be as simple as possible, but not more so."

—Albert Einstein

The objective of the main-effects model is to integrate the optimal models derived from the given subset of solution spaces into a unified multivariable model, which will serve as the foundation for a subsequent model-building process. In contrast to the preceding model-building processes, which yielded 21 model variants, this new process will commence with the automatic binning process (see **Figure 18-3**). Thereafter, the flow will be bifurcated into two principal branches. One of these branches will be connected directly to a scorecard node, employing the backward elimination only (BE-Only) multivariable selection method. The second branch will be connected to an iterative collinearity reduction method (ICRM) node, which will in turn be divided into two branches; one branch will utilize the BE-Only method, while the other will employ the all-subsets multivariable selection method. Subsequently, the main-effects building process will yield three variants:

CHAPTER 18   THE MAIN-EFFECTS MODEL

1. A main-effects IG, BE-Only model
2. A main-effects IG, ICRM, BE-Only model
3. A main-effects IG, ICRM, all-subsets model

It should be noted that at this juncture, it is not our intention to recreate the 21 previous variants using only the optimal variables identified during the hyperoptimization process. Instead, our objective is to ascertain the outcome of applying a range of multivariable selection *strategies*. It should be noted that the term *"multivariable selection methods"* was deliberately avoided in favor of *"multivariable selection strategies."* This is because, technically, only two multivariable methods are employed: all-subsets and BE-Only. The primary distinction is the incorporation of the ICRM method following the Interactive Grouping (IG) node and preceding the actual multivariable selection method. It is important to note that the recommendation so far has been to reduce collinearity before the IG node, rather than after. This prompts the question of the rationale behind this decision.

To address this question, it is necessary to recall that the set of variables constituting the main effects was derived from disparate collinearity reduction methods, including the ICRM, the covariance varclus, and the correlation varclus. Consequently, the set of variables representing the main effects has never been subjected to a single multivariable selection process simultaneously. It follows then that the introduction of this new set of variables may well give rise to the emergence of new collinearity issues. Consequently, the application of a collinearity reduction method prior to the transformation would merely eliminate the very effects that are to be combined, thus defeating the purpose of the rationale behind the creation of the multivariable main-effect model.

The objective is therefore to ascertain the impact of the combined main effects on the multivariable model. It is then to be expected that the collinearity will yield different results when reduced after the WoE transformation than when reduced before it. This is due to the fact that the transformation may affect the collinearity by selecting different group cutoffs for highly correlated variables, which may have a significant impact on the risk distribution of the resulting WoE.

18.1 The Variable Select Node

The modeling process will now be continued by the addition of a custom code node to the diagram. The node in question can be connected directly after the Univariable Procedure node or directly before it, depending on whether the initial rejection of variables[1] is to be tested. **Figure 18-1** illustrates an example of an EM diagram in which a new code node has been connected immediately preceding the Univariable Procedure node. The rationale for placing the new code node before the Univariable Procedure is to explicitly define the subset of variables to be selected from the initial set of explanatory variables.

---

[1] It is also important to bear in mind that the Univariable Procedure carries out an automatic rejection process, and consequently, a variable select node will have no effect over variables that have been previously rejected. This consideration assumes particular significance when adhering to the Hosmer and Lemeshow methodology, which advocates for the utilization of previously rejected variables in the main-effects model.

CHAPTER 18   THE MAIN-EFFECTS MODEL

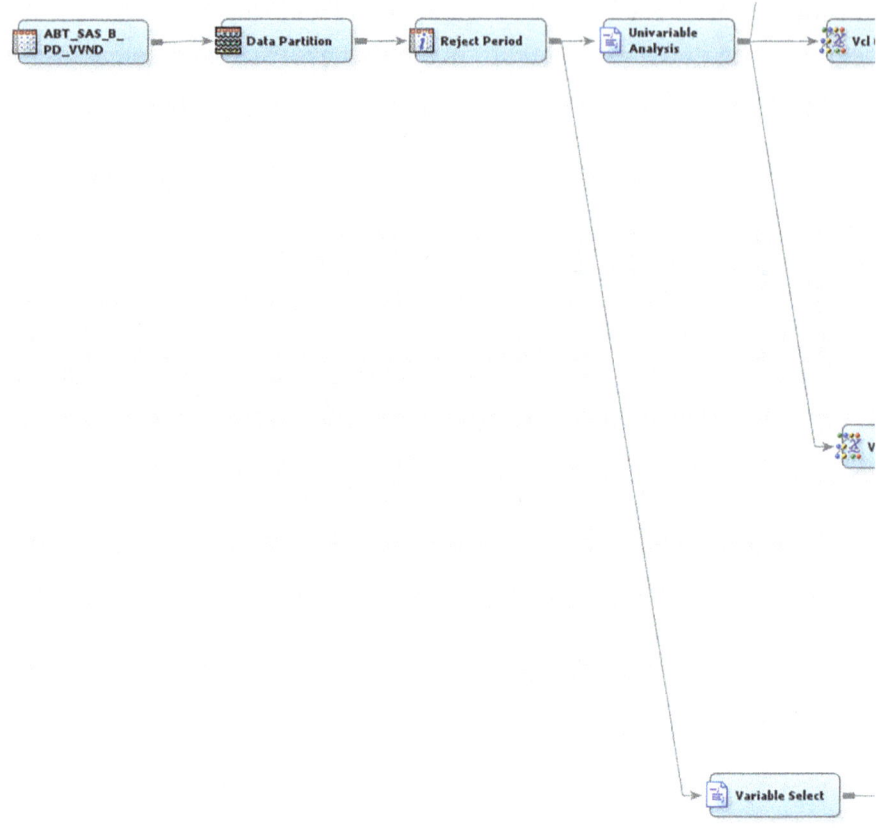

Figure 18-1. *EM diagram illustrating the location of a new code node called Variable Select, on the upper right side of the figure.*

Listing 18-1 illustrates the Variable Select code. It is noteworthy that the list of variables to be selected is explicitly defined by the macro-variable **varlist**, which is then recursively redefined by the macro **%enquote**; and finally, the listed variables are selected by the **%metadata** macro using the logical expression:

```
newrole=Rejected,where=(NAME not in (&varlist) and
upcase(ROLE)='INPUT')
```

Listing 18-1. Example of a variable select customized node code.

```
Training Code
%LET varlist=
N_OCUPATION
I_AGE_YRS
I_LOAN_AGE_YRS
I_MX_DYSPSTDUE_LON_L12M
I_MX_DYSPSTDUE_CCR_L09M
I_RISK_LEVEL
I_NUMDEPEND
I_R_CREDIT_TASSTS
I_R_PAYA_INST_CLC_OOMB
I_R_TOPEXP_TASSTS
I_SDC_R_CHRG_CRRA_CPM_L12M
I_SIN_R_BALT_APPR_CLC_L13_24
;

%LET varlist=%enquote(&varlist,separator=);

%metadata(newrole=Rejected ,where=(NAME NOT IN (&varlist) and upcase(ROLE)='INPUT'));
```

It should be noted that the logical expression rejects any variable that is not present in the **&varlist** macro-variable, and that currently has a role of `'INPUT'`.

An alternative approach would have been to add the Variable Select node immediately following the Univariable Procedure node. However, the placement of the Variable Select node before the Univariable Procedure would allow for the incorporation of any available variable from the original set of explanatory variables.

18.2 Selecting the Best Full Main-Effects Model

The model-building process will be initiated once more, but on this occasion, the Interactive grouping (IG) step will be reached directly, thus bypassing the Univariable and Collinearity Reduction steps. As illustrated in **Figure 18-2**, the main-effects building process section is derived from the primary hyperoptimization diagram. The main-effects section is

CHAPTER 18   THE MAIN-EFFECTS MODEL

located in the lower right portion of **Figure 18-2**, and it is enclosed by a red rectangle. **Figure 18-3** provides a detailed examination of the main-effects section.

It is noteworthy that in this instance, a single IG node is employed, with a configuration of 20 bins and 6 groups. The objective is to increase the complexity (resolution) of the WoEs by increasing the number of cutoff points, thereby facilitating a more comprehensive visualization of the potential cutoff points that can be adjusted during the manual WoE refinement process.

As illustrated in **Figure 18-3**, the initial main-effects model does not employ any collinearity reduction techniques. This approach is justified by the assumption that the collinearity level will be insignificant, thereby not affecting the model's performance. Consequently, the multivariable selection algorithm (BE-Only with SBC) is permitted to identify the optimal variables from the *full* main-effects model. Additionally, **Figure 18-3** illustrates that the ICRM is divided into two multivariable selection techniques: the BE-Only method with the SBC criterion at a 0.0001 significance level and the all-subsets method. The rationale behind employing both methods is to facilitate a comparative analysis of the outcomes between the reduced model derived from the BE-Only approach and the extended model generated by the all-subsets method. It is important to note that the no-ICRM branch does not include an all-subsets selection method. This is due to the fact that the first main-effects model is, by definition, an extended model.

CHAPTER 18   THE MAIN-EFFECTS MODEL

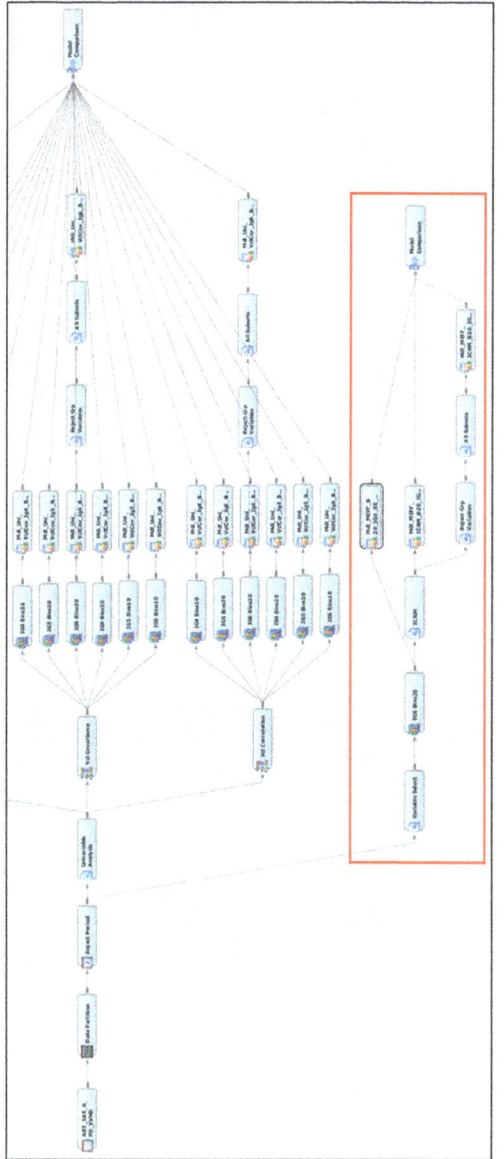

Figure 18-2. Example of a main-effects diagram.

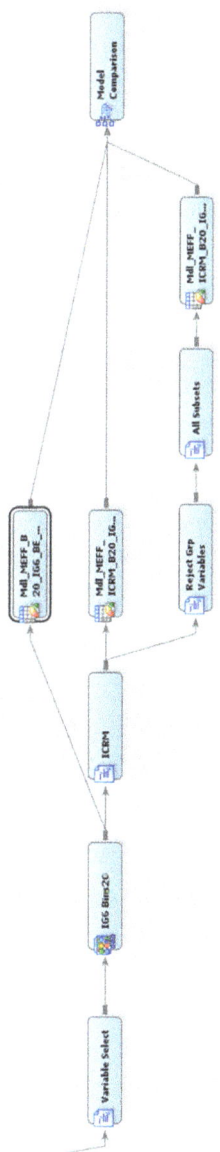

Figure 18-3. Zooming into the main-effects section.

573

CHAPTER 18   THE MAIN-EFFECTS MODEL

Figure 18-4 shows the results obtained from the diagram shown in **Figure 18-3**. It is notable that the three main-effects models exhibit a high degree of similarity in terms of performance, yet a lesser degree of similarity in terms of complexity. For instance, the three models exhibit a validation ROC value close to 0.9. However, the Mdl_MEFF_B20_IG6_BE_SBC model has 30 variables, the Mdl_MEFF_ICRM_B20_IG6_BE_SBC has 25, and the Mdl_MEFF_ICERM_B20_IG6_AllSub_Mcq has 28. Evidently, the negligible performance gain of the Mdl_MEFF_B20_IG6_BE_SBC model, of 0.006, does not justify its gain in complexity of 5 and 2 variables compared with the Mdl_MEFF_ICRM_B20_IG6_BE_SBC and the Mdl_MEFF_ICERM_B20_IG6_AllSub_Mcq models, respectively. Therefore, in this particular case, it is evident that the optimal option is the Mdl_MEFF_ICRM_B20_IG6_BE_SBC model.

Model Description	Selection Criterion: Valid: Roc Index	Train: Model Degrees of Freedom
Mdl_MEFF_B20_IG6_BE_SBC	0.901	30K
Mdl_MEFF_ICRM_B20_IG6_BE_SBC	0.895	25K
Mdl_MEFF_ICERM_B20_IG6_AllSub_Mcq	0.895	28K

Figure 18-4. Results of the main-effects models.

CHAPTER 18   THE MAIN-EFFECTS MODEL

Figure 18-5, on the other hand, shows the resulting models from four examples of main-effects multivariable selection processes. In examples (A), (B), and (C), the criteria for selecting the best model can be based on the principle of parsimony as the difference in performance does not vary greatly between the three selection strategies. In contrast, example (D) demonstrates an improvement of one ROC point, achieved at the cost of a single additional variable. Therefore, the Main_Effects_B20_IG6_BE_SBC model is the preferred choice in this instance. Furthermore, it is also notable in this instance that all models are relatively concise, comprising a total of between 10 and 12 variables. This renders the selection of an 11-variable model, a relatively straightforward process.

CHAPTER 18   THE MAIN-EFFECTS MODEL

(A)

Model Description	Selection Criterion: Valid: Roc Index	Train: Model Degrees of Freedom
Main_Effects_B20_IG6_BE_SBC	0.84	11
Main_Effects_ICRM_B20_IG6_AllSub_Mcq	0.832	13
Main_Effects_ICRM_B20_IG6_BE_SBC	0.832	9

(B)

Model Description	Selection Criterion: Valid: Roc Index	Train: Model Degrees of Freedom
Main_Effects_B20_IG6_BE_SBC	0.845	25
Main_Effects_ICRM_B20_IG6_AllSub_Mcq	0.843	23
Main_Effects_ICRM_B20_IG6_BE_SBC	0.843	19

(C)

Model Description	Selection Criterion: Valid: Roc Index	Train: Model Degrees of Freedom
Main_Effects_B20_IG6_BE_SBC	0.815	18
Main_Effects_ICRM_B20_IG6_AllSub_Mcq	0.812	20
Main_Effects_ICRM_B20_IG6_BE_SBC	0.811	17

(D)

Model Description	Selection Criterion: Valid: Roc Index	Train: Model Degrees of Freedom
Main_Effects_B20_IG6_BE_SBC	0.884	12
Main_Effects_ICRM_B20_IG6_AllSub_Mcq	0.878	13
Main_Effects_ICRM_B20_IG6_BE_SBC	0.875	11

Figure 18-5. Four selected examples of the multivariable selection process for the main-effects models.

18.3 Results and Discussion of the Best Full Main-Effects Model

The main-effects model, Mdl_MEFF_ICRM_B20_IG6_BE_SBC, will continue to be utilized in the subsequent sections, having been selected following an assessment of the performance statistics illustrated in **Figure 18-4**.

Table 18-1 (a detailed description of each variable is presented in **Table 18-2**) presents a summary of the results obtained from the Mdl_MEFF_ICRM_B20_IG6_BE_SBC model. The upper right section of the table displays the performance statistics KS, ROC, and Gini, while the lower section shows the final subset of 24 main effects selected by the BE-Only method following the ICRM. Respectively, **Figure 18-6** presents the assessment charts of the model, where chart (A) displays a line chart of the empirical odds vs. the average scorecard points for the training partition (left) and the validation partition (right) and chart (B) presents a bar chart with the absolute value of the effects' coefficients.

CHAPTER 18 THE MAIN-EFFECTS MODEL

Based on the performance statistics presented in **Table 18-1**, it can be concluded that the Mdl_MEFF_ICRM_B20_IG6_BE_SBC model has an excellent performance, as evidenced by the ROC value of the validation partition, which falls between 0.8 and 0.9 (**Table 17-1**). Furthermore, an examination of **Table 18-1** reveals that the total number of variables cannot be reduced any further based on statistical significance alone, as all of them are found to be extremely significant, with p-values <0.001. In contrast, the charts of the empirical odds (**Figure 18-6** (A)) indicate a linear and monotonic negative correlation between the observed default and the score estimated by the model, thereby suggesting that the present model provides an adequate fit to the empirical data. Finally, the bar chart of the absolute value of the coefficients of the main-effects model reveals two noteworthy observations. Firstly, there is a considerable discrepancy in the magnitude of the effects as one progresses from left to right on the chart in **Figure 18-6** (B). This is evidenced by the fact that the absolute value of the coefficient for the variable I_LOAN_AGE_YRS is 1.0384, while that of the variable I_QV_R_PAYT_APPR_CLC_L12_09 is 0.093, which represents a deviation of −91%. This discrepancy may suggest that some variables can be excluded without compromising the overall performance of the model. Secondly, the presence of an inverted coefficient, i.e., I_SV_R_PAYA_INST_CLC_L12_06, indicates that the model continues to exhibit collinearity issues.

It is important to note that, based on statistical performance alone, the Mdl_MEFF_ICRM_B20_IG6_BE_SBC model would have been accepted as the *"champion model,"* concluding the model-building process without addressing the issues identified in the previous assessment. These issues include the substantial discrepancy in magnitude between the coefficients of the variables, the coefficient inversion, and the inherent collinearity that results from it. These observations suggest that the model could benefit from reducing the total number of variables, despite their statistical significance.

CHAPTER 18 THE MAIN-EFFECTS MODEL

Table 18-1. *Summary results of the Md1_MEFF_ICRM_B20_IG6_BE_SBC model. The performance statistics KS, ROC, and Gini are presented on the top part, and the final subset of main effects that were selected by the BE-Only method, with collinearity reduction, are presented on the bottom part.*

Partition	KS	ROC	Gini
Train	0.657	0.897	0.795
Validate	0.653	0.895	0.791
Over Training	0.006	0.002	0.004

No.	Parameter	DF	Estimate	Error	χ^2	p-value	Std. Estimate	Odds
	Intercept	1	-2.3213	0.0063	136486.9	<.0001	0.098	
1	WOE_N_OCCUPATION	1	-0.354	0.0148	574.49	<.0001	-0.1242	0.702
2	WOE_I_AGE_YRS	1	-0.39	0.0141	765.11	<.0001	-0.0954	0.677
3	WOE_I_LOAN_AGE_YRS	1	-1.0384	0.0492	445.02	<.0001	-0.0838	0.354
4	WOE_I_MX_DYSPSTDUE_LON_L12M	1	-0.415	0.0084	2446.6	<.0001	-0.1564	0.66
5	WOE_I_MX_DYSPSTDUE_CCR_L09M	1	-0.183	0.0232	62.28	<.0001	-0.0199	0.833
6	WOE_I_RISK_LEVEL	1	-0.8641	0.0038	52876.57	<.0001	-0.4964	0.421
7	WOE_I_NUMDEPEND	1	-0.328	0.0571	33.03	<.0001	-0.0185	0.72
8	WOE_I_R_CREDIT_TASSTS	1	-0.5114	0.0409	156.65	<.0001	-0.0389	0.6
9	WOE_I_R_PAYA_INST_CLC_00MB	1	-0.4654	0.038	149.68	<.0001	-0.0368	0.628
10	WOE_I_R_TOPEXP_TASSTS	1	-0.4858	0.0457	113.11	<.0001	-0.0384	0.615
11	WOE_I_SDC_R_CHRG_CRRA_CPM_L12M	1	-0.4709	0.0393	143.61	<.0001	-0.0471	0.624
12	WOE_I_SIN_R_BALT_APPR_CLC_L13_24	1	-0.2671	0.0177	228.77	<.0001	-0.0504	0.766
13	WOE_I_SV_R_PAYA_INST_CLC_L12_06	1	0.1187	0.0224	28.18	<.0001	0.0144	1.126
14	WOE_I_QV_R_PAYT_APPR_CLC_L09_06	1	-0.1545	0.0135	130.99	<.0001	-0.044	0.857
15	WOE_I_QV_R_PAYT_BALT_CLC_L06_03	1	-0.36	0.0117	945.47	<.0001	-0.1108	0.698
16	WOE_N_HOMESTATUS	1	-0.3044	0.0297	105.17	<.0001	-0.0336	0.738
17	WOE_N_COLLATERAL_TYPE	1	-0.1624	0.0154	111.46	<.0001	-0.0392	0.85
18	WOE_N_PROFESSION	1	-0.1137	0.0306	13.77	0.0002	-0.0148	0.893
19	WOE_I_QV_R_PAYT_APPR_CLC_L12_09	1	-0.093	0.0142	43.14	<.0001	-0.0242	0.911
20	WOE_N_BRANCH	1	-0.4318	0.0103	1763.78	<.0001	-0.1632	0.649
21	WOE_I_AV_R_BALT_APPR_CLC_L06M	1	-0.1318	0.0112	137.74	<.0001	-0.0392	0.877
22	WOE_I_SIN_R_PAYT_APPR_CLC_L13_24	1	-0.2683	0.0123	476.53	<.0001	-0.0914	0.765
23	WOE_I_SIN_R_BALT_APPR_CLC_L03M	1	-0.1428	0.022	42	<.0001	-0.0213	0.867
24	WOE_I_MV_R_BALT_APPR_CLC_L02_01	1	-0.3012	0.0227	175.48	<.0001	-0.0331	0.74

CHAPTER 18 THE MAIN-EFFECTS MODEL

It is evident that the modeling process cannot be concluded without addressing the aforementioned issues. However, prior to this, it is necessary to undertake one additional task: namely, the visual inspection of the WoE functions to assess their coherence in terms of business logic.

CHAPTER 18   THE MAIN-EFFECTS MODEL

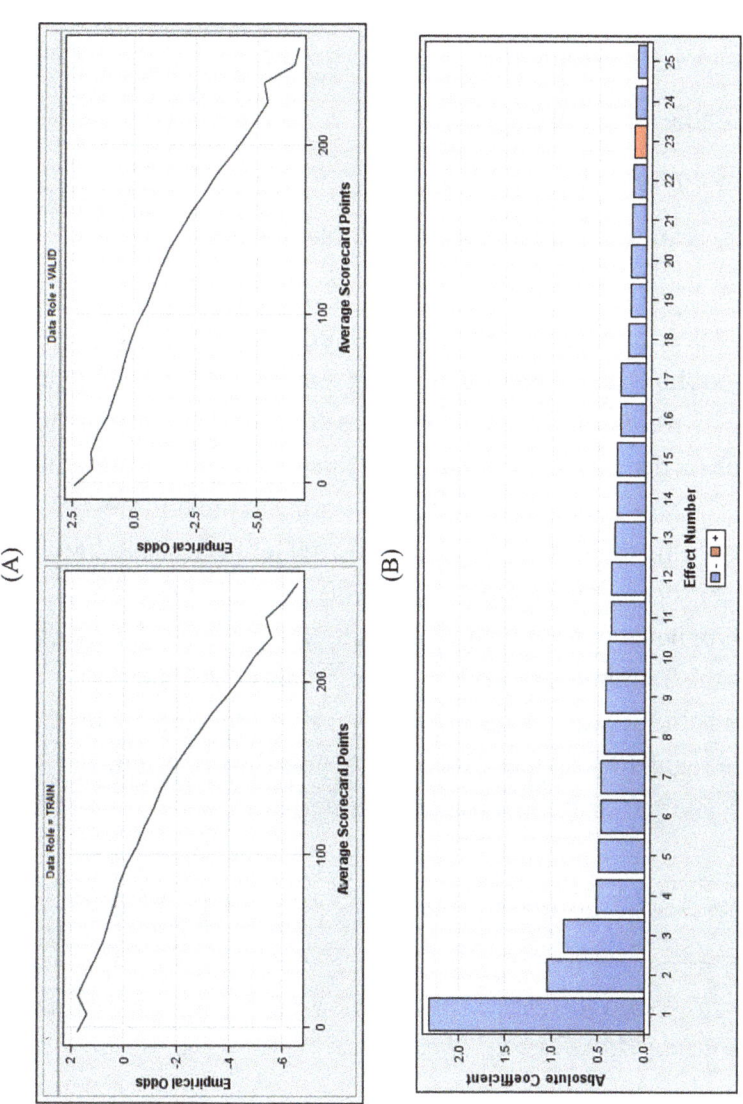

Figure 18-6. *Assessment charts for model* Md1_MEFF_ICRM_B20_IG6_BE_SBC. *(A) Line chart of the empirical odds vs. the average scorecard points for the training partition (left) and the validation partition (right). (B) Bar chart with the absolute value of the effects' coefficients.*

581

18.4 The WoE Refinement Process

The final selection of main effects will be isolated by the addition of an additional branch comprising three nodes:

1. A variable select node
2. An IG node
3. A scorecard node with no multivariable selection method applied

Figure 18-7 illustrates the resulting three-node branch in EM. The creation of an additional branch is justified by two main reasons. Firstly, it is imperative to avoid modifying any of the results obtained during the preceding steps, as this is a common mistake during the model-building process. This is because the *"accidental"* overwriting of previous results can have far-reaching consequences, including the loss of traceability of the modeling process, the loss of crucial information that may be challenging to replicate or recover, and the potential for more significant issues, such as the inability to reproduce the modeling process or provide evidence to support the methodology used to develop the final model during a regulatory review. This underscores the importance of maintaining a separate branch for ongoing refinement.

Figure 18-7. Illustrative example of a three-node branch created to isolate the final selection of main effects.

The second rationale for establishing an additional branch is to isolate and freeze the variables of the selected main-effects model. It is important to note that although the scorecard node is capable of automatically selecting the variables of the best model selected by the BE-Only algorithm, it is

CHAPTER 18 THE MAIN-EFFECTS MODEL

always preferable to connect the additional branch to the initial nodes of the diagram for the sake of clarity. The Variable Select node is therefore a convenient tool for this purpose, as it allows the user to select the 24 variables from **Table 18-1** (a detailed description of each variable is presented in **Table 18-2**) by inputting their names as a list into the **varlist** macro-variable, as demonstrated in **Listing 18-2**. This is in contrast to the EM's metadata node, which does not permit the selection of variables in this manner.

Listing 18-2. Example of the macro-variable `varlist` listing the 24 variables presented in **Table 18-1**.

```
%LET varlist=
N_OCUPATION
I_AGE_YRS
I_LOAN_AGE_YRS
I_MX_DYSPSTDUE_LON_L12M
I_MX_DYSPSTDUE_CCR_L09M
I_RISK_LEVEL
I_NUMDEPEND
I_R_CREDIT_TASSTS
I_R_PAYA_INST_CLC_OOMB
I_R_TOPEXP_TASSTS
I_SDC_R_CHRG_CRRA_CPM_L12M
I_SIN_R_BALT_APPR_CLC_L13_24
I_SV_R_PAYA_INST_CLC_L12_06
I_QV_R_PAYT_APPR_CLC_L09_06
I_QV_R_PAYT_BALT_CLC_L06_03
N_HOMESTATUS
N_COLLATERAL_TYPE
N_PROFESSION
I_QV_R_PAYT_APPR_CLC_L12_09
N_BRANCH
I_AV_R_BALT_APPR_CLC_L06M
I_SIN_R_PAYT_APPR_CLC_L13_24
I_SIN_R_BALT_APPR_CLC_L03M
I_MV_R_BALT_APPR_CLC_L02_01
;
```

CHAPTER 18 THE MAIN-EFFECTS MODEL

18.4.1 Defining the Expected Trend

In Section 3.4, the inductive reasoning is defined as a type of reasoning guided exclusively by data. This category of reasoning is exemplified by scientific research, particularly statistical analysis. In statistical analysis, conclusions are drawn based on a sufficient amount of data, allowing hypotheses to be proven or rejected according to a predefined level of significance. Accordingly, from a purely statistical perspective, it is not possible to predetermine the trend or relationship that our explanatory variables must have with the target variable, as the relationship must be supported by the data and not the other way around. However, in contrast to a purely statistical perspective, from a scorecard perspective, it is indeed possible to do so.

One of the fundamental distinctions between the scorecard and the classical approach is that the former is proactive and deductive, whereas the latter is passive and inductive. Consequently, in contrast to the methodologies described in other publications, this book presents a comprehensive methodology based on a thorough examination of the influencing factors affecting the system of interest, with the objective of designing specific variables for the problem at hand. This is the key aspect that distinguishes the scorecard approach from the classical approach and allows us to set the relationship between explanatory and target variables. This is due to the fact that the relationship between these variables is to some extent predetermined from the beginning of the analysis, during the conceptual variable design stage.

It must be emphasized that the conceptual variable design is capable of anticipating the logical trends of most explanatory variables, but not all. This is not an inherent defect but rather an inherent aspect of the modeling process. To illustrate this point, **Table 18-2** provides a comprehensive account of the 24 primary effects presented in **Table 18-1**, along with their anticipated patterns with respect to the target variable. It should be noted that the 24 variables are conveniently ordered according to their degree of similarity. For example, the first five variables are nominal in nature, and therefore, no trend can be deduced in advance (or at least not apparently, since professions and occupations associated with a higher income, for instance, may also be negatively correlated with the risk of default).

The subsequent two variables are discrete sociodemographic variables, specifically the age of the customer and the number of economic dependents. The utilization of age as a variable is a common practice in PD models. However, it is not essential to be aware of its prevalence in PD models to conclude that as individuals age, they transition into adulthood, which is associated with traits such as conscientiousness, responsibility, financial literacy, and higher income (Thomas Sowell [64]). Conversely, the traits associated with youthfulness are frequently linked with impulsivity, lack of financial experience, and lower income. At this juncture, it is important to acknowledge that a counterargument to the previous reasoning may be that in countries where opportunities for the younger demographic are limited or starting salaries are low, financial concerns may outweigh age as an underlying factor. Nevertheless, this line of reasoning fails to acknowledge the fact that age, salary, and risk are *confounded* (as evidenced by **Figure 3-5**). This is because age affects both the independent and dependent variables (such as salary and marital status in the case of age). Therefore, it is reasonable to assume that a negative correlation will be observed between the default event and age.

Conversely, the number of economic dependents is not as straightforward to ascertain. One might posit that an increase in the number of dependents will inevitably result in an elevated risk of default, given that the financial obligations associated with supporting multiple dependents may reach a critical point, potentially leading to an inability to fulfill these obligations. However, this assertion is not supported by empirical observations, as previously discussed in Section 3.2, where the phenomenon known as the *"Married with Children Effect"* was examined. In this section, the available data demonstrate that, contrary to popular belief, the presence of additional dependents, such as children, is associated with a reduced probability of default.

CHAPTER 18 THE MAIN-EFFECTS MODEL

Table 18-2. *Detailed description of the 24 explanatory variables presented in Table 18-1 and their expected trend with respect to the target variable.*

No.	Variable	Description	Expected Trend
1	N_BRANCH	Application branch	N/A
2	N_COLLATERAL_TYPE	Type of collateral or security	N/A
3	N_OCCUPATION	Customer's occupation	N/A
4	N_PROFESSION	Customer's profession	N/A
5	N_HOMESTATUS	Present home status	N/A
6	I_AGE_YRS	Customer's age (in years)	Upward trend, reduces risk
7	I_NUMDEPEND	Economic dependents	Upward trend, reduces risk
8	I_LOAN_AGE_YRS	Loan age (time since origination)	unknown
9	I_MX_DYSPSTDUE_LON_L12M	Loan's maximum days past due in the last 12 months	Upward trend, increases risk
10	I_MX_DYSPSTDUE_CCR_L09M	Credit card's maximum days past due in the last 9 months	Upward trend, increases risk
11	I_RISK_LEVEL	Internal risk level (ordinal)	Upward trend, increases risk
12	I_R_CREDIT_TASSTS	Total credits (all active loans and credits) divided by total assets (current account, collaterals, investments)	Upward trend, increases risk
13	I_R_TOPEXP_TASSTS	Total expenses (all purchases, cash withdrawals, charges) divided by total assets (current account, collaterals, investments)	Upward trend, increases risk
14	I_R_PAYA_INST_CLC_00MB	Monthly payment/instalment of the last month	Upward trend, decreases risk
15	I_SV_R_PAYA_INST_CLC_L12_06	Six-month variation of the ratio of monthly payment/instalment from 12 months back to 6 months back	Upward trend, decreases risk
16	I_QV_R_PAYT_BALT_CLC_L06_03	Quarterly variation of the ratio of cumulative payment/outstanding amount from 6 months back to 3 months back	Upward trend, decreases risk
17	I_QV_R_PAYT_APPR_CLC_L09_06	Quarterly variation of the ratio of cumulative payment/approved amount from 9 months back to 6 months back	Upward trend, decreases risk
18	I_SIN_R_PAYT_APPR_CLC_L13_24	Increments of the ratio of cumulative payment/approved amount from the last 13 to 24 months	Upward trend, decreases risk
19	I_QV_R_PAYT_APPR_CLC_L12_09	Quarterly variation of the ratio of cumulative payment/approved amount from 12 months back to 9 months back	Upward trend, decreases risk
20	I_SIN_R_BALT_APPR_CLC_L13_24	Increments of the ratio of outstanding amount/approved amount from the last 13 to 24 months	Upward trend, increases risk
21	I_AV_R_BALT_APPR_CLC_L06M	Average ratio of outstanding amount/approved amount for the last 6 months	Upward trend, increases risk
22	I_SIN_R_BALT_APPR_CLC_L03M	Increments of the ratio of outstanding amount/approved amount for the last 3 months	Upward trend, increases risk
23	I_MV_R_BALT_APPR_CLC_L02_01	Monthly variation of the ratio of outstanding balance/approved amount from 2 months back to 1 month back	Upward trend, increases risk
24	I_SDC_R_CHRG_CRRA_CPM_L12M	Decrements of the ratio of charges/current amount for the last 12 months	Upward trend, decreases risk

Variable 8, the age of the loan (time since origination), provides an additional illustrative example of a variable for which the trend is not immediately apparent. Nevertheless, there may be an intuitive sense that the age of the loan could possess significant predictive power. The question then becomes: How might this be the case? As demonstrated in **Table 18-1**, the selection of loan age as a primary predictor is attributable to its substantial contribution to the multivariable model (by statistical means). However, this does not provide a satisfactory explanation for its inclusion in the initial training set of variables.

It is important to bear in mind that the modeling process involves both creativity and logical deduction. Consequently, it is not uncommon to include certain variables based solely on intuition rather than purely deductive reasoning. It is therefore important to trust one's intuition and consider including variables that may not have an immediate logical explanation or business judgment. It is important to acknowledge that the absence of an immediate rationale for including a variable does not necessarily diminish its importance. Instead, it may be indicative that its importance will become apparent through subsequent analysis, particularly if it is selected as one of the main effects.

The next three variables—maximum loan delinquency in the last 12 months (9), maximum credit card delinquency in the last 9 months (10), and the internal risk level in ordinal units (11)–are all directly correlated with the number of days past due. Therefore, a positive correlation is anticipated between these variables and the target variable. It is crucial to acknowledge that for variables directly associated with the number of days past due, a positive linear correlation with the target variable is both anticipated and required. Any observed deviation from this anticipated outcome may indicate deficiencies in the data processing or data quality.

The remaining 13 variables can be grouped into six different groups, according to their base ratio or conceptual meaning:

CHAPTER 18 THE MAIN-EFFECTS MODEL

Financial Ratios: The financial ratios presented in **Equation 18-1** have a negative implication. They represent the ratio of the customer's total credits, which include all existing loans and credits, to their total expenses, which include recorded purchases, charges, and cash withdrawals. These figures are then compared with the customer's total assets, which include the combined value of savings accounts, current accounts, investment instruments, and collateral holdings. Consequently, it is anticipated that a positive correlation with the target variable will be observed when the numerator increases or the denominator decreases, and a negative correlation when both the numerator and denominator move in opposite directions.

$$\frac{Credits_{Total}}{Assets_{Total}} \quad \frac{Expenses_{Total}}{Assets_{Total}}$$

Equation 18-1. Financial ratios.

Payment-to-Installment Ratio: The correlation between the monthly payment-to-installment ratio, and the target variable, is expected to be negative when the ratio increases, and positive when the ratio decreases.

$$\frac{Monthly\ Payment}{Instalment}$$

Equation 18-2. Monthly payment over installment.

Cumulative Payment-to-Outstanding, or Approved Amount Ratio: The cumulative payment represents the total debt covered up-to-date; consequently, the correlation between the cumulative payment-to-outstanding or approved amount ratio and the target variable is expected to be negative as the ratio increases, and positive otherwise.

$$\frac{Cumulative\ Payment}{Outstanding\ Amount} \quad \frac{Cumulative\ Payment}{Approved\ Amount}$$

Equation 18-3. Cumulative payment over outstanding amount and approved amount.

Outstanding Amount-to-Approved Amount Ratio: The ratio of outstanding amount-to-approved amount can also be expressed as

$$1 - \frac{Cumulative\ Payment}{Approved\ Amount}$$

This formulation has two implications: First, this ratio will have an inverse correlation with the cumulative payment-to-approved amount ratio, and second, it suggests that one of the two ratios may be redundant, given their complementary nature.

It is worth noting that linear combinations among our design variables are not uncommon, especially when creating hundreds or thousands of them.

$$\frac{Outstanding\ Amount}{Approved\ Amount}$$

Equation 18-4. Outstanding amount over approved amount.

Charges-to-Savings and Current Accounts Ratio: The correlation between the monthly ratio of charges-to-savings and current accounts and the target variable is expected to be positive as the ratio increases, and negative otherwise.

$$\frac{Charges}{savings - current\ accounts}$$

Equation 18-5. Charges over savings-current account.

18.4.2 Interactive Grouping Feature

The IG node [13] incorporates a functionality that enables the interactive examination of the automatically calculated groups for each of the main effects listed in **Table 18-2**. As illustrated in **Figure 18-8**, the initial window

CHAPTER 18   THE MAIN-EFFECTS MODEL

of the IG feature presents a summary of the 24 main effects, listed in descending order according to their Gini statistic, as presented in **Table 18-1**. The summary displays the following information:

- **Variable/Label**: Display the physical name of the variable.
- **Predefined grouping**: Shows whether the current WoE was calculated or precalculated in a previous IG node. Notice that in this case, the predefined flag is set to "Y" for all variables, this is because the WoEs calculated in **Figure 18-3** were imported into the IG node presented in **Figure 18-7** instead of being reestimated.
- **Level**: Shows the level of the variable.
- **Calculated Role**: Indicates whether the current variable has been accepted or rejected according to the predefined Gini cutoff of 5.
- **New role**: Allows the user to change the automatically estimated role for the desired role.
- **Original Gini**: Displays the automatically estimated Gini statistic for the current variable.
- **Original Information Value**: Displays the automatically estimated IV for the variable at hand.
- **Gini Statistic**: Shows the resulting Gini statistic after applying manual changes to the automatic estimates.
- **Information Value**: Shows the resulting IV after applying manual changes to the automatic estimates.

CHAPTER 18 THE MAIN-EFFECTS MODEL

Label	Pre-Defined Grouping	Level	Calculated Role	New Role	Original Gini	Original Information Value	Gini Statistic	Information Value
I_RISK_LEVEL	Y	INTERVAL	Input	Default	58.937	1.837	58.937	1.837
I_MX_DYSPSTDUE_LON_L12M	Y	INTERVAL	Input	Default	37.842	0.684	37.842	0.684
I_SIN_R_PAYT_APPR_CLC_L13_24	Y	INTERVAL	Input	Default	32.412	0.354	32.412	0.354
N_BRANCH	Y	NOMINAL	Input	Default	32.254	0.404	32.254	0.404
I_AV_R_BALT_APPR_CLC_L06M	Y	INTERVAL	Input	Default	31.03	0.383	31.03	0.383
I_OV_R_PAYT_BALT_CLC_L06_03	Y	INTERVAL	Input	Default	29.304	0.35	29.304	0.35
I_OV_R_PAYT_APPR_CLC_L09_06	Y	INTERVAL	Input	Default	27.482	0.298	27.482	0.298
I_OV_R_PAYT_APPR_CLC_L12_09	Y	INTERVAL	Input	Default	25.8	0.248	25.8	0.248
I_AGE_YRS	Y	INTERVAL	Input	Default	25.155	0.206	25.155	0.206
N_COLLATERAL_TYPE	Y	NOMINAL	Input	Default	21.355	0.179	21.355	0.179
N_OCUPATION	Y	NOMINAL	Input	Default	20.703	0.253	20.703	0.253
I_SIN_R_BALT_APPR_CLC_L13_24	Y	INTERVAL	Input	Default	18.877	0.128	18.877	0.128
I_SIN_R_BALT_APPR_CLC_L03M	Y	INTERVAL	Input	Default	12.817	0.083	12.817	0.083
N_HOMESTATUS	Y	NOMINAL	Input	Default	10.483	0.042	10.483	0.042
N_PROFESSION	Y	NOMINAL	Input	Default	8.445	0.044	8.445	0.044
I_SV_R_PAYA_INST_CLC_L12_06	Y	INTERVAL	Input	Default	7.937	0.067	7.937	0.067
I_LOAN_AGE_YRS	Y	INTERVAL	Input	Default	7.408	0.021	7.408	0.021
I_R_CREDIT_TASSTS	Y	INTERVAL	Input	Default	7.088	0.02	7.088	0.02
I_MX_DYSPSTDJE_CCR_L06M	Y	INTERVAL	Input	Default	7.009	0.053	7.009	0.053
I_MV_R_BALT_APPR_CLC_L02_01	Y	INTERVAL	Input	Default	6.049	0.055	6.049	0.055
I_R_PAYA_INST_CLC_00M6	Y	INTERVAL	Input	Default	5.908	0.024	5.908	0.024
I_R_TOPEXP_TASSTS	Y	INTERVAL	Input	Default	5.486	0.019	5.486	0.019
I_NUMDEPEND	Y	INTERVAL	Input	Default	5.409	0.011	5.409	0.011
I_SDC_R_CHRG_CRRA_CPM_L12M	Y	INTERVAL	Input	Default	5.328	0.026	5.328	0.026

Figure 18-8. Summary of the 24 main effects, shown in Table 18-1, presented in descending order according to their Gini statistic.

CHAPTER 18 THE MAIN-EFFECTS MODEL

It is important to note that the decision tree (DT) algorithm, employed by the IG node, operates as a constrained optimization engine. As a result, the grouping and splitting operations performed by the DT algorithm are based solely on the numerical maximization or minimization of the objective function. Consequently, the formation of meaningless groupings and/or misclassification of categories is to be expected.

Awareness of this fact makes it clear why the refinement process is necessary and also explains why sophisticated algorithms cannot be a substitute for visual and commonsense inspection by an experienced human. It should be evident that optimization algorithms lack the capacity to comprehend the intricacies of the problem at hand, which is why they require a multitude of constraints to produce a result that is, at best, only partially meaningful.

This leads to the next section, where a series of real-world and applied examples will be used to explain how to perform a visual inspection, followed by the corresponding diagnosis and the recommended actions needed to improve (where possible) the consistency of the WoE function.

18.4.3 Applied Examples of Visual Assessment, Diagnosis, and Manual Adjustments

We will now proceed to the Groupings tab for a visual inspection of the main effects. This will allow us to assess the consistency of the WoEs and determine whether manual adjustments are necessary. **Table 18-3** presents a summary of the diagnostics and subsequent actions taken for each of the 24 main effects, which can be further categorized into four groups based on their diagnostics. Group one comprises eight variables that were excluded on the grounds of insufficient consistency. Group two comprises seven variables that demonstrate a monotonic linear behavior in the anticipated direction and thus do not necessitate adjustment. Group three comprises six variables that necessitate some degree of adjustment prior to their selection. Finally, group four consists of three variables with ambiguous results that may require adjustment. It is recommended that these variables be included in the multivariable model for subsequent evaluation of their impact on the overall performance of the model. The aforementioned groups may be referred to as follows:

1. Inconsistent and rejected
2. Consistent, no adjustments required
3. Consistent, but manual adjustments required
4. Ambiguous, may require manual selection

CHAPTER 18   THE MAIN-EFFECTS MODEL

Table 18-3. *WoE diagnostics and actions taken after visual inspection.*

No.	Variable	Description	WoE Diagnostics
1	N_BRANCH	Application branch	Adjustment needed
2	N_COLLATERAL_TYPE	Type of collateral or security	Adjustment needed
3	N_OCCUPATION	Customer's occupation	Adjustment needed
4	N_PROFESSION	Customer's profession	No adjustment needed
5	N_HOMESTATUS	Present home status	Adjustment needed
6	I_AGE_YRS	Customer's age (in years)	No adjustment needed
7	I_NUMDEPEND	Economic dependents	Adjustment needed
8	I_LOAN_AGE_YRS	Loan age (time since origination)	Adjustment needed
9	I_MX_DYSPSTDUE_LON_L12M	Loan's maximum days past due in the last 12 months	No adjustment needed
10	I_MX_DYSPSTDUE_CCR_L09M	Credit card's maximum days past due in the last 9 months	Rejected, lack of consistency
11	I_RISK_LEVEL	Internal risk level (ordinal)	No adjustment needed
12	I_R_CREDIT_TASSTS	Total credits (all active loans and credits) divided by total assets (current account, collaterals, investments)	Rejected, lack of consistency
13	I_R_TOPEXP_TASSTS	Total expenses (all purchases, cash withdrawals, charges) divided by total assets (current account, collaterals, investments)	Rejected, lack of consistency
14	I_R_PAYA_INST_CLC_00MB	Monthly payment/instalment of the last month	Rejected, lack of consistency
15	I_SV_R_PAYA_INST_CLC_L12_06	Six-month variation of the ratio of monthly payment/instalment from 12 months back to 6 months back	Inconclusive, assess after regression
16	I_QV_R_PAYT_BALT_CLC_L06_03	Quarterly variation of the ratio of cumulative payment/outstanding amount from 6 months back to 3 months back	Rejected, lack of consistency
17	I_QV_R_PAYT_APPR_CLC_L09_06	Quarterly variation of the ratio of cumulative payment/approved amount from 9 months back to 6 months back	Rejected, lack of consistency
18	I_SIN_R_PAYT_APPR_CLC_L13_24	Increments of the ratio of cumulative payment/approved amount from the last 13 to 24 months	No adjustment needed
19	I_QV_R_PAYT_APPR_CLC_L12_09	Quarterly variation of the ratio of cumulative payment/approved amount from 12 months back to 9 months back	Rejected, lack of consistency
20	I_SIN_R_BALT_APPR_CLC_L13_24	Increments of the ratio of outstanding amount/approved amount from the last 13 to 24 months	No adjustment needed
21	I_AV_R_BALT_APPR_CLC_L06M	Average ratio of outstanding amount/approved amount for the last 6 months	No adjustment needed
22	I_SIN_R_BALT_APPR_CLC_L03M	Increments of the ratio of outstanding amount/approved amount for the last 3 months	Questionable, assess after regression
23	I_MV_R_BALT_APPR_CLC_L02_01	Monthly variation of the ratio of outstanding balance/approved amount from 2 months back to 1 month back	Rejected, lack of consistency
24	I_SDC_R_CHRG_CRRA_CPM_L12M	Decrements of the ratio of charges/current amount for the last 12 months	adjustment needed, assess after regression

18.4.3.1 Examples of Inconsistent WoEs

Figures 18-9 through **18-16** present the WoE's comprehensive report on the variables within group one. The upper section of the document presents two bar-and-line charts. On the left side, the event rate (primary y axis) and the calculated WoE (secondary y axis) are plotted against the cutoff point (Bin). On the right side, the event frequency (primary y axis) and the calculated WoE (secondary y axis) are plotted against the estimated group. It is notable that in all cases, the event rate, and consequently the WoE, does not exhibit a linear monotonic behavior with respect to the grouping variable or the risk level. Moreover, the variables I_QV_R_PAYT_BALT_CLC_L06_03, I_QV_R_PAYT_APPR_CLC_L09_06, and I_QV_R_PAYT_APPR_CLC_L12_09 not only fail to demonstrate linearity, but they also exhibit a zigzag-like pattern. It is noteworthy that the variable I_MX_DYSPSTDUE_CCR_L09M was rejected not only due to its lack of linearity, but also because 86% of the observations exhibited null values.

This is the list of the eight rejected variables:

1. I_MX_DYSPSTDUE_CCR_L09M
2. I_R_CREDIT_TASSTS
3. I_R_TOPEXP_TASSTS
4. I_R_PAYA_INST_CLC_OOMB
5. I_QV_R_PAYT_BALT_CLC_L06_03
6. I_QV_R_PAYT_APPR_CLC_L09_06
7. I_QV_R_PAYT_APPR_CLC_L12_09
8. I_MV_R_BALT_APPR_CLC_L02_01

CHAPTER 18 THE MAIN-EFFECTS MODEL

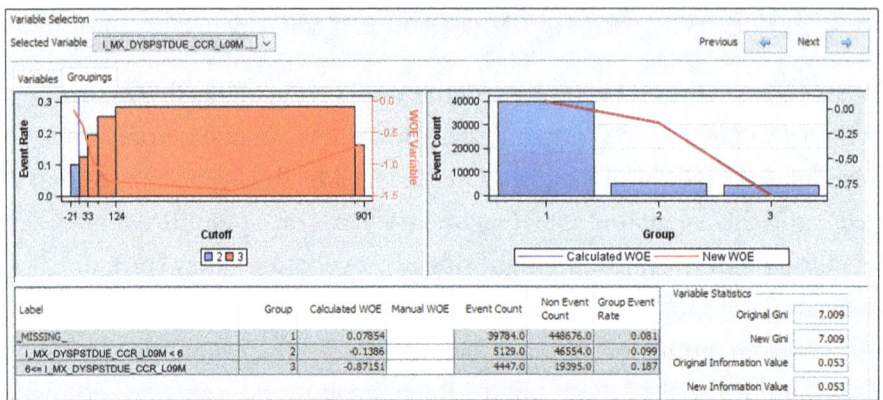

Figure 18-9. WoE diagnostics for I_MX_DYSPSTDUE_CCR_L09M.

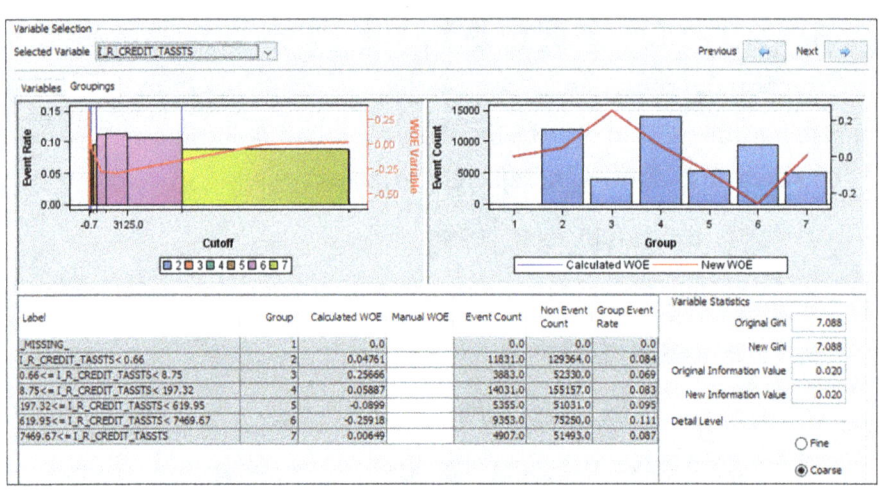

Figure 18-10. WoE diagnostics for I_R_CREDIT_TASSTS.

CHAPTER 18 THE MAIN-EFFECTS MODEL

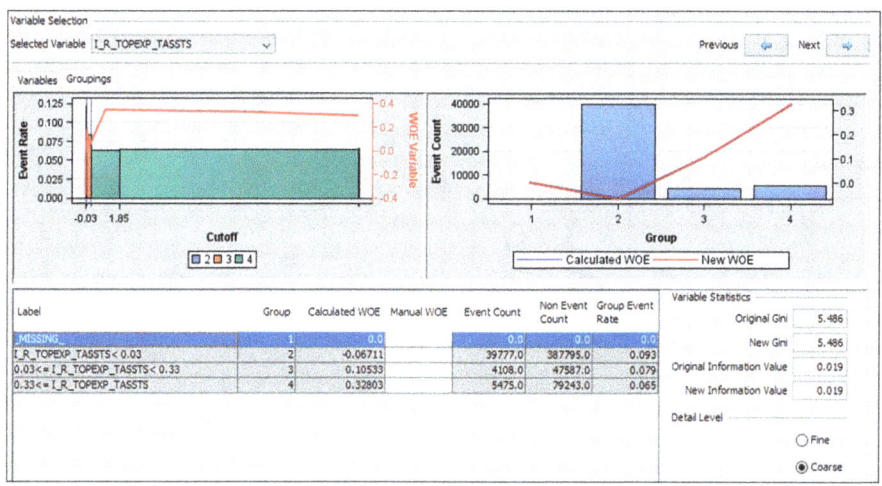

Figure 18-11. WoE diagnostics for I_R_TOPEXP_TASSTS.

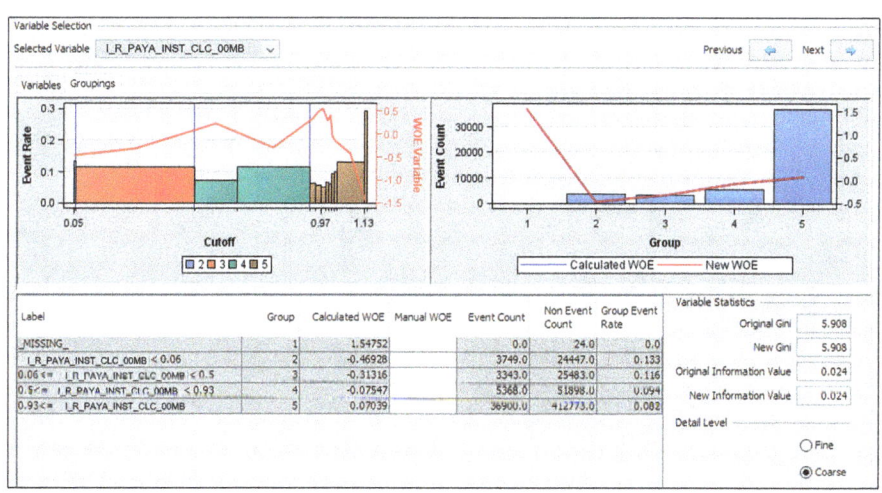

Figure 18-12. WoE diagnostics for I_R_PAYA_INST_CLC_00MB.

597

CHAPTER 18 THE MAIN-EFFECTS MODEL

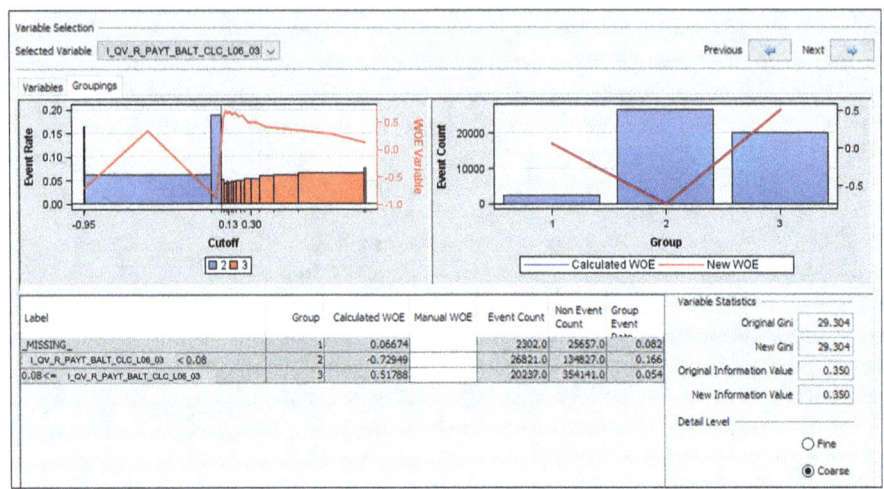

Figure 18-13. WoE diagnostics for I_QV_R_PAYT_BALT_CLC_L06_03.

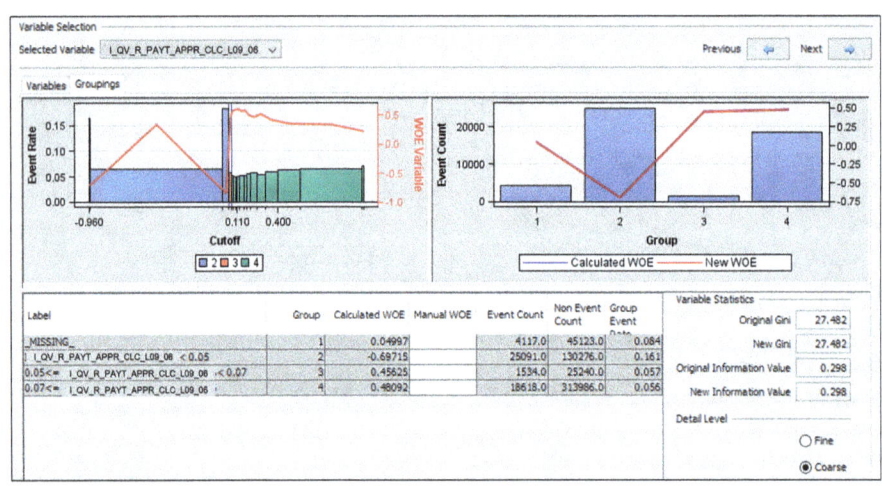

Figure 18-14. WoE diagnostics for I_QV_R_PAYT_APPR_CLC_L09_06.

CHAPTER 18 THE MAIN-EFFECTS MODEL

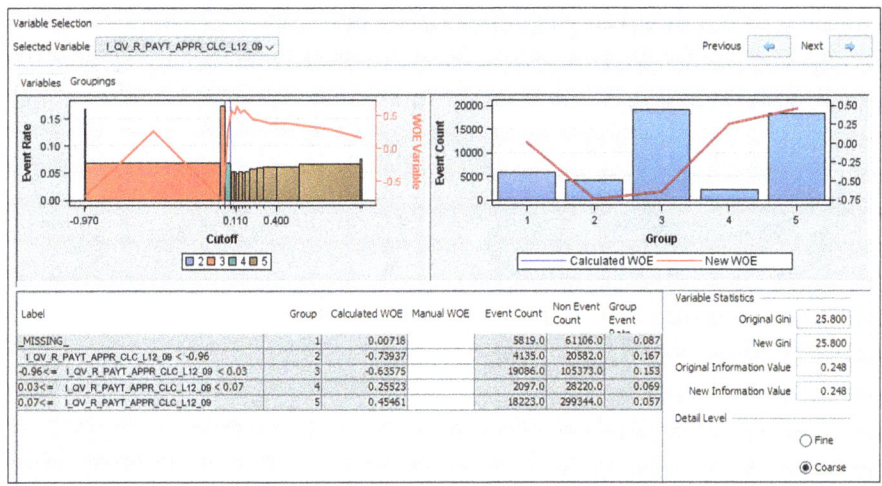

Figure 18-15. WoE diagnostics for I_QV_R_PAYT_APPR_CLC_L12_09.

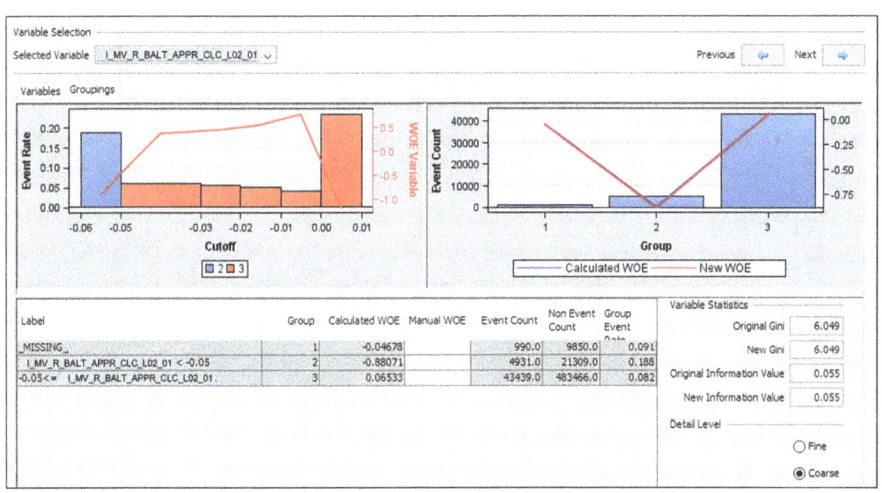

Figure 18-16. WoE diagnostics for I_MV_R_BALT_APPR_CLC_L02_01.

18.4.3.2 Examples of Ideal Monotonic Linear WoEs

Figures 18-17 through **18-23** present the detailed WoE reports for the variables in group two. It is noteworthy that these reports demonstrate a nearly ideal monotonic linear behavior of the event rate across the cutoff

599

points (top left), which consistently follows the expected trend for all variables in this group. This observation not only explains why no manual adjustments were required for the variables in group two, but also provides a rationale for rejecting the variables in group one.

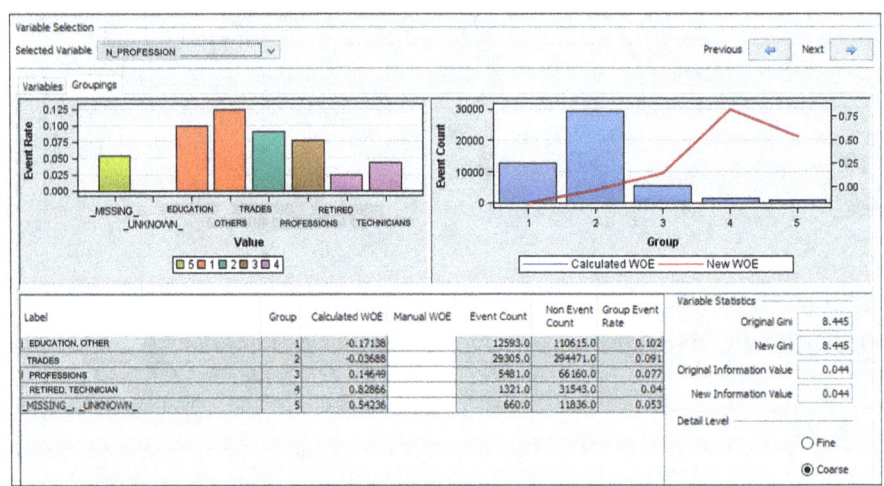

Figure 18-17. WoE diagnostics for N_PROFESSION.

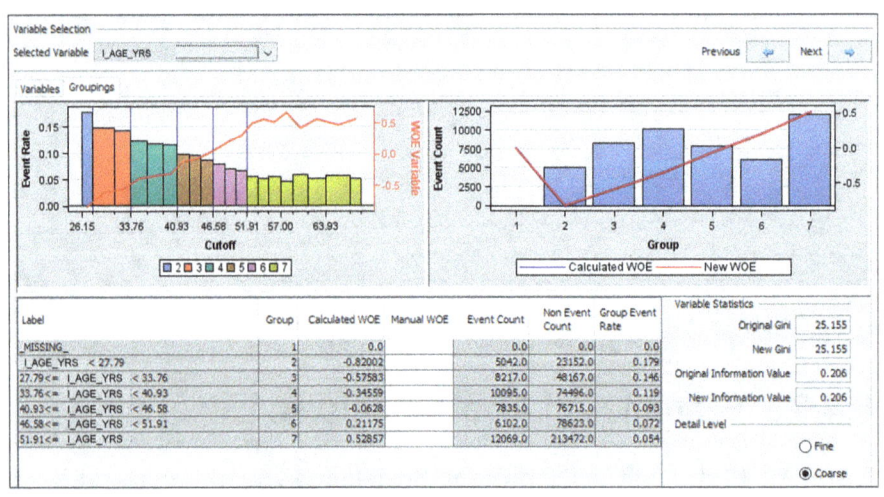

Figure 18-18. WoE diagnostics for I_AGE_YRS.

CHAPTER 18   THE MAIN-EFFECTS MODEL

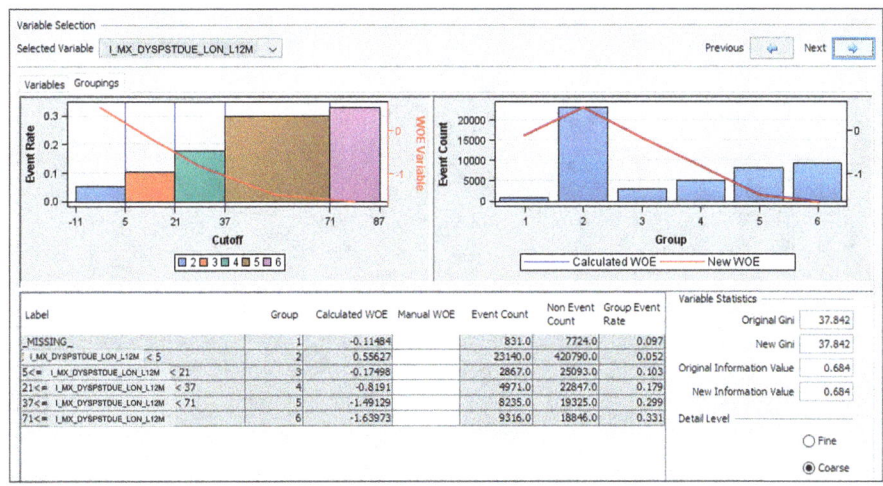

Figure 18-19. WoE diagnostics for I_MX_DYSPSTDUE_LON_L12M.

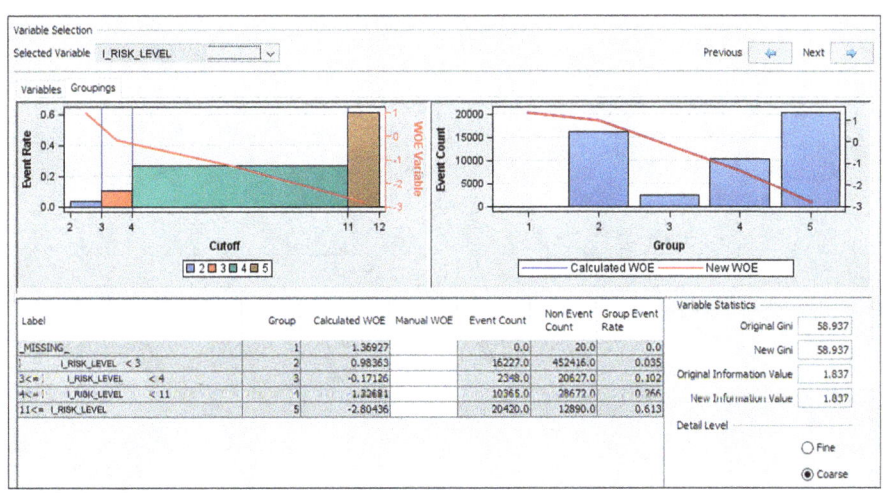

Figure 18-20. WoE diagnostics for I_RISK_LEVEL.

CHAPTER 18 THE MAIN-EFFECTS MODEL

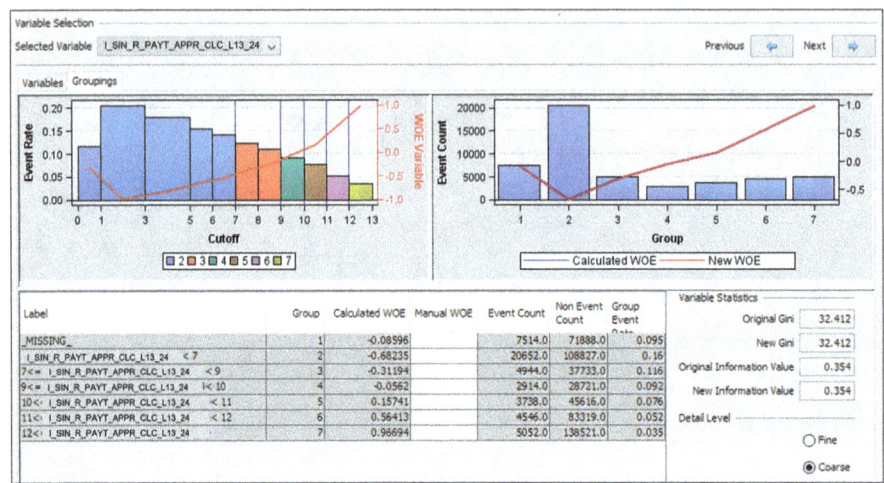

Figure 18-21. WoE diagnostics for I_SIN_R_PAYT_APPR_CLC_L13_24.

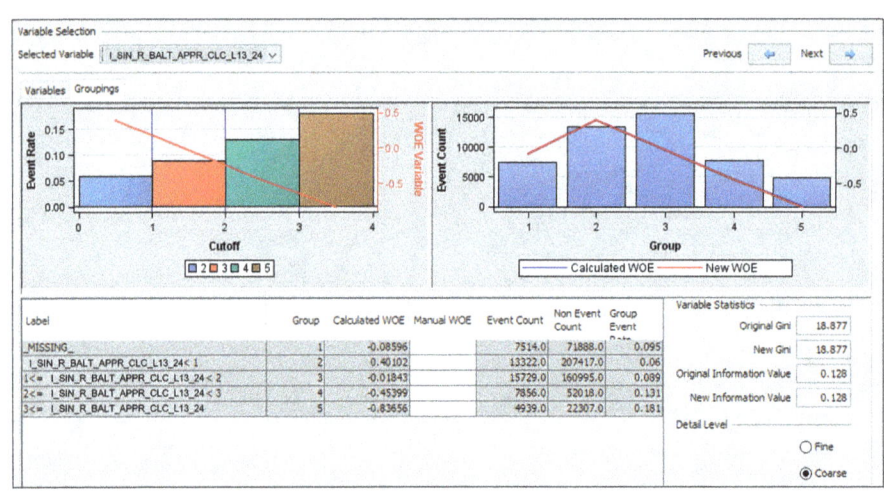

Figure 18-22. WoE diagnostics for I_SIN_R_BALT_APPR_CLC_L13_24.

CHAPTER 18 THE MAIN-EFFECTS MODEL

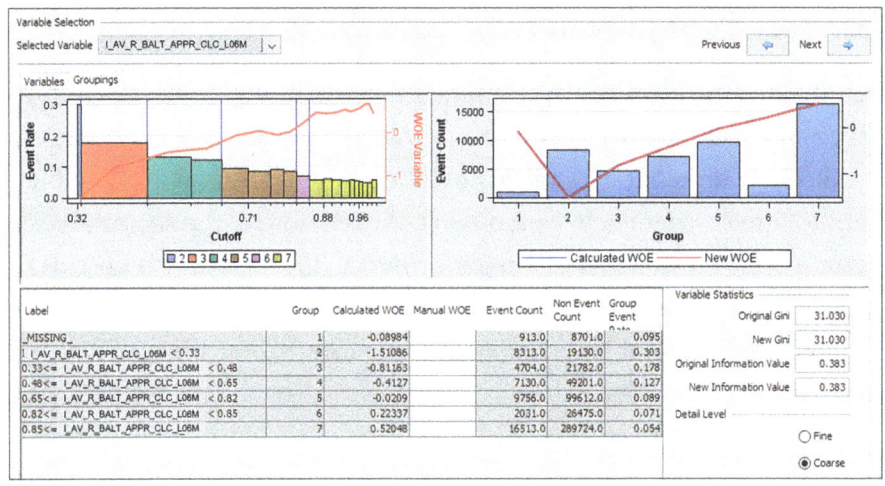

Figure 18-23. *WoE diagnostics for* I_AV_R_BALT_APPR_CLC_L06M.

From these results, one might argue that the estimated WoEs for the group two variables could be further refined after examining the detailed tabular summary report (lower part). Alternatively, one might even be tempted to collapse some of the risk levels based on the summary reports alone. Nevertheless, it is my recommendation that readers refrain from making any adjustments at this stage. Instead, it is recommended that readers consider this as an initial step toward the development of the final model.

The list of variables in group two is as follows:

1. N_PROFESSION
2. I_AGE_YRS
3. I_MX_DYSPSTDUE_LON_L12M
4. I_RISK_LEVEL
5. I_SIN_R_PAYT_APPR_CLC_L13_24
6. I_SIN_R_BALT_APPR_CLC_L13_24
7. I_AV_R_BALT_APPR_CLC_L06M

603

18.4.3.3 Examples of Refinable WoEs

In contrast to the variables in group two, those in group three necessitated refinement to enhance or rectify their projected trajectory. Nevertheless, this necessity for refinement does not inherently signify that the variables in group three are ineffective predictors. Indeed, four of the six variables in group three are among the top performers in the summary report, as illustrated in **Figure 18-8**, based on their Gini coefficient. Notably, these high-performing variables are all nominal, a feature that aligns with the observed challenges of the IG decision tree algorithm in accurately categorizing variables based on their risk level. This observation may provide a rationale for the classification of these variables within group three.

Figure 18-24 presents the detail report of the variable N_HOMESTATUS (current housing status). It is notable that the event rates for Risk-Groups[2] 2 and 3 are nearly identical, with values of 10.3% and 9.9%, respectively. The optimal course of action in this case is to merge Risk-Groups 1 and 3, given that their event rates exhibit minimal discrepancy.

Figure 18-25 depicts the detailed report subsequent to the merging of Risk-Groups 2 and 3. It is also notable that the difference between the Gini coefficient before and after the adjustment is minimal, with a difference of 0.103. Furthermore, the risk difference between the Risk-Groups has been enhanced, with the figures now standing at 12.7% for the Risk-Group 1, 10.2% for the Risk-Group 2, and 7.4% for the Risk-Group 3.

---

[2] To avoid confusion, we will refer to the WoE's classification groups as "Group" and to the groups pertaining to a specific WoE function as "Risk-Group."

CHAPTER 18   THE MAIN-EFFECTS MODEL

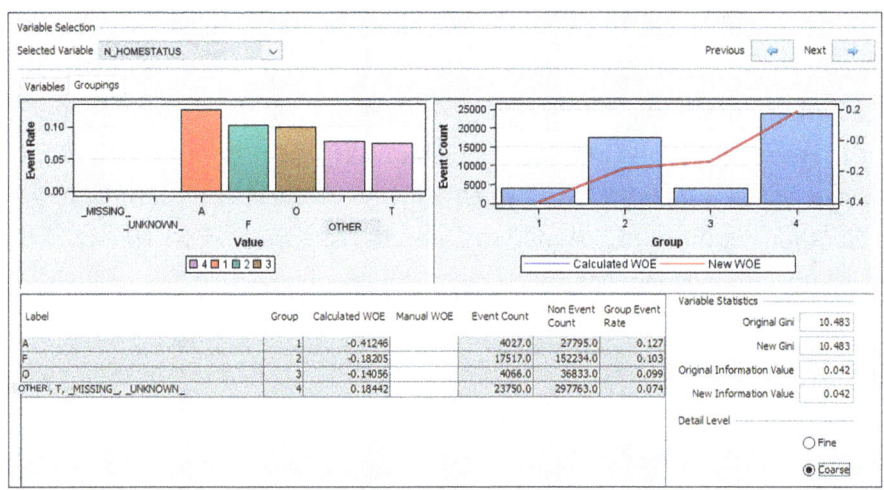

Figure 18-24. WoE diagnostics for N_HOMESTATUS.

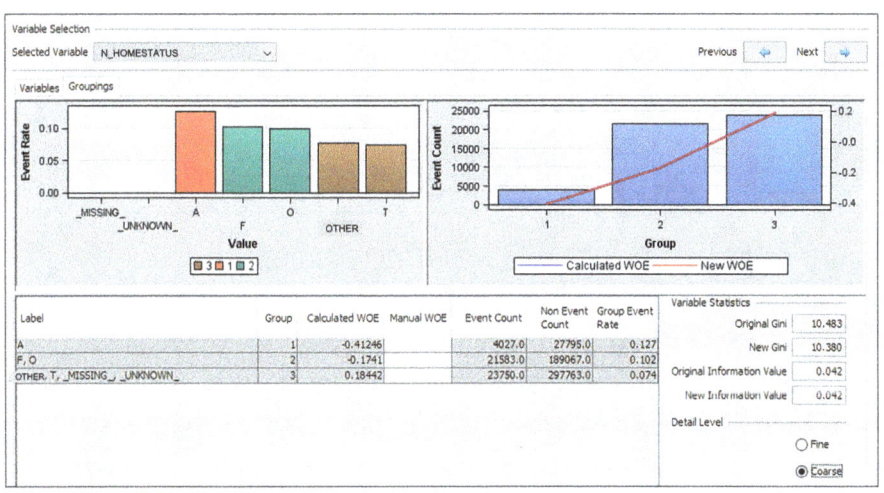

Figure 18-25. WoE diagnostics for N_HOMESTATUS after merging grouping—groups 2 and 3.

Figure 18-26 illustrates the detailed report for N_OCCUPATION (customer's occupation), which has been set to a fine level of detail. This configuration permits the visualization of event rates for each category

605

CHAPTER 18 THE MAIN-EFFECTS MODEL

within N_OCCUPATION. Upon closer examination, it becomes evident that the categories *"TRADES," "PUBLIC OFFICIALS,"* and *"PROFESSIONS"* are situated in close proximity, yet they are classified within distinct risk levels, with corresponding event rates of 11.4% and 10.8%, respectively. Similarly, the categories *"OTHERS"* and *"EDUCATION"* are in close proximity to each other, yet they are assigned to different risk levels, with event rates of 6.7% and 6.6%, respectively.

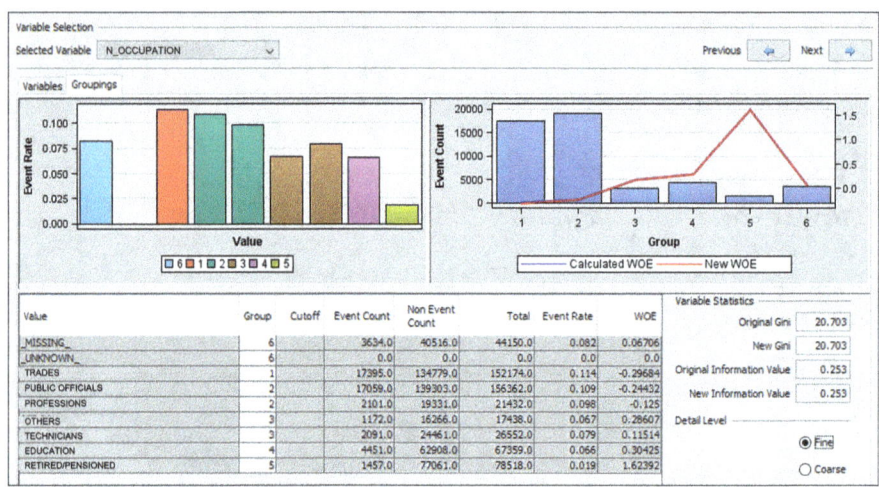

Figure 18-26. WoE diagnostics for N_OCCUPATION.

Figure 18-27 presents the detailed report for N_OCCUPATION, rearranged in a more consistent manner according to the individual event rates observed for each category. As with N_HOMESTATUS, the adjusted version shows a minimal discrepancy between the original (20.7) and the revised (19.9) Gini coefficient. Similarly, the disparity in event rates between risk groups is more pronounced after adjustment than before.

CHAPTER 18   THE MAIN-EFFECTS MODEL

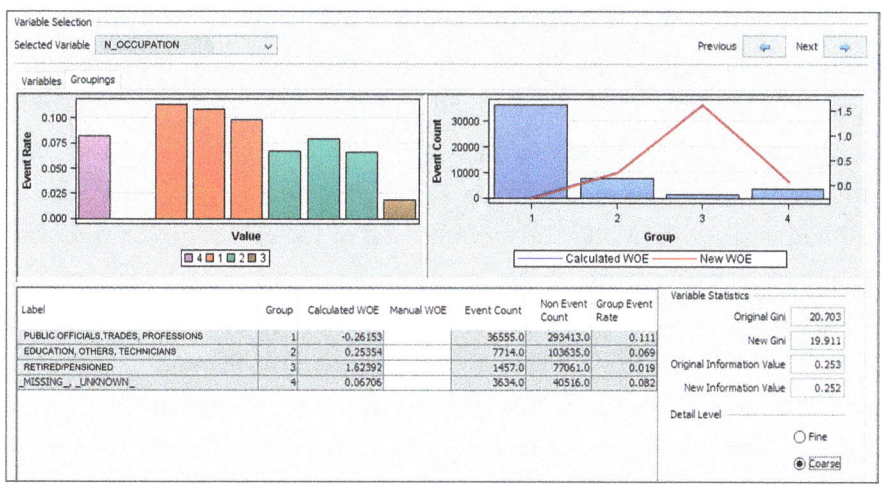

Figure 18-27. WoE diagnostics for N_OCCUPATION after manual adjustments.

It is noteworthy that N_OCCUPATION (customer's occupation) and N_PROFESSION (customer's profession) have similar categories, which makes one wonder why both are included in the model. While occupation refers to an individual's employment status, profession refers to their educational background, thus delineating two distinct variables.

However, it is important to note that the inclusion of these variables in the main-effects model was determined by statistical tests conducted throughout the modeling process. Surprisingly, both occupation and profession remained in the main-effects model even after the collinearity reduction (ICRM), after the IG node (see **Figure 18-3**).

This observation strongly suggests the presence of a significant interaction between occupation and profession, indicating that their inclusion is not only nonredundant but could also be beneficial in improving the overall performance of the model.

CHAPTER 18   THE MAIN-EFFECTS MODEL

The refinement process of the variable N_COLLATERAL_TYPE[3] will be omitted, as the adjustments made to it are largely similar to those made to previous nominal variables. Instead, the focus will be directed toward the variable N_BRANCH (application branch), which will allow us to see an example of how the IG algorithm deals with nominal variables of high cardinality, since the N_BRANCH is comprised of 121 categories or, rather, 121 application branches.

Figure 18-28 shows the coarse version of the detailed summary report, where it can be seen that there is no clear trend in the event rate on the x axis, and yet the Gini coefficient remains remarkably high at 32.25. In addition, **Figure 18-29** presents the detailed list of branches in descending order by event rate. Notably, the branches in Risk-Groups 1 (A) and 2 (B) exhibit consistent event rate trends, moving from lower to higher risk levels. However, an examination of the top and bottom five branches in Risk-Group 3 (C) and the top five branches in Risk-Group 4 (D) reveals inconsistent event rate trends that overlap across all Risk-Groups. In addition, **Table 18-4** shows that two-thirds (66.6%) of the branches were classified within Risk-Group 3 by the IG decision tree algorithm, encompassing nearly 44% of the observations in the training sample.

---

[3] The detailed report for N_COLLATERAL_TYPE, before and after adjustments, can be seen in **Figure 15-40** at the end of this section. It is worth noting that the improvements were related to the consistency in the classification of the categories, in terms of their event rate, and not so to the improvement of discriminatory power.

CHAPTER 18   THE MAIN-EFFECTS MODEL

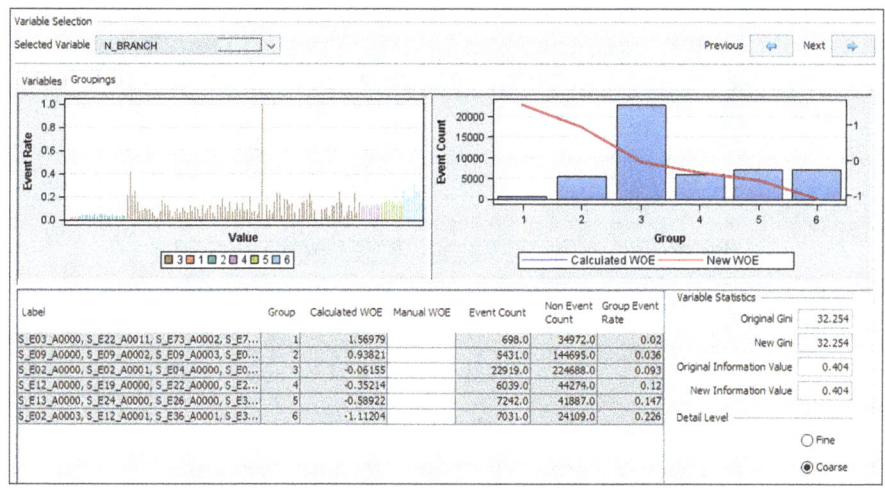

Figure 18-28. WoE diagnostics for N_BRANCH before adjustments.

Subsequently, each branch was manually reclassified based on its event rate, resulting in **Table 18-5**, which demonstrates a more balanced distribution of branches and training observations across risk groups. Moreover, the fit markedly enhanced the linear correlation between the average event rate and the Risk-Groups, as illustrated in **Figure 18-30** (A) and (B), which show the average event rate before and after the fit, respectively.

609

CHAPTER 18 THE MAIN-EFFECTS MODEL

(A)

	A	B	C	D	E	F	G
1	Value	Group	Events	No-Events	Total	Rate	WoE
4	S_E03_A0000	1	74	5,766	5,840	0.0130	2.0114
5	S_E22_A0011	1	93	3,569	3,662	0.0250	1.3031
6	S_E73_A0002	1	156	6,746	6,902	0.0230	1.4226
7	S_E77_A0000	1	375	18,891	19,266	0.0190	1.5752

(B)

	A	B	C	D	E	F	G
1	Value	Group	Events	No-Events	Total	Rate	WoE
8	S_E09_A0000	2	182	5,584	5,766	0.0320	1.0794
9	S_E09_A0002	2	145	3,374	3,519	0.0410	0.8028
10	S_E09_A0003	2	158	3,996	4,154	0.0380	0.8862
11	S_E09_A0004	2	236	6,118	6,354	0.0370	0.9109
12	S_E11_A0000	2	536	13,798	14,334	0.0370	0.9039

(C)

	A	B	C	D	E	F	G
1	Value	Group	Events	No-Events	Total	Rate	WoE
26	S_E02_A0001	3	18	26	44	0.4090	-1.9766
27	S_E05_A0000	3	616	1,915	2,531	0.2430	-1.2101
28	S_E44_A0001	3	461	1,472	1,933	0.2380	-1.1833
29	S_E74_A0006	3	213	690	903	0.2360	-1.1689
30	S_E77_A0002	3	13	43	56	0.2320	-1.1481

(D)

	A	B	C	D	E	F	G
1	Value	Group	Events	No-Events	Total	Rate	WoE
114	S_E09_A0005	3	22	2,248	2,270	0.0100	2.2825
117	S_E37_A0002	3	0	8	8	0.0000	0.4889
118	S_E52_A0001	3	0	44	44	0.0000	2.1443
119	S_E62_A0002	3	0	11	11	0.0000	0.7912
120	S_E62_A0005	3	0	2	2	0.0000	-0.7349

(E)

	A	B	C	D	E	F	G
1	Value	Group	Events	No-Events	Total	Rate	WoE
122	S_E58_A0001	4	518	3,405	3,923	0.1320	-0.4613
123	S_E74_A0000	4	593	4,173	4,766	0.1240	-0.3931
124	S_E22_A0000	4	970	6,992	7,962	0.1220	-0.3691
125	S_E36_A0000	4	958	6,949	7,907	0.1210	-0.3628
126	S_E19_A0000	4	670	5,020	5,690	0.1180	-0.3304

Figure 18-29. Detailed report for the variable N_BRANCH exported to a spreadsheet and ordered in descending order by event rate. (A) The four branches included in Risk-Group 1. (B) The top five branches for Risk-Group 2. (C) The top five branches for Risk-Group 3. (D) The bottom five branches for Risk-Group 3. (E) The top five branches for Risk-Group 4.

CHAPTER 18 THE MAIN-EFFECTS MODEL

Table 18-4. Summary statistics for N_BRANCH before adjustments

Group	Branches	% Branches	Observations	% Obs.	Avg. event rate	Max event rate	Min event rate	Std. dev. event rate	95% CI
1	4	2.78%	35,670	6.32%	0.020	0.025	0.013	0.005	0.005
2	19	13.19%	150,126	26.62%	0.037	0.049	0.031	0.006	0.003
3	**96**	**66.67%**	**247,607**	**43.90%**	**0.116**	**1.000**	**0.000**	**0.113**	**0.023**
4	8	5.56%	50,313	8.92%	0.122	0.132	0.108	0.008	0.006
5	9	6.25%	49,129	8.71%	0.148	0.161	0.135	0.010	0.007
6	8	5.56%	31,140	5.52%	0.219	0.295	0.162	0.043	0.030
	144	**100.00%**	**563,985**	**100.00%**	**0.111**	**1.000**	**0.000**	**0.102**	

CHAPTER 18 THE MAIN-EFFECTS MODEL

Table 18-5. Summary statistics for N_BRANCH after adjustments.

Group	Branches	% Branches	Observations	% Obs.	Avg. event rate	Max event rate	Min event rate	Std. dev. event rate	95% CI
1	12	8.33%	42,067	7.46%	0.014	0.028	0.000	0.012	0.007
2	23	15.97%	157,588	27.94%	0.037	0.049	0.031	0.005	0.002
3	**39**	**27.08%**	**179,121**	**31.76%**	**0.071**	**0.099**	**0.000**	**0.021**	**0.007**
4	37	25.69%	106,755	18.93%	0.129	0.152	0.106	0.013	0.004
5	15	10.42%	39,902	7.08%	0.162	0.179	0.152	0.008	0.004
6	18	12.50%	38,552	6.84%	0.279	1.000	0.186	0.187	0.086
	144	**100.00%**	**563,985**	**100.00%**	**0.111**	**1.000**	**0.000**	**0.102**	

CHAPTER 18   THE MAIN-EFFECTS MODEL

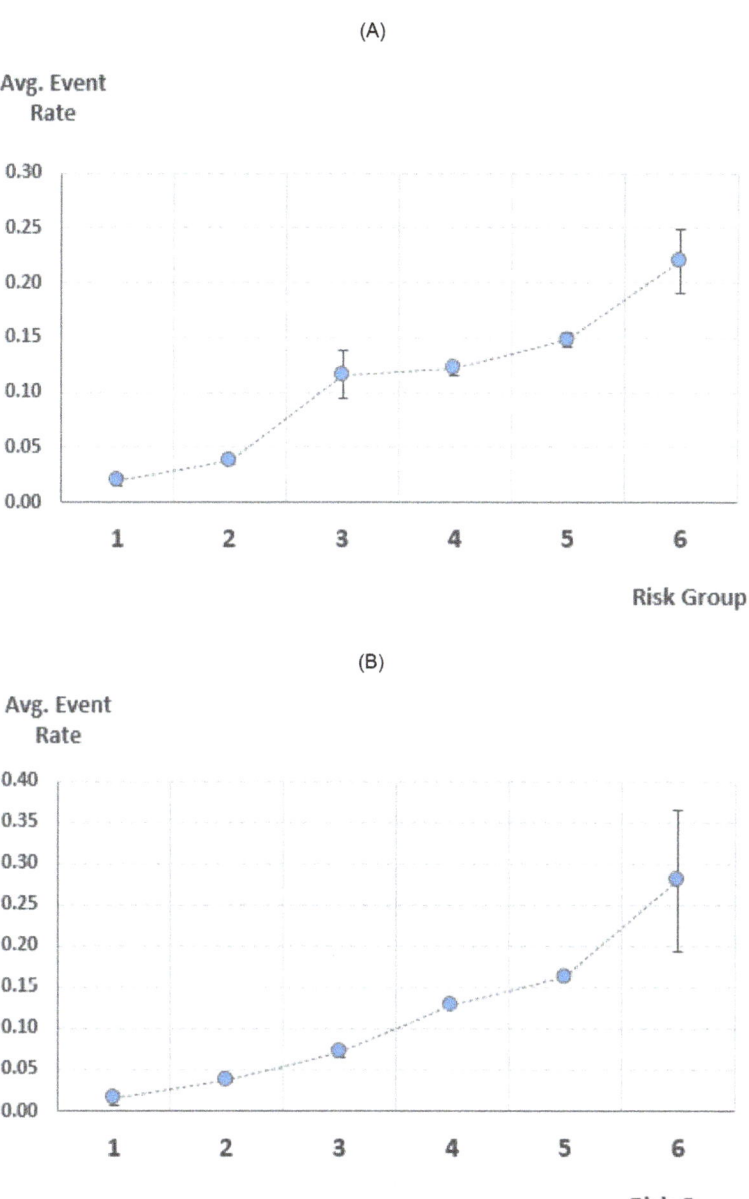

Figure 18-30. Line charts showing the average event rate by Risk-Group and their corresponding 95% confidence interval (A) before adjustment and (B) after adjustment.

CHAPTER 18 THE MAIN-EFFECTS MODEL

These modifications not only enhanced the upward trajectory of the mean event rate but also removed the overlap of the confidence intervals between Risk-Groups 3 and 4. This substantial improvement is accompanied by an increase in the discriminatory power, as reflected in the Gini coefficient, which increased from 32.25 to 37.59 after the adjustments, as illustrated in **Figure 18-31**.

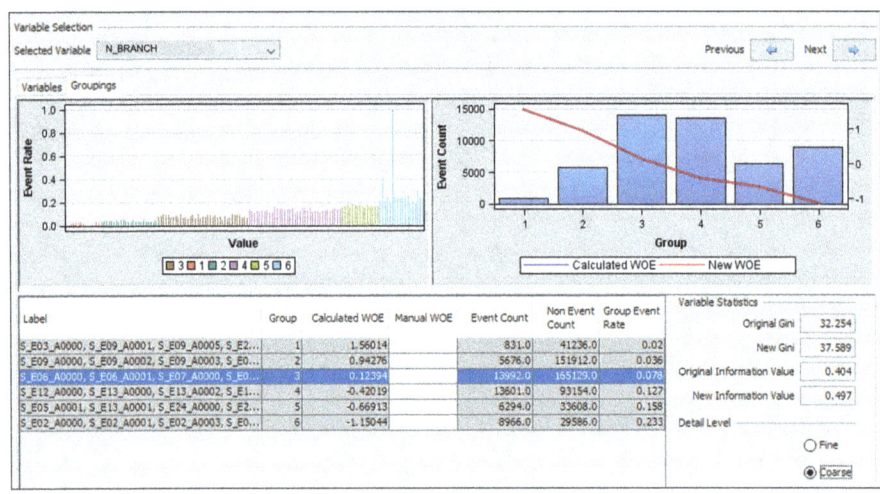

Figure 18-31. WoE diagnostics for N_BRANCH after adjustments.

It is important to note that while the confidence interval of Risk-Group 6 widened, there was no overlap with other groups. This widening is expected due to the proper reclassification of branches, according to their event rate. In addition, the widening is evident, by observing the range of event rates of Risk-Group 6, which was between 0.162 and 0.295, before the adjustments (see **Table 18-4**), and was reset to 0.186 and 1.0, after the adjustments (see **Table 18-5**).

As discussed in Section 9.2, there are alternative approaches to grouping the N_BRANCH categories other than the IG node, such as

CHAPTER 18 THE MAIN-EFFECTS MODEL

1. Greenacre's method [26, 47].
2. Decision trees
3. Weight of evidence (WoE)

The WoE calculation embedded in the IG node has already been employed; thus, Greenacre's method[4] [26, 47] will be utilized as an additional process to the IG node, rather than as a replacement.

Decision tree algorithms have been observed to become trapped in local solutions when confronted with high-cardinality variables. To circumvent this issue, the objective is to facilitate the grouping process of the IG node by *pregrouping* the categories present in N_BRANCH through the utilization of Greenacre's method [26, 47] and subsequently grouping the resulting clusters through the decision tree of the IG node.

Figure 18-32 presents the main-effects mining diagram, incorporating Greenacre's method [26, 47] branch, while **Table 18-6** presents the list of the 23 clusters generated by Greenacre's method [26, 47], arranged in descending order according to their event rate.

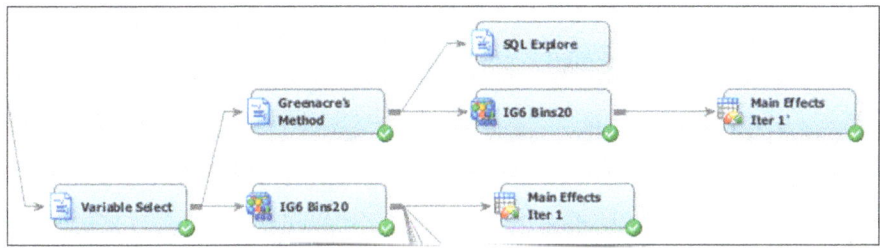

Figure 18-32. *Main-Effects mining diagram including Greenacre's method [26, 47] branch.*

---

[4] A bonus code of Greenacre's method can be downloaded from https://github.com/saulzapiain/ConceptualVariableDesign. File: Bonus Greenacre's_Method.sas.

CHAPTER 18 THE MAIN-EFFECTS MODEL

Table 18-6. *List of the 23 clusters created by Greenacre's method [26, 47], ordered in descending order by their event rate.*

CLUS_N_BRANCH	OBS	% OBS	EVENT_RATE	EVENTS
23	4	0.00%	100.00%	4
20	44	0.01%	40.91%	18
21	6,646	1.18%	29.49%	1,960
22	3,607	0.64%	25.40%	916
14	12,443	2.21%	23.49%	2,923
15	9,454	1.68%	20.70%	1,957
19	6,354	1.13%	18.70%	1,188
7	18,886	3.35%	16.29%	3,077
5	21,510	3.81%	15.30%	3,292
6	24,390	4.32%	14.28%	3,482
16	24,753	4.39%	13.27%	3,285
8	27,692	4.91%	12.24%	3,390
4	29,426	5.22%	11.45%	3,369
17	15,802	2.80%	9.50%	1,501
12	53,307	9.45%	8.73%	4,653
18	17,280	3.06%	8.06%	1,392
2	59,971	10.63%	7.35%	4,407
13	31,713	5.62%	6.27%	1,987
11	29,777	5.28%	4.53%	1,350
9	38,229	6.78%	3.73%	1,425
3	90,630	16.07%	3.26%	2,953
10	33,892	6.01%	2.17%	735
1	8,175	1.45%	1.17%	96

CHAPTER 18 THE MAIN-EFFECTS MODEL

Likewise, **Figures 18-33** and **18-34** present the detailed report for the new segmentation variable, CLUS_N_BRANCH, at both the detailed and aggregated levels, respectively. It is noteworthy that the Gini coefficient has reached a value close to 38 (37.66) in this instance, whereas in the preceding example, the initial Gini was of 32.254 and was increased to a value of 37, only after performing the manual reclassification of the 121 categories (branches). It is also noteworthy that most of the clusters seem to have been correctly classified, according to their risk level. However, clusters 20 and 23 represent exceptions to this, but their contribution is so negligible that the automatic classification proposed by the IG node can be disregarded, and the focus can move on to the next variable.

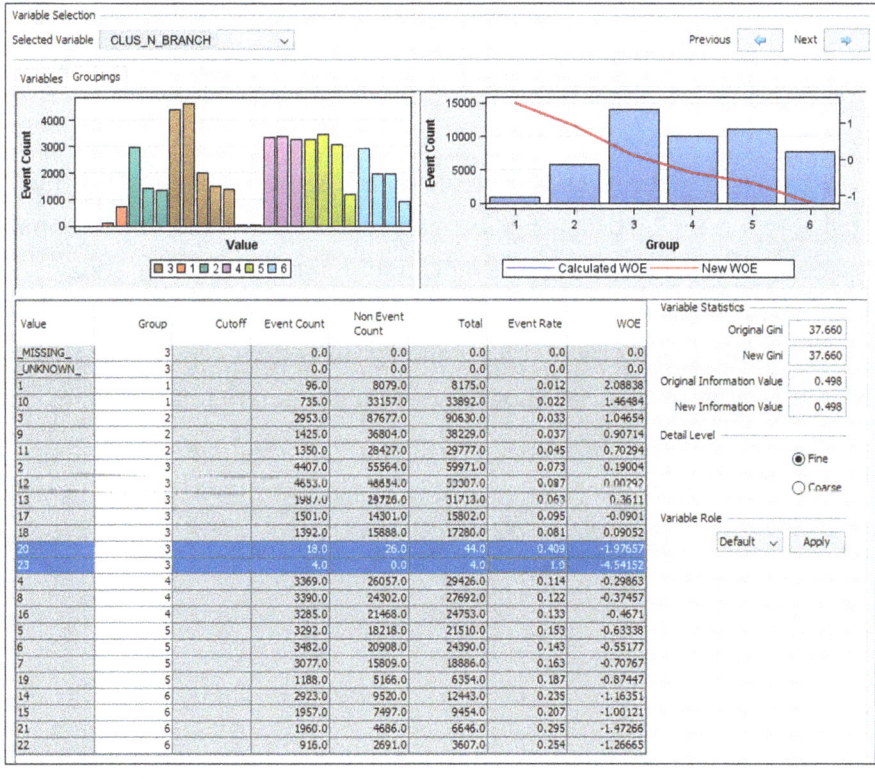

Figure 18-33. *WoE diagnostics for CLUS_N_BRANCH, set at fine data level.*

617

CHAPTER 18 THE MAIN-EFFECTS MODEL

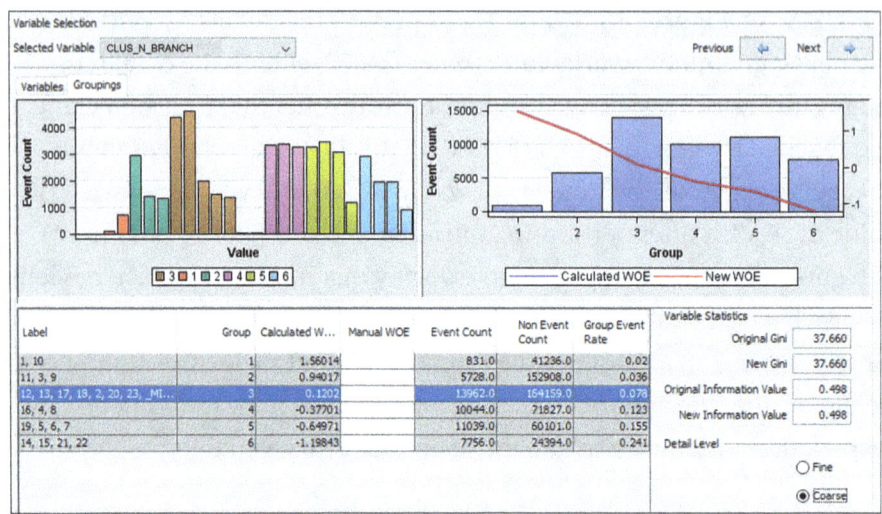

Figure 18-34. *WoE diagnostics for CLUS_N_BRANCH, set at coarse data level.*

With regard to the interval variables side, **Figure 18-35** presents the detailed report for I_LOAN_AGE_YRS (loan age, since origination). The main difference from the previous interval variables is the noticeable slide-like behavior, observed on the event rate chart (top left side). While this deviation from a linear trend might initially suggest the rejection of the variable, the interpretation is not immediately apparent in this instance.

CHAPTER 18 THE MAIN-EFFECTS MODEL

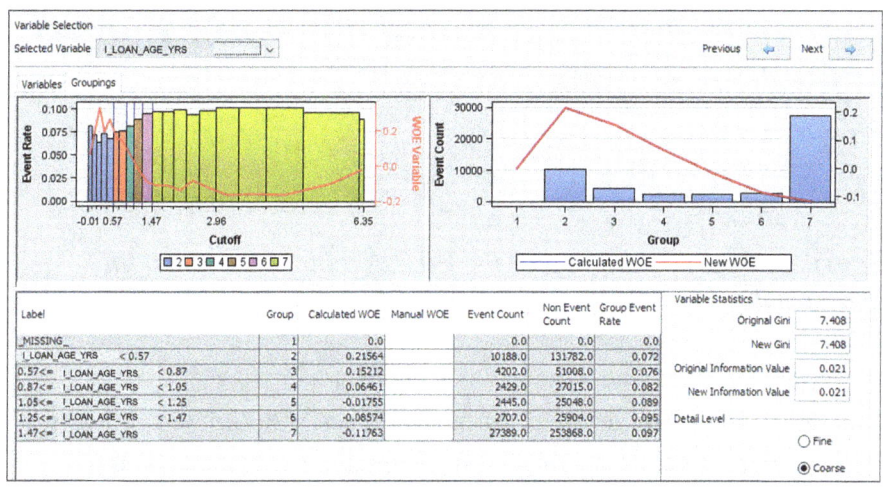

Figure 18-35. *WoE diagnostics for* I_LOAN_AGE_YRS *before adjustments.*

The occurrence of a slide-like behavior, whether inverted or not, could be interpreted as representing a two- or three-stage transitory state. As illustrated in the event rate chart, the initial steady state can be observed to extend from Risk-Group 2 to Risk-Group 3 (Risk-Group 1 corresponds to missing values), followed by an exponential decay that progresses through Risk-Groups 4, 5, and 6, reaching a second steady state at Risk-Group 7.

Figure 18-36 shows the lower tabular report for I_LOAN_AGE_YRS at a fine level of detail, highlighting the existence of a three-stage transient state. This is evidenced by the observation that Risk-Groups 2 and 3 have virtually the same event rate level, with values of 0.072 and 0.076, respectively. Similarly, the difference between Risk Groups 4 and 5 and Risk Groups 6 and 7 is negligible, with event rates of 0.082 and 0.089 and 0.095 and 0.097, respectively.

619

CHAPTER 18 THE MAIN-EFFECTS MODEL

Label	Group	Calculated WOE	Manual WOE	Event Count	Non Event Count	Group Event Rate
MISSING	1	0.0		0.0	0.0	0.0
I_LOAN_AGE_YRS < 0.57	2	0.21564		10188.0	131782.0	0.072
0.57<= I_LOAN_AGE_YRS < 0.87	3	0.15212		4202.0	51008.0	0.076
0.87<= I_LOAN_AGE_YRS < 1.05	4	0.06461		2429.0	27015.0	0.082
1.05<= I_LOAN_AGE_YRS < 1.25	5	-0.01755		2445.0	25048.0	0.089
1.25<= I_LOAN_AGE_YRS < 1.47	6	-0.08574		2707.0	25904.0	0.095
1.47<= I_LOAN_AGE_YRS	7	-0.11763		27389.0	253868.0	0.097

Figure 18-36. *Tabular report for* I_LOAN_AGE_YRS *set to a fine detail level.*

Figure 18-37 presents the detailed report for I_LOAN_AGE_YRS following to the aforementioned group mergers, resulting in a reduction from seven groups to three. It is noteworthy that the Gini coefficient exhibited a slight reduction (going from 7.4 to 7.16) following the implementation of the adjustments, indicating that no substantial loss of discriminatory power occurred during the merging process. It is also important to note that the objective here is not to gain or lose discriminatory power; rather, it is to adjust the WoE function (if possible) to align with expected commonsense behavior.

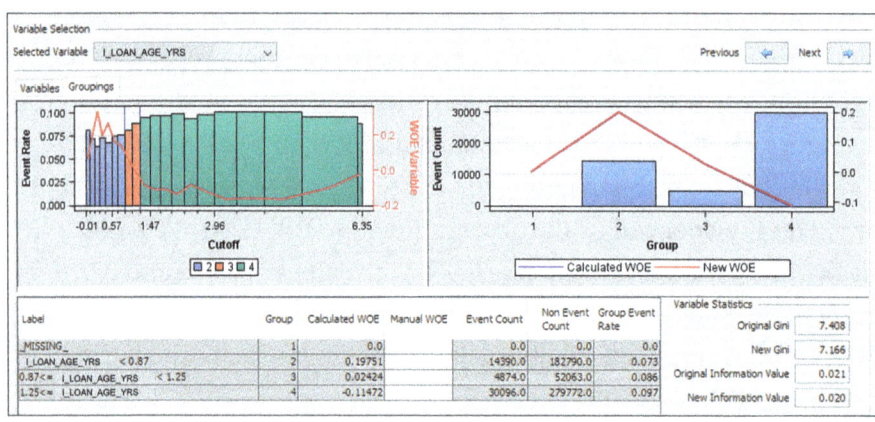

Figure 18-37. *WoE diagnostics for* I_LOAN_AGE_YRS *after adjustments.*

CHAPTER 18 THE MAIN-EFFECTS MODEL

We will now continue with the second interval variable and the last main effect that needs a slight adjustment. As demonstrated in **Figure 18-38**, which presents the detailed report for I_NUMDEPEND (economic dependent), the event rates of Risk-Group 3 and Risk-Group 4 (upper left corner) are nearly identical, suggesting a possible merging of the two groups.

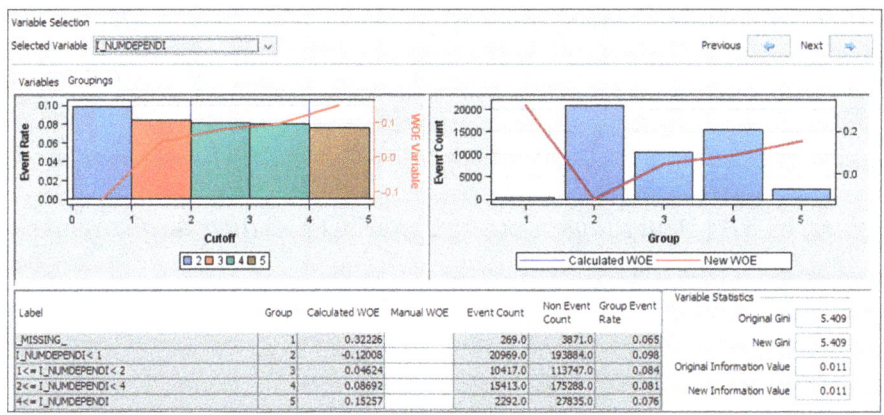

Figure 18-38. WoE diagnostics for I_NUMDEPEND before adjustments.

Figure 18-39 shows the detailed report for I_NUMDEPEND after merging Risk-Groups 3 and 4, based on the above observations. It is notable that the Gini coefficient demonstrates a negligible difference between the pre- and post-adjustment periods, similar to the observation made for the variable I_LOAN_AGE_YRS. It is also important to highlight the fact that the Gini coefficient for I_NUMDEPEND was already low and close to the rejection threshold of 5. However, the number of economic dependents has an important conceptual meaning, as explained in Section 3.2, through the *"Married with Children effect,"* which supports its inclusion in the final main-effects model. It should also be noted that, strictly speaking, the variable I_NUMDEPEND is still above the rejection threshold, no matter by how much.

621

CHAPTER 18 THE MAIN-EFFECTS MODEL

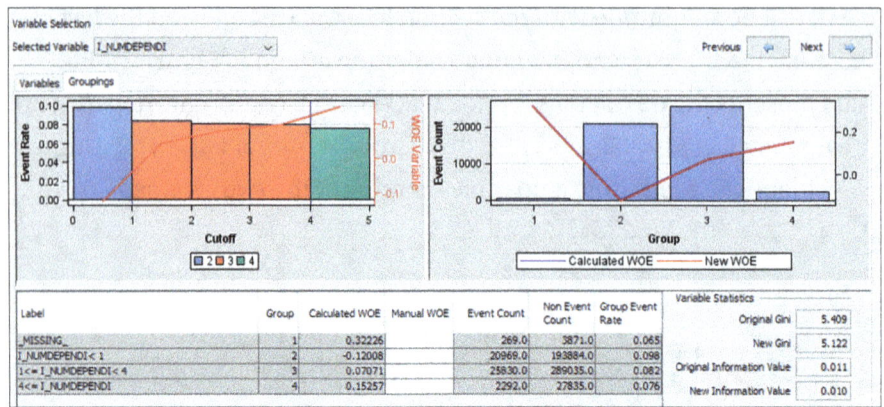

Figure 18-39. *WoE diagnostics for* I_NUMDEPEND *after adjustments.*

CHAPTER 18 THE MAIN-EFFECTS MODEL

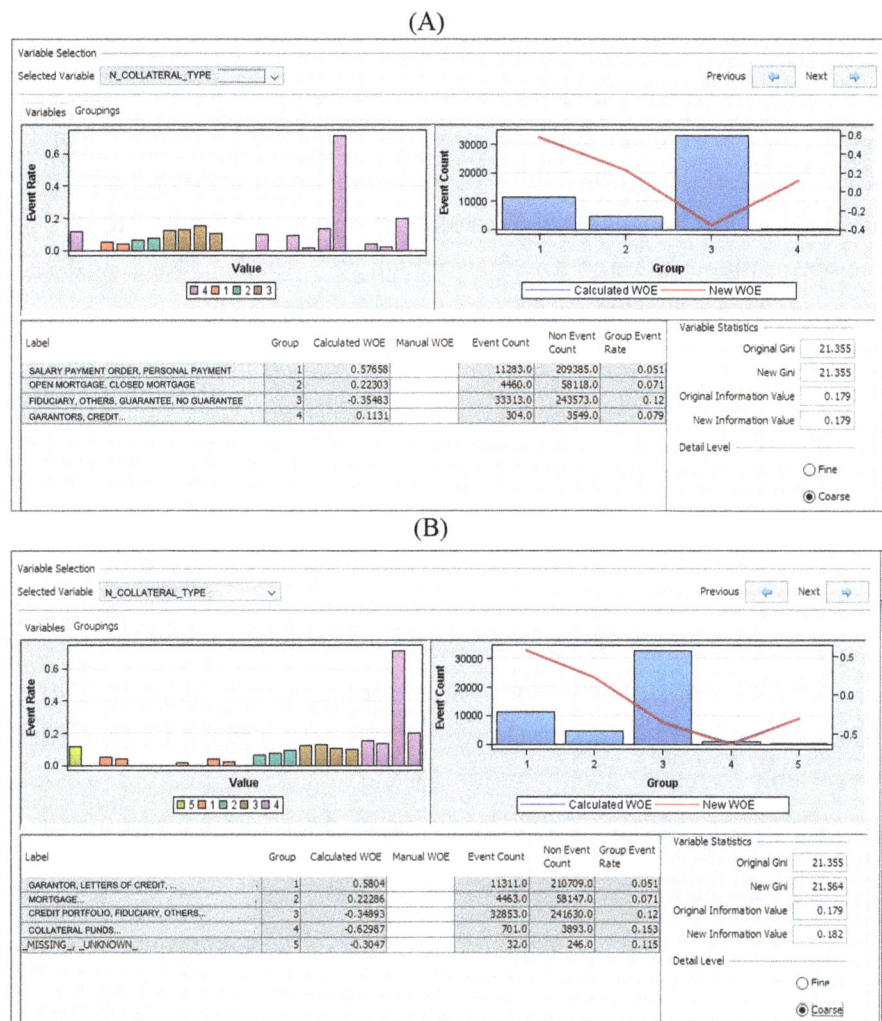

Figure 18-40. Detailed report for N_COLLATERAL_TYPE. (A) Before adjustments. (B) After adjustments.

18.4.3.4 Examples of Ambiguous WoEs

An ambiguous WoE is characterized by a relatively low Gini coefficient and may necessitate minor adjustments to enhance the consistency of its observed trends. However, the general event rate function of such a variable does not appear to exhibit a strong correlation with the cutoff value assigned to the Risk-Group. The first ambiguous variable in **Table 18-3** is I_SV_R_PAYA_INST_CLC_L12_06 (six-month variation of the ratio of monthly payments/installments from 12 months back to 6 months back), and its detailed report is shown in **Figure 18-41**. It is notable that this variable, with a value of 7.937, is marginally above the rejection threshold, by a margin of approximately three points. It is also noteworthy that there are only three Risk-Groups (the fourth one belonging to the null values), and their event rates are also reasonably separated from each other in descending order, with values of 0.196, 0.109, and 0.081, corresponding to Risk-Groups 2, 3, and 4, respectively. However, the event rate function does not exhibit the anticipated monotonic linear behavior, with over 80% of observations classified in Risk-Group 4. It is therefore recommended that this variable be retained in the model for the time being, in order to assess its contribution to the main-effects model following the regression.

CHAPTER 18   THE MAIN-EFFECTS MODEL

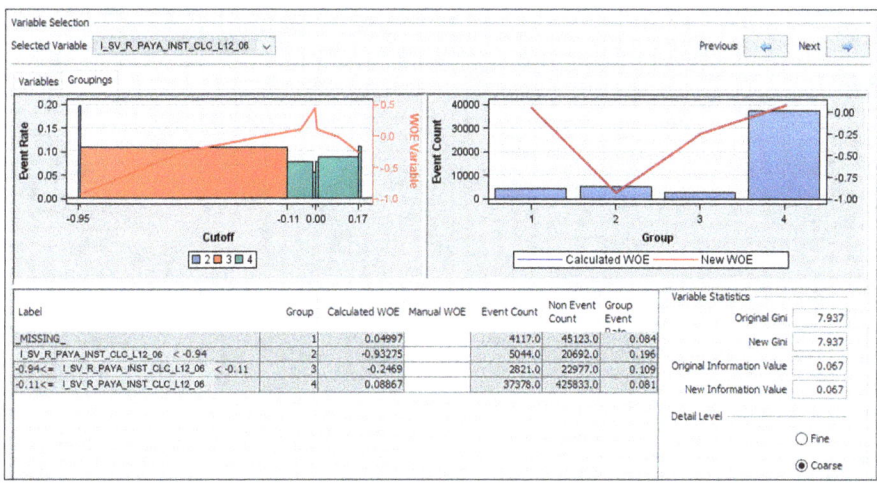

Figure 18-41. *WoE diagnostics for* I_SV_R_PAYA_INST_CLC_L12_06.

As demonstrated in **Figure 18-42**, the variable I_SIN_R_BALT_APPR_CLC_L03M (increments of the ratio of outstanding balance amount to approved amount in the last 3 months) exhibits a Gini coefficient that exceeds even that of the previous variable, with a value of 12.817. Moreover, a marked disparity in event rate was evident among the distinct Risk-Groups, with values of 0.074 for Risk-Group 2 and 0.135 for Risk-Group 3, while Risk-Group 1 belongs to the null values. These observations suggest the presence of a variable with a fairly good discriminatory power. What stands out in this case is the dichotomic nature of the resulting Risk-Groups. This variable will be maintained in the model to assess its contribution subsequently.

625

CHAPTER 18   THE MAIN-EFFECTS MODEL

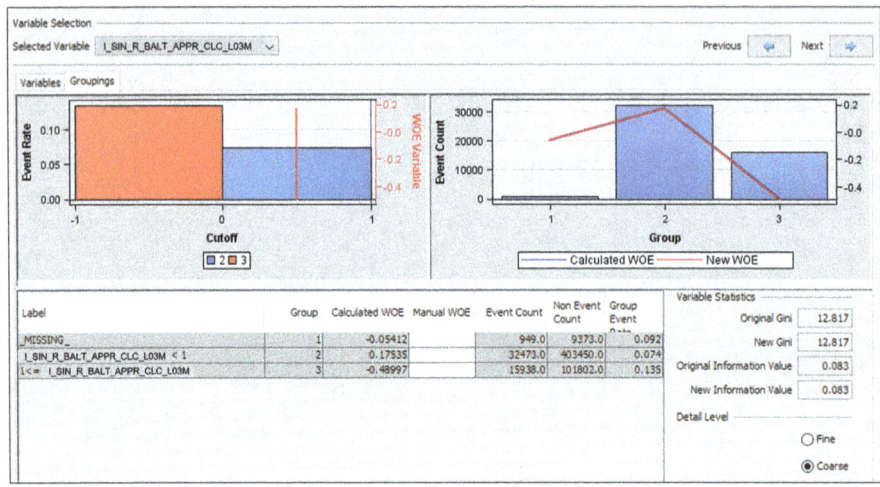

Figure 18-42. *WoE diagnostics for* I_SIN_R_BALT_APPR_CLC_L03M.

The third and final ambiguous interval variable is I_SDC_R_CHRG_CRRA_CPM_L12M (decrements in the ratio of charges/savings and current amount in the last 12 months), shown in **Figure 18-43**. It is important to note that, in contrast to the two preceding variables, some adjustments will be necessary. Specifically, Risk-Group 3 exhibits no discernible difference from Risk-Group 4 with respect to event rate; therefore, it should be merged with Risk-Group 4.

CHAPTER 18 THE MAIN-EFFECTS MODEL

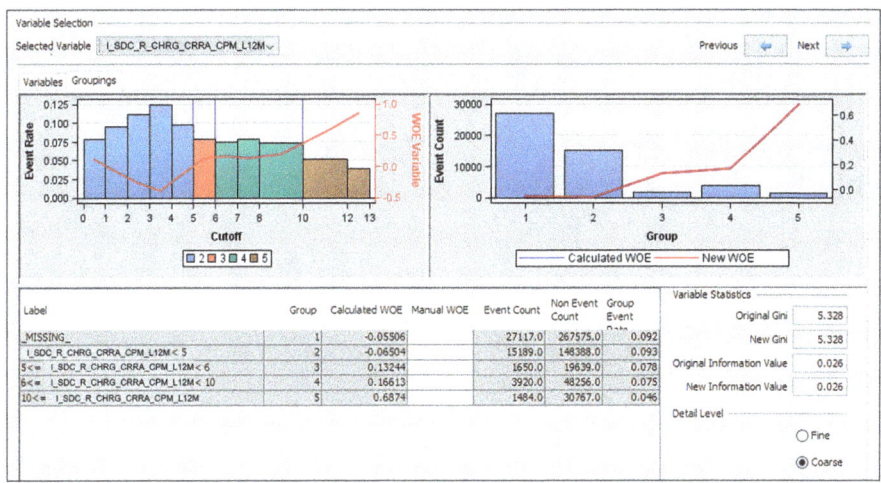

Figure 18-43. WoE diagnostics for I_SDC_R_CHRG_CRRA_CPM_L12M before adjustments.

Figure 18-44 shows the detailed report for the variable I_SDC_R_CHRG_CRRA_CPM_L12M after adjustments. The most noticeable aspects of these adjustments pertain to the correction of the WoE trend (see top right chart), and the insignificant difference between the Gini coefficient, before (5.328) and after (5.318) adjustments. It should be noted, however, that the Gini coefficient for this particular variable was barely above the rejection threshold. Furthermore, the event rate function (top left) demonstrates a tail-like behavior on the left side of the chart that does not correspond with the anticipated monotonic linear behavior, despite the fact that the entirety of the tail-like section was engulfed by Risk-Group 2. As previously indicated, this variable will be maintained in the model and its contribution to the multivariable model will be evaluated following the regression analysis.

CHAPTER 18 THE MAIN-EFFECTS MODEL

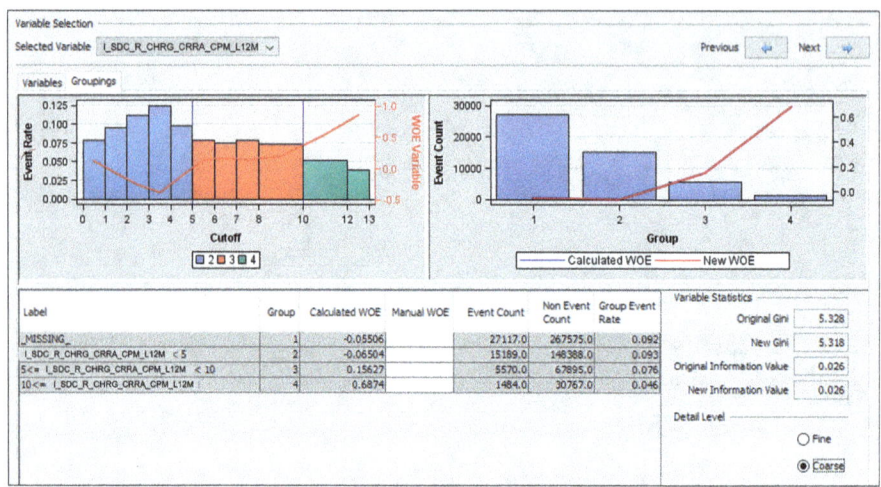

Figure 18-44. WoE diagnostics for I_SDC_R_CHRG_CRRA_CPM_L12M after adjustments.

18.4.4 Conclusion of the WoE Refinement Process

In this section, we have presented a methodology that departs from the classical statistical analysis, thereby allowing us to anticipate the expected relationship between the explanatory and target variables based on their conceptual meanings. The preceding discussion should now make evident why the methodology presented here differs from the conventional passive modeling process, which typically involves the indiscriminate inclusion of variables in the hope that some of them might significantly predict the event under study.

In accordance with the anticipated relationships between the explanatory and target variables, a visual inspection of the detailed WoEs reports, provided by the IG node, was conducted. Subsequently, the 24 main effects were enumerated and classified into four distinct categories based on their distinguishing characteristics. The four groups, comprising a total of 24 variables, were as follows:

1. Inconsistent and rejected, eight variables
2. Consistent, no adjustment needed, seven variables
3. Consistent, but required manual adjustments, six variables
4. Ambiguous, may require manual selection, three variables

The variables in group one were excluded primarily because they did not exhibit monotonic linear behavior. It was determined that the empirical data exhibited a significant deviation from the anticipated trend, rendering any adjustment ineffective. In contrast, group two variables exhibited characteristics that were the opposite of those in group one, demonstrating not only the anticipated trend but also an almost-ideal monotonic linear behavior. Consequently, no further action was required for the selection of variables in group two.

The variables in group three demonstrated an acceptable trend that exhibited slight deviations from the anticipated monotonic linear behavior. It is noteworthy that four of the six variables included in group three were nominal variables. It has been argued that this result is not uncommon, based on empirical observations indicating that the decision tree (DT) algorithm of the IG node tends to exhibit suboptimal performance when dealing with categorical variables. From this group, the variable N_BRANCH was of particular interest, as it provides a practical illustration of the IG node's approach when dealing with high-cardinality variables. The efficacy of the IG node in handling high-cardinality variables was evidenced by a comparison of the summary statistics presented in **Table 18-4** (before improvement) and **Table 18-5** (after improvement). This enhancement was further substantiated by the event rate charts shown in **Figures 18-28** and **18-31**, before and after the enhancements, respectively.

CHAPTER 18 THE MAIN-EFFECTS MODEL

In addition, an alternative approach was proposed for dealing with the N_BRANCH variable, i.e., Greenacre's method [26, 47]. However, it was emphasized that this method was not being used as a replacement for the IG node, but rather as a contributing process. In this section, the argument is put forth that the DT utilized by the IG node is susceptible to becoming stuck in local solutions. It is therefore recommended that the grouping process performed by the IG node be facilitated by pregrouping the highly cardinal variables. In this example, Greenacre's method [26, 47] pregrouped the 121 categories into 23 clusters, according to their event rate, which were subsequently collapsed into six groups by the IG node. The outcome of this process was a WoE with a Gini coefficient of 37.66, indicating that no adjustment was required for the automatic groups created by the IG node.

Another notable case from group three is the variable I_LOAN_AGE_YRS, which illustrated an instance of a variable exhibiting a three-stage transition state. In this example, it was proposed that not all variables must exhibit monotonic linear behavior to be included in the model. Some variables may be treated as trichotomous or dichotomous variables. It is noteworthy that I_LOAN_AGE_YRS has been designated as a trichotomous variable for the time being, but a subsequent examination will be conducted following the regression analysis. In the event of overfitting on the empirical odds charts, the I_LOAN_AGE_YRS variable will be subjected to collapsing, with the objective of transforming it from a trichotomous variable to a dichotomous one.

The variables in group four exhibit a combination of the characteristics observed in the preceding three cases, displaying a notable degree of discriminatory power. However, they also exhibit a suspicious trend in their event rate charts, which is worthy of further investigation. The most illustrative cases in this group are the variable I_SDC_R_CHRG_CRRA_CPM_L12M, which exhibited a tail-like behavior on the left side of the event rate chart, and the variable I_SIN_R_BALT_APPR_CLC_LO3M, which appears to be dichotomous in nature. Therefore, the variables in group 4 were classified as ambiguous due to their inability to adhere to the anticipated monotonic linear behavior, yet the available evidence is insufficient to exclude them from the multivariable model.

18.5 Construction of the Final Main-Effects Model

Table 18-7 shows the resulting model after conducting the WoE refinement process (version without Greenacre's method [26, 47]). The first thing to notice, when comparing the model presented in **Table 18-7** with the initial main-effects model presented in **Table 18-1**, is the removal of eight variables that were eliminated by visual inspection of the event rate charts alone, leaving us with a model of 16 variables. The second observation is that surprisingly, the performance of the model is not reduced even after eliminating eight variables. The third observation is that there are no inverted coefficients in the new refined model (**Figure 18-45** (B)). Finally, the only element affected was the empirical odds (**Figure 18-45** (A)), which now shows a pronounced deviation from the observed event rate at high score values. This is in contrast to the monotonic linear behavior obtained from the full main-effects model (**Figure 18-6** (B)).

CHAPTER 18 THE MAIN-EFFECTS MODEL

Table 18-7. Summary results of the refined main-effects model.

					Partition	KS	ROC	Gini
					Train	0.653	0.895	0.790
					Validate	0.649	0.893	0.786
					Over Training	0.005	0.002	0.004

Num	Parameter	DF	Estimate	Error	χ^2	p-value	Std. Estimate	Odds
	Intercept	1	-2.3327	0.0063	138496	<.0001	0.097	
1	WOE_N_OCCUPATION	1	-0.3345	0.0148	512.57	<.0001	-0.1172	0.716
2	WOE_I_AGE_YRS	1	-0.3932	0.014	793.83	<.0001	-0.0962	0.675
3	WOE_I_LOAN_AGE_YRS	1	-0.4108	0.0454	82.04	<.0001	-0.0327	0.663
4	WOE_I_MX_DYSPSTDUE_LON_L12M	1	-0.5152	0.0077	4482.56	<.0001	-0.1942	0.597
5	WOE_I_NUMDEPEND	1	-0.2985	0.0574	27.09	<.0001	-0.0167	0.742
6	WOE_I_SIN_R_BALT_APPR_CLC_L13_24	1	-0.242	0.0175	190.46	<.0001	-0.0457	0.785
7	WOE_N_HOMESTATUS	1	-0.2626	0.0296	78.75	<.0001	-0.029	0.769
8	WOE_N_COLLATERAL_TYPE	1	-0.1661	0.015	121.91	<.0001	-0.0404	0.847
9	WOE_N_PROFESSION	1	-0.0728	0.0305	5.69	0.0171	-0.00951	0.93
10	WOE_I_RISK_LEVEL	1	-0.8712	0.0037	54331.63	<.0001	-0.5005	0.418
11	WOE_N_BRANCH	1	-0.4535	0.0092	2457.3	<.0001	-0.1845	0.635
12	WOE_I_AV_R_BALT_APPR_CLC_L06M	1	-0.2649	0.0105	633.88	<.0001	-0.0787	0.767
13	WOE_I_SDC_R_CHRG_CRRA_CPM_L12M	1	-0.69	0.0389	314.47	<.0001	-0.069	0.502
14	WOE_I_SIN_R_PAYT_APPR_CLC_L13_24	1	-0.3395	0.0117	837.33	<.0001	-0.1157	0.712
15	WOE_I_SIN_R_BALT_APPR_CLC_L03M	1	-0.1774	0.0215	68.11	<.0001	-0.0264	0.837
16	WOE_I_SV_R_PAYA_INST_CLC_L12_06	1	-0.0724	0.0213	11.55	0.0007	-0.00881	0.93

CHAPTER 18 THE MAIN-EFFECTS MODEL

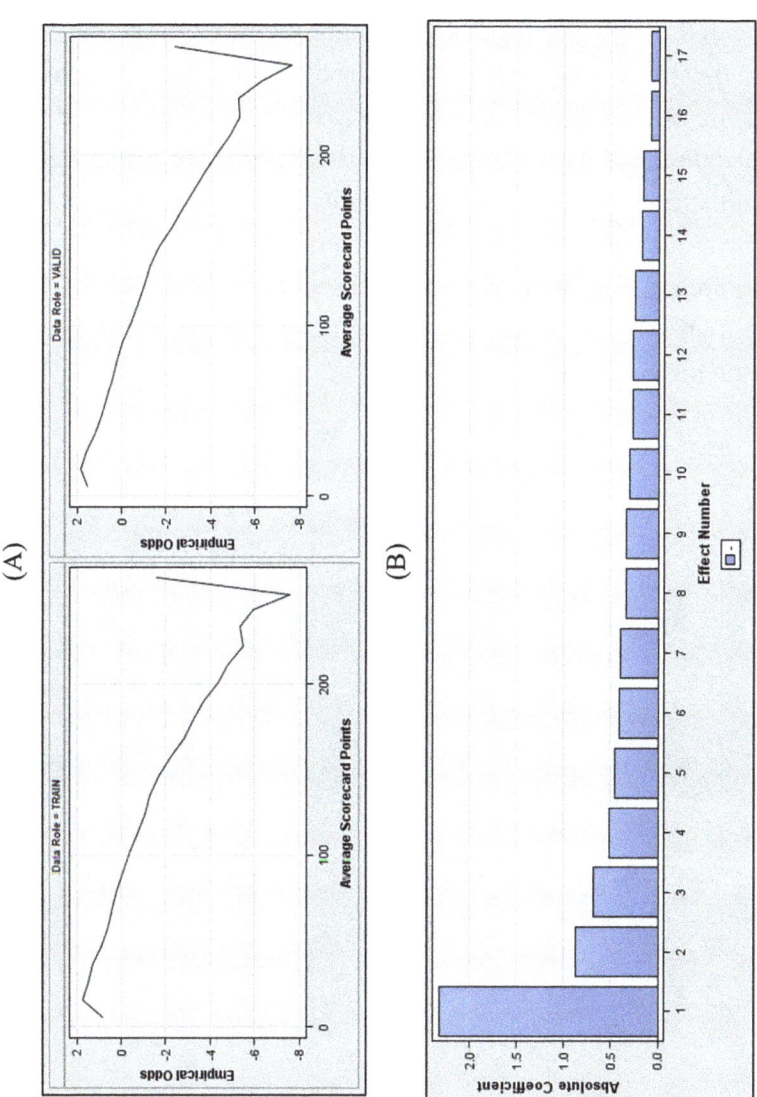

Figure 18-45. Assessment charts for the refined main-effects model. (A) Line chart of the empirical odds vs. the average scorecard points for the training partition (left) and the validation partition (right). (B) Bar chart of the absolute value of the effect coefficients.

CHAPTER 18 THE MAIN-EFFECTS MODEL

18.5.1 Manual Elimination Based on the χ^2 Statistic

The objective of the present study is to further reduce and refine the model. This objective will be accomplished by continuing to eliminate variables that do not contribute significantly to the model. **Table 18-8** shows the 16 refined main effects ordered in descending order, according to their $\chi 2$ statistic. It is evident that the statistical significance is unable to assist in the reduction of the number of effects, given that all of the variables are already extremely significant (except for N_PROFESSION), with *p*-values less than 0.001. Nevertheless, the number of variables can be reduced on the basis of their $\chi 2$ statistic.

CHAPTER 18 THE MAIN-EFFECTS MODEL

Table 18-8. *Refined main effects ordered in descending order according to their χ^2 statistic.*

No.	Parameter	DF	Estimate	Error	χ^2	p-value	Std. Estimate	Odds
	Intercept	1	-2.3327	0.0063	138,496	<.0001	0.097	
1	WOE_I_RISK_LEVEL	1	-0.8712	0.0037	54,332	<.0001	-0.5005	0.418
2	WOE_I_MX_DYSPSTDUE_LON_L12M	1	-0.5152	0.0077	4,483	<.0001	-0.1942	0.597
3	WOE_N_BRANCH	1	-0.4535	0.0092	2,457	<.0001	-0.1845	0.635
4	WOE_I_SIN_R_PAYT_APPR_CLC_L13_24	1	-0.3395	0.0117	837	<.0001	-0.1157	0.712
5	WOE_I_AGE_YRS	1	-0.3932	0.014	794	<.0001	-0.0962	0.675
6	WOE_I_AV_R_BALT_APPR_CLC_L06M	1	-0.2649	0.0105	634	<.0001	-0.0787	0.767
7	WOE_N_OCCUPATION	1	-0.3345	0.0148	513	<.0001	-0.1172	0.716
8	WOE_I_SDC_R_CHRG_CRRA_CPM_L12M	1	-0.69	0.0389	314	<.0001	-0.069	0.502
9	WOE_I_SIN_R_BALT_APPR_CLC_L13_24	1	-0.242	0.0175	190	<.0001	-0.0457	0.785
10	WOE_N_COLLATERAL_TYPE	1	-0.1661	0.015	122	<.0001	-0.0404	0.847
11	WOE_I_LOAN_AGE_YRS	1	-0.4108	0.0454	82	<.0001	-0.0327	0.663
12	WOE_N_HOMESTATUS	1	-0.2626	0.0296	79	<.0001	-0.029	0.769
13	WOE_I_SIN_R_BALT_APPR_CLC_L03M	1	-0.1774	0.0215	68	<.0001	-0.0264	0.837
14	WOE_I_NUMDEPEND	1	-0.2985	0.0574	27	<.0001	-0.0167	0.742
15	WOE_I_SV_R_PAYA_INST_CLC_L12_06	1	-0.0724	0.0213	12	0.0007	-0.00881	0.93
16	WOE_N_PROFESSION	1	-0.0728	0.0305	6	0.0171	-0.00951	0.93

CHAPTER 18   THE MAIN-EFFECTS MODEL

As indicated in **Table 18-8**, the variables N_PROFESSION and I_SV_R_PAYA_INST_CLC_L12_06 exhibit the lowest χ^2 statistic. It would be possible to remove both variables simultaneously and assess the model again; however, from this point forward, variables will be removed one at a time, with careful observation of the effect of each removal, until no further variables can be removed from the model without compromising its performance in terms of ROC points or empirical odds fit.

Table 18-9 illustrates trial 2 of the refined main-effects model, in which the variable N_PROFESSION was excluded. In contrast to **Table 18-8**, **Table 18-9** illustrates the discrepancy between the variables and their preceding iteration. It is notable that the overall performance of the model did not undergo a significant change between iteration 1 and iteration 2. It is also evident that no variable deviates by more than one percent. This finding suggests that the eliminated variable did not function as a confounder, thereby enabling us to continue with the manual elimination process.

CHAPTER 18 THE MAIN-EFFECTS MODEL

Table 18-9. *Refined main effects after iteration 2, in descending order according to their χ^2 statistic. Version without the N_PROFESSION variable.*

Partition	KS	ROC	Gini
Train	0.653	0.895	0.790
Validate	0.648	0.893	0.786
Overtraining	0.005	0.002	0.004

No.	Parameter	DF	Estimate	Error	χ^2	p-value	Std. Estimate	Deviation
	Intercept	1	-2.333	0.006	138,514	<.0001	0.097	0.000
1	WOE_I_RISK_LEVEL	1	-0.872	0.004	54,409	<.0001	-0.501	0.000
2	WOE_I_MX_DYSPSTDUE_LQN_L12M	1	-0.515	0.008	4,481	<.0001	-0.194	0.000
3	WOE_N_BRANCH	1	-0.455	0.009	2,484	<.0001	-0.185	-0.003
4	WOE_I_SIN_R_PAYT_APPR_CLC_L13_24	1	-0.340	0.012	839	<.0001	-0.116	-0.001
5	WOE_I_AGE_YRS	1	-0.398	0.014	825	<.0001	-0.097	-0.011
6	WOE_I_AV_R_BALT_APPR_CLC_L06M	1	-0.266	0.011	637	<.0001	-0.079	-0.002
7	WOE_N_OCCUPATION	1	-0.336	0.015	517	<.0001	-0.118	-0.004
8	WOE_I_SDC_R_CHRG_CRRA_CPM_L12M	1	-0.690	0.039	315	<.0001	-0.069	0.000
9	WOE_I_SIN_R_BALT_APPR_CLC_L13_24	1	-0.241	0.018	190	<.0001	-0.046	0.002
10	WOE_N_COLLATERAL_TYPE	1	-0.168	0.015	125	<.0001	-0.041	-0.010
11	WOE_I_LOAN_AGE_YRS	1	-0.412	0.045	82	<.0001	-0.033	-0.003
12	WOE_N_HOMESTATUS	1	-0.264	0.030	80	<.0001	-0.029	-0.006
13	WOE_I_SIN_R_BALT_APPR_CLC_L03M	1	-0.177	0.022	68	<.0001	-0.026	0.003
14	WOE_I_NUMDEPEND	1	-0.299	0.057	27	<.0001	-0.017	-0.001
15	WOE_I_SV_R_PAYA_INST_CLC_L12_06	1	-0.072	0.021	11	0.0007	-0.009	0.004

CHAPTER 18 THE MAIN-EFFECTS MODEL

Table 18-10 summarizes the entire refinement process, which encompasses a total of seven iterations.[5] The summary displays, from left to right, the number of effects remaining in the model, the type of refinement performed, the manually rejected variable, the initial WoE diagnosis, the training and validation ROC statistic, the overfit level (training minus validation), and the model name at each iteration. Similarly, **Figure 18-46** presents the EM nodes that gave rise to the results presented in **Table 18-10**, arranged in an iterative fashion, and **Figure 18-47** shows the validation ROC progression at each iteration step. Lastly, **Table 18-11** presents the final model, and **Figure 18-48** presents the progression of the empirical odds charts, for both training and validation, over the seven refinement iterations.

---

[5] The main-effects details at each iteration are shown at the end of this section.

CHAPTER 18 THE MAIN-EFFECTS MODEL

Table 18-10. Summary of the iterative refinement process.

Iteration	Effects in model	Refinement type	Rejected variable	Diagnostics	Train ROC	Validation ROC	Overtraining	Model
0	24				0.8973	0.8953	0.0020	Full model
1	16	WoE adjustment	Multiple		0.8950	0.8931	0.0019	Reduced model
2	15	Manual elimination	N_PROFESSION	No adjustment	0.8950	0.8931	0.0019	Reduced model
3	14	Manual elimination	I_SV_R_PAYA_INST_CLC_L12_06	Ambiguous	0.8950	0.8930	0.0019	Reduced model
4	13	Manual elimination	I_NUMDEPEND	Adjustment	0.8949	0.8930	0.0019	Reduced model
5	12	Manual elimination	I_SIN_R_BALT_APPR_CLC_L03M	Ambiguous	0.8948	0.8929	0.0019	Reduced model
6	11	Manual elimination	I_LOAN_AGE_YRS	Adjustment	0.8947	0.8928	0.0019	Reduced model
7	**10**	**Manual elimination**	**N_HOMESTATUS**	**Adjustment**	**0.8945**	**0.8927**	**0.0018**	**Final model**

CHAPTER 18   THE MAIN-EFFECTS MODEL

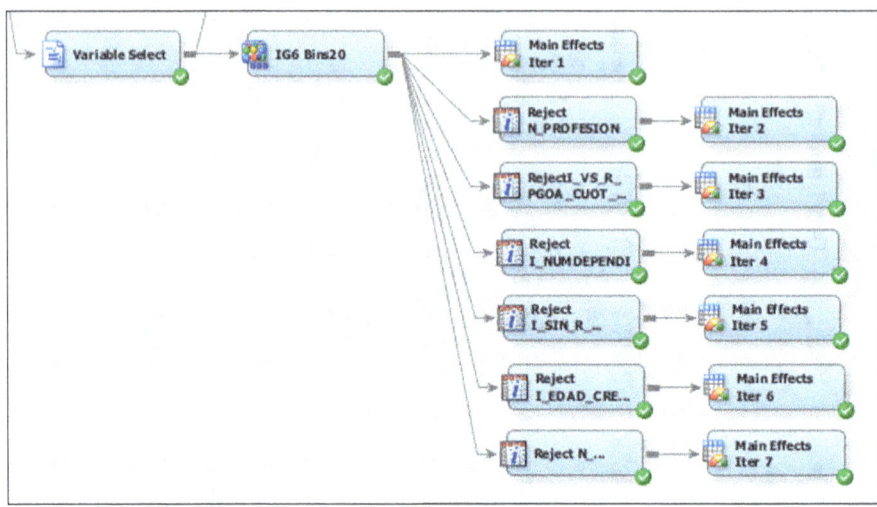

Figure 18-46. *EM nodes arranged as a refinement iteration process.*

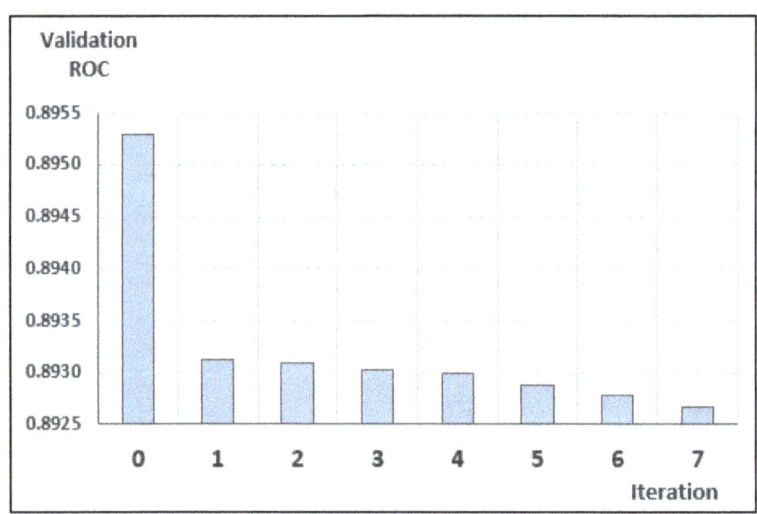

Figure 18-47. *Validation ROC progression at each iterative step.*

CHAPTER 18 THE MAIN-EFFECTS MODEL

Table 18-11. Final main-effects model.

Partition	KS	ROC	Gini
Train	0.653	0.895	0.789
Validate	0.648	0.893	0.785
Over Training	0.005	0.002	0.004

No.	Parameter	DF	Estimate	Error	χ^2	p-value	Std. Estimate	Deviation
	Intercept	1	-2.335	0.006	138,922	<.0001	0.097	0.000
1	WOE_I_RISK_LEVEL	1	-0.873	0.004	54,714	<.0001	-0.502	-0.001
2	WOE_I_MX_DYSPSTDUE_LON_L12M	1	-0.532	0.007	5,063	<.0001	-0.201	0.002
3	WOE_N_BRANCH	1	-0.460	0.009	2,557	<.0001	-0.187	-0.005
4	WOE_I_AGE_YRS	1	-0.425	0.013	1,042	<.0001	-0.104	-0.087
5	WOE_I_AV_R_BALT_APPR_CLC_L06M	1	-0.303	0.010	938	<.0001	-0.090	0.004
6	WOE_I_SIN_R_PAYT_APPR_CLC_L13_24	1	-0.327	0.012	794	<.0001	-0.111	-0.002
7	WOE_N_OCCUPATION	1	-0.334	0.015	517	<.0001	-0.117	0.007
8	WOE_I_SDC_R_CHRG_CRRA_CPM_L12M	1	-0.676	0.039	303	<.0001	-0.068	0.009
9	WOE_I_SIN_R_BALT_APPR_CLC_L13_24	1	-0.265	0.017	234	<.0001	-0.050	0.008
10	WOE_N_COLLATERAL_TYPE	1	-0.177	0.015	139	<.0001	-0.043	-0.024

CHAPTER 18   THE MAIN-EFFECTS MODEL

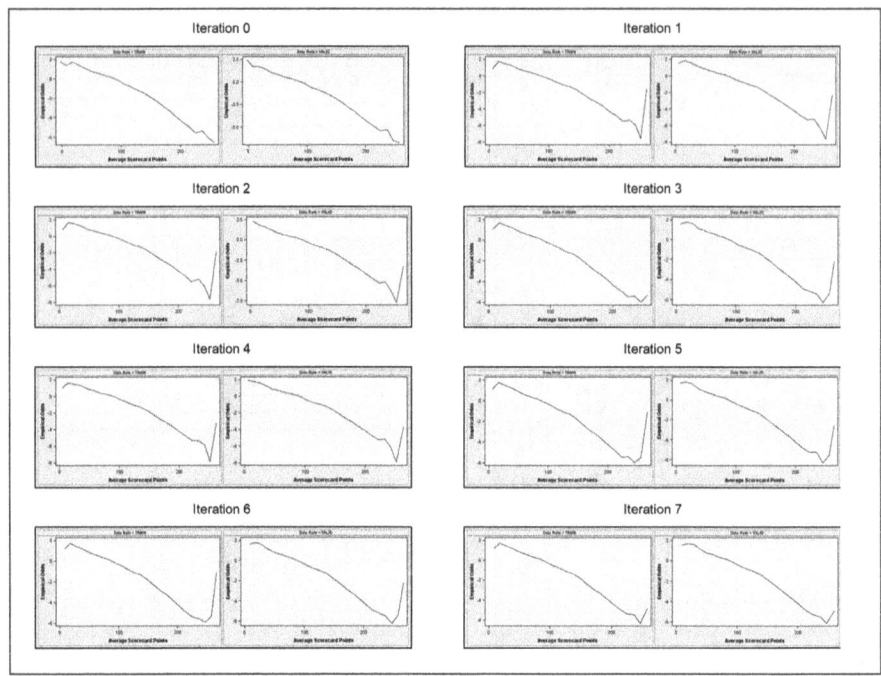

Figure 18-48. Progression of the empirical odds charts, for training and validation, over the seven iterations.

The first element to notice from these results is the substantial reduction in effects of almost 60% of the initial set of variables, which decreased from 24 effects in the initial full main-effects model to 10 effects in the final main-effects model.[6] Secondly, the model's performance demonstrated remarkable resilience to the aforementioned 60% reduction in the total number of explanatory variables, exhibiting a marginal decline in performance metrics. This was evidenced by a slight decrease in the ROC value from 0.8973 to 0.8945, representing a negligible difference of 28 thousandths in ROC units.

Another striking result, in addition, to the negligible difference between the initial and final validation ROC values, is the fact that the

---

[6] It is worth mentioning that the largest elimination of effects was done via the WoE refinement process that rejected eight effects in one single iteration.

CHAPTER 18 THE MAIN-EFFECTS MODEL

overtraining was not only insignificant at all times, but that it was further reduced throughout the refinement process.

From the results presented in **Table 18-10**, it is also worth noting that most of the manually rejected effects belonged to effects that were previously classified as ambiguous or that needed adjustment. This outcome is not merely coincidental, particularly in light of the recommended course of action for ambiguous variables, which was to incorporate them into the regression and subsequently evaluate their contribution. This finding underscores the importance of the WoE refinement process, which has been instrumental in directing efforts toward identifying potential variables that might have been rejected, even before the regression analysis has been conducted.

Similarly, it is unsurprising the fact that three of the adjusted variables that were rejected throughout the refinement process already exhibited a low discriminatory power, close to the rejection threshold of a Gini of 5. Furthermore, the necessity for adjustment prior to the regression analysis is indicative of their initial lack of consistency.

It is noteworthy that the only variable that does not seem to belong to the group of rejected variables as presented in **Table 18-10** is N_PROFESSION. In addition, as mentioned in Section 18.4.3.3, there is a possibility of interaction between N_PROFESSION and N_OCCUPATION, which is evidenced by the inclusion of both variables in the full main-effects model. In light of the aforementioned considerations, a decision was taken to construct an alternative model by reintroducing N_PROFESSION back into the final model, thereby resulting in an 11-variable model.

The final main-effects model and the corresponding empirical odds are presented in **Table 18-12** and **Figure 18-49**, respectively. It is evident that the incorporation of an additional variable does not seem to have a significant impact on the model's performance and coefficient size. However, the enhancement in the empirical odds is remarkable, thereby further substantiating the existence of an N_PROFESSION-N_OCCUPATION interaction. In light of these findings, we will proceed to consider the model presented in **Table 18-12** as our definitive final main-effects model.

CHAPTER 18 THE MAIN-EFFECTS MODEL

Table 18-12. *Final main-effects model after bringing back the N_PROFESSION variable.*

Partition	KS	ROC	Gini
Train	0.653	0.895	0.789
Validate	0.648	0.893	0.785
Over Training	0.005	0.002	0.004

No.	Parameter	DF	Estimate	Error	χ^2	p-value	Std. Estimate	Deviation
	Intercept	1	-2.335	0.006	138,904	<.0001	0.097	0.000
1	WOE_I_RISK_LEVEL	1	-0.873	0.004	54,629	<.0001	-0.501	0.000
2	WOE_I_MX_DYSPSTDUE_LON_L12M	1	-0.532	0.007	5,066	<.0001	-0.201	0.000
3	WOE_N_BRANCH	1	-0.458	0.009	2,528	<.0001	-0.186	0.003
4	WOE_I_AGE_YRS	1	-0.420	0.013	998	<.0001	-0.103	0.011
5	WOE_I_AV_R_BALT_APPR_CLC_L06M	1	-0.302	0.010	934	<.0001	-0.090	0.002
6	WOE_I_SIN_R_PAYT_APPR_CLC_L13_24	1	-0.327	0.012	792	<.0001	-0.111	0.001
7	WOE_N_OCCUPATION	1	-0.333	0.015	512	<.0001	-0.117	0.004
8	WOE_I_SDC_R_CHRG_CRRA_CPM_L12M	1	-0.676	0.039	303	<.0001	-0.068	0.000
9	WOE_I_SIN_R_BALT_APPR_CLC_L13_24	1	-0.266	0.017	235	<.0001	-0.050	-0.003
10	WOE_N_COLLATERAL_TYPE	1	-0.175	0.015	135	<.0001	-0.043	0.011
11	WOE_N_PROFESSION	1	-0.079	0.031	7	0.0101	-0.010	

CHAPTER 18   THE MAIN-EFFECTS MODEL

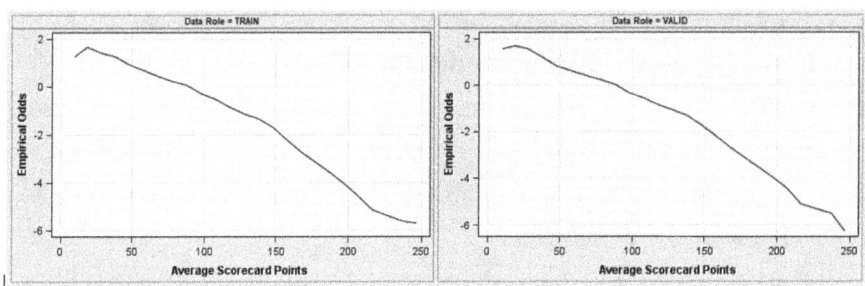

Figure 18-49. Empirical odds for the final main-effects model after bringing back the N_PROFESSION variable.

18.6 Conclusion

Chapter 18 provides an example of the conventional and most common steps needed to obtain a satisfactory final main-effects model. It is important to acknowledge that, as previously mentioned, machine-generated models are inherently incapable of generating results that are comparable to, or even approximating those obtained through the methodology outlined here. To illustrate this point, we may consider the model presented in **Table 18-1**, which constituted the initial input utilized in the construction of the final main-effects model presented in **Table 18-12**. It could be argued that the model presented in **Table 18-1** could be considered a machine-generated model. Nevertheless, it should be noted that the construction of this model would not have been possible without the elimination of the sign-inverted coefficients that were identified during the manual assessment of the coefficient-variant association matrix.

For the sake of illustration, let us assume that the model presented in **Table 18-1** was indeed 100% machine-generated. This prompts the question of whether there is a significant difference between the model presented in **Table 18-1** and the one presented in **Table 18-12**. This is particularly pertinent given that both models comprise variables with a *p*-value of less than 0.001 and thus possess excellent discriminatory power, with validation ROCs of 0.89.

CHAPTER 18 THE MAIN-EFFECTS MODEL

It is my hope that by the conclusion of this book, the reader will be in a position to answer this question in the affirmative. Indeed, there is a notable distinction between the model presented in **Table 18-1** and the one presented in **Table 18-12**. While both models exhibit comparable statistical performance, they diverge significantly in terms of simplicity, coherence, and business logic.

As has been demonstrated, statistical performance is not a definitive criterion, as models may become indistinguishable from one another at a certain point, thereby rendering automatic model selection an arbitrary process. Moreover, the observation that all variables in the model presented in **Table 18-1** are extremely significant is merely coincidental and attributable to the fact that we are dealing with large training datasets comprising over 100,000 observations. As previously discussed in Section 16.2.1.2, the standard error, denoted as $SE(\hat{\beta})$, is inversely proportional to the sample size, n. Consequently, too many meaningless variables will become highly or extremely significant, simply due to the large sampling size, rather than because they actually have a meaningful relationship with the target variable (not to mention their conceptual meaning).

Therein lies the problem with the machine-generated models approach. On the one hand, there are sophisticated algorithms capable of detecting the most insignificant effects; and on the other hand, there are huge datasets that magnify the effect of meaningless variables. Furthermore, the use of automatic selection procedures is also questionable due to the inherent limitations of the data itself. Data is often of poor quality, containing errors and inconsistencies. Additionally, there are legal, ethical, and business considerations that must be taken into account when determining which variables should or should not be included in a model—but who is going to separate the wheat from the chaff...the analyst or the machine?

CHAPTER 18 THE MAIN-EFFECTS MODEL

18.7 Summary

This chapter provides a comprehensive guide to building a robust and conceptually sound main-effects model, starting with the reduction of effects derived from the hyperspace in the previous chapter (*Experimental Design and Hyperoptimization*) and culminating in the refinement of the *weight of evidence* (WoE) transformations. The chapter not only outlines the multivariable selection strategies used to create the main-effects model but also delves deeply into the *WoE refinement process*, offering some of the most valuable practical examples in the book.

Key highlights of the chapter:

1. **Multivariable Selection Strategies for Main Effects**

   The chapter begins by addressing the challenge of reducing the hyperspace of effects using three distinct strategies:

   1. IG, BE-Only model
   2. IG, ICRM, BE-Only
   3. IG, ICRM, all-subsets

   A critical innovation in these strategies is the placement of the ICRM (iterative collinearity reduction method) after the IG (Interactive Grouping) node. This placement ensures that collinearity issues introduced by combining variables from disparate reduction methods are addressed after transformations like weight of evidence (WoE), which can influence collinearity by

CHAPTER 18   THE MAIN-EFFECTS MODEL

selecting different group cutoffs for highly correlated variables.

One strategy notably omits the ICRM, based on the assumption that collinearity levels will not significantly affect the model's performance. The chapter explains the rationale behind contrasting BE-Only and all-subsets methods, offering insights into the trade-offs between reduced and extended models. A set of criteria is then provided for selecting the *best full main-effects model*.

2. **The WoE Refinement Process**

   The *WoE refinement process* is one of the most valuable and revealing sections of the book, offering real-world examples that are indispensable for practitioners, particularly those new to scorecard development. These examples, meticulously detailed and nuanced, demonstrate how to refine WoE transformations to align with conceptual expectations and business logic.

   The process begins by defining *expected trends* for each variable—an essential step that highlights the need for a deep understanding of the system being modeled. This proactive, deductive approach distinguishes the scorecard methodology from the passive, inductive methods of classical modeling. As the chapter explains:

   "The relationship between explanatory and target variables is to some extent predetermined from the

CHAPTER 18   THE MAIN-EFFECTS MODEL

beginning of the analysis, during the conceptual variable design stage."

However, not all variables conform to these expected trends, which is an inherent aspect of the modeling process. To address this, the chapter introduces a classification system based on visual inspection of the WoE functions:

1. Inconsistent and rejected
2. Consistent, no adjustments required
3. Consistent, but manual adjustments required
4. Ambiguous, may require manual selection

Each category is addressed in detail, with practical guidance on how to refine variables to ensure consistency and logical coherence. The examples provided in this section offer a level of depth and clarity that is rarely available in other publications. For those beginning their journey in scorecard development, these examples are a dream come true.

3. **Reducing Variables Through WoE Insights**

   By examining the WoE charts alone, the chapter demonstrates how the total number of main-effects variables can be reduced from 24 to 16. This reduction is achieved through a combination of conceptual understanding, visual inspection, and manual refinement. The chapter also presents a methodology for further reducing variables when all are highly significant, emphasizing the importance of the χ^2 statistic and conceptual meaning.

CHAPTER 18 THE MAIN-EFFECTS MODEL

A standout example is the interaction between N_OCCUPATION and N_PROFESSION, which perfectly illustrates the importance of retaining meaningful interactions. This real-life example aligns beautifully with theoretical expectations, showcasing the power of manual intervention in variable selection.

4. **Machine-Generated Models vs. Manual Refinement**

The chapter concludes by contrasting the manually refined main-effects model with a hypothetical machine-generated model. While both may exhibit similar statistical performance, the manually refined model stands out for its simplicity, coherence, and business logic. The discussion highlights the pitfalls of relying solely on automated methods, particularly in the presence of large datasets where meaningless variables may appear highly significant due to sampling size effects.

The chapter ends with a thought-provoking question:

"Who is going to separate the wheat from the chaff— the analyst or the machine?"

Key takeaways:

1. **Multivariable Selection Strategies:**
   - The chapter outlines three strategies for refining the hyperspace of effects, with a focus on addressing collinearity and optimizing variable selection.

CHAPTER 18 THE MAIN-EFFECTS MODEL

2. **Weight of Evidence (WoE) Refinement:**
   - The WoE refinement process is a cornerstone of the chapter, offering invaluable real-world examples that demonstrate how to align variables with expected trends and refine their transformations.

3. **Proactive, Deductive Approach:**
   - The scorecard methodology emphasizes a proactive, deductive approach to modeling, setting it apart from the passive, inductive methods of classical approaches.

4. **Variable Classification:**
   - Variables are classified into four categories based on their WoE functions, with detailed guidance provided for addressing each category.

5. **Reduction of Variables:**
   - The chapter demonstrates how to reduce the number of main-effects variables from 24 to 16 through visual inspection and manual refinement alone.

6. **Manual Elimination and Interactions:**
   - A methodology for manually eliminating variables is presented, with a standout example illustrating the importance of retaining meaningful interactions.

CHAPTER 18   THE MAIN-EFFECTS MODEL

7. **Limitations of Automation:**
   - The chapter critiques the limitations of machine-generated models, highlighting the importance of manual refinement for achieving simplicity, coherence, and business logic.

In the next chapter, *The Scoring Process*, we move from model construction to implementation. This transition marks a pivotal step where the insights and transformations derived from the main-effects model are translated into actionable scoring rules. Here, we will explore how to assign scores to variables, calibrate the model for real-world use, and ensure that the scoring system is both effective and easy to exploit for decision-making. The scoring process is where the model truly comes to life, transforming data into meaningful, quantifiable outcomes that drive business strategies.

CHAPTER 19

The Scoring Process

"In theory there is no difference between theory and practice. In practice there is."

—Yogi Berra

One of the topics most frequently overlooked in contemporary machine learning (ML) and data mining (DM) literature, along with those of variable design and data processing, is the scoring process. This is because the majority of inexperienced analysts have been led to believe that the model-building process ends with the obtaining of the final model. However, even excellent models can ultimately be discarded if they are unable to be successfully deployed and executed on a recurring basis. It is unfortunate that I have observed on numerous occasions throughout my professional career that models that were developed with significant effort and resources have ultimately been discarded due to a lack of understanding of the scoring process. It is easy to understand that the loss of all the work invested in the construction of a model would be a significant tragedy. It is therefore crucial to have a plan in place to move forward with the model in question, should it be accepted for production. However, it is imperative to first comprehend the intricacies of the scoring process.

CHAPTER 19 THE SCORING PROCESS

19.1 Principal Elements of the Scoring Process

Figure 19-1 presents the nine principal elements that comprise the scoring process:

1. Official data sources
2. ETLs for the foundation mart
3. Foundation mart
4. CVIC (ETL for the construction of the scoring ABT)
5. Scoring ABT
6. Scoring model
7. Scored ABT
8. Scoring snapshot table
9. Follow-up statistics

CHAPTER 19 THE SCORING PROCESS

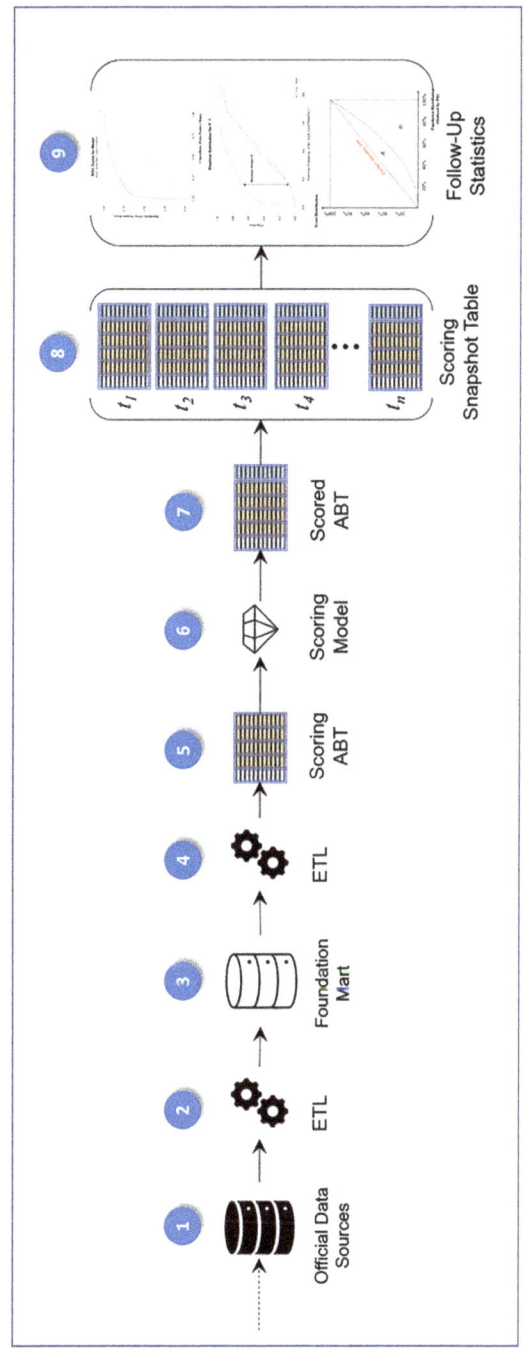

Figure 19-1. Principal elements comprising the scoring process.

CHAPTER 19 THE SCORING PROCESS

19.1.1 Official Data Sources

The official data sources comprise all the original data sources (predominantly tables) that were employed as inputs for the construction of the inputs, prior to the construction of the ABT. These inputs are typically the official tables that provide data to the overall corporation, such as the snapshot statement tables, which may pertain to loans, credit cards, checking accounts, or other types of accounts. The transactional tables encompass those pertaining to credit, debit, or any other kind of event. In addition, dimension and catalog tables for slowly changing characteristics or descriptions, as well as any other pertinent source, should be included.

19.1.2 ETLs for the Foundation Mart

As previously discussed in Section 10.4, it is highly probable that your initial data sources will necessitate some degree of customization prior to their utilization as viable inputs for the construction of your ABT. This encompasses the following:

1. Extraction of the building window and/or the target population

2. Computation of the sequential synthetic keys (Section 9.5) for new records and addition of the corresponding synthetic key for already existing records

3. Addition of grouping columns to the transactional tables (Section 10.4.1.1), which includes

    a. Transaction classification labels (Section 3.3.1)

    b. Higher granularity levels, such as customer number or CUSTOMER_RK (Section 10.4.1.2)

4. Conducting the corresponding grouping operations of the transactional tables (Section 10.4.1.3)

5. Construction of the TBTs for the computation of base metrics with the use of Oracle's analytical functions (OAFs) (Sections 10.4.1.4 and 10.4.1.5)

6. Recurring loading and accumulation for historical table, e.g., snapshot and TBTs

19.1.3 Foundation Mart

As discussed in Section 10.8.1, the recommended strategy is to construct an *ad hoc* data mart in conjunction with the model. This will enable the model to be continuously fed during the scoring process once the final model has been accepted and approved for production. Additionally, Section 10.8.1 indicated that the *ad hoc* data mart could serve not only as the input for the recently approved model but also as the foundation for the development of numerous future models. It is therefore evident that the foundation mart should form an integral part of the scoring process.

19.1.4 CVIC

As outlined in Section 10.5, the continuous variable integration cycle (CVIC) facilitates the systematic and sequential development of the training ABT. It is evident that the CVIC can also be employed to construct our scoring ABT. Indeed, it is anticipated that the CVIC will execute considerably more rapidly for scoring ABTs than for training ABTs, given that the number of variables and records to be incorporated into the scoring ABT is significantly smaller.

19.1.5 Scoring ABT

In contrast to the training ABT, the scoring ABT comprises solely the final set of variables selected for the final model. Additionally, the target population typically encompasses only the accounts that were active

CHAPTER 19   THE SCORING PROCESS

during the reference period. As an illustration, **Table 19-1** depicts the typical columns that would be incorporated into the scoring ABT, based on the final model presented in **Table 18-12**. It should be noted that only four additional columns have been included for identification purposes: PERIOD, DATE, ACCOUNT_ID, and ACCOUNT_RK. However, the inclusion of additional columns is optional.

Table 19-1. *Typical columns included in the scoring ABT according to the final model presented in* ***Table 18-12****.*

No.	Column	Brief description
1	PERIOD	Scoring date in YYYYMM format
2	DATE	Scoring date in mm/dd/yyyy format
3	ACCOUNT_ID	Natural business key
4	ACCOUNT_RK	Synthetic key
5	I_RISK_LEVEL	Final interval explanatory variable
6	I_MX_DYSPSTDUE_LON_L12M	Final interval explanatory variable
7	N_BRANCH	Final nominal explanatory variable
8	I_AGE_YRS	Final interval explanatory variable
9	I_AV_R_BALT_APPR_CLC_LO6M	Final interval explanatory variable
10	I_SIN_R_PAYT_APPR_CLC_L13_24	Final interval explanatory variable
11	N_OCCUPATION	Final nominal explanatory variable
12	I_SDC_R_CHRG_CRRA_CPM_L12M	Final interval explanatory variable
13	I_SIN_R_BALT_APPR_CLC_L13_24	Final interval explanatory variable
14	N_COLLATERAL_TYPE	Final nominal explanatory variable
15	N_PROFESSION	Final nominal explanatory variable

CHAPTER 19 THE SCORING PROCESS

19.1.6 Scoring Model

The scoring model is a *SAS Score Code* that performs the necessary calculations to determine the WoE functions and the score for the final variables. Additionally, it calculates the scorecard points and the PD for each record in the scoring ABT. The execution of the scoring model is what is commonly referred to as *"scoring."* It is important to note that this is the only stage of the scoring process where the term *"scoring"* is accurately applied.[1]

The complete SAS Score Code can be copied from the *Model Package*. **Figure 19-2** shows a series of screens that illustrate the process to generate and open the Model Package:

1. Right-click on the scorecard node of your final main-effects model.

2. Click on the "Create Model Package..." option.

3. Name your Model Package.

4. Once your Model Package has been generated, click on the "Open" button.

Figure 19-3 shows the series of options needed to access the SAS Score Code:

1. Select the View option from the toolbar.

2. Select the "Scoring" option, and then, click on SAS Score Code.

---

[1] This observation is important because the term "*scoring*" is loosely used in the industry to refer to the entire scoring process. As can be seen, this process encompasses a multitude of activities in addition to the mere act of *scoring* the scoring ABT. It is therefore recommended that the reader employs the term "*Scoring Process*" in lieu of "*scoring*" to avoid perpetuating the underestimation of the scoring process.

CHAPTER 19   THE SCORING PROCESS

The SAS Score Code should appear in a new emergent window.

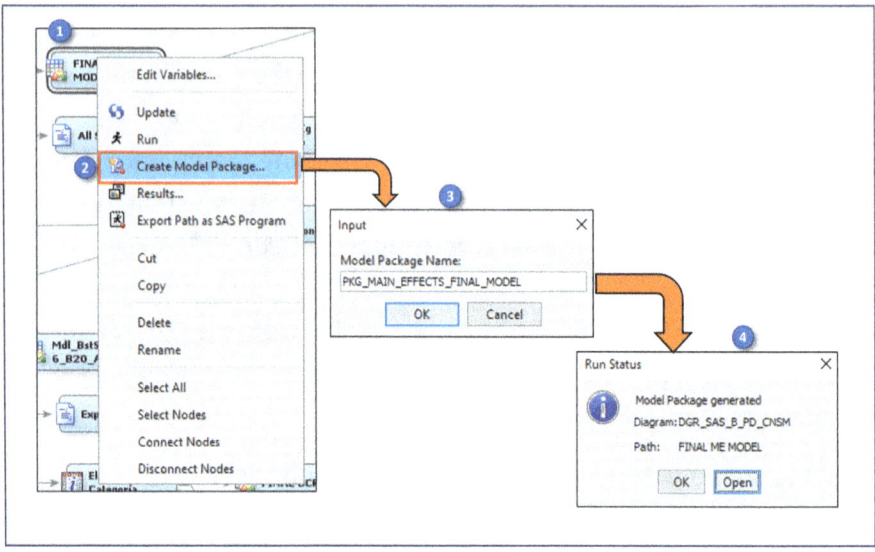

Figure 19-2. Illustrated example of the four steps needed to generate and open the Model Package.

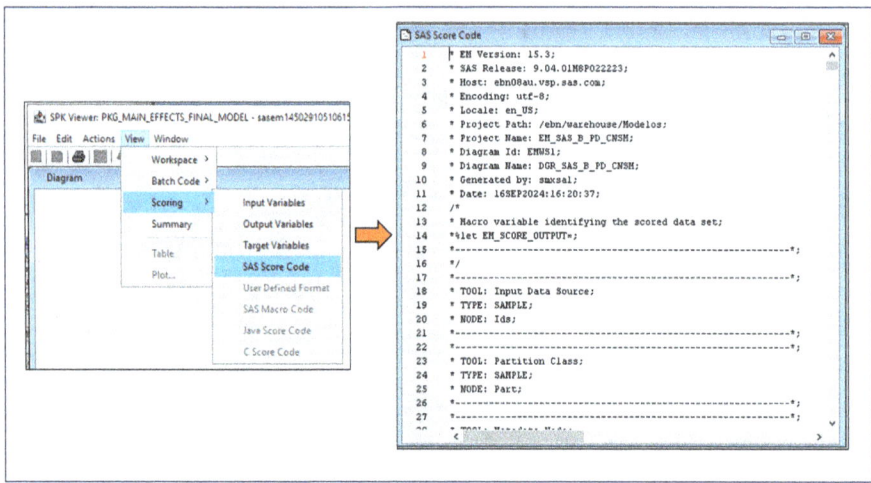

Figure 19-3. Illustrated example showing the location of the SAS Score Code within the Model Package.

Henceforth, the SAS Score Code can be transferred to a more appropriate programmer interface, such as Enterprise Guide (EG), for the purpose of further editing.

It should be noted that the SAS Score Code generated by EM is in fact a SAS DATA step code. Consequently, it must first be placed within a SAS DATA step clause before it can be utilized. **Listing 19-1** illustrates the utilization of a SAS Score Code within a SAS DATA step. It should be noted that two additional lines have been incorporated at the outset of the code, in lines 1 and 2. Line 2 illustrates the SET clause, incorporating the input dataset designated as SCR_ABT_B_PD_CRD_202301. This dataset corresponds to the scoring ABT, as detailed in **Table 19-1**. Line 1 illustrates the DATA clause with the output dataset, designated as SCR_MDL_B_PD_CRD_202301, which corresponds to the *scored ABT*.

CHAPTER 19 THE SCORING PROCESS

Listing 19-1. Extract of the SAS Score Code placed in a SAS DATA step.

```sas
DATA    SCR_MDL_B_PD_CRD_202301;
        SET SCR_ABT_B_PD_CRD_202301;

        *----------------------------------------;
        * Variable: I_AGE_YRS;
        *----------------------------------------;
        LABEL GRP_I_AGE_YRS =
        "Grouped: I_AGE_YRS";
        LABEL WOE_I_AGE_YRS =
        "Weight of Evidence: I_AGE_YRS";

        if MISSING(I_AGE_YRS) then do;
        GRP_I_AGE_YRS = 1;
        WOE_I_AGE_YRS =          0;
        end;
        else if NOT MISSING(I_AGE_YRS) then do;
        if I_AGE_YRS < 27.79 then do;
        GRP_I_AGE_YRS = 2;
        WOE_I_AGE_YRS = -0.820019759;
        end;
        else
        if 27.79 <= I_AGE_YRS AND I_AGE_YRS < 33.76 then do;
        GRP_I_AGE_YRS = 3;
        WOE_I_AGE_YRS =   -0.57582914;
        end;
        else
        if 33.76 <= I_AGE_YRS AND I_AGE_YRS < 40.93 then do;
        GRP_I_AGE_YRS = 4;
        WOE_I_AGE_YRS =   -0.34559292;
        end;
        else
        if 40.93 <= I_AGE_YRS AND I_AGE_YRS < 46.58 then do;
        GRP_I_AGE_YRS = 5;
        WOE_I_AGE_YRS =   -0.06280172;
        end;
        else
        if 46.58 <= I_AGE_YRS AND I_AGE_YRS < 51.91 then do;
        GRP_I_AGE_YRS = 6;
        WOE_I_AGE_YRS = 0.2117495907;
        end;
        else
        if 51.91 <= I_AGE_YRS then do;
        GRP_I_AGE_YRS = 7;
        WOE_I_AGE_YRS = 0.5285673944;
        end;
        end;
```

Furthermore, it should be noted that the SAS Score Code contains the rules that assign the group levels and their corresponding weight values for each of the final variables presented in **Table 19-1**. Consequently, each variable will yield two additional columns. In the case of the variable shown in **Listing 19-1**, I_AGE_YRS, its corresponding derive variables are GRP_I_AGE_YRS and WOE_I_AGE_YRS.

Listing 19-2 presents a different section of the SAS Score Code with the rules that assign the scorecard points[2] for the variable I_AGE_YRS, the resulting derive variable is SCR_I_AGE_YRS.

Listing 19-2. Example of the SAS Score Code rules that assign the scorecard points for the variable I_AGE_YRS.

```
790     *------------------------------------------------------------*;
791     * Variable: I_AGE_YRS;
792     *------------------------------------------------------------*;
793     if MISSING(I_AGE_YRS) then do;
794        SCORECARD_POINTS = SCORECARD_POINTS + 35;
795        SCR_I_AGE_YRS= 35;
796     end;
797     else if NOT MISSING(I_AGE_YRS) AND I_AGE_YRS < 27.79 then do;
798        SCORECARD_POINTS = SCORECARD_POINTS + 23;
799        SCR_I_AGE_YRS = 23;
800     end;
801     else if NOT MISSING(I_AGE_YRS) and 27.79 <= I_AGE_YRS AND I_AGE_YRS < 33.76 then do;
802        SCORECARD_POINTS = SCORECARD_POINTS + 27;
803        SCR_I_AGE_YRS = 27;
804     end;
805     else if NOT MISSING(I_AGE_YRS) and 33.76 <= I_AGE_YRS AND I_AGE_YRS < 40.93 then do;
806        SCORECARD_POINTS = SCORECARD_POINTS + 30;
807        SCR_I_AGE_YRS = 30;
808     end;
809     else if NOT MISSING(I_AGE_YRS) and 40.93 <= I_AGE_YRS AND I_AGE_YRS < 46.58 then do;
810        SCORECARD_POINTS = SCORECARD_POINTS + 34;
811        SCR_I_AGE_YRS = 34;
812     end;
813     else if NOT MISSING(I_AGE_YRS) and 46.58 <= I_AGE_YRS AND I_AGE_YRS < 51.91 then do;
814        SCORECARD_POINTS = SCORECARD_POINTS + 38;
815        SCR_I_AGE_YRS = 38;
816     end;
817     else if NOT MISSING(I_AGE_YRS) and 51.91 <= I_AGE_YRS then do;
818        SCORECARD_POINTS = SCORECARD_POINTS + 42;
819        SCR_I_AGE_YRS = 42;
820     end;
```

---

[2] For more information about the calculation of the scorecard points, see Siddiqi [63], pp 240-244 or visit https://go.documentation.sas.com/doc/en/emref/15.2/n181vl3wdwn89mn1pfpqm3w6oaz5.htm.

CHAPTER 19 THE SCORING PROCESS

The SAS Score Code adds seven additional columns to the scoring ABT:

1. **_WARN_:** Warning column

2. **I_KGB_GE60D_N12M:** Nominal unnormalized estimated target[3] variable

3. **U_KGB_GE60D_N12M:** Interval unnormalized estimated target variable

4. **P_KGB_GE60D_N12M1:** Estimated probability of event

5. **P_KGB_GE60D_N12M0:** Complementary probability of no event

6. **SCORECARD_POINTS:** Scaled total score

7. **SCORECARD_BIN (Listing 19-4):** A 25-level bin of the scaled total score used for charts (such as the empirical odds chart) or for the calculation of performance statistics, like the *system stability index* (SSI).

It is noteworthy that the coefficients of the logistic regression estimated during the execution of the Scorecard node are also present in the SAS Score Code with 16 digits of precision (see **Listing 19-3**).

---

[3] The target variable in this example is the KGB_GE60D_N12M variable, and the default event occurs when the maximum number of days past due in the next 12 months is greater or equal to 60 days.

Listing 19-3. *Fragment of the SAS Score Code showing the estimates of the logistic regression.*

```
691         ***   Effect:  WOE_N_OCCUPATION ;
692         _TEMP = WOE_N_OCCUPATION ;
693         _LP0 = _LP0 + (    -0.50336293490518 *  _TEMP);
694
695         ***   Effect:  WOE_I_AGE_YRS ;
696         _TEMP = WOE_I_AGE_YRS ;
697         _LP0 = _LP0 + (    -0.47691413825136 *  _TEMP);
698
699         ***   Effect:  WOE_I_LOAN_AGE_YRS ;
700         _TEMP = WOE_I_LOAN_AGE_YRS ;
701         _LP0 = _LP0 + (    -0.91554624555803 *  _TEMP);
702
703         ***   Effect:  WOE_I_MX_DYSPSTDUE_LON_L12M ;
704         _TEMP = WOE_I_MX_DYSPSTDUE_LON_L12M ;
705         _LP0 = _LP0 + (    -0.57725162181871 *  _TEMP);
706
707         ***   Effect:  WOE_I_SIN_R_BALT_APPR_CLC_L13_24 ;
708         _TEMP = WOE_I_SIN_R_BALT_APPR_CLC_L13_24 ;
709         _LP0 = _LP0 + (    -0.31783380514361 *  _TEMP);
710
711         ***   Effect:  WOE_N_HOMESTATUS ;
712         _TEMP = WOE_N_HOMESTATUS ;
713         _LP0 = _LP0 + (    -0.32113449961759 *  _TEMP);
```

CHAPTER 19 THE SCORING PROCESS

Listing 19-4. Fragment of the SAS Score Code showing an example of the SCORECARD_BIN variable.

```
1191        * Assign SCORECARD_BIN values;
1192        *;
1193        if SCORECARD_POINTS < 412 then SCORECARD_BIN = 1;
1194        else if SCORECARD_POINTS < 423 then SCORECARD_BIN = 2;
1195        else if SCORECARD_POINTS < 434 then SCORECARD_BIN = 3;
1196        else if SCORECARD_POINTS < 445 then SCORECARD_BIN = 4;
1197        else if SCORECARD_POINTS < 456 then SCORECARD_BIN = 5;
1198        else if SCORECARD_POINTS < 467 then SCORECARD_BIN = 6;
1199        else if SCORECARD_POINTS < 478 then SCORECARD_BIN = 7;
1200        else if SCORECARD_POINTS < 489 then SCORECARD_BIN = 8;
1201        else if SCORECARD_POINTS < 500 then SCORECARD_BIN = 9;
1202        else if SCORECARD_POINTS < 511 then SCORECARD_BIN = 10;
1203        else if SCORECARD_POINTS < 522 then SCORECARD_BIN = 11;
1204        else if SCORECARD_POINTS < 533 then SCORECARD_BIN = 12;
1205        else if SCORECARD_POINTS < 544 then SCORECARD_BIN = 13;
1206        else if SCORECARD_POINTS < 555 then SCORECARD_BIN = 14;
1207        else if SCORECARD_POINTS < 566 then SCORECARD_BIN = 15;
1208        else if SCORECARD_POINTS < 577 then SCORECARD_BIN = 16;
1209        else if SCORECARD_POINTS < 588 then SCORECARD_BIN = 17;
1210        else if SCORECARD_POINTS < 599 then SCORECARD_BIN = 18;
1211        else if SCORECARD_POINTS < 610 then SCORECARD_BIN = 19;
1212        else if SCORECARD_POINTS < 621 then SCORECARD_BIN = 20;
1213        else if SCORECARD_POINTS < 632 then SCORECARD_BIN = 21;
1214        else if SCORECARD_POINTS < 643 then SCORECARD_BIN = 22;
1215        else if SCORECARD_POINTS < 654 then SCORECARD_BIN = 23;
1216        else if SCORECARD_POINTS < 665 then SCORECARD_BIN = 24;
1217        else SCORECARD_BIN = 25;
```

19.1.7 Scored ABT

Table 19-2 shows the expected layout of the scored ABT (which in this example is SCR_MDL_B_PD_CRD_202301) after executing the scoring model, i.e., the SAS Score Code. It is notable that there has been an increase in the number of columns from 15 to 55. The increase in the number of columns can be attributed primarily to the incorporation of the group, WoE, and score blocks, which quadruple the number of explanatory variables presented in **Table 19-1** from 11 to 44.

CHAPTER 19 THE SCORING PROCESS

Table 19-2. *List of expected columns in the scored ABT.*

No.	Column	Brief Description
1	PERIOD	Identification Columns
2	DATE	Identification Columns
3	ACCOUNT_ID	Identification Columns
4	ACCOUNT_RK	Identification Columns
5	I_RISK_LEVEL	Base Predictive Variable
6	I_MX_DYSPSTDUE_LON_L12M	Base Predictive Variable
7	N_BRANCH	Base Predictive Variable
8	I_AGE_YRS	Base Predictive Variable
9	I_AV_R_BALT_APPR_CLC_L06M	Base Predictive Variable
10	I_SIN_R_PAYT_APPR_CLC_L13_24	Base Predictive Variable
11	N_OCCUPATION	Base Predictive Variable
12	I_SDC_R_CHRG_CRRA_CPM_L12M	Base Predictive Variable
13	I_SIN_R_BALT_APPR_CLC_L13_24	Base Predictive Variable
14	N_COLLATERAL_TYPE	Base Predictive Variable
15	N_PROFESSION	Base Predictive Variable
16	GRP_I_RISK_LEVEL	Group Level of the Predictive Variable
17	GRP_I_MX_DYSPSTDUE_LON_L12M	Group Level of the Predictive Variable
18	GRP_N_BRANCH	Group Level of the Predictive Variable
19	GRP_I_AGE_YRS	Group Level of the Predictive Variable
20	GRP_I_AV_R_BALT_APPR_CLC_L06M	Group Level of the Predictive Variable
21	GRP_I_SIN_R_PAYT_APPR_CLC_L13_24	Group Level of the Predictive Variable
22	GRP_N_OCCUPATION	Group Level of the Predictive Variable
23	GRP_I_SDC_R_CHRG_CRRA_CPM_L12M	Group Level of the Predictive Variable
24	GRP_I_SIN_R_BALT_APPR_CLC_L13_24	Group Level of the Predictive Variable
25	GRP_N_COLLATERAL_TYPE	Group Level of the Predictive Variable
26	GRP_N_PROFESSION	Group Level of the Predictive Variable
27	WOE_I_RISK_LEVEL	WoE of the Predictive Variable
28	WOE_I_MX_DYSPSTDUE_LON_L12M	WoE of the Predictive Variable
29	WOE_N_BRANCH	WoE of the Predictive Variable
30	WOE_I_AGE_YRS	WoE of the Predictive Variable
31	WOE_I_AV_R_BALT_APPR_CLC_L06M	WoE of the Predictive Variable
32	WOE_I_SIN_R_PAYT_APPR_CLC_L13_24	WoE of the Predictive Variable
33	WOE_N_OCCUPATION	WoE of the Predictive Variable
34	WOE_I_SDC_R_CHRG_CRRA_CPM_L12M	WoE of the Predictive Variable
35	WOE_I_SIN_R_BALT_APPR_CLC_L13_24	WoE of the Predictive Variable
36	WOE_N_COLLATERAL_TYPE	WoE of the Predictive Variable
37	WOE_N_PROFESSION	WoE of the Predictive Variable
38	SCR_I_RISK_LEVEL	Score of the Predictive Variable
39	SCR_I_MX_DYSPSTDUE_LON_L12M	Score of the Predictive Variable
40	SCR_N_BRANCH	Score of the Predictive Variable
41	SCR_I_AGE_YRS	Score of the Predictive Variable
42	SCR_I_AV_R_BALT_APPR_CLC_L06M	Score of the Predictive Variable
43	SCR_I_SIN_R_PAYT_APPR_CLC_L13_24	Score of the Predictive Variable
44	SCR_N_OCCUPATION	Score of the Predictive Variable
45	SCR_I_SDC_R_CHRG_CRRA_CPM_L12M	Score of the Predictive Variable
46	SCR_I_SIN_R_BALT_APPR_CLC_L13_24	Score of the Predictive Variable
47	SCR_N_COLLATERAL_TYPE	Score of the Predictive Variable
48	SCR_N_PROFESSION	Score of the Predictive Variable
49	_WARN_	Complementary Variable
50	I_KGB_GE60D_N12M	Complementary Variable
51	U_KGB_GE60D_N12M	Complementary Variable
52	P_KGB_GE60D_N12M1	Estimated Probability
53	P_KGB_GE60D_N12M0	Estimated Probability
54	SCORECARD_POINTS	Score
55	SCORECARD_BIN	25-Level Scorecard Bin

CHAPTER 19   THE SCORING PROCESS

It should be noted that the scored ABT represents the primary output of the scoring process, which encompasses both the probability of default and the scorecard points. In the subsequent phase of the scoring process, the accumulated scoring results will be integrated into a single snapshot table by appending the scored ABTs one on top of the other. However, as will be demonstrated in the following section, the objective is not to create a distinct scoring snapshot table for each model that is constructed. Instead, the intention is to incorporate all of the scoring models in a single master snapshot table. In order to achieve this, it is essential that all models adhere to the same layout, which necessitates that all models possess the same variables.

It is evident that no two models, let alone all of them, will have identical variables. Therefore, the scoring snapshot table will only store fundamental and shared information between the scoring models, such as the scoring period, the account number, the PD, and the scorecard points. This raises the question of how to store the predictive variables of each scoring model.

The utilization of a master snapshot table, which stores the primary results of all scoring models, is a highly advantageous approach for data processing and utilization. This is due to the fact that long (many rows) and narrow (few columns) tables are always preferable to long (many rows) and wide (many columns) tables for data management. However, each scoring model presents a distinct challenge due to the inherent differences in their respective layouts.

There are two approaches to address this issue. The first method involves the creation of a sequential set of tables, designated by their code name and their respective scoring period. To illustrate this, we can consider the previously mentioned scored ABT example SCR_MDL_B_PD_CRD_202301. In this case, we would create and store a separate table for each scoring period, such as SCR_MDL_B_PD_CRD_202301, SCR_MDL_B_PD_CRD_202302, SCR_MDL_B_PD_CRD_202303,..., and so on, where

"SCR_MDL_B_PD_CRD" corresponds to the model's code, and it reads as *"scored PD behavioral model for credit card,"* with the scoring period represented in the format YYYYMM.

However, this approach presents a challenge in analyzing the evolution of predictive variables over time, as the results are dispersed across multiple tables. An alternative approach would be to create a separate snapshot table for each scoring model in the form of SCR_MDL_B_PD_CRD_SNP, which would read as *"Snapshot table for the scored PD behavioral model for credit card."* At a minimum, one additional column would need to be added to identify each scoring period. It should be noted that this option is more convenient in the long run, as it allows for the aggregation of results across multiple periods with a simple GROUP BY clause.

It is also noteworthy that the scoring ABTs could be appended and stored; however, their information is redundant, and thus, it is recommended that they be dropped after the termination of each scoring cycle.

19.1.8 Scoring Snapshot Table

As previously discussed, the purpose of the scoring snapshot table is to accommodate the historical results of scoring models in a single table. **Table 19-3** presents the list of suggested columns for the scoring snapshot table. It is important to note that only a limited number of compatible columns are included in the scoring snapshot table, which allows for the accurate identification and incorporation of virtually any scoring model.

Table 19-3. List of suggested columns for the scoring snapshot table.

No.	Column	Brief description
1	MODELING_ABT	Name of the modeling ABT
2	MODEL_CD	Scoring model code
3	MODEL_DESC	Scoring model description
4	VERSION_NO	Scoring model version
5	STATUS	Scoring model status
6	PERIOD	Scoring period (YYYYMM)
7	DATE	Scoring date (mm/dd/yyyy)
8	CUSTOMER_RK	Customer retained key
9	CUSTOMER_ID	Customer natural key
10	ACCOUNT_RK	Account retained key
11	ACCOUNT_ID	Account natural key
12	PROBABILITY	Probability of default
13	SCORE	Scorecard points
14	SCORECARD_BIN	Scorecard bin
15	EVENT	Observed event

For instance, the table includes the ABT from which the scoring model was derived. This facilitates the determination of the number of scoring models per ABT. This information is important to consider in light of the fact that a single ABT can be utilized as an input for multiple scoring models with different targets simultaneously.

Furthermore, the snapshot table comprises a unique code for each scoring model, accompanied by a description and versioning information, as well as the status of the scoring model. This is due to the possibility of there being multiple versions of the same model, such as the primary scoring model and some alternatives that may be employed for comparison purposes. The status indicator enables the differentiation between active and inactive models, particularly in the context of a snapshot table that may encompass historical data spanning multiple years. However, the most significant attribute of the snapshot table is the incorporation of the EVENT column, which encompasses the *observed default event*.

19.1.9 The Observed Default Event and the Validity Window

It is important to note that the observed default event is not part of the scoring process per se; however, it must be included at some point in the process if the performance of the model is to be estimated during the production phase.

In contrast to the modeling phase, the observed default event, also known as the actual event, serves as the target variable in the production phase. However, unlike the modeling phase, the outcome of the target variable is not known until 12 months after the last scoring process. This is due to the fact that it is only possible to ascertain whether an event has occurred *if the maximum number of days past due is greater than or equal to 60 days over the subsequent 12 months*. Consequently, the follow-up performance statistics will invariably lag behind the most recent scoring period by a period of 12 months.

Table 19-4 provides an illustrative example of the validity of various activities and performance statistics over a period from January of 2020 to December of 2022. It should be noted that this example assumes the construction of a scoring model based on a 12-month stacked ABT, spanning the period from January 2020 to December 2020. It is evident that for the initial two years of the specified timeframe, there is no issue

CHAPTER 19 THE SCORING PROCESS

in estimating the default event and the performance statistics for each individual period. The scoring of each period utilized for training the model is referred to as backtesting.[4] However, from January of 2022, it is no longer possible to estimate the actual event, nor can any of the related performance statistics, such as the ROC, the KS, or the Gini index, be estimated.

---

[4] The backtesting process is similar to the scoring process, except that in the backtesting, the scored periods correspond to the same periods used to build the training ABT, or even periods before.

CHAPTER 19 THE SCORING PROCESS

Table 19-4. *Validity of time windows for training, backtesting, scoring, and the actual event estimation, alongside the validity for the performance statistics of SSI, ROC, KS, and Gini.*

Period	Training	Backtesting	Scoring	Event	SSI	ROC	KS	Gini
202001	x	x		x	x	x	x	x
202002	x	x		x	x	x	x	x
202003	x	x		x	x	x	x	x
202004	x	x		x	x	x	x	x
202005	x	x		x	x	x	x	x
202006	x	x		x	x	x	x	x
202007	x	x		x	x	x	x	x
202008	x	x		x	x	x	x	x
202009	x	x		x	x	x	x	x
202010	x	x		x	x	x	x	x
202011	x	x		x	x	x	x	x
202012	x	x		x	x	x	x	x
202101			x	x	x	x	x	x
202102			x	x	x	x	x	x
202103			x	x	x	x	x	x
202104			x	x	x	x	x	x
202105			x	x	x	x	x	x
202106			x	x	x	x	x	x
202107			x	x	x	x	x	x
202108			x	x	x	x	x	x
202109			x	x	x	x	x	x
202110			x	x	x	x	x	x
202111			x	x	x	x	x	x
202112			x	x	x	x	x	x
202201				x		x		
202202				x		x		
202203				x		x		
202204				x		x		
202205				x		x		
202206				x		x		
202207				x		x		
202208				x		x		
202209				x		x		
202210				x		x		
202211				x		x		
202212				x		x		

CHAPTER 19   THE SCORING PROCESS

It is evident that the scorecard points can be calculated at any given time with the scoring model, provided that the scoring ABT is in place. However, this raises the question of how the performance of the scoring model can be evaluated without having to wait a full year.

19.1.10 The System Stability Index

The system stability index (SSI) is, in fact, an alternative representation of the IV. It quantifies the extent of deviation in the current score distribution relative to the final development model distribution, which is attained at the conclusion of the modeling process.

The definition of the SSI is identical to that of the IV, where the distribution of observations of the modeling population is given by

$$d_i^M = \frac{n_i^M}{N^M}$$

Equation 19-1. Distribution of the modeling population.

And the distribution of the actual scoring population is given by

$$d_i^A = \frac{n_i^A}{N^A}$$

Equation 19-2. Distribution of the actual scoring population.

Where d_i^M and d_i^A are the proportion of observations for the modeling and the actual scoring populations of the *ith* scorecard bin and N^M and N^A are the total number of observations for the modeling and the actual scoring populations.

The WoE in this case is defined as

$$WoE_i = ln\left[\frac{d_i^A}{d_i^M}\right]$$

Equation 19-3. WoE.

And the difference between the modeling and the actual scoring distributions of the *ith* scorecard bin is given by

$$dd_i = d_i^A - d_i^M$$

Equation 19-4. Difference between the modeling and the actual score distributions.

The system stability index (SSI) has then the same definition presented in **Equation 15-11**:

$$SSI_i = WoE_i \cdot dd_i$$

Equation 19-5. System stability index.

Table 19-5 provides an illustrative example of the calculation of the SSI. In this example, we have a 25-scorecard bin SSI, where DM represents the distribution of the modeling population and DA represents the distribution of the actual scoring population. It should be noted that the calculations follow precisely the same logic as that shown in **Table 15-1** for the WoE, except that in this case, the sum of the IVs is referred to as the SSI.

CHAPTER 19 THE SCORING PROCESS

Table 19-5. *Illustrative example that shows how to perform the SSI calculation.*

Scorecard Bin	DM	DA	WoE ln[DA/DM]	DA-DM	IV
Score< 77	0.0001	0.0000	0.0000	-0.0001	0.0000
77<=Score< 87	0.0002	0.0000	-1.7068	-0.0002	0.0003
87<=Score< 97	0.0006	0.0007	0.0709	0.0000	0.0000
97<=Score< 107	0.0020	0.0022	0.1062	0.0002	0.0000
107<=Score< 117	0.0042	0.0049	0.1687	0.0008	0.0001
117<=Score< 127	0.0067	0.0123	0.6087	0.0056	0.0034
127<=Score< 137	0.0117	0.0237	0.7017	0.0120	0.0084
137<=Score< 147	0.0212	0.0255	0.1880	0.0044	0.0008
147<=Score< 157	0.0284	0.0269	-0.0561	-0.0016	0.0001
157<=Score< 167	0.0396	0.0385	-0.0285	-0.0011	0.0000
167<=Score< 177	0.0490	0.0490	0.0004	0.0000	0.0000
177<=Score< 187	0.0533	0.0621	0.1533	0.0088	0.0014
187<=Score< 197	0.0721	0.0900	0.2211	0.0178	0.0039
197<=Score< 207	0.1060	0.1161	0.0910	0.0101	0.0009
207<=Score< 217	0.1492	0.1376	-0.0806	-0.0116	0.0009
217<=Score< 227	0.1928	0.1802	-0.0671	-0.0125	0.0008
227<=Score< 237	0.1287	0.1120	-0.1384	-0.0166	0.0023
237<=Score< 247	0.0679	0.0614	-0.1005	-0.0065	0.0007
247<=Score< 257	0.0333	0.0285	-0.1539	-0.0047	0.0007
257<=Score< 267	0.0191	0.0169	-0.1214	-0.0022	0.0003
267<=Score< 277	0.0090	0.0070	-0.2457	-0.0020	0.0005
277<=Score< 287	0.0037	0.0035	-0.0520	-0.0002	0.0000
287<=Score< 297	0.0011	0.0005	-0.8135	-0.0006	0.0005
297<=Score< 307	0.0001	0.0002	0.3851	0.0001	0.0000
307<=Score	0.0000	0.0000	0.0000	0.0000	0.0000
Total	1	1			SSI = 0.0262

Based on the rules of thumb presented in **Table 15-2**, we can define the respective rules of thumb for the SSI as in Table 19-6.

Table 19-6. Rules of thumb for the system stability index.

System stability index	Distribution shift
Less than 0.02	Negligible shift
0.02 to 0.1	Small shift
0.1 to 0.3	Fair shift
0.3 to 0.5	Significant shift
>0.5	Extreme shift

Note that unlike the IV, the bigger the SSI, the worse the distribution shift.

19.1.11 Follow-Up Statistics

It can be argued that the follow-up statistics do not form part of the scoring process. Nevertheless, it is evident that no scoring process would be complete without some method of measuring the performance and stability of the model. This enables the assessment of the necessity for recalibration of the scoring model, or the estimation of the time at which recalibration would be required, by extrapolating the trends of the performance statistics.

As previously discussed in Section 17.1.1, the most effective statistical methods for evaluating the performance of our model, in terms of discriminatory power, are the ROC curve, the Gini coefficient, and the KS test statistic. However, it should be noted that all performance statistics can only be measured retrospectively. Therefore, the SSI should also be included in the set of follow-up statistics.

CHAPTER 19 THE SCORING PROCESS

The incorporation of these follow-up statistics will facilitate the monitoring of the model's performance over time, in a manner analogous to a process control chart. This will assist in the identification of any unexpected fluctuations and/or trends in the scoring model. It is important to note that a sudden fluctuation in the model's performance does not necessarily indicate that the model is uncalibrated. Rather, it could also be indicative of potential issues with the input data. The follow-up statistics, therefore, serve as an excellent response variable, not only for the model itself but for the entire scoring process.

Figures 19-4, **19-5**, and **19-6** illustrate the expected behavior for the follow-up statistics, the ROC curve, the KS test statistic, and the SSI, respectively. It is noteworthy that all three statistics exhibit minimal fluctuations over time, across multiple scoring periods. Moreover, the ROC curve and the KS even show a slight upward trend. It is also noteworthy that the aforementioned metrics fall within the optimal performance range for the specified scoring window, with values between 0.83 and 0.91 for the ROC curve, values between 0.54 and 0.72 for the KS, and finally values between 0.04 and 0.06 for the SSI.

CHAPTER 19   THE SCORING PROCESS

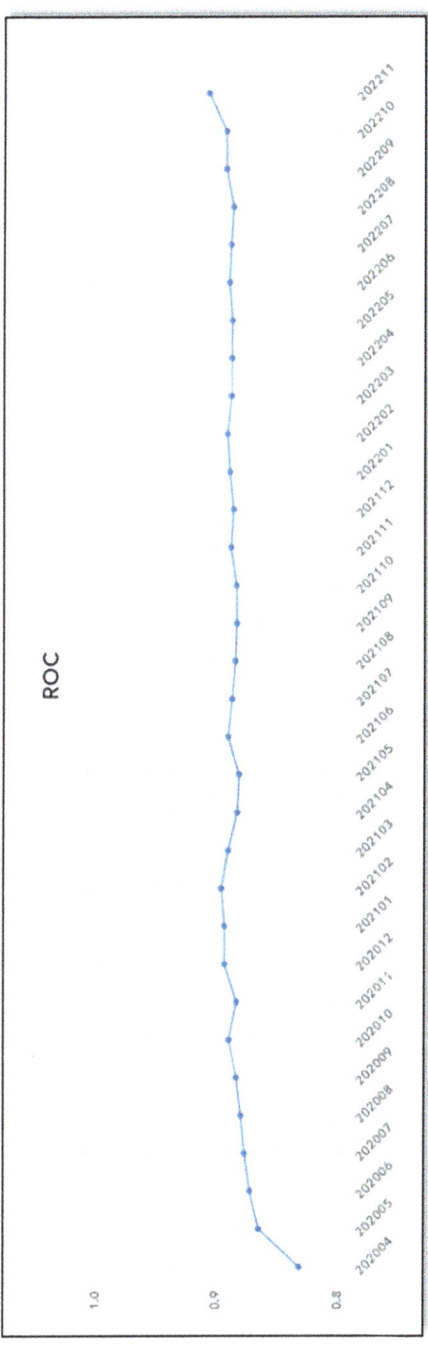

Figure 19-4. Monitoring chart for the ROC curve.

CHAPTER 19 THE SCORING PROCESS

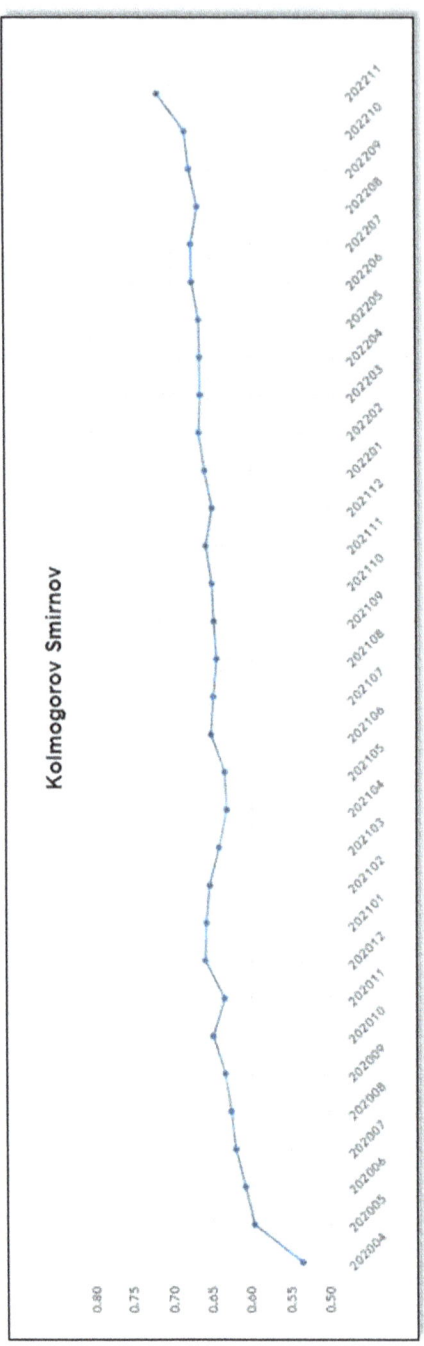

Figure 19-5. Monitoring chart for the KS test statistic.

CHAPTER 19 THE SCORING PROCESS

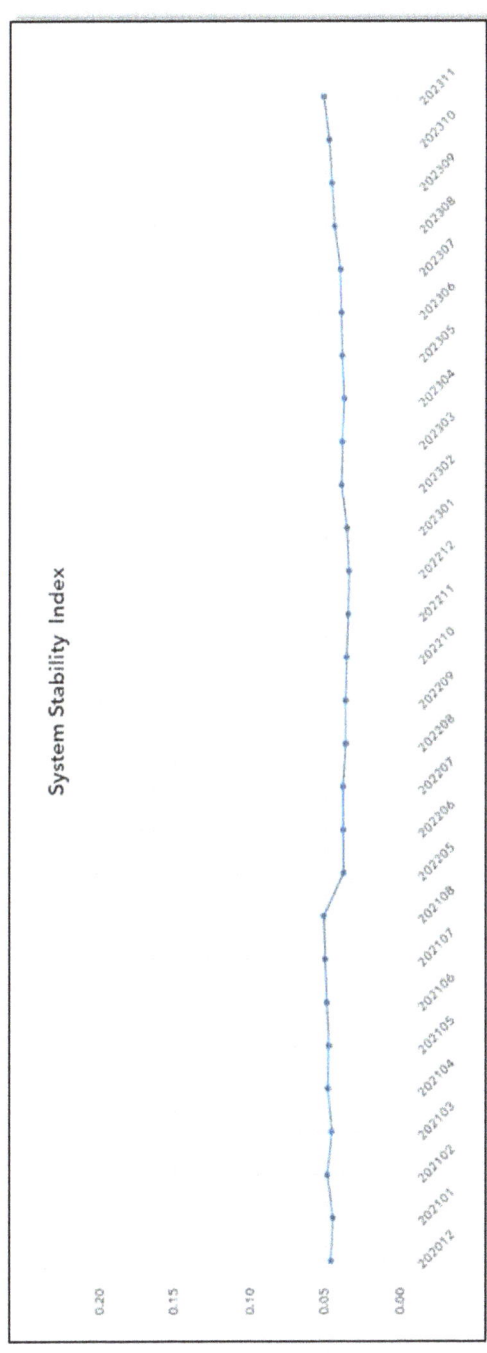

Figure 19-6. Monitoring chart for the SSI.

CHAPTER 19   THE SCORING PROCESS

It is therefore recommended that the scoring process be expanded to include the computation, updating, and storage of the aforementioned statistics, thereby enabling the monitoring of the scoring model.

19.2 Summary

In this chapter, we saw the detailed steps required to implement the scoring process effectively, ensuring that all the efforts invested in experimental design, hyperoptimization, and main-effects modeling are not wasted. The scoring process is far more than just assigning scores to data—it is about creating a sustainable, repeatable, and efficient system that enables the storage, monitoring, and exploitation of results in a way that drives actionable insights. Without proper implementation of this process, all prior efforts risk being rendered meaningless.

We explored the *nine critical steps* of the scoring process:

1. **Official Data Sources**: Identifying and accessing validated, reliable data sources to ensure the foundation of the process is robust.

2. **ETLs for the Foundation Mart**: Extracting, transforming, and loading data to create a clean and structured foundation for scoring.

3. **Foundation Mart**: Building a centralized repository that serves as the base for all subsequent models and processes.

4. **CVIC (ETL for the Construction of the Scoring ABT)**: Constructing the analytical base table (ABT) specifically for scoring, ensuring it aligns with the model's requirements.

5. **Scoring ABT**: Preparing the scoring dataset by applying transformations and ensuring all variables are aligned with the scoring model.

6. **Scoring Model**: The actual set of rules and parameters that serve to estimate the probability of default and the corresponding score.

7. **Scored ABT**: Applying the predictive model to generate scores based on the variables defined in the scoring ABT.

8. **Scoring Snapshot Table**: Storing the scored data in an efficient and structured manner, enabling long-term usability and facilitating performance monitoring.

9. **Follow-Up Statistics**: Measuring the performance of the model over time to ensure it remains effective and relevant.

Implementing a *sustainable process* capable of being executed on a recurrent basis is critical. A well-designed scoring process ensures that the model can be run consistently, with minimal manual intervention, and that results are stored efficiently for future use. The key challenge is not simply scoring the data but storing the results in a way that allows for easy exploitation. This includes creating a *scoring snapshot table* that enables seamless access to the scores for analysis, monitoring, and decision-making.

We also emphasized the importance of the *validity window* in measuring model performance. The validity window, which depends entirely on the event definition, ensures that the model's predictions are evaluated over a consistent and meaningful timeframe. For example, if the event is defined as *"maximum number of days past due is greater than or equal to 60 days over the subsequent 12 months,"* the validity window must align with this definition to accurately measure performance.

To monitor and follow up on the scoring model's performance, three key statistics are recommended:

1. **ROC (Receiver Operating Characteristic):** Measures the model's ability to distinguish between positive and negative outcomes retrospectively

2. **KS (Kolmogorov-Smirnov):** Identifies the maximum separation between the cumulative distributions of positive and negative outcomes, also retrospective in nature

3. **SSI (Score Stability Index):** Unlike the ROC and KS, the SSI measures the model's actual performance over time by analyzing shifts in the score's distribution, providing insight into potential data drift or changes in the population

It is important to distinguish between the retrospective nature of the **ROC** and **KS** statistics, which evaluate how well the model performed on historical data, and the **SSI**, which infers the model's ongoing validity by identifying changes in score distributions. The SSI is particularly valuable for assessing whether the model remains relevant in a dynamic environment, ensuring it continues to deliver reliable predictions.

In conclusion, this chapter highlighted that the scoring process is the final and critical step in operationalizing predictive models. Its success depends on proper implementation, efficient storage of results, and regular monitoring of performance. By following the nine steps outlined and leveraging the recommended statistics, we can ensure the scoring process is both sustainable and impactful, enabling data-driven decisions that stand the test of time.

CHAPTER 20

Closing Thoughts

"I may not have gone where I intended to go, but I think I have ended up where I intended to be."

—Douglas Adams

"You can't always get what you want...But if you try, sometimes...well, you might find you get what you need."

—Mick Jagger

A few years ago, I read a two-volume book called *The Structure of Magic* [4], written by two clinical psychologists, Richard Bandler and John Grinder. The book is about how to conduct a successful therapy that can actually produce change in their clients, by analyzing the *meta-structure* of language itself. But what really caught my eye about this book was its intention to unravel what really lies behind the charismatic superstars of clinical psychology, whom Bandler and Grinder describe as

> "These people seemingly perform the task of clinical psychology with the ease and wonder of therapeutic magician...Through the approaches they bring to this task seem varied and as different as day and night, they all seem to share a unique wonder and potency."

CHAPTER 20 CLOSING THOUGHTS

This description strike me as familiar, and I thought that these people are not exclusively in the realm of clinical psychology, but in fact, they are ubiquitous in our daily lives and throughout all human history—we also call these people gurus. Gurus are the experts in our organizations, and they are the people we turn to when there is a problem that everyone thinks only they can solve.

But the work of gurus seems truly magical to everyone else but other gurus, which begs the question, what makes a guru a guru?

20.1 The Structure of Magic

In their book, Bandler and Grinder [4] refer to the structure of magic as the meta-structure of the phenomenon being studied, which in their case was the language used during therapy that was capable of producing real change in their clients and in our case is the modeling process that is capable of producing successful models in a structured and standardized manner—but what is the meta-structure of the modeling process?

If you remember Section 2.1.2, we mentioned that a meta-structure can also be called an archetype, that is, what is true across all possible examples, what remains the same regardless of the methodology or the author.

There is not much new about the individual elements that I have presented in this book, such as the WoE refinement process, manual variable elimination, or the multivariable selection methods. And the truth is that while experienced analysts are well aware of these methods and they use them in their daily work, their methods remain a mystery, and their knowledge is rarely, if ever, passed on to the next generation of analysts.

But it is also true that the lack of knowledge transfer is not entirely due to a lack of willingness on the part of experienced analysts, but to the fact that many of them, like me, have not been able to articulate what they are doing when they build models, let alone formally document their process in a proper sequence of logical steps, so that their process can be standardized and replicated.

It is interesting to note that the inability to articulate or to explain what one is doing and what everyone else perceives as unique is not uncommon and is also mentioned by Satir, in Bandler and Grinder's book.

> "Looking back, I see that, although I was aware that change was happening, I was unaware of the specific elements that went into the transaction which made change possible."

This is what originally motivated me to write this book—to try to explain to everyone else, and even to myself, what it was that I was doing when I was building models—or in other words, to explain the structure of magic.

20.2 The End of a Journey

Our journey through the various steps of the modeling process has come to an end, and we have come a long way from understanding the system under study to the final steps needed to put our model into production. So I hope that after reading this book, you will leave with a more realistic understanding of what the modeling process actually is and all that it entails. But what I would really like to see is a significant change in the present perception that model building is still a mystery reserved only for the privileged, or that the building process should be a painful and exhausting process in which the attainment of an excellent model seems to be the product of sheer luck, rather than the product of a structured, ordered series of logical steps. So what I would like to see is a new generation of analysts who are well aware of the fact that the modeling process can and should be standardized and who are also aware that

CHAPTER 20 CLOSING THOUGHTS

while the creative process will always be present in the modeling process, it should always be present within the confines of a preestablished methodology.

As a final thought, I would like to invite my dear readers to reflect on this idea: *if you have made it this far, perhaps it is because you have some of a magician in you too.*

References

[1] **Altman Edward I**. Financial Ratios, Discriminant Analysis and The Prediction of Corporate Bankruptcy. The Journal of Finance, Sep 1968, Vol. 23, No. 4, pp. 589-609.

[2] **Altman Edward I. and Sabato Gabriele**. Modelling Credit Risk for SMEs: Evidence from the U.S. Market. Abacus, 43, pp. 332–357, 2005.

[3] **Altman Edward I. and Sabato Gabriele Wilson Nicholas**. The value of non-financial information in SME risk management. June 2010, The Journal of Credit Risk 6(2):95-127.

[4] **Bandler Richard and Grinder John**, The structure of magic. 1975, Science and Behavioral Books Inc. United States of America.

[5] **Basel Committee on Banking Supervision**. Working Paper No. 14 Studies on the Validation of Internal Rating Systems, May 2005.

[6] **Becker Gerhold K**. Integrity as Moral Ideal and Business Benchmark. Journal of International Business Ethics, Vol.2 No.2 2009.

[7] **Bellini Tiziano**. IFRS 9 and CECL Credit Risk Modelling and Validation: A Practical Guide with Examples Worked in R and SAS, 2019. Academic Press.

REFERENCES

[8] **Borghans Lex, Duckworth Angela Lee, Heckman James J. and Weel Bas Ter**. The Economics and Psychology of Personality Traits. The Journal of Human Resources, Fall, 2008, Vol. 43, No. 4, Noncognitive Skills and Their Development (Fall, 2008), pp. 972-1059.

[9] **Brown Iain.** Developing Credit Risk Models Using SAS Enterprise Miner and SAS/STAT. December 2014, SAS Institute, United States of America.

[10] **Campbell Anne**. A Mind of Her Own: The evolutionary psychology of women, 2nd Edition, 2013. Oxford University Press, United States of America.

[11] **Chapra Steven C. and Canale Raymond P.** Numerical Methods for Engineers, 5th Edition, 2006. McGraw Hill.

[12] **Chatterjee Samprit and Hadi Ali S**. Regression Analysis by Example, 5th Edition, 2012, John Wiley & Sons, Inc, United States of America.

[13] **Costa Paul T. Jr., Terracciano Antonio, and McCrae Robert R**. Gender differences in personality traits across cultures: Robust and surprising findings. Journal of Personality and Social Psychology, 2001, Vol. 81, No. 2, 322-331.

[14] **Dorfman Paul M**. Table Look-Up by Direct Addressing: Key-Indexing -- Bitmapping -- Hashing, SUGI 26 conference in Long Beach, California, April 23, 2001, Paper 8-26.

REFERENCES

[15] **Dorfman Paul M and Shajenko Lessia S.** Crafting Your Own Index: Why, When, How, SUGI 30 Proceedings, Philadelphia Pennsylvania, April 10-13, 2005, Paper 232-31.

[16] **Dorfman Paul M and Vyverman Koen S.** DATA Step Hash Objects as Programming Tools, SUGI 30 Proceedings, Philadelphia Pennsylvania, April 10-13, 2005, Paper 241-31.

[17] **Eberhardt Peter.** The SAS® Hash Object: It's Time To .find() Your Way Around – Tutorial. Pharma SUG 2010 – Paper TU02.

[18] **Ekman Paul.** Emotions Revealed, Second Edition: Recognizing Faces and Feelings to Improve Communication and Emotional Life, 2nd Edition, March 2007. St Martin's Griffin, United States of America.

[19] **Fernandez Gorge Et al.** Time Series Modeling Essentials Course Notes, 2016, SAS Institute Inc., Cary, NC, USA.

[20] **Field Andy.** Discovering Statistics Using SPSS, 3rd Edition, 2009. SAGE Publications, United States of America.

[21] **Fleeson William, M. Gallagher Patrick.** The Implications of Big-Five Standing for the Distribution of Trait Manifestation in Behavior: Fifteen Experience-Sampling Studies and a Meta-Analysis. J Pers Soc Psychol. 2009 December ; 97(6): 1097–1114. doi:10.1037/a0016786.

REFERENCES

[22] **Furnival George M. and Wilson Robert W**. Regressions by Leaps and Bounds, Technometrics, Nov., 1974, Vol. 16, No. 4 (Nov., 1974), pp. 499-511

[23] **Georges Jim Et al**. Advanced Predictive Modeling Using SAS® Enterprise MinerTM 6.1, Course Notes, 2009, SAS Institute Inc. Cary, NC, USA.

[24] **Georges Jim Et al**. Applied Analytics Using SAS® Enterprise MinerTM Vol. I, Course Notes, 2011, SAS Institute Inc. Cary, NC, USA.

[25] **Gete Pedro And Reher Michael**. Two Extensive Margins of Credit and Loan-to-Value Policies. Journal of Money, Credit and Banking, Vol. 48, No. 7 (October 2016), pp. 1397-1438.

[26] **Greenacre Michael**. Correspondence Analysis in Practice, 3rd Edition, 2017, CRC Press, Boca Raton, Florida, United States of America.

[27] **Haidt Jonathan**. The Righteous Mind: Why Good People Are Divided by Politics and Religion. February 2013. Vintage Books, United States of America.

[28] **Harari Yuval Noah**. Sapiens: A Brief History of Humankind. May 2018. HarperCollins Publishers, United States of America.

[29] **Hawkins Jeff, Blakeslee Sandra**. On Intelligence: How a New Understanding of the Brain Will Lead to the Creation of Truly Intelligent Machines. August 2005, St. Martin's Griffin New York.

REFERENCES

[30] **Himmelblau, Lasdon Edgar**. Optimization of Chemical Processes. 2nd Edition, April 2001, McGraw Hill, United States of America.

[31] **Hoffmaster Barry and Hooker Cliff**. The Nature of Moral Compromise: Principles, Values, and Reason. Social Theory and Practice, January 2017, Vol. 43, No. 1 (January 2017), pp. 55-78.

[32] **Hosmer David W. Jr., Lemeshow Stanley, Sturdivant Rodney X**. Applied Logistic Regression. 3rd Edition, April 2013, Wiley.

[33] **Koehn Daryl**. Integrity as a Business Asset. Journal of Business Ethics 58 (1/2/3)(2005): 125-136.

[34] **Kozlowski Jan and Gawelczyk Adam. T**. Why are species' body size distributions usually skewed to the right?. Functional Ecology 2002 16, 419 – 432.

[35] **Krebs Dennis L**. Morality: An Evolutionary Account. Perspectives on Psychological Science, May 2008, Vol. 3, No. 3 (May 2008), pp. 149-172.

[36] **Kuhnen Camelia M. and Melzer Brian T**. Noncognitive Abilities and Financial Delinquency: The Role of Self Efficacy in Avoiding Financial Distress. The Journal of Finance, Vol. 73, No. 6 (December 2018), pp. 2837-2869.

[37] **Lakoff George, Johnson Mark**. Metaphors We Live By. University of Chicago Press; 1st edition (April 15, 2003).

[38] **Loren Judy**. How Do I Love Hash Tables? Let Me Count The Ways!, SAS Global Forum, March 16-19, 2008, San Antonio, Texas, Paper 029-2008.

REFERENCES

[39] **Mallows C. L.** Some Comments on C_p, Technometrics, Vol. 15, No. 4 (Nov., 1973), pp. 661-675.

[40] **Mazar Nina, Amir On and Ariely Dan.** The Dishonesty of Honest People: A Theory of Self-Concept Maintenance. Journal of Marketing Research, Dec. 2008, Vol. 45, No. 6 (Dec. 2008), pp. 633-644.

[41] **Modlin Danny Et al.** Statistics 1 Introduction to ANOVA, Regression, and Logistic Regression Vol 1 and 2, Course Notes, 2012, SAS Institute Inc. Cary, NC, USA.

[42] **Motulsky Harvey.** Intuitive Biostatistics: A Nonmathematical Guide to Statistical Thinking. 4th Edition, March 2017, Oxford University Press.

[43] **Neisser Ulric.** Intelligence: Knowns and Unknowns. Article in American Psychologist · February 1996, Vol. 51, No. 2, 77-101.

[44] **Nisbet Robert, Elder John, Miner Gary.** Handbook of Statistical Analysis and Data Mining Applications. May 2009, Academic Press an Imprint of Elsevier Science.

[45] **Ohlson James A.** Financial Ratios and the Probabilistic Prediction of Bankruptcy. Journal of Accounting Research, Spring, 1980, Vol. 18, No. 1 (Spring, 1980), pp. 109-131.

[46] **Oyama S. Ted.** Membrane Science and Technology Chapter 1 - Correlations. Elsevier, Volume 14, 2011, Pages 1-24.

[47] **Patetta Mike Et al**. Predictive Modeling Using Logistic Regression Course Notes, April 2010, SAS Institute Inc. Cary, NC, USA.

[48] **Perline Rich.** Development of Credit Scoring Applications Using SAS Enterprise Miner Course Notes. SAS Institute.

[49] **Peterson Jordan B**. Maps of Meaning: The Architecture of Belief. March 1999, Routledge.

[50] **Peterson Jordan B**. 12 Rules for Life an Antidote to Chaos, 2018, Allen Lane, New Delhi India.

[51] **Petrick, J. A., & Quinn**, J. F. (2000). The Integrity Capacity Construct and Moral Progress in Business. Journal of Business Ethics, 23, 3–18.

[52] **Qi Min and Yang Xiaolong**. Loss Given Default of High Loan-to-Value Residential Mortgages. Office of the Comptroller of the Currency OCC Economics Working Paper 2007-4 August 2007.

[53] **Quiroga Rodrigo Quian**. The Forgetting Machine: Memory, Perception, and the Jennifer Aniston Neuron. October 2017, Ben Bella Books Inc.

[54] **Ratner Bruce**. Statistical and Machine-Learning Data Mining :: Techniques for Better Predictive Modeling and Analysis of Big Data. 3rd Edition, June 2020, CRC Press.

[55] **Rhoades Stephen & Dodoo Lee**. Another Helping of Hash, NESUG 2008. Cary, NC: SAS Institute Inc

[56] **Rischert Alice**. Oracle SQL by Example, 4th Edition, August 2009. Prentice Hall.

REFERENCES

[57] **Rousseeuw Peter J. and Leroy Annick M**. Robust Regression and Outlier Detection, 2003, John Wiley & Sons, Inc, United States of America.

[58] **Royston Patrick, Sauerbrei Willi**. Multivariable Model - Building: A Pragmatic Approach to Regression Analysis Based on Fractional Polynomials for Modelling Continuous Variables. June 2008, Wiley.

[59] **Sarma Kattamuri S., PhD**. Predictive Modeling with SAS® Enterprise Miner™ Practical Solutions for Business Applications, 2nd Edition, December 2013. Cary, NC: SAS Institute Inc.

[60] **SAS Institute Inc 2011**. SAS® Enterprise Miner™ 7.1 Extension Nodes: Developer's Guide. Cary, NC: SAS Institute Inc.

[61] **Schmitt David P**. Why Can't a Man Be More Like a Woman? Sex Differences in Big Five Personality Traits Across 55 Cultures. Journal of Personality and Social Psychology, 2008, Vol. 94, No. 1, 168–182.

[62] **Shmueli Galit, Patel Nitin R., Bruce Peter C**. Data Mining For Business Intelligence: Concepts, Techniques, and Applications in Microsoft Office Excel® with XLMiner®. 2nd Edition, October 2010, Wiley.

[63] **Siddiqi Naeem**. Intelligent Credit Scoring: Building and Implementing Better Credit Risk Scorecards. 2nd Edition, January 2017, Wiley.

[64] **Sowell Thomas**. Discrimination and Disparities, 2019, Basic Books, United States of America.

[65] **Stango Victor and Zinman Jonathan**. Exponential Growth Bias and Household Finance. The Journal of Finance, Dec. 2009, Vol. 64, No. 6 (Dec. 2009), pp. 2807-2849.

[66] **Stroud K.A**. Engineering Mathematics. 5th Edition, 2001, Palgrave Macmillan.

[67] **Taha Hamndy A**. Operations Research: Introduction: An Introduction. 8th Edition, 2007, Pearson Prentice Hall.

[68] **Wagenaar William A., Sagaria Sabato D**. Misperception of exponential growth. Perception & Psychophysics 1975, Vol. 18(6), 416-422.

[69] **Warner-Freeman Jennifer K**. I cut my processing time by 90% using hash tables – You can do it too!, NESUG 2007.

[70] **Weisberg Yanna J., DeYoung Colin G. and Hirsh Jacob B**. Gender differences in personality across the ten aspects of the Big Five. Frontiers in Psychology, August 2011 | Volume 2 | Article 178.

[71] **Wilson Edward O**. Sociobiology: The New Synthesis, 1975, Harvard University Press, Belknap Press.

[72] **Zhang Yanan, Ji Lu, and Liu Fei**. Local Housing Market Cycle and Loss Given Default: Evidence from Sub-Prime Residential Mortgages. IMF Working Paper, International Monetary Fund.

Online Resources

(1) A Brief History of Loans: Business Lending Through the Ages	https://www.become.co/blog/a-brief-history-of-loans-business-lending-through-the-ages/
(2) Online Etymology Dictionary	https://www.etymonline.com/word/conceive
(3) Britannica	https://www.britannica.com/
(4) Ebaqdesign	https://www.ebaqdesign.com/blog/bank-logos
(5) Kohlberg's Stages of Moral Development	www.simplypsychology.org/kohlberg.html
(6) Merriam-Webster Dictionary	https://www.merriam-webster.com/
(7) ZIPPIA	https://www.zippia.com/research/earnings-vs-hours-worked/
(8) National Center for Education Statistics	https://nces.ed.gov/programs/raceindicators/indicator_reg.asp
(9) Credit Card Delinquency Statistics	https://www.creditcards.com/statistics/credit-card-delinquency-statistics-1276/
(10) The LOGISTIC Procedure	https://support.sas.com/documentation/cdl/en/statug/63347/HTML/default/viewer.htm#logistic_toc.htm

(*continued*)

ONLINE RESOURCES

(11) The REG Procedure	https://support.sas.com/documentation/cdl/en/statug/63033/HTML/default/viewer.htm#reg_toc.htm
(12) The OPTMODEL Procedure	https://support.sas.com/documentation/cdl/en/ormpug/63975/HTML/default/viewer.htm#optmodel_toc.htm
(13) Interactive Grouping Node	https://documentation.sas.com/doc/en/emref/14.3/p1qzwz7onopjqcn11uc04i18urg7.htm
(14) Example 73.2 Aerobic Fitness Prediction	https://support.sas.com/documentation/cdl/en/statug/63033/HTML/default/viewer.htm#statug_reg_sect055.htm
(15) Varclus	https://go.documentation.sas.com/doc/en/emref/14.3/p19e837tepmjz0n1hjt2gdk3sqfg.htm.
(16) SAS® Enterprise Miner™ 14.3 Reference Guide	https://go.documentation.sas.com/doc/en/emref/14.3/n1jqzz8cssr9m2n1ktx2iyv87q56.htm
(17) Weight of Evidence (WOE) and Information Value (IV) Explained	https://www.listendata.com/2015/03/weight-of-evidence-woe-and-information.html
(18) Gini, Cumulative Accuracy Profile, AUC	https://www.listendata.com/2019/09/gini-cumulative-accuracy-profile-auc.html
(19) Scorecard Node	https://go.documentation.sas.com/doc/en/emref/15.2/n181vl3wdwn89mn1pfpqm3w6oaz5.htm

Abbreviations and Acronyms

Abbreviation	Description
##MB	Number of months back
ABT	Analytical base table
AI	Artificial intelligence
AIC	Akaike's information criterion
APPV	Average performance per variable
ATM	Cash withdrawals
BE	Backward elimination
CC	Credit card
CECL	Current expected credit loss
CLTV	Current loan to value
CRC	Costa Rica colones
CS	Credit scoring
CVIC	Continuous variable integration cycle
DB	Database
DC	Decrement
DE	Distribution of events
DM	Data mining
DNE	Distribution of no events
DO	Default option

(*continued*)

ABBREVIATIONS AND ACRONYMS

Abbreviation	Description
DR	Default rate
DS	Date-specific
EAD	Exposure at default
EG	Enterprise Guide
EM	Enterprise Miner
EMWS	Enterprise Miner Workspace
ESS	Equipment and specialized supplies and services
ETL	Extraction, transformation, and loading
FL	Factor level
FP	Fixed parameter
FS	Forward selection
GDP	Gross domestic product
GES	Governmental expenses and services
HLE	Hobbies, luxuries, and entertainment
ICRM	Iterative collinearity reduction method
IG	Interactive Grouping
IGN	Interactive Grouping node
IN	Increment
IT	Information technology
IV	Information value
KGB	Known good and bad
KS	Kolmogorov-Smirnov
L##M	Last number of months

(*continued*)

ABBREVIATIONS AND ACRONYMS

Abbreviation	Description
LGD	Loss given default
LTV	Loan to value
MCA	Married by community of acquired assets
MCC	Merchant category code
MFB	Metabolic flux balancing
ML	Machine learning
MPR	Matrimonial property regime
MSG	Married by separation of goods
MV	Monthly variation
MWR	Mid-West Region
NDS	Non-date-specific
OAF	Oracle's analytical function
OLS	Ordinary least squares
OR	Odds ratio
OTH	Others
PCA	Principal component analysis
PD	Probability of default
PHE	Personal care, health, and education
PR	Pacific Region
QP	Quadratic programing
QV	Quarterly variation
Re	Reynolds number
RFF	Restaurants and fast food

(continued)

ABBREVIATIONS AND ACRONYMS

Abbreviation	Description
RMR	Rocky Mountain Region
ROC	Receiver operator characteristic
RSH	Retail stores, supermarket, and home
SBC	Schwarz's Bayesian criterion
SCD2	Slowly changing dimension type 2
SDC	Sum of decrements
SER	Southeast Region
SES	Socioeconomic status
SIN	Sum of increments
SME	Small and medium enterprises
SQL	Structured query language
SSI	System stability index
STEM	Science, technology, engineering, and mathematics
SV	Semestral variation
SWR	Southwest Region
TBT	Transactional base table
TE	Trial-and-error
TRN	Transportation
TRX	Transaction
UI	User interface
USD	United States dollar
VIF	Variance inflation factor
VTR	Vacations and trips
WoE	Weight of evidence

Index

A

Abductive reasoning, 47–49
Absolute values, 406
Abstraction, 3
Accumulation factor, 52
Accuracy, 141
ADD_MONTHS function, 132
Aerobic Fitness Prediction, 364
Akaike's information criterion (AIC), 473, 476, 477
Algebraic notation, 444
All-subsets method, 515
All-subsets procedure, 466, 470, 491–493
   brand, 479
   by-product, 478
   complexity, 488
   Cq's optimization problem, 485–494
   GRP variables, 479
   scorecard/regression node, 478
   SCORE option, 479
   score test approximation, 481–485
   stepwise procedure, 490, 491
   top models, 488
All-subsets variants, 552
Alternative factors, 521–524
ALTER TABLE statement, 173
Altman Z-Score Model, 4–9, 122
Amazon, 89
Amount-to-approved amount ratio, 589
Analytical base tables (ABTs), 116, 119, 127, 153, 217, 269, 288, 319, 451, 495, *See also* Merchant category code (MCC)
   ATM/EXP, 194, 195
   common mistakes, 171
   data processing, 122, 123
   diagramatic representation, 257–264
   event definition, 238–243
   hypotheses, 121
   input tables, 166
   proactive/reactive approaches, 254, 255
   recommended set of tables, 255
   scoring, 253, 657
   SQL, 123, 124
   stacked, 245–253
   transactional tables, 121
ANOVA, 415

INDEX

Application process, 35, 64, 69, 108
Archetype, 686
Archetypical definition, 16
Artificial intelligence (AI), 100, 326
Association matrix, 541, 553–556
Attributes, 105
Automated multivariable selection algorithms, 415
Automation, 263
Average performance per variable (APPV), 505
AVG aggregation, 132

B

Backtesting, 672
Backward elimination (BE) algorithm, 468, 469, 497
Backward elimination only (BE-Only), 468, 476, 515, 548, 551, 567
Balance equation, 9–12, 30, 31, 507
    contributing and subtracting forces, 32
    main-effects multivariable model, 31
    married with children effects, 33–39, 41
Bar charts, 408, 409, 546
Behavioral patterns
    out-calls/in-calls, 88
    payments and balances, 88
    periodicities, 88
    timestamps, 89

Best model, 317
Better-fitting model, 515
Bridge tables, 171
Building plan
    code, 132–136
    errors and inconsistencies, 130, 131
    identification, 130
    inconsistency, 130
    increment/decrement flags, 130
    modeling process, 127
    nomenclature and consistency, 135–137
    variables, 128–130
Business rules, 149–153

C

CALL MISSING statement, 181–183
Catalog tables, 170, 171
Categorical lumping, 415
Category lumping, 144
Cause-and-effect terms, 407
Champion model, 317, 533, 578
Charges-to-savings, 589
Chi-squared distribution, 354
CLASS statement, 350
CLUS_N_BRANCH, 618
Code Editor user interface (UI), 307, 308
Coefficient inversion, 553, 557
Collinearity, 361, 472, 510–512, 514, 515, 518, 524, 526, 558
    conflicting variables, 377

INDEX

diagnostics, 369–372
issues, 375–377
methods, 371, 372
optimization issues, 363–368
parameters, 539
variance inflation factor (VIF), 361–363
Collinearity reduction method, 495, 497, 568
Commitments, 32, 41, 42, 61, 100, 119
Common sense, 46, 47, 71
Communication, 123
Conceptualization process, 2, 3
Altman Z-Score Model, 4–9
balance equation, 8–11
Conceptual modelling, 52
borrower, 20, 21
contract, 17, 18
evolutionary psychology, 24, 25
external factors, 28, 29
integrity, 21
lender, 17–20
loan, 14, 15
modeling process, 13
moral values, 24–26
PD moral, 26–28
trust, 15, 16
Conditional grouping functions, 217–226
Confidence intervals (CI), 332
Conflicting variables, 387
Consistency, 135–137
Constrained Optimal Options, 523

Continuous variable integration cycle (CVIC), 166, 255, 657
account-customer relationship, 233–238
interval variables, 230–232
nominal variables, 232
Contractual agreements, 16
Correlation, 112, 132
Correlation matrix, 372
Costa Rica Colones (CRC), 65
CREATE TABLE clause, 311, 312
Credit card (cc) utilization, 66, 90, 92, 114, 116, 196, 237, 398–404, 413, 420, 535–538
Credit line's utilization, 237
Credit scoring (CS), 14
Cross-referencing process, 173
Cross-validation error, 474
Cross-validation misclassification, 474
Cumulative payment, 588
Customer loyalty, 420
Customer's personal information, 108–114

D

Data manipulation languages, 124
Data mining diagram
probability, 313
scorecard model, 313
six basic steps, 315, 316
Data mining (DM), 562, 653
Data processing, 122, 123

INDEX

Data proportion, 328, 329
Data sets, 178
Data validation, 349–352
Date-specific (DS) chronological sequence, 246
Debt restructuring, 156
Decision-making process, 476
Decision tree (DT) algorithm, 416, 418, 421, 592, 604, 608, 615, 629
Decomposition, 32
Deductive reasoning, 47–49
Default event, 239
Default option (DO), 528
Default rate (DR), 35, 115, 453
    children, lifelong responsibility, 41
    dependent children numbers, 40, 41
    incomes, 38
    marital status, 33
    married and single people, 38
    married persons by age, 37
    matrimonial property regimes, 35
    single category, 36
    singles by choice, 39
Degrees of freedom (df), 535
DELETE clause, 375
Denarii, 16
Dimensional tables, 170, 171
Dimensionless numbers, 73–78
Dollar bill, 15, 16
DOSUBL function, 184

E

Economic indicators, 115
Education, 110
Egalitarian, 501
Elements of daily life, 43–47
EM code node, 306
Employment, 111
ENQUOTE macro, 276–278
Enterprise Guide (EG), 186, 345, 428, 661
Enterprise Miner (EM), 395
Equation Balancing, 8–11
Error function, 363, 364
Event definition, 238–245
Evolutionary psychology, 25, 26
Exclusive models, 533
Expected variables, 145
Experimental design, 130
    applied example, 524–528
    collinearity, 497
    defined, 496
    factors and factor levels, 507–515
    model's assessment, 498–507
    multivariable model, 496
    number of variants, 515–520
    root factors, 517–519
    selection criterion factors and BE, 497
    symmetry, 518
    treatments/variants, 516–518
Explanatory variables, 135, 136, 146, 160, 302, 585

INDEX

Exploration panel, 271, 272
Exportgroup contents, 429

F

Factor levels, 521–524
False positive rate (FPR), 499
Financial institutions, 17, 117
Financial obligations, 105–108
Financial products, 151
Financial ratios, 76, 77, 588
Fixed parameter (TE), 528
Flux directionality, 51–58
Folding time, 95–100
Follow-up statistics, 677–681
Foreign keys, 171
Forward-looking approach, 254
Foundation Mart, 656, 657
FROM clause, 189

G

Generalized Algebraic Modeling System (GAMS), 442
Geolocation, 112
Gini coefficient, 501–503, 535, 604, 608, 614, 617, 620, 625, 627, 672, 677
Gini index, 426, 436–439
Global factors, 508
Greenacre's method, 418, 615, 616, 630, 631
Gross domestic product (GDP), 114

GROUP BY clause, 669
Grouping options, 423
Grouping process, 433

H

Hands-on approach, 122
Hash DATA step statement, 179
Hash inner joins, 187–189
Hash outer join DATA step statement, 180–193
   DOSUBL function, 184
   SELECT correlated subqueries, 188–193
Histograms, 350
Historical consistency, 160
Historical inconsistencies, 155
Human error, 293
Hyperoptimization, 563, 568
   association matrix, 534, 535
   experimental design phase, 532
   fictional example, 532
   graphical representation, 533
Hyperparameters, 544, 566
Hyperresponse
   collinearity reduction techniques, 539
   credit card PD Model, 535–538
   retail loan, 538, 539
Hyperspace, 541, 544
   credit card PD model, 541–553
   optimal model, 562–563
   retail loan PD model, 553–563

Hypothetical customer-level
    account, 255
Hypothetical illustration, 145

I, J

Identification, 32
Independent variable, 363
Inductive reasoning, 47–49
Information technology (IT), 124
Information value (IV), 402, 403,
    409, 426, 436–439
Input flux, 51
Input tables
    dimensional and catalogs,
        170, 171
    snapshot, 167–169
    transaction, 169, 170
Input variables, 509
Integrity, 21, 39, 41, 101
Interactive grouping (IG), 568,
    571, 589–592
Interactive grouping node (IGN)
    binning and grouping
        options, 423–426
    grouping data, 427–432
    predefined groupings, 432, 433
    tree-based grouping
        options, 421–423
Interpretability, 113
Interval grouping method, 423
Interval targets, 451–456
    resolution issues, 453–456
Interval variables, 67, 407–415, 423

Irrelevant variables, 315
Iterative collinearity reduction
    method (ICRM) node,
    393–395, 459, 496, 567, 568,
    572, 607, 648
    applied example, 385–389
    collinearity issues, 375–377
    competing variable, 380–385
    higher proportion of
        variance, 380–385
    initial collinearity
        diagnostics, 373–375
    main loop, 377–382
    maximum VIF variable, 378–642
    rejected variable, 384, 385
    steps, 372
Iterative refinement, 638–642
Iterator object, 180

K

Kolmogorov-Smirnov (KS) test,
    503, 672, 677, 678

L

Leverage points, 406
Line-and-box workflow
    diagram, 302
Linearization, 404
Line charts, 362, 363, 409, 412, 614
Line graphs, 334
Link functions, 116
Loan-to-value ratio, 76, 105

INDEX

Logit units, 406, 407
Log-likelihood function, 477
Log rerouting, 344, 345
Long-term commitments, 35
Lorenz curves, 501, 502
Low-resolution analysis, 14, 15
Low-resolution conceptual
    representation, 49, 50
  balance equation, 50
  flux directionality, 51–58
L-shaped behavior, 63
Lumping categories, 52, 53

M

Machine learning (ML), 100, 122,
    326, 562, 653
Macroeconomic
    information, 114–117
Macro-system, 10, 11
Macro-variables, 302
Magnitudes, 59, 95, 105
  definition, 70
  dimensionless numbers,
    73–78
  frequencies, 70
  intrinsic benefits, 78, 79
  payroll payment, 71
  ratio history, 71–73
  ratio lists, 79–84
  ratio types, 73–78
Main-effects model, 566, 631, 632
  ambiguous
    characterization, 624–628

assessment charts, 581
collinearity reduction,
    568, 580–581
descending order, 634, 635
detailed report, 624
diagnostics, 622
economic dependents, 621
event rate function, 624
final models, 641–643
manual elimination, χ^2
    statistic, 634–645
model-building process,
    567, 571
model's performance, 642
multivariable selection
    processes, 575, 577
regression, 624
variable select node, 569–571
visual inspection, 631
WoE refinement process, 631
Main loop, 349–352
Mantel-Haenszel χ^2 test, 35
Marital status, 58, 65, 67, 68, 112
Mathematical expressions, 87
Mathematical models, 2, 121
Matrimonial property
    regime (MPR), 33
Maximum clusters, 510, 522
Maximum number of groups, 433
  Gini Index, 436–439
  grouping options, 433
  parameters, 436
  variable, 435
MaxPulse variable, 365–368, 370

Merchant category code (MCC), 42, 43, 170
   and ABTs, 196–199
      cursor, 225–227
      customer ID, 200, 201
      desired level grouping, 201
      empty rows deletion, 216
      grouping columns, 199
      input variables, 217
      new variables, 216
      PL/SQL, 227
      transactional base table, 202
   groupings, 169
Meta-analysis, 128
Metabolic flux balancing (MFB), 1
Metadata, 282–284
   advanced option, 285–295
   advisor options screen, 287
   basic option, 288
   column metadata screen, 289
   logistic regression, 288
   properties screen, 287
   source screen, 286
   standardized nomenclature, 290
   target variable, 302
   typographical error, 290
Meta-structure, 685
Mid-West Region (MWR), 116
Minimal discrepancy, 604, 606
Minimum WOE Difference, 523
Mining diagram, 317, 318
Misclassification rate, 474
Missing values, 405, 418, 419, 422

Model-building process, 149, 293, 319, 340, 350, 495, 496, 521, 563, 564, 653
Modeling process, 687
Model's assessment (response)
   complexity, 505
   cumulative accuracy ratio, 501–503
   Kolmogorov-Smirnov (KS) test, 503
   overfitting, 506–508
   performance, 499
   predictive power, 509
   ROC curve, 499–501
Model selection criterion, 473–478
Monetary balance, 103, 104
Monetary variables
   effective approach, 62
   linear and logistic regression, 62
   scale, dimensionality, and units, 64–69
   wealth distribution, 63, 64
Monotonic event rate, 423
Monotonic linearization, 407–415
Monotonic linear WoEs, 599–603
Monotonic relationship, 426
Monthly variation (MV), 203
Moral values, 22–26, 41
Mortgage, 150
Mortgage models, 145
Multi-model subset, 532
Multiplicative combination, 148
Multivariable models, 530, 538

Multivariable selection methods, 317, 460, 465, 468, 472, 473, 476, 479, 491, 514, 515, 527, 528
  AIC, 466
  preferable use, 469–472
  second strategy, 466
  sequential selection, 466
  stepwise procedures, 467–469
  variable, 470–472

N

Naming your variants, 530–534
Native keys, 154
Natural logarithm, 401
Neighborhood, 113
Nested rules, 153
Nested subqueries, 208
New variables, 216
N_OCCUPATION, 606
Nodes panel, 272
Nomenclature, 135–137
Nominal variables, 404, 415–418
Non-date-specific (NDS) chronological sequence, 246
Non-significant variables, 355–357
Northeast Region (NR), 116
N_PROFESSION variable, 644
Null model, 469
Numerical stability issues, 66

O

Object-oriented programming
  explanatory variables, 302
  interval variables, 299
  macro-variables, 302
  nominal variables, 299
  properties, 297
  selected list, 297–298
  single-process workflow, 303, 306
  two-node workflow, 306, 307
Observed default event, 371–674
One-to-many relationship, 170
One-to-one relationships, 302
Optimal criterion, 423
Optimal multivariable models, 535, 536, 539, 540
Optimal variables, 534
Optimization issues, 363–368
Optimization methods, 67
Optimization problem, 441–453
Optimization program, 447
OPTMODEL output, 451, 452
Oracle cursor, 238
  begin statement, 176, 177
  cursor statement, 175, 176
  DECLARE statement, 175
  PL/SQL cursors, 174
  query performance, 177
  sections, 174, 175
Oracle's analytical functions (OAF), 208

INDEX

ORDERED argument, 180
Outliers, 404–406
Output Delivery System (ODS), 353, 373
Overfitting, 326, 437, 506–508, 522, 537, 630
Overtraining, 326, 327

P

Pacific Region (PR), 116
Parameter initialization, 345, 346
PARAMETERSESTIMATES report, 375
Partitioning
   data proportion, 328, 329
   representativeness, 324
   stratified sample analysis, 329
   stratified sampling, 324
   training, validation, and test partitions, 328, 329
Payment-to-installment ratio, 588
Penalization, 477
Periodicity, 88, 90, 105, 106, 130, 168, *See also* Time series
PL/SQL server, 227
Point-in-time variables, 232
Polychotomous variables, 477
Polynomial transformations, 406
Predefined groupings, 432, 433
Predictive error, 486
Predictive power, 509

Predictor, 361
Predictor strength, 472
Preferable use, 469–472
PREFIXLIST, 276, 277
Prerequisite parameters, 342
Preselection method, 338
Primary key, 173
Primary loop, 350
Principal component analysis (PCA), 361, 395
Probability, 89
Probability of default (PD) model, 12, 14, 26–28, 39, 57, 89, 92, 93, 115, 501, 535–538, 668, 683
PROC REG for model, 367, 370, 371
Product information, 144
Programming-friendly interface, 313
Programming language, 123
Python, 124

Q

Quarterly variation (QV), 203
Quasi-linear behavior, 546

R

Ratio, 59, 95, 102, 105, 112, 119
   ATM/EXP, 194
   and deviations, 97
   financial modeling, 61

lists from elements, 79–84
vs. magnitudes, 70–72
monetary variables, 62–70
payment-to-installment, 588
Receiver operating
      characteristics (ROC)
   curve, 535, 677, 678
   overfitting, 506
   statistical curve, 499–501
   training, 506
   validation, 537
   value, 672
RECODE macro, 279–282
Rectangular wave pattern, 455
Redundant variables, 317
REG procedure, 375
Regression, 352–357
Regression analysis, 643
Regressor coefficients, 368
Regressors, 367
REMOVE macro, 278, 279
Resolution, 113
Retail loan PD model, 553–563
Risk-groups, 608
Robustness, 45
Rocky Mountain
      Region (RMR), 116
ROC value, 642
Roll rate analysis, 240–243
Root factors, 516–518
Round-off error, 68
Row multiplication
      of ABTs, 172
ROW_NUMBER function, 211

Row-to-column ratio, 170
RunPulse variable, 365–368, 370
RunTime, 369

S

Sample size, 509
SAS DATA step, 661, 663
SAS DATA Step Hash, 178, 179
   DOSUBL function, 184
   hash inner join
         statement, 186–188
   hash outer join DATA step
         statement, 180–185
   hash table, 186
SAS® Enterprise Miner™ (EM), 475
   code editor user
         interface (UI), 307–308
   description, 270–272
   diagram/canvas panel, 273
   exploration panel, 271, 272
   nodes panel, 272
   object-oriented programming,
         297–317
   programming exapmles,
         309–314
   properties panel, 272
   quick help panel, 272
   standardized methodology,
         292, 293
   startup code, 273, 274
   utility macros, 274–284
SAS Score Code, 659–666
SCD2-type tables, 171

INDEX

Schwarz's Bayesian criterion (SBC), 473, 515, 528
Science, technology, engineering, and mathematics (STEM), 110, 111
Scorecard models, 108, 451
Scored ABT, 666–669
Score test approximation, 481–494
Scoring process
    ABTs, 657–659
    ETLs, 656, 657
    follow-up statistics, 677–681
    Foundation Mart, 656, 657
    model package, 659–666
    official data sources, 656
    principal elements, 654
    snapshot table, 669
    system stability index (SSI), 674–677
Second dominant variable, 382–387
Segmentation, 142–146
SELECT clause, 188–193
Selection criteria, 475
Self-employment, 111
Semestral variation (SV), 203
Sequential relationship, 285
SET statement, 181
SET clause, 661
Skewness, 62
Slide-like behavior, 619
Slowly Changing Dimension Type 2 (SCD2), 158–159

Snapshot table, 167–169, 668
    scoring process, 669
Sociodemographic variables, 585
Solver engine, 460
Southwest Region (SWR), 116
Spatial dimension, 324
Squared errors, 363
Stacked ABT, 245
    average period n, 249
    conceptual representation, 246, 247
    timeline handling, 249–253
Standardization, 69, 404
Standard units, 406, 407
Startup code, 273, 274
Statement tables, 167
Statistical representativeness, 146–149, 509
Stay significance value, 472
Stepwise procedures, 467–469, 490, 491
    model selection criterion, 473–478
    preferable use, 469–472
    stay significance value, 472
Stratification, 327
Stratified sample analysis, 329–337
Stratified sampling
    overtraining, 326, 327
    risk dimension, 325
    spatial dimension, 324
    time dimension, 325
    ZIP codes, 327

Structured Query Language (SQL), 123, 124
Subject level of analysis, 153
Subqueries, 134, 188
Substantial evidence, 28
Substantial scale, 64
Sumerian civilization, 15
Sum of squares (SSE), 473
Survival rate, 464
Symmetrical mining diagram, 497
Synthetic keys, 154, 155
   historical inconsistencies, 155, 156
   slowly changing dimension type 2 (SCD2), 158, 159
   switching keys, 159
   system-generated key, 155
   Telcos, 160–162
System-generated keys, 154
System stability index (SSI), 664, 672, 674–677, 682

T

Tabular report, 620
Tail-like behavior, 630
Tantamount, 326
Target population, 141, 165, 168, 243–245
   business rules, 149–153
   CLTV variable, 142
   credit cards, 142
   defined, 141, 142
   PD model, 141
   segmentation, 142–146
   statistical representativeness, 146–149
   subject level of analysis, 153
   synthetic keys, 154, 155
Target variable, 302, 584, 586
Telecommunication, 88
Telecommunications companies (telcos), 160–162
Temporary storage tables, 349–351
Three-stage transition, 630
Time dimension, 325
Time series, 116
   economic conditions, 92
   increment/decrement flags, 97–101
   patterns, 90
   periodicities, 93
   periodic pattern, 92
   point-in-time, 93
   scorecard, 90
   timeframe, 90, 92
   utilization level, 93
Timestamps, 89
Training Code section, 310
Transactional base table (TBT), 205
Transactional data, 202
Transaction tables, 169, 170
Transformations, 408
Travel and entertainment (T&E), 43
Tree-based grouping options, 421–423

INDEX

Trial and error (TE), 46, 528
True positive rate (TPR), 499
Trust, 15, 16, 32, 35

U

Unconventional information, 104
Unicity, 172
UNION ALL clause, 312
Unit-length scaling, 69
Univariable analysis, 144, 339
    execution parameters
        setting, 341–344
    hypothesis testing, 339
    log rerouting, 344, 345
    main loop and data
        validation, 349–352
    next variable, 355
    non-significant
        variables, 355–357
    parameter initialization,
        345, 346
    regression, 352–357
    selection criterion, 340
    significance, 353–355
    step-by-step
        explanation, 340–357
    temporary storage
        tables, 349–351
Univariable procedure, 459, 464,
    495, 509
Univariable procedure
    node, 569–571

Universal community property, 34
Utility macros
    ENQUOTE, 276–278
    METADATA, 282–392
    NLIST macro counts, 274, 275
    PREFIXLIST, 276, 277
    RECODE, 279–282
    REMOVE macro, 278, 279

V

Validation error, 473
Validation partition, 323, 578
Validity window, 671–674
Varclus, 389–394, 526
Variable clustering, 389–393
Variables, 103, 235, 459
    business judgment, 587
    categorization, 103
    code, 132–136
    controversy, 117, 118
    customer's banking
        information, 114
    customer's personal
        information, 108–114
    data quality, 587
    explanatory, 302, 585
    financial obligations, 105–108
    independent and
        dependent, 485
    integration cycle, 237
    interval, 299–232
    lists, 105–106

maximum number of groups, 435
multivariable model, 593
multivariable selection
    methods, 470–472
nomenclature, 138
nominal, 232, 299
select node, 569–571
spreadsheets, 128–130
target, 302
transformation, 512–514
unconventional
    information, 104
Variable selection options, 426, 427
Variable selection process, 495
Variable survival rate
  modeling process, 460
  PD models, 461
  rejection rate, 463
  scorecard models, 460
  survival rate, 462–464
Variance inflation factor (VIF),
    361–363, 370
  collinearity issues, 369
  conflicting variables, 377
  cutoff value, 386
  diagnostic tool, 394
  infinity, 362
  second variable, 381
  staggering value, 379
  substitute, 372
  variables, 378–379
Variation proportion, 510, 522
Visual assessment, 130, 592–595

W, X, Y

Wealth distribution, 63, 64
Weight of evidences (WoEs), 62,
    460, 464, 480, 495, 523, 557
  advantages, 404, 405
  definition and
    calculation, 398–404
  dependent variable, 397
  fictitious example, 409
  independent variable, 397
  interactive grouping
    node (IGN), 421–433
  interval targets, 451–456
  missing values, 418, 419
  nominal variables, 415–418
  optimization problem, 441–453
  OPTMODEL procedure,
    442–451
  outliers, 405, 406
  profession, 416, 417
  regression model, 398
  standard units, 406, 407
WHERE clause, 191
WoE-refinement process, 686
  BE-Only algorithm, 582
  categories, 614
  characteristics, 629
  classical approach, 584
  conceptual variable design, 584
  diagnostics, 598, 607, 619
  Greenacre's method, 615, 616
  ideal monotonic linear, 599–603
  inconsistent, 595–599

INDEX

WoE-refinement process (*cont.*)
    interactive grouping (IG) feature, 589–592
    interval variables, 618, 621
    line charts, 614
    macro-variable, 583
    main-effects mining, 615
    model-building process, 582
    N_BRANCH, 611
    nodes, 582
    pre-and post-adjustment periods, 621
    reclassification, 614
    refinable, 604–624
    refinement process, 608
    rejected variables, 595
    risk groups, 604, 609, 611
    statistical perspective, 584
    summary statistics, 612–614
    tabular report, 619, 620
    target variables, 628
    traceability, 582
    visual assessment, 592–595

Z

ZIP code, 287
Z-Score Model, 4–9

GPSR Compliance

The European Union's (EU) General Product Safety Regulation (GPSR) is a set of rules that requires consumer products to be safe and our obligations to ensure this.

If you have any concerns about our products, you can contact us on

ProductSafety@springernature.com

In case Publisher is established outside the EU, the EU authorized representative is:

Springer Nature Customer Service Center GmbH
Europaplatz 3
69115 Heidelberg, Germany